LAGRANGIAN & HAMILTONIAN DYNAMICS

Lagrangian & Hamiltonian Dynamics

Peter Mann
University of St Andrews

OXFORD
UNIVERSITY PRESS

Great Clarendon Street, Oxford, OX2 6DP,
United Kingdom

Oxford University Press is a department of the University of Oxford.
It furthers the University's objective of excellence in research, scholarship,
and education by publishing worldwide. Oxford is a registered trade mark of
Oxford University Press in the UK and in certain other countries

Published in the United States of America by Oxford University Press
198 Madison Avenue, New York, NY 10016, United States of America

British Library Cataloguing in Publication Data
Data available

Library of Congress Control Number: 2017960667

ISBN 978–0–19–882237–0 (hbk.)
ISBN 978–0–19–882238–7 (pbk.)

Printed and bound by
CPI Group (UK) Ltd, Croydon, CR0 4YY

DOI 10.1093/oso/9780198822370.001.0001

For my beautiful daughter Hallie.

Contents

Preface

The purpose of this book is to introduce and explore the subject of *Lagrangian and Hamiltonian dynamics* to science students within a relaxed and self-contained setting for those unacquainted with mathematics or university-level physics. Lagrangian and Hamiltonian dynamics is the continuation of Newton's classical physics into new formalisms, each highlighting novel aspects of mechanics that gradually build in complexity to form the basis for almost all of theoretical physics. Lagrangian and Hamiltonian dynamics also acts as a gateway to more abstract concepts routed in differential geometry and field theories and can be used to introduce these subject areas to newcomers.

In this book we journey in a self-contained manner from the very basics, through the fundamentals and onwards to the cutting edge of the subject at the forefront of research. Along the way, the reader is supported by all the necessary background mathematics, fully worked examples and thoughtful and vibrant illustrations, as well as an informal narrative and numerous fresh, modern and interdisciplinary applications. For example, the subject is rarely discussed within the scope of chemistry, biology or medicine, despite numerous applications and an *absolute* relevance.

The book contains a very detailed and explicit account of Lagrangian and Hamiltonian dynamics that ventures above and beyond most undergraduate courses in the UK. Alongside this, there are some unusual topics for a classical mechanics textbook. The most notable examples include the "classical wavefunction" Koopman-von Neumann theory, classical density functional theories, the "vakonomic" variational principle for non-holonomic constraints, the Gibbs-Appell equations, classical path integrals, Nambu brackets, Lagrangian-Hamilton-Jacobi theory, Dirac brackets and the full framing of mechanics in the language of differential geometry, alongside many more unique features!

The book features many fully worked examples to complement the core material; these can be attempted as exercises beforehand. It is here that many modern applications of mechanics are investigated, such as protein folding, atomic force microscopy, medical imaging, cell membrane modelling and surface adsorbate analysis, to name but a few! Key equations are highlighted throughout the text, and colour is used in derivations that could otherwise be hard to follow. At the back of the book there are vast mathematical chapters that cover all the material that is required to understand classical mechanics, so that the book is a self-contained reference.

The motivation for writing this text was generated during my undergraduate degree in chemistry. Often, I would try to research further information on my quantum mechanics courses, only to be inundated with insurmountable walls of equations and complex ideas which resulted in many opened Wikipedia tabs! Out of frustration, I began compiling a set of notes to aid my understanding; naturally, it got completely out of hand and resulted in the current work! I found myself referring back to my own notes in order to understand new concepts and, very quickly, I fell in love with mechanics, appreciating its beauty more and more with each new formulation I studied. It is hence for those without formal training but who wish to understand more about physics that I wrote this book. It is for this reason that every mathematical step is *fully* detailed, and plenty of diagrams and examples have been packed into the book to help along the way! A brief biography of the characters who developed the ideas we present is given at the back of the book, for those interested in the history of physics.

I don't think I could pick a "favourite equation" out of the book, or even a favourite formulation - each bring their own elegance and unique features which stand out among the rest! Koopman-von Neumann theory is extremely interesting and is useful for probing the structural features of mechanics but, on the other hand, d'Alembert's principle and the Gibbs-Appell equations are simply beautiful! I happen to think however that Newton and Euler are possibly the two cleverest individuals in the history of natural philosophy, as, without them, we would have no mechanics!

The book can be used in many ways and for different audiences; for an interested individual, it can provide further information on a particular topic; alternatively, it can serve as a support for undergraduate courses in classical mechanics. There are perhaps three types of courses this book would suit: (i) courses on Newtonian mechanics and classical mathematical physics; (ii) introductory Lagrangian and Hamiltonian dynamics courses at the undergraduate level; (iii) advanced courses on mechanics at the postgraduate level.

I would like to thank Duncan Stewart for his illustrations, which far exceeded my expectations and crude drawings. I would like to thank Elizabeth Farrell for correcting my ferocious lack of punctuation and terrible grammar. I would also like to thank Ania Wronski and Lydia Shinoj for their help in the final stages of the book. Thank Renald Schaub and Collin Bleak for their kind encouragements and John Mitchell, Bernd Braunecker and Jonathan Keeling for their supportive feedback. In addition I would like to thank my tutors at Liverpool College and my lecturers at the University of St Andrews, whose passion and enthusiasm for science I have inherited. I would like to thank all my friends for their encouragement and for always being up for a kick about with the football at any spontaneous moment—in particular Giorgio and Josh. Above all, I would like to thank my parents and my beautiful wife, whose love, support and encouragements, from the opening lines to the final edits, made this book a reality!

So here we go—I hope you enjoy reading it as much as I enjoyed writing it!

Peter Mann

Part I

Newtonian Mechanics

The aim of this section is to introduce the reader to the principles of Newtonian mechanics. This chapter covers Newton's laws of motion and subsequently builds up a framework of classical physics in preparation for our treatment of analytical dynamics. It has been kept brief in order to move quickly to the main focus of the book, Lagrangian and Hamiltonian mechanics, but covers the necessary foundations upon which we build.

Newtonian mechanics was the first successful description of dynamics in history. There are previous works and theories some from the ancient Greek philosophers or Renaissance mathematicians; however, none were as successful as Newton's approach. The theory we develop will be useful in all branches of modern physics, as it can be thought of as the very first application of the scientific method we understand today. We begin by defining the basic ideas of positions, velocities and accelerations as well as two types of spaces in classical mechanics, "configuration space" and "phase space". Newton's three laws are explored and conservation laws are defined. We then move on to discuss the work-energy theorem, angular momentum and the harmonic oscillator before closing with some key results of classical mathematical physics. In subsequent chapters, we will utilise the configuration and phase spaces of mechanical systems for more advanced descriptions and see the hidden structure of classical mechanics!

1
Newton's Three Laws

In 1687 the natural philosopher Sir Isaac Newton (see figure 1.1) published the *Principia Mathematica* and with it sparked the revolutionary ideas key to all branches of classical physics. In the late seventeenth and early eighteenth century Newton developed his findings into three fundamental laws of the motion of matter; however, before delving into them let's first discuss some general concepts.

Dynamics is concerned with the propagation of systems in a smooth, continuous and therefore mathematically differentiable evolution of time. The time can be divided up into intervals to consider the motion of particles between **boundary conditions**, e.g. t_1 and t_2. The **system** is our object of interest and for this chapter is considered to be either a single or a collection of generic particles that are not governed by quantum mechanics, for quantum systems do not follow these laws explicitly.

Fig. 1.1: Sir Isaac Newton (1643–1727).

We can use a **coordinate system** to give the location of a single particle meaning in a **configuration space** by specifying three **Cartesian coordinates** (x, y, z). Figure 1.2 illustrates this for a cartesian coordinate system; the parameters that describe the location of the particle (coordinates) in the three-dimensional space (configuration space) are its positions (time-dependent vectors) along each axis. Specifying a configuration for a system of particles is specifying their positions at an instant. For two free particles the configuration space is six-dimensional; for N free particles, a point in the $3N$-dimensional configuration space fully describes the system's positions, as there is a single dimension for each dimension in which the system is free to move. The coordinates of configuration space don't have to be Cartesian; any valid set of parameters that identify the location of a particle can be used to span the space.

As the particle evolves through time we aim to characterise its behaviour with an **equation of motion**, which when applied to the system will result in **deterministic** differential equations that give complete predictability for the scenario. When a system is deterministic, this means that, if the current state of a system is known and the

Lagrangian & Hamiltonian Dynamics. Peter Mann, Oxford University Press (2018).
© Peter Mann. DOI: 10.1093/oso/9780198822370.001.0001

equations of motion that describe the evolution of the system are known, past and future states can be determined with complete predictability. For a system of n dimensions to be deterministic, we need to know $2n$ pieces of information about the system. So, for a free particle that can move in three-dimensional space, we might choose the position and velocities along each axis x, y and z or the position and momenta.

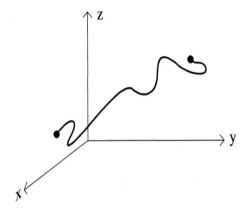

Fig. 1.2: A Cartesian coordinate system and a particle's trajectory between two times.

The **position** of a particle $\vec{r}$ was chosen by Newton to be the fundamental dynamical variable and can be divided into three vector **components**; almost all theories of classical mechanics describe the evolution of position in some way or other. A **trajectory** $\gamma(t)$ is a particle's path through space as a function of time. It maps two positions together via a collection of intermediate states. It is therefore the goal of classical mechanics to identify the trajectory of the system. **Velocity** $\vec{v}$ is the time derivative of position, the rate of change of position with respect to time, and can similarly be split into components in each direction. It is a vector quantity sometimes signified by a dot above the position character: $v_i = \dot{r}_i$. **Speed** is the magnitude of velocity.

$$\vec{r} = (x, y, z), \qquad \vec{v} = (\dot{x}, \dot{y}, \dot{z}), \qquad \vec{a} = (\ddot{x}, \ddot{y}, \ddot{z}) \qquad (1.0.1)$$

The **acceleration** $\vec{a}$ of a particle is the time derivative of its velocity, the rate at which the velocity is changing. If the particle is stationary or is travelling at constant speed *and direction* then there is no acceleration. Since position, velocity and acceleration are vector quantities, a change in the direction of the velocity constitutes an acceleration; hence, particles undergoing circular motion are constantly accelerating even if the speed is constant.

We can recover velocity from acceleration by integration and likewise recover position from velocity. This is perhaps the most important concept and is mostly overlooked;

when we talk of solutions to an equation of motion, we are trying to obtain the accelerations, velocities and positions as functions of time, therefore giving us determinism over the system.

Exercise 1.1 To illustrate these relations let's suppose that we have an equation of motion for a particle in one dimension as a function of time $\vec{r}(t)$:

$$\vec{r}(t) = a + bt + ct^2$$

Differentiating this with respect to time gives the velocity $\dot{r}(t) = b + 2ct$, and a second differentiation gives the acceleration $\ddot{\vec{r}}(t) = 2c$. At $t = 0$ the particle is at position a, with a velocity of b. The particle is under a uniform acceleration of $2c$. It is easy to integrate with respect to time to obtain the velocity and again to obtain an expression for the particle's position.

A particle's **inertia** is its resistance to change in its state of motion, whether it be at rest or at a certain velocity. It is quantitatively measured by the particle's mass; the more massive an object is, the larger the **force** required for a given acceleration. Forces are influences that can change the state of motion of a system, provided they overcome other forces that may act in opposition on the body.

Newton's three laws of motion are neatly presented below:

- An object will remain at rest or constant velocity unless acted upon by an external force, when viewed in an inertial reference frame.
- The vector sum of the forces on an object is equal to the rate of change of velocity multiplied by the mass of the object.
- When a body exerts a force on a second body, the second body exerts a force equal in magnitude and opposite in direction on the first body.

An inertial reference frame is one in which Newtonian physics is valid. The third law tells us straight away to ignore those forces that do not conserve total momentum, even before we have considered their dynamics. Newton's second law is where his famous equation originates: the net force on a body is the product of its mass and acceleration. The acceleration is the response to a force in a given direction.

$$\boxed{\vec{F} = m\vec{a} \quad \textbf{Newton's second law}} \tag{1.0.2}$$

When the net force is zero the system is in **equilibrium**. It may appear that the first law is simply a special case of the second law when the resultant force on the body is zero, setting $\vec{F} = 0$ and substituting $\vec{a}$ for $d\vec{v}/dt$:

$$m\frac{d\vec{v}}{dt} = 0 \tag{1.0.3}$$

We can therefore see that if the time derivative of velocity is zero then the velocity will have the same value it did at $t = 0$, as predicted by and qualifying the first law.

This does not negate the need for the first law, however, since the first law gives us the notion of inertial frames. We can now imagine a scenario of a constant applied force, F_x, for a particle free to move in one dimension; we aim to arrive at a formula for the motion of a particle as a function of time and we do this as follows. We rearrange Newton's second law to give us an expression for acceleration in terms of the applied force:

$$\frac{dv_x}{dt} = \frac{F_x}{m}$$

We now integrate both sides with respect to time, between the limits of t and 0:

$$\int_0^t \frac{dv_x}{dt} dt = \int_0^t \frac{F_x}{m} dt$$

$$v_x \Big|_0^t = \left(\frac{F_x}{m} t + c\right)\Big|_0^t$$

Evaluation of the limits gives us an expression we can rearrange to equal the velocity at time t:

$$v_x(t) - v_x(0) = \left(\frac{F_x}{m} t + c\right) - \left(\frac{F_x}{m} 0 + c\right)$$

Another integration with respect to t and subsequent solving for $r_x(t)$ gives us,

$$\int_0^t v_x(t) dt = \int_0^t \left(v_x(0) + \frac{F_x}{m} t\right) dt$$

$$r_x(t) = r_x(0) + v_x(0)t + \frac{F_x}{2m} t^2$$

Remembering $F_x = ma_x$ we can substitute this into our equation to give,

$$r_x(t) = r_x(0) + \dot{r}_x(0)t + \frac{1}{2}\ddot{r}_x t^2$$

This result is the **Verlet algorithm**, which is the basic way to integrate an equation of motion. This demonstrates the predictability of Newton's formalism: by knowing a few parameters of our system we can predict exactly what will happen under a known applied force. Please note, however, we are assuming that the mass is constant and that we do, indeed, understand the force we are applying.

Some equations of motion generally have more than one solution; it is the choice of the **initial conditions** which provide the specific solution. In this way we can follow a unique path between two points in configuration space from any number of feasible trajectories that would also provide a solution.

1.1   Phase Space

Momentum $\vec{p}$ is the product of the particle's mass and its velocity; it is therefore a vector quantity, even if we later become careless with the assignation of the vectorial notation in following chapters:

$$\vec{p} = m\vec{v} \qquad (1.1.1)$$

If we now take the time derivative of both sides we obtain

$$\frac{d}{dt}\vec{p} = m\frac{d}{dt}\vec{v} \qquad (1.1.2)$$

We use Newton's second law to see that the time derivative of the momentum is equal to the resultant force on the system. We can therefore rewrite Newton's second law by saying *force is the rate of change of momentum*. We now have two very important relations that form the backbone of the Newtonian formalism:

$$\vec{p} = m\dot{\vec{r}}, \qquad \vec{F} = \dot{\vec{p}} \qquad (1.1.3)$$

Together, they incorporate the essence of Newton's formalism and allow us to make a statement:

> *Knowing the position and momentum provides us with complete Laplacian determinism of the system, provided we know the equations of motion and have chosen initial conditions.*

We compile this into the idea of **phase space**. Consider a particle free to move in one dimension. It can move back and forth but let's say that its location is within some specific area at a given time along the x-axis. We can also consider the particle's momentum. We know that it's going to be within certain values and that there's going to be a maximum, a minimum and a sort of average momentum. If we make a plot of x-dimensional momentum on the vertical axis versus x-dimensional position on the horizontal we are going to end up with a region, a patch, where all the actual values of both p_x and r_x are contained. This two-dimensional patch is called a **phase plane** and, for our one-dimensional particle on a line, it contains all the configurations of position and possible values of momentum for that dimension. For motion in one dimension the phase plane is a two-dimensional space and we can talk of 'the area of the phase plane the system occupies' as it traces out trajectories.

Now, imagine the same plot for each dimension x, y and z in Cartesian coordinates and their respective momentum coordinates. The particle will now occupy some **volume of phase space** which is interpreted as containing all of its possible configurations in momentum and position over the six-dimensional space (position and momenta for each dimension). Trying to imagine graphically of what this look like is impossible; rather, we just accept that it is a space of states. Each point in phase space describes the **state** of the particle. The state of the system is described not only by its position but also by its momentum (we could also choose a velocity phase space, which is position versus velocity). The state of a system of n degrees of freedom is described by a

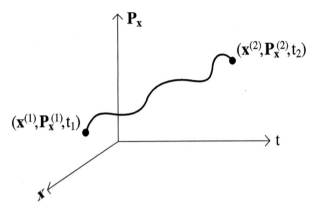

Fig. 1.3: The extended phase space of a particle evolving during the time interval $[t_1, t_2]$.

volume in $2n$-dimensional phase space, a degree of freedom being x, y, z, etc. Because we can't draw such a diagram for more than one dimension, we can use bold symbolled characters to represent all the dimensions into a two-dimensional representation; and then plot that two-dimensional plane against a time axis to represent time evolution. Such a phase space with a time axis is an **extended phase space**, as depicted in figure 1.3.

So, configuration space is all the possible positions of the system, and phase space is all the possible positions *and* all the possible momenta the system can take on. We will come back to the idea of phase space later when discussing **Hamiltonian mechanics** and then give it a deeper meaning when we discuss statistical entropy, but for now our definition is enough. You see that knowing the point in phase space a system occupies satisfies the conditions necessary to be deterministic. A point in configuration space does not satisfy this. Think of it this way: is it enough to know the position of a moving body to know where it is going? No; we need to know its velocity (or momentum), in addition to the position, to determine where it will be.

1.2   Systems of Particles

Newton's third law is tricky at first but if we take some time with it we can see it is rather intuitive, especially for a system of particles, which we now turn to consider. We use the notation i and j to represent different particles, and the subscript ij under F reads *the force on i due to j*, while ji reads *the force on j due to i*. For particles i and j the third law can be stated as

$$\vec{F}_{ij} = -\vec{F}_{ji} \tag{1.2.1}$$

The force on particle i due to particle j is equal yet opposite, hence the negative sign, to the force on j due to i. This begs the question, why, if forces are seemingly always balanced in these pairs acting in opposite directions, do we get resultant forces on bodies? The answer is because they act on different particles, so each individual particle experiences a resultant force. Before we move on we must clarify what we mean by **internal forces** and **external forces** for a system of particles. If we consider a gas in a box as our system of N particles then an *external force* acts on the entire system (i.e the entire box), whilst *internal forces* occur within the system (i.e molecular collisions or interactions).

The total internal force on a particle is the sum of all the forces acting on it due to all the other particles it is influenced by:

$$\vec{F}_i = \sum_j \vec{F}_{ij} \tag{1.2.2}$$

where $\vec{F}_i$ is the total internal force on particle i and $\vec{F}_{ij}$ is the force on i due to particle j. The summation is over all $j \neq i$, which just means over every other particle that acts on i except for the particle itself, as you can't consider a particle's collision with itself. The total rate of change of momentum for all particles is given by the sum over i of the change of momentum of the ith particle $\dot{\vec{p}}_i$ and is equal to the sum of all the internal forces due to all of the particle collisions:

$$\sum_i \dot{\vec{p}}_i = \sum_{i,j} \vec{F}_{ij} \tag{1.2.3}$$

Since we are now counting over i as well, we are including all the reactive forces as well (i.e the force on j due to i). By Newton's third law the sum over both of these will be zero, since every force is counted twice, but one is + and the other -. This leaves us with the **conservation of momentum**: the total momentum of an isolated system never changes:

$$\frac{d}{dt} \sum_i \vec{p}_i = 0 \tag{1.2.4}$$

Conservation laws state that the system retains information about the value of a quantity as the dynamics plays out. We can test to see if a given quantity is a conserved quantity by taking its time derivative; if it is zero then the quantity does not change with time and will have the same value over a time interval $\forall \ t \in [t_0, t_1]$ between t_1 and t_2 ($\forall$ means 'for all', t 'times', $\in$ 'in', the 'interval' $[t_0, t_1]$).

We now consider external forces too. The second law for such a system is written in equation 1.2.5, where $\dot{p}_i$ is now the total rate of change in momentum of the individual particle. This is the sum of internal and external forces, where $F_i^{(e)}$ is the net external force acting on particle i, and $\sum F_{ij}$ is the total internal force.

$$\sum_j \vec{F}_{ij} + \vec{F}_i^{(e)} = \dot{\vec{p}}_i \tag{1.2.5}$$

Using the third law and summing over all i particles we can write an expression for the total rate of change of momentum of the system as follows. The left-hand side is written in equation 1.2.6, but note that $j \neq i$, since it is silly to think of the force a particle is exerting on itself via a collision with itself; so, the sum is from $j = 1$ to $j = N$, excluding $j = i$:

$$\sum_i \sum_{j,j\neq i} \vec{F}_{ij} + \sum_i \vec{F}_i^{(e)} = \vec{F}_{\text{total}}^{(e)} \tag{1.2.6}$$

Since we know by the third law that

$$\sum_i \sum_{j,j\neq i} \vec{F}_{ij} = 0 \tag{1.2.7}$$

assuming the mass is constant we can sum over the right-hand side of equation 1.2.5:

$$\sum_i \dot{\vec{p}}_i = \sum_i \frac{d\vec{p}_i}{dt} = \sum_i m_i \frac{d\vec{v}_i}{dt} = \sum_i m_i \frac{d}{dt}\left(\frac{d\vec{r}_i}{dt}\right) = \frac{d^2}{dt^2}\left(\sum_i m_i \vec{r}_i\right) \tag{1.2.8}$$

For a system with a distribution of mass we can define a point where the *weighted* position of the mass all added up is equal to zero; it is a sort of mass balance point if you like. For N identical particles with coordinates r_i where the centre of mass *relative* to the origin of the coordinate system is given by $\vec{R}$:

$$\vec{R} = \frac{\sum_i m_i \vec{r}_i}{\sum_i m_i} = \frac{\sum_i m_i \vec{r}_i}{M} \tag{1.2.9}$$

where M is total mass, the sum of all the particles of the system. This coordinate is defined from mass balance theory or **moments**. The coordinates are defined by solving for $\vec{R}$ in the following relation:

$$\sum_i m_i(\vec{r}_i - \vec{R}) = 0 \tag{1.2.10}$$

The vector $\vec{R}$ is a mass-weighted average of the positions. We now substitute this into equation 1.2.8 to obtain an expression for the total external force acting on the system in terms of the coordinates of the centre of mass and the total mass of the particles.

$$\vec{F}_{\text{total}}^{(e)} = \frac{d^2}{dt^2}\left(\sum_i m_i \vec{r}_i\right) = M\frac{d^2\vec{R}}{dt^2} \tag{1.2.11}$$

This result tells us that the centre of mass moves like a point particle with a mass equal to the sum of all the particles and with all the external forces acting directly on it. This is very useful for a system with a large number of degrees of freedom where we wish to characterise the dynamics of the overall system without all the computational cost.

We come back to the conservation of linear momentum using the expression for the total momentum and sum the right-hand side over N particles to give the total rate of change of momentum of the system:

$$\vec{p}_{\text{total}} = \sum_i p_i = \sum_i m_i \frac{d\vec{r}_i}{dt} = \frac{d}{dt} \sum_i m_i \vec{r}_i = M \frac{d\vec{R}}{dt} \qquad (1.2.12)$$

If the total external force on a system is zero then the total linear momentum is conserved (since the time derivative is zero) which can be seen by simple integration of the equation below, it is a constant:

$$\vec{F}^{(e)}_{\text{total}} = \dot{\vec{p}}_{\text{total}} \qquad (1.2.13)$$

When we attempt to solve problems in Newtonian mechanics we start by solving the second law for the trajectory and once we have the equation of motion it is simply a mathematical exercise to find a solution. The general problem is written below for the ith particle. The sum over all particles cancels the first term by the third law:

$$\sum_j \vec{F}_{ij} + \vec{F}_i^{(e)} = m_i \ddot{\vec{r}}_i \qquad (1.2.14)$$

In general, however, when **constraints** are present we may not fully understand the form of the external forces. We will not consider constraints in the Newtonian formalism but will cover them extensively in analytical frameworks. You must agree that the above proof of the conservation of momentum is a bit contrived but, alas, there is no other way in Newtonian mechanics to find conservation laws than brutish calculations; this is not so of Lagrangian mechanics!

1.3 The N-body Problem

We conclude this section by looking at a system of N particles that have a central force $\vec{F}$ acting between them. The force has constant coefficients κ_{ij} such that

$$\vec{F}_{ji} = -\kappa_{ij} \frac{\vec{r}_i - \vec{r}_j}{|\vec{r}_i - \vec{r}_j|^n}, \qquad j \neq i \qquad (1.3.1)$$

The equations of motion are formed from Newton's second law and can be written as a system of N-coupled equations with $i \neq k$:

$$m_i \ddot{\vec{r}}_i = -\sum_{j \neq i}^N \kappa_{ij} \frac{\vec{r}_i}{r_{ij}^n} + \sum_{j \neq i,k}^N \kappa_{ij} \frac{\vec{r}_j}{r_{ij}^n} + \kappa_{ik} \frac{\vec{r}_k}{r_{ik}^n} \qquad (1.3.2)$$

This is known as the **N-body problem** for a set of initial positions and velocities. Motivated by Newton's *Principia*, the solution to the problem was the subject of a prize in honour of the sixtieth birthday of King Oscar II of Sweden in 1889. Although never solved in the general case even to this day, the prize was awarded to Henri Poincaré, a mathematician whose works we will investigate later in the book! This system has $6N$ variables $(r_x, r_y, r_z, \dot{r}_x, \dot{r}_y, \dot{r}_z)$ for each particle and, using the **Cauchy-Lipschitz theorem**, it is known that there does exist a unique solution.

There are certain quantities of a dynamical system called **first integrals** of the motion that allow us to reduce the dimensionality of the system.[1] For the N-body system there are three for the centre of mass, three for the linear momentum, three for the angular momentum and one for the energy. These integrals reduce the problem to a dimension of $6N - 10$. A further two integrals can be found, as shown by Jacobi (another mathematician we will investigate).

In the case that $N = 1, 2$ the N-body problem can be solved analytically as we have enough integrals of motion to reduce the system sufficiently. In the case that $N = 3$ there is still a six-dimensional system of equations. Further, Poincaré proved that the system contains no further integrals of the motion other than those above. This means that the system cannot be solved using the method of first integrals.

Due to this problem new methods were developed that are widely used today in all aspects of science, most notably **perturbation theory**. We will investigate Poincaré's canonical perturbation theory later in the book and discover what **KAM theory** has to say about the stability of perturbative solutions. We will also look at **distribution functions**, in particular, the Liouville probability density function $\rho(\boldsymbol{r}, \boldsymbol{p}, t)$ in phase space. We also note that many numerical methods have been developed to brute force a solution, such as the Runge-Kutta integration scheme.

The N-body problem is still an open unsolved problem in mechanics and, although much headway has been made in the case that $N = 3$, solutions are usually restrictive, without collisions or fixed inter-particle distances.

---

[1] Don't worry about these for now; we will come back to them when they are needed.

Chapter summary

- The position data of N particles is encoded by a point in a $3N$-dimensional **configuration space**. The **state** of the system is the specification of both the positions and the momenta (or velocity).

- The state of a system with n degrees of freedom is specified by a point in the $2n$-dimensional **phase space**.

- A **conservation law** describes the invariance of a given quantity under time evolution. In practice we identify conserved quantities by the vanishing of their time derivatives.

- We first introduced single-particle dynamics before generalising to consider a collection of particles.

Exercise 1.2 For a system of $i = 1, \ldots, N$ particles Newton's second law is a system of coupled second-order ordinary differential equations, (ODEs):

$$m_i \frac{d^2}{dt^2} \vec{r}_i = \vec{F}_i(\vec{r}_1, \ldots, \vec{r}_N; \frac{d\vec{r}_1}{dt}, \ldots, \frac{d\vec{r}_N}{dt}, t) \qquad (1.3.3)$$

The coordinate of the ith particle as a function of time $\vec{r}_i(t)$ is the trajectory we aim to find. As with any second-order dim(N) system we must specify exactly $2N$ quantities to solve for the trajectories of all the particles. In practice these quantities are often the initial coordinates and the initial velocities at the initial time. We would then say that Newton's second law becomes an **initial value problem**.

We may simplify this system somewhat by rewriting it as a system of $2N$ coupled first order systems:

$$\frac{d\vec{v}_i}{dt} = \frac{1}{m_i} \vec{F}_i(\vec{r}_1, \ldots, \vec{r}_N, \vec{v}_1, \ldots, \vec{v}_N, t), \qquad \frac{d\vec{r}_i}{dt} = \vec{v}_i \qquad (1.3.4)$$

Such a system may be solved by **Picard's iteration scheme** via integral equations:

$$\vec{v}_i(t) = v_i(t_0) + \int_{t_0}^{t} \frac{1}{m_i} \vec{F}_i(\vec{r}_1(t'), \ldots, \vec{r}_N(t'), \vec{v}_1(t'), \ldots, \vec{v}_N(t'), t') dt' \qquad (1.3.5)$$

$$\vec{r}_i(t) = \vec{r}_i(t_0) + \int_{t_0}^{t} v_i(t') dt' \qquad (1.3.6)$$

As for the one-dimensional case we find the solution to the jth-order in terms of the $j - 1$th order.

Exercise 1.3 Newtonian mechanics is valid in an inertial coordinate system. In general, a transformation of the coordinate frame will not preserve the form of the equations of motion. To see this, we suppose that the original coordinates of the inertial frame are given by $\vec{x}_i$ and that there are new coordinates $\vec{y}_i$ of a non-inertial frame related by the **transformation**

equation $\vec{y}_i = \vec{y}_i(\vec{x}_i, t)$. This equation tells us how to convert from the $\vec{x}$ frame to the $\vec{y}$ frame. We know that the dynamical equation of motion will contain second derivatives of the coordinate so we will have to compute the second time derivative in the $\vec{y}_i$ frame:

$$\frac{d\vec{y}_i}{dt} = \sum_j \frac{\partial \vec{y}_i}{\partial x_i^j} \frac{dx_i^j}{dt} + \frac{\partial \vec{y}_i}{\partial t} \tag{1.3.7}$$

$$\frac{d^2 \vec{y}_i}{dt^2} = \sum_j \frac{\partial \vec{y}_i}{\partial x_i^j} \frac{d^2 x_i^j}{dt^2} + \sum_{j,k} \frac{\partial^2 \vec{y}_i}{\partial x_i^j \partial x_i^k} \frac{dx_i^j}{dt} \frac{dx_i^k}{dt} + 2 \sum_j \frac{\partial^2 \vec{y}_i}{\partial x_i^j \partial t} \frac{dx_i^j}{dt} + \frac{\partial^2 \vec{y}_i}{\partial t^2} \tag{1.3.8}$$

If we then multiply this by m_i and use Newton's equation in the $\vec{x}$ frame, $\vec{F}_i = m_i(d^2 \vec{x}_i / dt^2)$, we have the transformed form of Newton's second law:

$$m_i \frac{d^2 \vec{y}_i}{dt^2} = \sum_j \frac{\partial \vec{y}_i}{\partial x_i^j} \vec{F}_i + m_i \sum_{j,k} \frac{\partial^2 \vec{y}_i}{\partial x_i^j \partial x_i^k} \frac{dx_i^j}{dt} \frac{dx_i^k}{dt} + 2m_i \sum_j \frac{\partial^2 \vec{y}_i}{\partial x_i^j \partial t} \frac{dx_i^j}{dt} + m_i \frac{\partial^2 \vec{y}_i}{\partial t^2} \tag{1.3.9}$$

The first term on the right is the transformed inertial force. There are, however, three additional forces that all contain m_i and, hence, we call them **inertial forces**, or **d'Alembert forces**. If we set the force $\vec{F}_i$ equal to zero in the inertial $\vec{x}_i$ frame these d'Alembert forces are still present in the non-inertial frame.

There are a set of transformations from an inertial frame to another inertial frame that preserve the form of Newton's second law. These are called **Galilean transformations**. They include a shift of the origin of the original frame by a fixed position vector, a shift of the origin by a fixed velocity vector, and rotation of the origin of the frame or a shift in the time by a fixed value. If we perform these transformations to the coordinate frame the form of Newton's equations of motion does not change relative to the first coordinate frame: the two frames are said to be in *constant relative motion*.

2
Energy and Work

In this chapter we investigate the work-energy theorem and define the kinetic and potential energies of the system. While there is some vector calculus involved it has been kept to the bare minimum and the reader should not require in-depth knowledge to understand the salient points.

If there is a net force on the particle, it accelerates in the direction of the unbalanced force. The **work done** W on the particle is the result of an unbalanced force acting on the particle, which as a result, moves some distance. It is given by the dot product of the net external force vector and the **displacement vector** $\vec{s}$ (just the difference in the two position vectors) integrated over the trajectory, connecting two points in configuration space. In the language of vector calculus, we would call it a **contour integral**, as depicted in figure 2.1.

$$W_{12} = \int_1^2 \vec{F} \cdot d\vec{s} \tag{2.0.1}$$

Work is a scalar quantity and it is dependent on the path taken; if the particle remains stationary during the applied force, then no work has been done. Work and energy are closely related in sort of cyclic definition:

Energy is the ability to do work, while work is the transfer of energy.

The following two relations can be substituted into the above expression:

$$\vec{F} = m\frac{d\vec{v}}{dt}, \qquad d\vec{s} := \vec{v}dt$$

$$W_{12} = \int_1^2 m\frac{d\vec{v}}{dt} \cdot \vec{v}dt \tag{2.0.2}$$

If we now take a quick look at the following time derivative, which is utilising the product rule, we obtain

$$\frac{d}{dt}(v^2) = \frac{d}{dt}(\vec{v} \cdot \vec{v}) = \frac{d\vec{v}}{dt} \cdot \vec{v} + \vec{v} \cdot \frac{d\vec{v}}{dt} = 2\left[\frac{d\vec{v}}{dt} \cdot \vec{v}\right]$$

Lagrangian & Hamiltonian Dynamics. Peter Mann, Oxford University Press (2018).
© Peter Mann. DOI: 10.1093/oso/9780198822370.001.0001

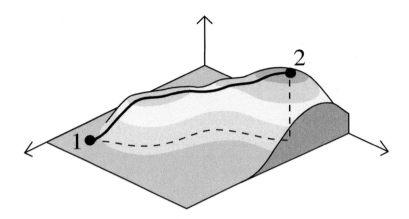

Fig. 2.1: The area under a path between two points in configuration space is evaluated using a contour integral.

$$\frac{1}{2}\frac{d}{dt}\left(v^2\right) = \frac{d\vec{v}}{dt} \cdot \vec{v} \tag{2.0.3}$$

$$W_{12} = \int_1^2 \frac{1}{2}m\frac{dv^2}{dt}dt = \int_1^2 \frac{1}{2}md(v^2) = \frac{1}{2}m\left(v_2^2 - v_1^2\right) \tag{2.0.4}$$

If we define a scalar quantity called the **kinetic energy** of a particle as $T = \frac{1}{2}mv^2$, this result is known as the **Work-Energy Theorem**, which states that the net work done is equal to the change in kinetic energy of a system:

$$W_{12} = T_2 - T_1 \tag{2.0.5}$$

Before, it was stated that, if we know the momentum and position of a particle, we have complete determinism of the system. To illustrate this, we can show that the kinetic energy can be considered to be a function of the momentum of the particle by simple rearrangement; remember, we are only considering one particle:

$$2mT = (mv)^2 \tag{2.0.6}$$

$$T = \frac{p^2}{2m} \tag{2.0.7}$$

So we see that we can *invert* the description from velocity to momenta. This simple concept will prove very useful later on in the book!

A **conservative force** is a force whose work done on going from initial to final positions depends only on the end points and not the path taken. Therefore, if we

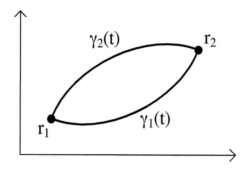

Fig. 2.2: A closed path in configuration space.

imagine a closed path (where there are no end points) then the net work done is zero. Let the two trajectories connecting two points in configuration space r_1 and r_2 equal $\gamma_1(t)$ and $\gamma_2(t)$, according to figure 2.2. We can therefore write:

$$\int_{\gamma_1} \vec{F} \cdot d\vec{s} = \int_{\gamma_2} \vec{F} \cdot d\vec{s}$$

Rearranging it, we obtain

$$\int_{\gamma_1} \vec{F} \cdot d\vec{s} - \int_{\gamma_2} \vec{F} \cdot d\vec{s} = 0$$

This result simplifies using slightly different notation for the integral sign.

$$\oint \vec{F} \cdot d\vec{s} = 0 \tag{2.0.8}$$

That is to say, the integral around a closed path and for any conservative force is exactly zero. We use **Stokes' theorem** to show that the integral around the closed path is equal to the curl of the force:

$$\oint \vec{F} \cdot d\vec{s} = \iint_S (\vec{\nabla} \times \vec{F}) \cdot d\vec{n} \tag{2.0.9}$$

There must be a scalar function such that the right hand side is zero (see chapter 34), thus allowing us to define the force up to an additive constant:

$$\vec{F} = -\vec{\nabla} U \tag{2.0.10}$$

since

$$\vec{\nabla} \times \vec{\nabla} U = 0 \tag{2.0.11}$$

We call U the **potential energy** of the system, and the negative sign is just for convention; it ensures that a given force accelerates the particle to a lower potential energy, so that the work is done against the force.

We say that the force is a **central force** if it depends only on the distance between the point the force acts and the coordinate origin. In general, for spherical coordinates, the potential can be written $\vec{\nabla} U(r, \theta, \phi)$,

$$\vec{\nabla} U = \frac{\partial U}{\partial r} \hat{r} + \frac{1}{r} \frac{\partial U}{\partial \theta} \hat{\theta} + \frac{1}{r \sin \theta} \frac{\partial U}{\partial \phi} \hat{\phi}$$

where characters with a hat symbol are the basis vectors in each dimension. A central potential or **spherically symmetric potential** is one where $U = U(r)$ such that the derivatives with respect to θ and ϕ are zero. In this case, the potential is independent of the angles θ and ϕ, and the gradient reduces to

$$\vec{\nabla} U = \frac{\partial U}{\partial r} \hat{r}$$

A central force will act in straight lines towards or pointing outwards from the coordinate frame origin. At any point on a sphere of constant radius, the magnitude of the force is constant; this means the force is **isotropic**.

A general **potential energy surface** is shown in figure 2.3, which depicts how the potential varies with position within the defined domain. We should note that, at stationary points, the gradient of the potential $\vec{\nabla} U$ is zero, meaning that there is no net force on the particle. The stationary points may be minima, maxima or saddle points.

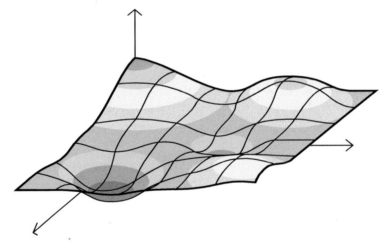

Fig. 2.3: A potential energy surface in $\mathbb{R}^3$.

It is important to note that a conservative potential is time independent; this is something which we will return to in later chapters. It should also be apparent that the work is a function of dU and so the absolute values of the potential are unimportant, rather the *difference* in potential; the **potential difference** is what determines the work done. Non-conservative forces cannot be derived from a scalar potential and they include velocity-dependant potentials (e.g. a particle in a charged field, and other non-inertial reference frames):

$$W_{12} = \int_1^2 \vec{F} \cdot d\vec{s} = -\int_1^2 dU = U_1 - U_2 \qquad (2.0.12)$$

Neither potential nor kinetic energy are conserved quantities on their own. Rather, it is the sum of the potential and kinetic contributions, in other words, the **total energy** E that is conserved for isolated systems. We can show that mechanical energy is conserved when a system is acted upon by conservative forces, using the work-energy theorem.

$$T_2 - T_1 = U_1 - U_2 \qquad (2.0.13)$$
$$T_1 + U_1 = T_2 + U_2 \qquad (2.0.14)$$

We can prove this result from Newton's second law and highlight the importance of the forces being conservative. We take the time derivative of the total energy to obtain

$$\frac{dE}{dt} = \frac{d}{dt}\left(\frac{1}{2}m\frac{dx}{dt} \cdot \frac{dx}{dt} + U(x)\right) = m\frac{dx}{dt} \cdot \frac{d^2x}{dt^2} + \vec{\nabla}U(x) \cdot \frac{dx}{dt} = \frac{dx}{dt} \cdot \left(m\frac{d^2x}{dt^2} + \vec{\nabla}U(x)\right) = 0$$

where the last term in brackets is Newton's second law. We now apply these ideas to a system of N particles. We should note that the work done by a system of particles is evaluated by considering every force acting in the motion of the system. We do this in the regular fashion of dividing the total net force up into internal and external contributions:

$$W = \sum_i \int_1^2 \vec{F}_i \cdot d\vec{s}_i$$

$$= \sum_i \int_1^2 \vec{F}_i^{(e)} \cdot d\vec{s}_i + \sum_i \sum_{j \neq i} \int_1^2 \vec{F}_{ij} \cdot d\vec{s}_i \qquad (2.0.15)$$

If we now assume that all the forces are conservative, we can express this as a potential gradient. The external forces are assumed to be independent of the positions of other particles; they are only a function of the position of the particle being considered. The internal force couples from the third law only depend on the difference between their positions (i.e. the magnitude of the displacement $\vec{r}_{ij} = |\vec{r}_i - \vec{r}_j|$ between particles i and j). Further, we must assume that the potentials are symmetric functions $U_{ij} = U_{ji}$ up to some constant and that they act along the same line connecting the particles:

$$\vec{F}_i^{(e)} = -\vec{\nabla}_i U_i(\vec{r}_1), \qquad \vec{F}_{ij} = -\vec{\nabla}_{ij} U_{ij}(|\vec{r}_i - \vec{r}_j|) \qquad (2.0.16)$$

where $\vec{\nabla}_i = \frac{\partial}{\partial \vec{r}_i}$,

$$W = \sum_i \int_1^2 -\frac{\partial}{\partial \vec{r}_i} U_i \cdot d\vec{r}_i + \frac{1}{2} \sum_i \sum_{j \neq i} \int_1^2 -\frac{\partial}{\partial \vec{r}_{ij}} U_{ij} \cdot d\vec{r}_{ij}$$

$$= -\sum_i \int_1^2 dU_i - \frac{1}{2} \sum_i \sum_{j \neq i} \int_1^2 dU_{ij} \tag{2.0.17}$$

The factor of a half comes from the sum over both i and j; we count the same particles twice, so must therefore, half the result. In the third step, we have cancelled the partial derivatives with the integration volumes, since the potential is only a function of $\vec{r}_i$, and dU_i is an exact differential, and similarly for $\vec{r}_{ij}$ and dU_{ij}.

If we can write both the external and internal forces as potential gradients, then we can write the total force as a potential gradient too:

$$V = \underbrace{\sum_i U_i}_{\text{external}} + \underbrace{\frac{1}{2} \sum_i \sum_{j \neq i} U_{ij}}_{\text{internal}} \tag{2.0.18}$$

For **rigid bodies** where the forces act along the line connecting the particles, the separation vector is (by definition) constant and, as such, $d\vec{r}_{ij}$ is zero; the internal forces do no work in these systems. This concept is useful for the rigid rotor model.

We can follow the same process of decomposition for the kinetic energy of a collection of particles. The position vector $\vec{r}_i$ can be decomposed into the coordinate of the centre of mass $\vec{R}$ and the distance from the centre of mass to the particle is $\hat{r}_i$ as shown in figure 2.4:

$$\vec{r}_i = \vec{R} + \hat{r}_i \tag{2.0.19}$$

The velocity can also be decomposed in a similar manner, simply by time differentiation:

$$\vec{v}_i = \vec{V} + \hat{v}_i \tag{2.0.20}$$

where $\hat{v}_i$ is the velocity of the ith particle relative to the centre of mass, and $\vec{V}$ is the velocity of the centre of mass. The total kinetic energy of the system is given by $\sum_i \frac{1}{2} m_i \vec{v}_i^2$:

$$\vec{v}_i \cdot \vec{v}_i = \left(\vec{V} + \hat{v}_i\right) \cdot \left(\vec{V} + \hat{v}_i\right)$$

$$= \vec{V}^2 + \left(\hat{v}_i\right)^2 + 2\left(\vec{V} \cdot \hat{v}_i\right)$$

$$T = \frac{1}{2} \sum_i m_i \vec{V}^2 + \frac{1}{2} \sum_i m_i \left(\hat{v}_i\right)^2 + \underbrace{\sum_i m_i \left(\vec{V} \cdot \hat{v}_i\right)}_{\vec{V} \cdot \left[\sum_i m_i \hat{v}_i\right] = 0}$$

Looking at the diagram, we can see that vectors $\vec{V}$ and $\hat{v}$ are orthogonal (the dot product of orthogonal vectors vanishes) and hence the end term disappears.

$$T = \frac{1}{2}M\vec{V}^2 + \frac{1}{2}\sum_i m_i(\hat{v}_i)^2 \qquad (2.0.21)$$

Indeed, what this result tells us is that the total kinetic energy of the system of particles is comprised of the translational motion of the centre of mass, and the kinetic energy about the centre of mass. It is clear that the total momentum of the system vanishes:

$$\sum_i \hat{p}_i = \sum_i m_i\hat{v}_i = \sum_i m_i(\vec{v}_i - \vec{V}) \qquad (2.0.22)$$

Then recall from chapter 1 that $\vec{V} = (\sum m\vec{v})/\sum m$:

$$\sum_i m_i\vec{v}_i - \sum_i m_i\frac{\sum_j m_j\vec{v}_j}{\sum_j m_j} = 0 \qquad (2.0.23)$$

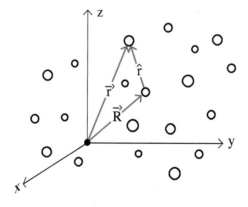

Fig. 2.4: Centre of mass coordinates for a system of particles.

Chapter summary

- The **work-energy theorem** is shown to lead to the **conservation of energy** in **conservative** mechanical systems.
- The **kinetic energy** T of a system is a scalar defined by $T = \frac{1}{2}mv^2 = p^2/2m$. It is the energy due to the movement of the system.
- The **potential energy** U of a conservative system is a scalar function defined by $\vec{F} = -\vec{\nabla}U(r)$. It is the energy due to a particle's position.
- The **total energy** of the system is the sum of the kinetic and potential energies $T + U$.

Exercise 2.1 Here, we would like to take a closer look at Stokes's theorem and understand the condition necessary for a force to be conservative:

$$\vec{\nabla} \times \vec{F} = 0 \qquad (2.0.24)$$

If we were to do this using the Cartesian form of $\vec{\nabla}$, then the theorem is a straightforward evaluation of $\vec{\nabla} \times (\vec{\nabla}U) = 0$.

Vector calculus was largely developed by the Scottish physicist Maxwell, who studied electromagnetism from a mechanical origin. He introduced the term 'curl' for two dimensions and 'twirl' for three dimensions! Although we look at Stokes' theorem in detail later on, from an exterior derivative point of view, it's good to get some intuition going, aside from all the maths!

So, what in layman's terms do all these funny looking symbols $\oint$, $\vec{\nabla}$ ... actually mean? Imagine we have some (oriented) surface with a vector field on it at each point; for instance, drop a piece of wood in a river: it doesn't just stay where it was dropped-it moves with the flow of the vector field. Although there is a general 'flow' downstream, there is also another kind of motion: the rotation at each point. If the river had no rotation anywhere, then the stick would remain perfectly straight down the river. If there is rotation, then we might expect the stick to spin around a little. In some cases, it may spin around a lot!

If we divide the surface up by a grid, then, at each point, quantify the 'amount of rotation' (the curl) or the 'circulation' of the field on that point of the grid:

$$\vec{\nabla} \times \vec{F} \qquad (2.0.25)$$

When we add this up over all the little grids the surface was divided into, we get

$$\iint_S (\vec{\nabla} \times \vec{F}) \cdot d\vec{n} \qquad (2.0.26)$$

It turns out that, at each grid inside the surface, for each little square, the opposite side actually cancels out the opposing contribution. So, the top will cancel out the bottom, and the left side will cancel out the right adjacent face. In fact, all we are left with is the contributions from the parts where there is no opposing side of the grid to cancel-the boundary of the surface!

If you draw a shape without sharp corners on a piece of paper, then the boundary is a line integral. In three dimensions, it is a surface integral. Therefore, the curl is equal to the integral around the boundary:

$$\oint \vec{F} \cdot d\vec{s} = \iint_S (\vec{\nabla} \times \vec{F}) \cdot d\vec{n} \tag{2.0.27}$$

Now, how do we know from all this that $\vec{F} = -\vec{\nabla}U$? Well, it is well known that $\vec{\nabla} \times \vec{\nabla}U = 0$; hence for Stokes' theorem to hold, for any generic closed curve, we must have that the force is equal to the gradient of a scalar function. The negative sign is something of a convention.

We can use $\vec{\nabla}$ therefore to denote the gradient of a scalar function, $\vec{\nabla}U$. The divergence of a vector field is $\vec{\nabla} \cdot \vec{F}$ and the curl of a vector field is $\vec{\nabla} \times \vec{F}$.

Exercise 2.2 Here, we will take a Lennard-Jones interaction force and compute the potential. This interaction is well known to physical chemists and is a model for describing intermolecular forces and bonding:

$$\vec{F} = \left(\frac{A}{r^{\alpha+1}} - \frac{B}{r^{\beta+1}} \right) \hat{r} \tag{2.0.28}$$

The potential is given by $U(r) = -\int dr F(r)$ and, hence,

$$U(r) = -\int \left(\frac{A}{r^{\alpha+1}} - \frac{B}{r^{\beta+1}} \right) dr$$
$$= \frac{C}{r^\alpha} - \frac{D}{r^\beta} \tag{2.0.29}$$

where $C = A/\alpha$, and $D = B/\beta$. In chemistry, this model is used to explain the origin of the thermal expansion in solids. As crystals heat up, the average intermolecular separation shifts to higher values in the potential well. Try to find the r value that corresponds to the minimum of the potential well for the $(12, 6)$ potential.

3

Introductory Rotational Dynamics

After considering the basic principles of linear motion, we are now going to talk about the importance of **circular motion** and rotations. This is our first example of a special case of motion, using the laws we have developed previously. We begin with uniform circular motion to develop some key ideas before moving on to some more advanced topics. It would be helpful to note that uniform rotation about an axis of symmetry or **principle axis** is much easier to consider and will be the focus of this section.

Consider a particle of mass m moving in a circle on the xy plane about the z axis with a fixed radius r and angle θ changing uniformly counterclockwise from the x axis as a function of time, according to figure 3.1. For this problem it is easiest to use polar coordinates, which are summarised below as $x(t) = r\cos\theta(t)$, $y(t) = r\sin\theta(t)$ and $\vec{r} = x\cos\theta + y\sin\theta$. The position of the particle around the circle, the **arc length**, s is given by the product of the radius r and the **angular displacement** θ,

$$s := r\theta \tag{3.0.1}$$

Define an **angular velocity** ω as the time derivative of the angular displacement.

$$\omega := \frac{d\theta}{dt} \tag{3.0.2}$$

In the case of uniform circular motion we can relate this to the linear velocity by considering the time derivative of equation 3.0.1, $v = r\omega$.

If the angular displacement is exactly 2π then *by definition*, the time interval dt is the period of the motion T. Angular velocity is a vector quantity so the rotating particle is constantly accelerating but, if we consider a constant angular speed, we can integrate our equation for angular velocity to give us the rotation angle as a function of time:

$$\int_0^t \omega \, dt = \int_0^t \frac{d\theta}{dt} dt \tag{3.0.3}$$

If we now define our coordinate system such that $\theta(0)$ is zero, then we can cancel it from our equations:

$$\theta(t) = \theta(0) + \omega t \tag{3.0.4}$$

Lagrangian & Hamiltonian Dynamics. Peter Mann, Oxford University Press (2018).
© Peter Mann. DOI: 10.1093/oso/9780198822370.001.0001

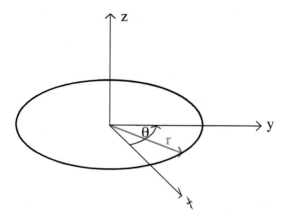

Fig. 3.1: Circular motion.

We now substitute this definition of θ at time t into the above equations for the $x(t)$ and $y(t)$ components:

$$x(t) = r \cos \omega t, \qquad y(t) = r \sin \omega t$$

The angular velocity components of the particle are now the time derivatives

$$v_x(t) = -\omega r \sin \omega t, \qquad v_y(t) = \omega r \cos \omega t$$

We can prove that the two vectors are perpendicular by taking their dot product using $\vec{A} \cdot \vec{B} = A_x B_x + A_y B_y$:

$$\vec{r} \cdot \vec{v} = x(t) v_x(t) + y(t) v_y(t) \tag{3.0.5}$$

$$= \left[r \cos \omega t \right] \left[-\omega r \sin \omega t \right] + \left[r \sin \omega t \right] \left[\omega r \cos \omega t \right] \tag{3.0.6}$$

$$= 0 \tag{3.0.7}$$

Therefore the direction of the angular velocity is perpendicular to the plane of rotation. The acceleration is the time derivative of the velocity components with respect to t:

$$a_x(t) = -r\omega^2 \cos \omega t, \qquad a_y(t) = -r\omega^2 \sin \omega t \tag{3.0.8}$$

Remembering that the magnitude of a vector is given by $|a| = (a_x^2 + a_y^2)^{1/2}$,

$$|a| = \sqrt{a_x^2 + a_y^2} \tag{3.0.9}$$

$$= \left[(-r\omega^2 \cos \omega t)^2 + (-r\omega^2 \sin \omega t)^2 \right]^{\frac{1}{2}} \tag{3.0.10}$$

$$= \left[\omega^4 r^2 (\cos^2 \omega t + \sin^2 \omega t) \right]^{\frac{1}{2}} = (r^2 \omega^4)^{\frac{1}{2}} \tag{3.0.11}$$

If this is our acceleration, then we can now plug that into Newton's second law to give us the force required to keep the particle in constant circular motion. The force is constant due to the constant acceleration, since we are changing the angle constantly:

$$F = mr\omega^2 = \frac{mv^2}{r} \tag{3.0.12}$$

From this we infer that the acceleration is given by $a = r\omega^2 = v^2/r$. Inspection of a_x and $x(t)$ shows that the vectors only differ by $-\omega^2$ (and for a_y and $y(t)$), meaning that the acceleration is in the opposite direction to the position for each component. By definition of our coordinate system, we defined the position to be pointing radially away from the origin of the circle, so the acceleration must, by definition, be pointing towards the centre of the circle. We have a centripetal acceleration and, by the second law, a **centripetal force**. We may, however, wonder what a centrifugal force is? By Newton's third law, we know there is an equal and opposite force acting on the particle as it turns. From the inertial frame of reference (the centre of the circle), the force is centripetal; however, the particle's reference frame is itself rotating and thus experiences a **fictitious force**, or a d'Alembert's force, if you prefer. A fictitious force is a force that appears to be acting in a non-inertial reference frame due to the inertial frames rotation itself, but disappears when observations are made in the inertial frame. There are three well-known fictitious forces to be aware of: **centrifugal**, **Coriolis** and **Euler forces**. Therefore, in Newtonian mechanics, the correct choice of coordinate frame upon which to stage the physics is important. We may, therefore, spend time selecting reference frames; for instance, in rotational-vibrational spectroscopy or particle accelerators, we are forced to consider the problem from a non-inertial frame and we will look at this in a minute.

Like the mass of a linear analogue, the rotating system has a quantitative measure of its intrinsic opposition to rotational motion, called its **moment of inertia**. The moment of inertia I is a measure of the resistance to the angular acceleration of the particle; it is a second-order or *degree* **tensor**, with nine components. However, for our discussion, it is assumed the particle is rotating about a fixed axis, which means that the **angular momentum** and the angular velocity are parallel vectors along the axis of rotation and we can just say that I is a scalar quantity. In general, this is an over-simplification and only holds for rotation about a symmetry axis angular momentum and angular velocity are not parallel if the axis of rotation is not symmetric or the axis is moving (precession), another reason why the dynamics of rotating bodies is fascinating.

The angular momentum $\vec{L}$ is a property of a rotating particle defined as the cross product of the distance from the centre of rotation $\vec{r}$ and the linear momentum $\vec{p}$:

$$\vec{L} := \vec{r} \times \vec{p} \tag{3.0.13}$$

For rotation about a fixed axis, $\vec{L}$ is given by the product of the particle's moment of inertia and its angular velocity. The expression for the moment of inertia depends on the nature of the system and, *in this case*, I is given by the product of the mass and the square of the shortest distance from the centre of the axis of rotation to the particle, "this case" being collinear motion about the principle axis:

$$\vec{L} = I\omega = (m\vec{r}^{\,2})(\dot{\theta}), \qquad \text{with} \qquad I = m\vec{r}^{\,2} \tag{3.0.14}$$

Therefore, in order to change the angular momentum of the particle, we need to change either I or ω. To change I or ω, we must supply a turning force. A **torque vector** $\vec{\tau}$ is defined as the cross product of the applied force and the distance from where the force is applied to the centre of the axis of rotation:

$$\vec{\tau} := \vec{r} \times \vec{F} \tag{3.0.15}$$

In order to give a torque a more applied definition, we can prove that it is equal to the rate of change of the angular momentum using the product rule:

$$\frac{d\vec{L}}{dt} = \left(\vec{r} \times \frac{d\vec{p}}{dt}\right) + \left(\frac{d\vec{r}}{dt} \times \vec{p}\right) = \vec{r} \times \vec{F} + \vec{v} \times \vec{p} \tag{3.0.16}$$

For our fixed axis rotating system the cross product of the parallel momentum and velocity vectors is zero.

$$\frac{d\vec{L}}{dt} = \vec{\tau} \tag{3.0.17}$$

When the applied force is zero, the torque is zero, which leads us neatly into the **conservation of angular momentum:** when no external torques act on the particle, the total angular momentum is constant. We conclude that both the torque and the angular momentum depend on the choice of the origin of the coordinate frame. We extend this to a system of particles by taking the cross product of $\vec{r}_i$ on both sides and summing over i for an aggregate of particles:

$$\sum_i \vec{r}_i \times \left[\sum_j \vec{F}_{ij} + \vec{F}_i^{(e)}\right] = \sum_i \vec{r}_i \times [\dot{\vec{p}}_i] \tag{3.0.18}$$

$$\underbrace{\sum_i \sum_j \left(\vec{r}_i \times \vec{F}_{ij}\right) + \sum_i \vec{r}_i \times \vec{F}_i^{(e)}}_{\sum_i \vec{\tau}_i} = \sum_i \vec{r}_i \times \frac{d}{dt}(m_i \vec{v}_i) \tag{3.0.19}$$

Realising that the cross product of two parallel vectors is zero, we obtain

$$\frac{d}{dt}(\vec{r} \times m\vec{v}) = \vec{r} \times \frac{d}{dt}(m\vec{v}) + [\vec{v} \times m\vec{v}] \tag{3.0.20}$$

which gives us

$$\sum_i \sum_j \left(\vec{r}_i \times \vec{F}_{ij} \right) + \sum_i \vec{\tau}_i = \sum_i \frac{d}{dt}(\vec{r}_i \times m_i \vec{v}_i) = \underbrace{\sum_i \frac{d\vec{L}_i}{dt}}_{\dot{\vec{L}}_{\text{total}}} \tag{3.0.21}$$

$$\overbrace{\sum_i \sum_j \left(\vec{r}_i \times \vec{F}_{ij} \right) + \sum_i \vec{\tau}_i}^{\text{sum of internal and external torques}} = \dot{\vec{L}}_{\text{total}} \tag{3.0.22}$$

The first term on the left is the total internal torque, while the second term on the left is the total external torque. We can now make the assumption that the internal forces $\vec{F}_{ij}$ and $\vec{F}_{ji}$ are not only equal and opposite to each other but act along a line joining the ith and jth particles; this is sometimes called the strong form of the third law. Under this condition, the sum over both i and j particles is zero. Therefore, the total rate of change of angular momentum is equal to the total torque of the external forces:

$$\frac{d}{dt}\vec{L}_{\text{total}} = \sum_i \vec{\tau}_i = \vec{\tau}_{\text{total}}^{(e)} \tag{3.0.23}$$

If the total external torque is zero, then the total angular momentum is conserved.

In chapter 2, we developed the idea of a centre of mass and relative coordinates. We can use this to separate the total angular momentum:

$$\vec{L}_{\text{total}} = \sum_i \vec{r}_i \times m_i \vec{v}_i$$

$$= \sum_i (\vec{R} + \hat{r}_i) \times m_i(\vec{V} + \hat{v}_i)$$

$$= \sum_i \left[(\vec{R} \times m_i \vec{V}) + (\vec{R} \times m_i \hat{v}_i) + (\hat{r}_i \times m_i \vec{V}) + (\hat{r}_i \times m_i \hat{v}_i) \right] \tag{3.0.24}$$

We can use the linearity of the cross product now to slightly alter this expression:

$$\vec{L}_{\text{total}} = \sum_i \left[(\vec{R} \times m_i \vec{V}) + \sum_i (\hat{r}_i \times m_i \hat{v}_i) + (\vec{R} \times \sum_i m_i \hat{v}_i) + (\sum_i m_i \hat{r}_i \times \vec{V}) \right] \tag{3.0.25}$$

Recalling the definition of $\vec{R}$ as a mass-weighted coordinate, we realise that $M\vec{R}$ can be written:

$$M\vec{R} = \sum_i m_i \vec{r}_i = \sum_i m_i(\vec{R} + \hat{r}_i) = M\vec{R} + \sum_i m_i \hat{r}_i \tag{3.0.26}$$

Therefore, $\sum_i m_i \hat{r}_i = 0$. If we take the time derivative of this, it still is zero, that is, $d/dt[\sum_i m_i \hat{r}_i] = 0$. Therefore, with these results, the total angular momentum of a system of particles is

$$\vec{L}_{\text{total}} = \vec{R} \times M\vec{V} + \sum_i (\hat{r}_i \times m_i \hat{v}_i) = \vec{L} + \vec{S} \qquad (3.0.27)$$

This shows that we can separate the total angular momentum into the angular momentum of the centre of mass $\vec{L}$ of the rotating system relative to the origin, plus the angular momentum about the centre of mass $\vec{S}$; these are **orbital** and **spin angular momenta**, respectively.

The kinetic energy of a single rotating particle is given by substituting the linear velocity for the angular velocity $\frac{1}{2}mv^2 = \frac{1}{2}m(r^2\omega^2)$. For a system of particles with equivalent angular velocities:

$$T = \frac{1}{2}\sum_i \omega(m_i r_i^2)\omega = \frac{1}{2}\omega I \omega = \frac{\vec{L}^2}{2I} \qquad (3.0.28)$$

We will discuss angular momentum periodically throughout the book, especially when considering the Poisson bracket. Although it is a long way off just now, we will see how effortlessly we can obtain fundamental results with minimum effort, something that Newtonian physics does not afford.

Lastly, we look at elements of the dynamics of rotations and rigid bodies through non-inertial frames, although it is not our intention to linger on this topic. In order to generalise out of the xy plane, we now write $\vec{\omega}$ rather than ω, signifying it is a vector quantity; along with this, some other generalisations include replacing $v = r\omega$ with $\vec{v} = \vec{\omega} \times \vec{r}$ and taking $a = \omega^2 r$ to $\vec{a} = \vec{\omega} \times (\omega \times \vec{r})$.

Consider the active[1] infinitesimal rotation $d\vec{\Omega}$ of a particle p such that $\vec{r} \to \vec{r} + d\vec{r}$, as shown in figure 3.2. The difference vector $d\vec{r}$ is given by the cross product of $d\vec{\Omega}$ and $\vec{r}$:

$$d\vec{r} = d\vec{\Omega} \times \vec{r} \qquad (3.0.29)$$

Dividing by dt,

$$\frac{d\vec{r}}{dt} = \frac{d\vec{\Omega}}{dt} \times \vec{r} = \vec{\omega} \times \vec{r} \qquad (3.0.30)$$

where we have taken $\vec{\omega} = \dot{\vec{\Omega}}$; this relation is a generalisation of the circular motion we considered previously. Newton's second law does not hold in non-inertial frames but, as mentioned earlier, there arise cases where we *must* base our description outside of an inertial frame. To illustrate this, we imagine two reference frames: the **fixed frame** (f) and the **particle frame** (p). The aim is to describe the position, velocity and

---

[1] An active rotation changes the vector $\vec{r}$, while a passive rotation changes the coordinates.

acceleration of the particle in (p) by vectors measured in (f) while the particle frame is undergoing rotation about a fixed axis. The key concept is that the measurement of the rate of change of a vector quantity is not consistent between the frames so we need some sort of *transformation formula* for converting between frames. This set-up is an example of a non-inertial frame.

As shown in figure 3.2, let us for the moment assume that the vector $\vec{R}$ is fixed and that the particle is at rest in (p) so that $\hat{r}$ is constant with respect to the particle frame. Despite this we know that $\hat{r}$ is, in fact, changing, due to the action of $d\mathbf{\Omega}$, as is apparent in the fixed frame. In the fixed frame, we can write

$$\left(\frac{d\hat{r}}{dt}\right)_{(f)} = \frac{d\mathbf{\Omega}}{dt} \times \hat{r} = \vec{\omega} \times \hat{r} \tag{3.0.31}$$

Now let us allow $\hat{r}$ to move inside the particle frame:

$$\left(\frac{d\hat{r}}{dt}\right)_{(f)} = \left(\frac{d\hat{r}}{dt}\right)_{(p)} + \vec{\omega} \times \hat{r} \tag{3.0.32}$$

The motion of $\vec{r}$ in the fixed frame can be written in terms of $\vec{R}$ and $\hat{r}$:

$$\left(\frac{d\vec{r}}{dt}\right)_{(f)} = \left(\frac{d\vec{R}}{dt}\right)_{(f)} + \left(\frac{d\hat{r}}{dt}\right)_{(f)} \tag{3.0.33}$$

Noticing that the final term is exactly what we derived earlier we can substitute our previous result in to this expression.

$$\left(\frac{d\vec{r}}{dt}\right)_{(f)} = \left(\frac{d\vec{R}}{dt}\right)_{(f)} + \left(\frac{d\hat{r}}{dt}\right)_{(p)} + \vec{\omega} \times \hat{r} = \vec{V} + \hat{v} + \vec{\omega} \times \hat{r} = \vec{v}_{(f)} \tag{3.0.34}$$

Reading this equation from left to right goes as follows. The velocity of the particle in the fixed-axis frame is equal to the velocity of the origin of the rotating particle frame, plus the velocity of the particle relative to the rotating particle frame, plus some extra term that originates from the rotation of the frame. In this way, there arise fictitious forces when the accelerations are computed, due to the rotation of (p) itself; within (p), however, these vanish. We generalise this into a **transformation operator**, which we can apply to any vector quantity $\hat{b}$ of (p) measured in (f):

$$\left(\frac{d\hat{b}}{dt}\right)_{(f)} = \left(\frac{d\hat{b}}{dt}\right)_{(p)} + \vec{\omega} \times \hat{b} \tag{3.0.35}$$

Taking the second time derivative of the velocities in the fixed frame (f):

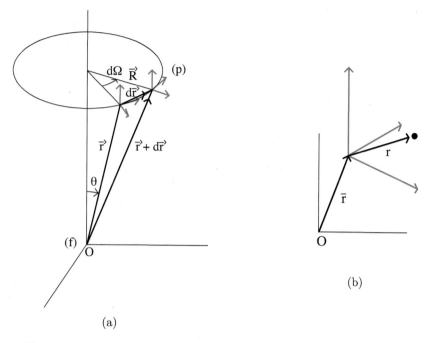

Fig. 3.2: The overall system (a) and a close up of the rotating frame (b).

$$\left(\frac{d\vec{v}_{(f)}}{dt}\right)_{(f)} = \left(\frac{d\vec{V}}{dt}\right)_{(f)} + \left(\frac{d\hat{v}}{dt}\right)_{(f)} + \vec{\omega} \times \left(\frac{d\hat{r}}{dt}\right)_{(f)} + \dot{\vec{\omega}} \times \hat{r} \tag{3.0.36}$$

However, since both $\hat{v}$ and $\hat{r}$ are measured in frame (p), we need to expand them using the transformation operator:

$$\left(\frac{d\vec{v}_{(f)}}{dt}\right)_{(f)} = \left(\frac{d\vec{V}}{dt}\right)_{(f)} + \left(\frac{d\hat{v}}{dt}\right)_{(p)} + \vec{\omega} \times \hat{v} + \vec{\omega} \times \left(\frac{d\hat{r}}{dt}\right)_{(p)} + \vec{\omega} \times \hat{r} + \dot{\vec{\omega}} \times \hat{r} \tag{3.0.37}$$

With the help of equation 3.0.30, we can write these accelerations formally and construct the second law in the fixed frame:

$$\vec{a}_{(f)} = \vec{A}_{\vec{R}} + \hat{a}_{\hat{r}} + 2(\vec{\omega} \times \hat{v}) + \vec{\omega} \times (\vec{\omega} \times \hat{r}) + \dot{\vec{\omega}} \times \hat{r} \tag{3.0.38}$$

This states that the acceleration of the particle in the fixed frame $\vec{a}_{(f)}$ is composed of the acceleration of the particle frame's origin $\vec{A}_{\vec{R}}$ (which is rotating) and the acceleration of the particle relative to the rotating frame $\hat{a}_{\hat{r}}$, plus additional terms that are dependent on the position $\hat{r}$ and velocity $\hat{v}$ of the particle in the particle frame. Newton's second law in the particles frame is simply $\vec{F} = m\hat{a}_{\hat{r}}$. Newton's second law for the net force $\vec{F}$

acting on the particle as observed in the **non-inertial frame** of reference (f) is now written as

$$\vec{F} = m\vec{a}_{(f)} = m\left[\vec{A}_{\vec{R}} + \hat{a}_{\hat{r}} + 2(\vec{\omega} \times \hat{v}) + \vec{\omega} \times (\vec{\omega} \times \hat{r}) + \dot{\vec{\omega}} \times \hat{r}\right] \qquad (3.0.39)$$

Or, more practically,

$$m\hat{a}_{\hat{r}} = \vec{F} - m\vec{A}_{\vec{R}} - 2m(\vec{\omega} \times \hat{v}) - m\vec{\omega} \times (\vec{\omega} \times \hat{r}) - m\dot{\vec{\omega}} \times \hat{r} \qquad (3.0.40)$$

We have the **non-inertial second law**. Reading this equation, the force on the left-hand side is the inertial force inside the particle frame $m\hat{a}_{\hat{r}}$; it is the "original bit" of Newton's second law. Next, $m\vec{A}_{\vec{R}}$ is a force arising due to the linear acceleration of the rotating frame from translation. The next term, $-2m(\vec{\omega} \times \hat{v})$, depends on the motion of the particle in the inertial frame; it is called the **Coriolis force**. The **centrifugal force** arises due to the rotation of the frame; we expect that it will point radially outwards since $(\vec{\omega} \times \hat{r})$ points in the azimuthal direction, $\vec{\omega} \times (\vec{\omega} \times \hat{r})$ points radially inwards and the negative of this points radially outwards. The **Euler force**, $-m\dot{\vec{\omega}} \times \hat{r}$, is present when the angular velocity changes. Apart from the inertial force, they are all fictitious forces that arise due to staging our observation from a non-inertial frame. If we include them in our analysis, then we can continue using Newton's laws in the fixed frame as if it were an inertial frame.

To simplify the equation, we could assume that the origin of the rotating particle frame has constant angular velocity such that $\dot{\vec{\omega}} = 0$, meaning that the Euler force disappears. We could assume that $\vec{R}$ is constant such that $\vec{A}_{\vec{R}} = 0$, too. Another simplification would be to assume that the origins of both coordinate frames coincide with one another such that $\vec{R} = 0$ and hence velocities $\vec{V}$ and accelerations $\vec{A}_{\vec{R}}$ cancel through the working.

Just as with linear dynamics, we can consider a system of particles using single-particle mechanics:

$$\sum_j \vec{F}_{ij} + \vec{F}_i^{(e)} = m_i\left[\vec{A}_{\vec{R}} + \hat{a}_{\hat{r}_i} + 2(\vec{\omega} \times \hat{v}_i) + \vec{\omega} \times (\vec{\omega} \times \hat{r}_i) + \dot{\vec{\omega}} \times \hat{r}_i\right] \qquad (3.0.41)$$

Summing over i particles we obtain

$$\sum_i \sum_j \vec{F}_{ij} + \sum_i \vec{F}_i^{(e)} = \sum_i m_i\vec{A}_{\vec{R}} + \sum_i m_i\hat{a}_{\hat{r}_i}$$
$$+ 2\left[\vec{\omega} \times \left(\sum_i m_i\hat{v}_i\right)\right] + \vec{\omega} \times \left[\vec{\omega} \times \left(\sum_i m_i\hat{r}_i\right)\right] + \dot{\vec{\omega}} \times \sum_i m_i\hat{r}_i$$

The obvious extension of our discussion is to treat non-collinear rotational dynamics, where the rotation is not about a principle axis or an axis that is itself free to move (precession). However, we will end our discussions here, as we have developed the necessary components required for our later work!

Chapter summary

- The **angular velocity** $\vec{\omega}$ is used to understand circular motion in the Newtonian framework. The **angular momentum** is defined as the cross product $\vec{L} = \vec{r} \times \vec{p}$ and a **torque vector** is the cross product of a force vector and the distance from the axis of rotation to the application of the force.

- The total angular momenta of a system of particles is found to be the sum of two contributions from **orbital** and **spin** angular momenta and we see that for a fixed axis system the time derivative of the angular momentum is given by the total external torque on the system.

- We see that **fictitious forces** appear due to non-inertial frames of reference causing us to generalise Newton's second law to afford non-inertial effects.

Exercise 3.1 Show that, for the motion generated by a central, isotropic force, the energy and angular momentum are conserved. To do this, first split the volume element into radial and perpendicular components:

$$\vec{F} \cdot d\vec{r} = F(r)\hat{r} \cdot (dr\hat{r} + d\vec{r}_{\perp}) \tag{3.0.42}$$

Then we note that $\hat{r} \cdot \hat{r} = 1$, since they are parallel, and that $\hat{r} \cdot d\vec{r}_{\perp} = 0$, since they are orthogonal:

$$d(T + U) = d \int F(r)dr - d \int F(r)dr \tag{3.0.43}$$

Hence, we have shown that energy is conserved. Similarly, we know that, if the component of a torque is zero, then the corresponding component of the angular momentum is conserved. To show that angular momentum is conserved, we can show the vanishing of the torque vector, due to $\vec{r} = r\hat{r}$ and $\hat{r}$ being parallel:

$$\vec{\tau} = \vec{r} \times \vec{F} = r\hat{r} \times F(r)\hat{r} = 0 \tag{3.0.44}$$

Note that, in showing the conservation of energy, we used that the dot product of the self vector is 1, while for the conservation of angular momentum, we used that the cross product of the self vector is 0.

4

The Harmonic Oscillator

The harmonic oscillator is a simple model ubiquitous to all branches of physics and is arguably the single most important system we can consider. The importance of the harmonic oscillator stems from the leading term of a Taylor expansion of a potential about a local minima, which are exactly harmonic oscillations. Equivalently small perturbations to a system are well approximated to low order by harmonic oscillations which we soon investigate in the Lagrangian section. The harmonic oscillator is a system with well-known solutions and has been fully investigated since it was first developed by Robert Hooke in the seventeenth century. These factors ensure that the harmonic oscillator is as relevant to a swinging pendulum as it is to a quantum field.

Consider a diatomic molecule consisting of two atoms of mass M_1 and M_2 connected by a massless spring free to oscillate. There is a trade-off between attractive and repulsive forces between the two atoms: when too close, the repulsive forces dominate; when too far, the attractive forces dominate. As such, there is an **equilibrium position** where a non-vibrating diatomic would rest. The unbalanced force is proportional and always acts in opposition to the displacement; it is easy to see in this case that it always acts towards the equilibrium position and hence is termed a **restoring force**. This is the origin of the minus sign in the definition of the restoring force $\vec{F}(x)$ and k, being a constant of proportionality:

$$\vec{F}(x) = -kx \tag{4.0.1}$$

The vector x is the difference between the position vectors of the two particles $x = \vec{r}_1 - \vec{r}_0$ from their equilibrium separation $\vec{r}_0$ but we drop the vectorial notation for brevity. This is **Hooke's law**. We call k the **spring constant**, it is a measure of the stiffness of the spring. A larger k means a much stiffer bond than a lower k, which can be seen as a larger force required to perturb the equilibrium state. Integration of Hooke's law with respect to x gives us the potential energy for the oscillator:

$$V(x) = \frac{1}{2}kx^2 \tag{4.0.2}$$

We introduce the concept of **reduced mass** m to simplify the problem by considering the motion of only a single mass attached via a spring to a fixed mount. Using a

Lagrangian & Hamiltonian Dynamics. Peter Mann, Oxford University Press (2018).
© Peter Mann. DOI: 10.1093/oso/9780198822370.001.0001

reduced mass reduces the complexity of the two-body problem to a one-body problem, and introducing the oscillator via a diatomic molecule is more for our intuition. The first task is to construct the equation of motion using Newton's second law and Hooke's law:

$$\frac{d^2x}{dt^2} = -\omega^2 x, \quad \text{or} \quad \ddot{x} + \omega^2 x = 0 \quad \text{with} \quad \omega = \sqrt{\frac{k}{m}} \tag{4.0.3}$$

This result is the equation of motion for the harmonic oscillator; it is a homogeneous, second-order, linear ODE. As with any problem in dynamics, we want to find an expression for the position as a function of time $x(t)$. A special type of solution is called a **mode** of the oscillator, expressed by

$$x(t) = a\cos(\omega t + \varphi), \quad \text{or} \quad x(t) = a\sin(\omega t + \varphi') \tag{4.0.4}$$

With amplitude a, $\omega = \sqrt{k/m}$ and, with φ taken as a **phase constant**, $\varphi' = \varphi - \pi/2$ represents a phase shift. To prove that this is indeed a solution to Newton's second law, it is necessary to take the second time derivative of $x(t)$, and use Hooke's law. We can see that the acceleration is not constant but a function of x so it matters where in the period of oscillation we evaluate it. Using a trigonometric identity $\cos(A \pm B) = \cos A \cos B \mp \sin A \sin B$, we can express another solution as

$$x(t) = A\cos(\omega t) - B\sin(\omega t) \tag{4.0.5}$$

with $A = a\cos\varphi$ and $B = a\sin\varphi$. It is a general result of differential equations that if x_1 and x_2 are both solutions then by the **superposition principle** so too is $x = c_1 x_1 + c_2 x_2$ where c_i are weightings coefficients. Therefore if we have more than one solution we can find lots more by taking taking linear combinations. The general solution therefore is a linear superposition of modes weighted by their amplitudes.

$$x(t) = \sum_i a_i \cos(\omega_i t + \varphi_i) \tag{4.0.6}$$

The total energy of the harmonic oscillator is the sum of the kinetic and potential energies. The expression of the potential is exactly $\frac{1}{2}kx^2$, which we rewrite in terms of ω for computation:

$$E = \frac{1}{2}m\dot{x}^2 + \frac{1}{2}m\omega^2 x^2 \tag{4.0.7}$$

We can take the time derivative to show that the energy is conserved using the general proof as a guideline, and the reader is invited to prove this. Looking at a solution to the equation of motion $x = a\sin(\omega t + \varphi')$, we can substitute the solution into the energy expression. The velocity is found by differentiating $x(t)$ with respect to time $v(t) = \omega a \cos(\omega t + \phi')$; however, we know that, when $|x| = a$ (at the full stretch of the spring), the velocity is exactly zero. Therefore, we obtain $E = \frac{1}{2}ka^2 = \frac{1}{2}m\omega^2 a^2$. Inverting this expression for the amplitude of the mode we obtain

$$x(t) = \frac{1}{\omega}\sqrt{\frac{2E}{m}}\sin(\omega t + \varphi') \qquad (4.0.8)$$

Since $\sin\theta$ is an oscillating function between ± 1, at the maximum amplitude it disappears and we can characterise the motion in intricate detail.

In the classical picture we have developed, we expect the particle to spend most of its time at the turning points, since it is here that the velocity is minimum. We can imagine a symmetric graph about the equilibrium position of probability versus position and rises to a maximum at both positive and negative amplitudes as in figure 4.1. In order to show this mathematically, lets define a new variable of the system $x'(t)$:

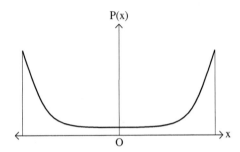

P(x)

Fig. 4.1: The probability density for the simple harmonic oscillator.

$$x' \equiv \omega\sqrt{\frac{\mu}{2E}}x \qquad (4.0.9)$$

Over one cycle, x will vary from a maximum to minimum value, then x' will vary between ± 1. After a straightforward substitution into equation 4.0.8, we obtain the following equation of motion for $x'(t)$:

$$x'(t) = \sin(\omega t + \varphi) \qquad (4.0.10)$$

The probability of finding the particle in between x' and $x' + dx'$ is

$$\int_{-1}^{1} \rho(x')dx' = 1 \qquad (4.0.11)$$

Note the form of the equation is an integral over a **probability density** $\rho(x')$ times a dimensional element (in the sense of length, area or volume), dx' gives a probability for that element. We must realise that, if we add all the volumes up (i.e., we integrate) over the entire range over which the particle is able to move, we must get unity (i.e. the particle is somewhere). This is a **normalisation condition** that is necessary for the probability density to give us results that are meaningful. The probability of the particle being in any given interval of dx' is also proportional to the time it spends in the region dt:

$$\rho(x')dx' = \lambda dt \qquad (4.0.12)$$

where the λ is a proportionality *constant*:

$$\rho(x') = \lambda \frac{dt}{dx'} \qquad (4.0.13)$$

We get the expression for the time from the equation for $x'(t)$, remembering that, if $\sin\theta = a$, then $\theta = \sin^{-1}a$:

$$t(x') = \frac{1}{\omega}\left(\sin^{-1}(x') - \varphi\right) \qquad (4.0.14)$$

which we can now differentiate with respect to time. Since φ/ω has no dependence on x' the problem becomes

$$\rho(x') = \lambda \frac{d}{dx'}\left[\frac{\arcsin(x')}{\omega}\right] = \frac{\lambda}{\omega}\frac{d}{dx'}\arcsin(x') \qquad (4.0.15)$$

The derivative of $\arcsin(x')$ is now shown below. Let

$$y = \arcsin x' \qquad \longrightarrow \qquad \sin y = x' \qquad (4.0.16)$$

Then use the chain rule to compute the derivative of this with respect to x':

$$\cos y \cdot \frac{dy}{dx'} = 1 \qquad (4.0.17)$$

Now, using the Pythagorean identity $\cos^2 y + \sin^2 y = 1$ and rearranging it so that $\cos y = \sqrt{1 - \sin^2 y}$, then squaring the expression for x' to give $\sin^2 y = x'^2$, we obtain

$$\frac{dy}{dx'} = \frac{1}{\sqrt{1 - x'^2}} \qquad (4.0.18)$$

Upon putting the λ and ω back in, we finally arrive at an equation expressing the probability density:

$$\rho(x') = \frac{\lambda}{\omega}(1 - x'^2)^{-1/2} \qquad (4.0.19)$$

If we use the normalisation requirement and the definition of ω as an angular frequency $\omega = \frac{2\pi}{T}$ for half a period (i.e. one oscillation from $x' = 1$ to $x' = -1$). this reduces to

$$\rho(x') = \frac{1}{2\pi}(1 - x'^2)^{-1/2} \qquad (4.0.20)$$

The amplitude a of x' is ± 1, so $a^2 = 1$; this can be used to generate an expression for the probability density in terms of the variables of the oscillator. Considering the term in brackets

$$\sqrt{a^2 - x'^2} = \sqrt{\frac{2E}{\omega^2 \mu} - x'^2} = \sqrt{2E - \mu \omega^2 x'^2} = \sqrt{2T} \qquad (4.0.21)$$

This indicates that the probability density is large when the kinetic energy T is small (i.e. at the turning points). Conversely, when the velocity is large, and hence T is large, the probability density is small. The particle spends more time at the turning points than at the equilibrium position, where the velocity is maximum. Interestingly, this is the exact opposite of the quantum harmonic oscillator, whose probability density is maximum about the equilibrium position.

We return to the harmonic oscillator time and again throughout this text, either to highlight aspects of physics or in worked problems. We treat oscillations in depth in part II.

Chapter summary

- **Hooke's law** is used to derive the equation of motion of the **harmonic oscillator**, which is a second-order homogeneous linear differential equation:

$$\ddot{x} + \omega^2 x = 0$$

- Solutions to this obey the **superposition principle** and take the general form

$$x(t) = \sum_i a_i \cos(\omega t + \varphi)$$

- The probability density of the classical oscillator is computed and shown to be inversely proportional to the kinetic energy.

Exercise 4.1 Here we will recognise that the harmonic oscillator is a linear second-order ODE with constant coefficients and solve the equation in more mathematical terms. We will see that our general solution is equivalent to the solution derived in text. The equation of motion for the system is written as before:

$$\frac{d^2 x}{dt^2} + \omega^2 x = 0 \qquad (4.0.22)$$

We attempt the solution of the form

$$x(t) = e^{\gamma t} \qquad (4.0.23)$$

with γ to be determined. We do this by putting this solution into the equation of motion to generate

$$(a^2 + \omega^2)x = 0 \qquad (4.0.24)$$

The coefficient γ is then found to be $\pm i\omega$. Therefore, there are *two* valid solutions; $e^{i\omega t}$ and $e^{-i\omega t}$. The general solution is a linear combination of them both:

$$x(t) = a_1 e^{i\omega t} + a_2 e^{-i\omega t} \qquad (4.0.25)$$

Using $e^{i\theta} = \cos\theta + i\sin\theta$ and $\cos(\theta + \phi) = \cos\theta\cos\phi - \sin\theta\sin\phi$, show that this solution is equivalent to our previous solution,

$$x(t) = A\cos(\omega t + \varphi) \qquad (4.0.26)$$

with,

$$A = (a^2 + b^2)^{1/2}, \qquad \varphi = -\tan^{-1}(b/a), \qquad a = a_1 + a_2, \qquad b = i(a_1 - a_2) \qquad (4.0.27)$$

Exercise 4.2 Use the expression for $x(t)$ and $\dot{x}(t)$ to show that the average kinetic energy over a cycle is equal to the average potential energy for the harmonic oscillator. Further, show that both of these equal half the average energy:

$$x(t) = a\cos(\omega t + \varphi), \qquad \dot{x}(t) = -a\omega\sin(\omega t + \varphi) \qquad (4.0.28)$$

Hence, show that,

$$\langle \tfrac{1}{2}m\dot{x}^2 \rangle = \langle \tfrac{1}{2}kx^2 \rangle = \tfrac{1}{2}\langle \tfrac{1}{2}m\dot{x}^2 + \tfrac{1}{2}kx^2 \rangle \qquad (4.0.29)$$

Exercise 4.3 Here, we would like to introduce a method for solving ODEs called the **series solution**. This method seeks to find a representative of the solution in a power series. It works nicely for some ODEs, and not for others. Without going into too much detail, we will introduce this method to the reader and use it to solve a very well-behaved second-order ODE - the harmonic oscillator.

So, the general problem is to solve the following equation for $f(x)$ and generate an expression in the form of a power series in x with coefficients a_j:

$$\frac{d^2 f(x)}{dx^2} + \omega^2 f(x) = 0, \qquad f(x) = \sum_{j=0}^{\infty} a_j x^j \qquad (4.0.30)$$

The first step is to compute some derivatives of the power series. We should always make sure that, before using them in our equation, we make the powers of x^j the same:

$$\frac{df(x)}{dx} = \sum_{j=0}^{\infty} j a_j x^{j-1} \qquad (4.0.31)$$

$$\frac{d^2 f(x)}{dx^2} = \sum_{j=0}^{\infty} j(j-1)a_j x^{j-2} = \sum_{j=0}^{\infty} (j+2)(j+1)a_{j+2} x^j \qquad (4.0.32)$$

We then stick this in the equation we are trying to solve:

$$\sum_{j=0}^{\infty} x^j \left\{ (j+2)(j+1)a_{j+2} + \omega^2 a_j \right\} = 0 \tag{4.0.33}$$

Now, for each element of the sum to vanish, we realise that the coefficients of each x^j must vanish independently:

$$(j+2)(j+1)a_{j+2} + \omega^2 a_j = 0 \tag{4.0.34}$$

It is from this that we generate the **recurrence relation** for the series. This basically gives us higher-order terms as a function of lower-order terms:

$$a_{j+2} = \frac{-\omega^2 a_j}{(j+2)(j+1)}, \qquad j = 0, 1, 2, \ldots \tag{4.0.35}$$

So, the next step is to actually feed in values of j and generate the first few coefficients of the series. For $j = 0, 1, 2$, we have

$$a_2 = -\frac{1}{2}\omega^2 a_0, \qquad a_3 = -\frac{1}{6}\omega^2 a_1, \qquad a_4 = \frac{\omega^4 a_0}{24} \tag{4.0.36}$$

When doing this, we see that the recurrence relation gives us only even or odd sequences, as the relation is a_{j+2} not the next adjacent term a_{j+1}. Since we always try to get as few base coefficients as possible, we must have two 'basis' coefficients, a_0 and a_1, which the others are functions of. The next stage is to write out the series, substitute in our terms and collect them in like coefficients:

$$\begin{aligned}
f(x) &= a_0 + a_1 x + a_2 x^2 + a_3 x^3 + a_4 x^4 + \cdots \\
&= a_0 + a_1 x - \frac{1}{2}\omega^2 a_0 x^2 - \frac{1}{6}\omega^2 a_1 x^3 + \frac{1}{24}\omega^4 a_0 x^4 + \cdots \\
&= a_0 \left\{ 1 - \frac{1}{2}\omega^2 x^2 + \frac{1}{24}\omega^4 x^4 + \cdots \right\} + a_1 \left\{ x - \frac{1}{6}\omega^2 x^3 + \cdots \right\} \\
&= a_0 \left\{ \sum_{j=0}^{\infty} (-1)^j \frac{\omega^{2j} x^{2j}}{(2j)!} \right\} + a_1 \left\{ \sum_{j=0}^{\infty} (-1)^j \frac{\omega^{2j} x^{2j+1}}{(2j+1)!} \right\} \\
&= a_0 \cos x + a_1 \sin x
\end{aligned}$$

Okay, so, in the last two steps, we happened to be able to write a nice closed form for the series (I did say this was a well-behaved equation)! From here, we have our solution to the system. We can, however, perform a little analysis on the recurrence relation, to gain insight into our series. In the limit of large j, the recurrence formula is dominated by the j^{-2} term:

$$\lim_{j \to \infty} a_{j+2} \approx -\frac{\omega^2}{j^2} a_j \tag{4.0.37}$$

We can then compute the ratio of two successive even terms in the series limit:

$$\frac{a_{j+2} x^{j+2}}{a_j x^j} = -\frac{\omega^2}{j^2} x^2 \tag{4.0.38}$$

Finally, we can find the condition to truncate the series at the next term by setting $a_{j+2} = 0$. Doing this implies that $\omega^2 a_j = 0$, and hence that $\omega = 0$ as a necessity.

You see this sort of thing cropping up all over physics, Hermite polynomials, Bessel equations, . . . it's jolly useful, and quite fun too! This problem was fairly straightforward and very well behaved. If you performed this for a quantum harmonic oscillator, you would see the appearance of a j in the numerator of the recurrence relation. Following the same procedure, we would generate energy eigenvalues with $(2j + 1)$ separation in the sequence! Elsewhere, we will take well-known differential equations and study their series solutions.

Exercise 4.4 Here, we explicitly solve the **Morse potential**, which is frequently used in chemical physics. This potential models anharmonic effects in diatomic molecules and can explain their rotational-vibrational spectra in a quantum setting. Like the earlier Lennard-Jones potential the Morse potential, is a combination of attractive and repulsive terms (see figure 4.2). This problem is rather difficult as it involves solving an **Abelian integral** (as usual, anything that has a name is usually difficult and I would like to thank Professor Lars Olsen of the University of St Andrews for his assistance)! In this exercise, we show that, while the general method of solving for the energy of a system is straightforward, solving for a trajectory is not always a well-defined process.

We begin by writing the total energy, E, of the one dimensional system.

$$E(x, \dot{x}) = \frac{1}{2}m\dot{x}^2 + D_e(1 - e^{-\beta x})^2 \qquad (4.0.39)$$

Here, D_e is the dissociation energy of the molecule; it is equal to the well depth at the minima. The x coordinate is the interatomic separation (average bond length at the minima), and β controls the well's width or curvature. We can see that, at short internuclear separations, the atoms repel each other, while, when the bond is stretched, they attract each other. At large distances, the molecule will dissociate.

In a quantum mechanical setting, if we plug this energy into the Schrödinger equation of motion, the solution generates a series of terms rather than a continuous value. We say that the system has energy **eigenstates**, or discrete values of the solution. In the classical sense, we show below that we can solve explicitly for the position coordinate during the motion.

We begin by rearranging the energy for the square of the velocity and substituting some values to tidy things up:

$$\dot{x}^2 = ay^2 + by + c \qquad (4.0.40)$$

where,

$$y = e^{-\beta x}, \qquad a = \frac{-2}{mD_e}, \qquad b = \frac{4}{m}D_e, \qquad c = \frac{2}{m}(E - D_e) \qquad (4.0.41)$$

We invert the first expression for x by taking the natural log of y and find dx/dy. We then integrate the velocity in time:

$$\int \frac{dx}{dt}dt = \int (\sqrt{ay^2 + by + c})dt \qquad (4.0.42)$$

resulting in the following Abelian integral I:

$$I = -\int \frac{dy}{y\sqrt{ay^2 + by + c}} = \int \beta dt \qquad (4.0.43)$$

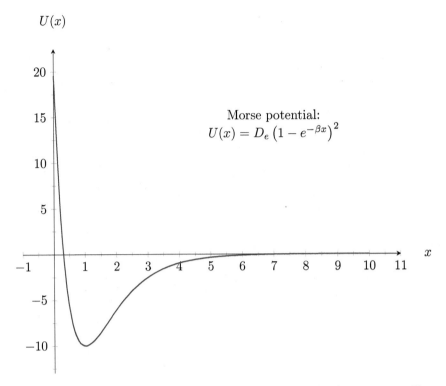

Fig. 4.2: The Morse potential $U(x)$ describing a classical anharmonic oscillator.

To solve this, we change the integration variable to $\gamma = 1/y$. We then complete the square:

$$
\begin{aligned}
I &= -\int \frac{1}{\sqrt{c\gamma^2 + b\gamma + a}}\,d\gamma \\
&= -\int \frac{1}{\sqrt{-c((\frac{D}{2c})^2 - (\gamma + \frac{b}{2c})^2)}}\,d\gamma \\
&= -\frac{1}{\sqrt{-c}} \int \frac{1}{\sqrt{\alpha^2 - (\gamma + \frac{b}{2c})^2}}\,d\gamma \\
&= -\frac{1}{\sqrt{-c}} \int \frac{1}{\sqrt{\alpha^2 - u^2}}\,du \\
&= -\frac{1}{\sqrt{-c}} \sin^{-1}\left(\frac{u}{\alpha}\right) + C
\end{aligned}
\tag{4.0.44}
$$

where $D = \sqrt{b^2 - 4ac}$, $\alpha = D/(2c)$ and $u = \gamma + b/(2c)$. This results in

$$I = -\frac{1}{\sqrt{-c}} \sin^{-1}\left(\frac{by + 2c}{y\sqrt{b^2 - 4ac}}\right) + C \qquad (4.0.45)$$

The trajectory is then found to be

$$x(t) = -\frac{1}{\beta} \ln\left(\frac{2c}{D \cdot \sin(\sqrt{-c}\beta t) - b}\right) \qquad (4.0.46)$$

The frequency of the oscillator ω is found by comparing the argument of the sine function to $\sin \omega t$. From this, we also obtain an expression for β:

$$\omega = \frac{\beta}{2\pi}\sqrt{\frac{2}{m}(D_e - E)} \qquad (4.0.47)$$

At the minimum of the well under a harmonic frequency, we can set E to zero to obtain $\beta = \sqrt{k/2D_e}$.

5

Wave Mechanics & Elements of Mathematical Physics

This short chapter considers the spherical wave solution to the wave equation and it's solutions in depth. Wave mechanics is useful in a wide area of physics, chemistry and geology, and the theory we develop is broad and general. One topic of particular importance to chemists and molecular physicists is the treatment of a wave on a sphere; we address this shortly. The problems for this chapter consider the series solution to classical differential equations that appear in physics; we show how to generate the Hermite and Legendre polynomials.

The one-dimensional wave equation was first written by d'Alembert in 1746; he is a character we will encounter later on in the text. This is a linear partial differential equation (PDE) whose solutions obey the superposition principle. We consider the three-dimensional wave equation first written by Euler:

$$\frac{1}{c^2}\frac{\partial^2}{\partial t^2}\psi(x,y,z,t) = \vec{\nabla}^2\psi(x,y,z,t) \tag{5.0.1}$$

One solution to such a wave equation $\psi(x,y,z,t)$ is a **plane wave**, so called since its wavefront or *surfaces of constant phase* are parallel two-dimensional planes propagating in an infinite uniform media. However, another class of solution exists to the wave equation - the **spherical wave**. A spherical wave consists of curved wavefronts propagating in the same direction.

A procedure of fundamental importance to the solution of PDEs is the **separation of variables** technique. Under certain conditions, the solution to the wave equation can be written as the product of terms with separated dependance on the independent variables. It is easiest to consider the one-dimensional wave equation at first with solution $\psi(x,t)$:

$$\psi(x,t) = X(x)T(t) \tag{5.0.2}$$

Substitution of this solution into the wave equation gives

$$\frac{\partial^2}{\partial x^2}\psi(x,t) = T(t)\frac{d^2X(x)}{dx^2} = \frac{1}{c^2}X(t)\frac{d^2T(t)}{dt^2} \tag{5.0.3}$$

Lagrangian & Hamiltonian Dynamics. Peter Mann, Oxford University Press (2018).
© Peter Mann. DOI: 10.1093/oso/9780198822370.001.0001

Therefore, we can collect variables to get two sides of the equation in different variables:

$$\frac{1}{X(x)}\frac{d^2 X(x)}{dx^2} = \frac{1}{c^2 T(t)}\frac{d^2 T(t)}{dt^2} = -\kappa^2 \tag{5.0.4}$$

The only way we can satisfy this relation is if both sides are equal to a constant. This is because the variables x and t are independent variables; changing one doesn't affect the other, and to have both sides be only a function of one variable only makes sense if they are constant. We call this constant the **separation constant** κ. We choose $-\kappa^2$ only for later convenience but the choice is irrelevant. This means we have separated the PDE into two ODEs:

$$\frac{d^2 X(x)}{dx^2} = -\kappa^2 X(x), \qquad \frac{d^2 T(t)}{dt^2} = -c^2 \kappa^2 T(t) \tag{5.0.5}$$

The nature of the separation constant is determined by the **boundary conditions** of the problem, and any solution to these equations must adhere to permissible allowed values. The boundary conditions force the separation constant to a limited number of solutions called **eigenvalues** to the corresponding **eigenfunction** solution. We do not discuss this further except to say that it is of the utmost importance to the physical result. With minor rearrangement (let $\omega = kc$), both of these results take the form

$$\left(\frac{d^2}{dx^2} + \kappa^2\right)X(x) = 0, \qquad \left(\frac{d^2}{dt^2} + \omega^2\right)T(t) = 0 \tag{5.0.6}$$

The first result is called the **Helmholtz equation**; it is a linear time-independent ODE. The right-hand equation is a second-order ODE in time. In the special case where $\kappa = 0$, we generate the **Laplace equation** $\nabla^2 X(x) = 0$. In the Cartesian frame, the solution to the Helmholtz equation is given by $X(x) = e^{ik \cdot x}$.

The more general result of this treatment is to consider the three-dimensional wave equation extending over x, y, z. Some readers will be familiar with this method when treating the Schrödinger equation for a hydrogen atom, as this equation is identical apart from a potential term. In the separations, the forms of the constants have been chosen to best suit the problem and are known from repeated experience. Returning to the three-dimensional wave equation in equation 5.0.1 we write the Laplacian in polar coordinates:

$$\nabla^2 \psi = \frac{\partial^2 \psi}{\partial r^2} + \frac{2}{r}\frac{\partial \psi}{\partial r} + \frac{1}{r^2 \sin\theta}\frac{\partial}{\partial \theta}\left(\sin\theta \frac{\partial \psi}{\partial \theta}\right) + \frac{1}{r^2 \sin^2\theta}\frac{\partial^2 \psi}{\partial \phi^2}$$

which, the reader is invited to prove. We now assume that ψ is a function of θ, ϕ, t and r and can be separated into $\psi(r, \theta, \phi, t) = R(r)\Theta(\theta)\Phi(\phi)T(t)$, where Θ and Φ

are angular dependant functions[1], R is a radial function and T is a time dependent function. We substitute this into the three-dimensional wave equation and divide by $R\Theta\Phi T$:

$$\underbrace{\frac{1}{R}\frac{d^2R}{dr^2} + \frac{2}{rR}\frac{dR}{dr}}_{\text{radial}} + \underbrace{\frac{1}{r^2}\frac{1}{\Theta\sin\theta}\frac{d}{d\theta}\left(\sin\theta\frac{d\Theta}{d\theta}\right) + \frac{1}{r^2\sin^2\theta}\frac{1}{\Phi}\frac{d^2\Phi}{d\phi^2}}_{\text{angular}} = \underbrace{\frac{1}{v^2}\frac{1}{T}\frac{d^2T}{dt^2}}_{\text{time}}$$

We are now going to systematically separate out all the functions until we have simplified the problem to a set of ODEs; when we separate off an equation, it will be highlighted for later use:

$$LHS(r,\theta,\phi) = RHS(t) = \text{constant} = -k^2$$

$$\boxed{\frac{1}{v^2}\frac{1}{T}\frac{d^2T}{dt^2} = -k^2} \tag{5.0.7}$$

This is the **time equation**. The next step is to separate the radial from the angular portions, which we do by multiplying through by r^2 and arranging to $LHS(\theta,\phi) = RHS(r) = \text{constant} = \lambda$:

$$-\frac{1}{\Theta\sin\theta}\frac{d}{d\theta}\left(\sin\theta\frac{d\Theta}{d\theta}\right) - \frac{1}{\sin^2\theta}\frac{1}{\Phi}\frac{d^2\Phi}{d\phi^2} = \frac{r^2}{R}\frac{d^2R}{dr^2} + \frac{2r}{R}\frac{dR}{dr} + k^2r^2 = \lambda$$

We then have the **radial equation**:

$$\boxed{r^2\frac{d^2R}{dr^2} + 2r\frac{dR}{dr} + \left[k^2r^2 - \lambda\right]R = 0} \tag{5.0.8}$$

To separate the final two variables, we multiply by $\sin^2\theta$, $LHS(\theta) = RHS(\phi) = \text{constant} = m^2$ giving us the **azimuth equation** in ϕ:

$$\boxed{\frac{d^2\Phi}{d\phi^2} = -m^2\Phi} \tag{5.0.9}$$

The final step is θ, and from the LHS we proceed by multiplying through by $\Theta/\sin^2\theta$:

$$\frac{\sin\theta}{\Theta}\frac{d}{d\theta}\left(\sin\theta\frac{d\Theta}{d\theta}\right) + \lambda\sin^2\theta = m^2$$

---

[1]We can impose boundary conditions on Φ so that it is a periodic function and insist that Θ is regular at the poles of the sphere such that $\theta = 0, \pi$. This type of problem then becomes a **Sturm-Liouville problem**.

giving us an **angular equation** in θ:

$$\frac{1}{\sin\theta}\frac{d}{d\theta}\left(\sin\theta\frac{d\Theta}{d\theta}\right) + \left[\lambda - \frac{m^2}{\sin^2\theta}\right]\Theta = 0 \tag{5.0.10}$$

There is now an ODE for each variable; these can now be either solved or converted to well-known forms. This is the result of the separation of variables procedure. The time and the azimuth equations have the easiest solutions:

$$T(t) = e^{\pm i(\omega t)}, \qquad \text{and} \qquad \Phi(\phi) = e^{\pm i(m\phi)} \tag{5.0.11}$$

Where $\omega = vk$, we can treat $T(t)$ as an eigenfunction of d^2/dt^2 with eigenvalue $-\omega^2$, and similarly for $\Phi(\phi)$: they are both harmonic oscillator ODEs. If we impose a **boundary condition** on ϕ such that, after every full rotation, the same point on the sphere has the same solution one period later then we are forcing ϕ to be single valued, and Φ to be a periodic function: $\Phi(\phi + 2\pi) = \Phi(\phi)$. We use Euler's formula $e^{i\pi} = -1$ to obtain

$$\Phi(\phi + 2\pi) = e^{im(\phi+2\pi)} = e^{im\phi}e^{2\pi im}$$

In order that $e^{im\phi} = e^{im\phi}e^{2\phi im}$ it is obvious that $e^{2\pi im} = 1$ and therefore m has to be an integer $m = 0, \pm 1, \pm 2, \ldots$, giving us our first separation constant.

The remaining two are a little more difficult and we will not solve them explicitly but express them in a form that has well-known solutions; the reader is then invited to investigate this at a later point. Let us now develop the radial equation by considering the following derivative and let $\rho = kr$:

$$r\frac{dR}{dr} = kr\frac{dR}{d(kr)} = \rho\frac{dR}{d\rho}, \qquad \longrightarrow \qquad r^2\frac{d^2R}{dr^2} = \rho^2\frac{d^2R}{d\rho^2} \tag{5.0.12}$$

Now let $\lambda \equiv l(l+1)$, where l is a non-negative integer $l = 0, 1, 2, \ldots$ (we are free to choose the separation constant):

$$\rho^2\frac{d^2R}{d\rho^2} + 2\rho\frac{dR}{d\rho} + \left[\rho^2 - l(l+1)\right]R = 0 \tag{5.0.13}$$

The radial equation is transformed into the **spherical Bessel equation**, which has well-known solutions called the **spherical Bessel functions** $j_l(\rho)$ and $y_l(\rho)$ named after Friedrich Bessel. The two solutions are known as Bessel functions of the **first kind**, $j_l(\rho)$, and the **second kind**, $y_l(\rho)$, respectively. Physically, we are only interested in the first kind, since they are finite at the origin of ρ; these are given by **Rayleigh's formula**:

$$j_l(\rho) = \rho^l\left(-\frac{1}{\rho}\frac{d}{d\rho}\right)^l\frac{1}{\rho}\sin\rho, \qquad \text{with} \qquad \lim_{\rho\to 0^+} j_l(\rho) = 0 \tag{5.0.14}$$

The final expression in Θ is a little more difficult to obtain. We begin by changing the variables of the angular equation using the chain rule; first, let $u = \cos\theta$:

$$\frac{d}{du} = \frac{d\theta}{du}\frac{d}{d\theta} = \left(\frac{du}{d\theta}\right)^{-1}\frac{d}{d\theta} = -\frac{1}{\sin\theta}\frac{d}{d\theta}$$

Then we obtain the following expression:

$$(1-u^2)\frac{d}{du} = -(1-\cos^2\theta)\frac{1}{\sin\theta}\frac{d}{d\theta} = -\sin\theta\frac{d}{d\theta}$$

Therefore,

$$\frac{d}{du}(1-u^2)\frac{d}{du} = \frac{1}{\sin\theta}\frac{d}{d\theta}\left(\sin\theta\frac{d}{d\theta}\right)$$

This term appears in the angular ODE we developed earlier:

$$\frac{d}{du}\left((1-u^2)\frac{d\Theta(u)}{du}\right) + \left(l(l+1) - \frac{m^2}{1-u^2}\right)\Theta(u) = 0$$

$$(1-u^2)\frac{d^2\Theta}{du^2} - 2u\frac{d\Theta}{du} + \left[l(l+1) - \frac{m^2}{1-u^2}\right]\Theta = 0 \qquad (5.0.15)$$

Again, this is a famous equation among physicists; it is called the **associated Legendre differential equation** and it has well-known solutions called the **associated Legendre polynomials** $\Theta(u) = P_l^m(u)$. The **Legendre equation** is the special case where $m = 0$ and the polynomials are found via the **Frobenius method** to find an infinite series solution, as we show in an exercise later on. The series is only convergent when $\lambda = l(l+1)$, giving us $P_l^m(u)$. For every l, there are $(2l+1)$ values of m between $\pm l$, and each associated polynomial crosses zero $(l-m)$ times, meaning there are $(l-m)$ **nodes** for that solution. Often, the angular and the azimuth solutions are considered together through $Y_l^m(\theta,\phi) = \Theta(\theta)\Phi(\phi)$, where Y_l^m is a **spherical harmonic function** of degree l and order m. With the addition of a normalisation constant $N_{l,m}$, the spherical harmonics become

$$Y_l^m(\theta,\phi) = N_{l,m}e^{im\phi}P_l^m(u) \qquad (5.0.16)$$

The general solution to the Helmholtz wave equation is $\psi(r,\theta,\phi) = R(r)\Theta(\theta)\Phi(\phi)$:

$$\psi(r,\theta,\phi) = \sum_{l=0}^{\infty}\sum_{m=-l}^{l} a_l^m j_l(\rho)Y_l^m(\theta,\phi) \qquad (5.0.17)$$

where the coefficient a_l^m is a collection of normalisation constants and boundary conditions heaped into one tidy symbol. It should be noted that the three-dimensional

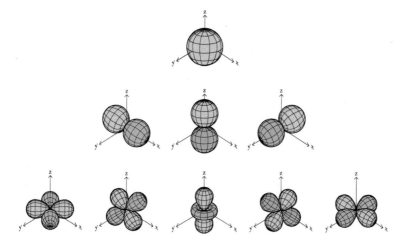

Fig. 5.1: The first few spherical harmonics.

Cartesian-frame analogue $\psi(x, y, z)$ is equal to this wavefunction; this result is known as the **Rayleigh plane wave expansion** and, in the literature, $\psi(x, y, z, t)$ is a plane wave, $\psi(r, \theta, \phi, t)$ is a spherical wave, and the time-independent analogues are referred to as plane or spherical **partial waves**, respectively.

The first few spherical harmonics are pictured in figure 5.1, and their similarity to **atomic orbitals** is no coincidence. The wave mechanics we have developed here is only the barest smattering of a rich theory of waves. There are plenty of physical applications of the separation of variables technique, ranging from plane wave diffraction and scattering experiments to pressure and sound waves, to disturbances propagating through solids and, of course, to atomic structures, each with their own wave equations. The choice of boundary conditions is a vast topic and is not something that can be covered in brief. The mathematics behind this topic is a bit horrific upon a first glance, with lots of complicated formulas which would be difficult to include; so, for further reading please see Boas (2005) book, *Mathematical Methods in the Physical Sciences*.

Exercise 5.1 Here I would like to continue our exploration of famous equations in classical physics by looking at **Hermite's equation**:

$$\frac{d^2 f(x)}{dx^2} - 2x\frac{df(x)}{dx} + 2\lambda f(x) = 0, \qquad f(x) = \sum_{j=0}^{\infty} a_j x^j \qquad (5.0.18)$$

We start by evaluating the first and second derivatives of the power series:

$$\frac{df(x)}{dx} = \sum_{j=1}^{\infty} j a_j x^{j-1}, \qquad \frac{d^2 f(x)}{dx^2} = \sum_{j=2}^{\infty} j(j-1) a_j x^{j-2} \qquad (5.0.19)$$

We then stick these into Hermite's equation:

$$\sum_{j=2}^{\infty} j(j-1)a_j x^{j-2} - 2t \sum_{j=1}^{\infty} j a_j x^{j-1} + 2\lambda \sum_{j=0}^{\infty} a_j x^j = 0 \qquad (5.0.20)$$

We now need to obtain equal powers in x^j:

$$\sum_{j=0}^{\infty} (j+2)(j+1)a_{j+2} x^j - \sum_{j=1}^{\infty} 2j a_j x^j + \sum_{j=0}^{\infty} 2\lambda a_j x^j = 0 \qquad (5.0.21)$$

The next step is to correct the summations to all start at the same index so that we can pull out each coefficient of x^j. To do this, we simply evaluate the zeroth term in the second sum. In this case, it is zero ($2 \cdot 0 \cdot a_0 \cdot x^0 = 0$), so we can simply start the middle sum at zero without any change (if it hadn't been zero, then we would have to correct for this):

$$\sum_{j=0}^{\infty} \left((j+2)(j+1)a_{j+2} - 2j a_j + 2\lambda a_j \right) x^j = 0 \qquad (5.0.22)$$

The recurrence relation is then found to be

$$a_{j+2} \frac{2(j-\lambda)}{(j+2)(j+1)} a_j, \qquad j = 0, 1, 2, \ldots \qquad (5.0.23)$$

All we need to do now is set the value of λ, a_0 and a_1 to generate the solutions. If we choose λ to be a positive integer, a_0 to be 1 (0) and a_1 to be 0 (1), then we generate polynomials of even (odd) order up to the value of λ. Evaluating terms greater than $j = \lambda$ will lead to zero coefficients, indicating that the series truncates to polynomials of order λ. These polynomials are called the **Hermite polynomials** and, amazingly, they appear all over mathematical physics! Try for yourself to compute the first few polynomials.

Exercise 5.2 Use the series solution technique to solve **Legndre's equation** by generating a recurrence relation:

$$(1-x^2)\frac{d^2 f(x)}{dx^2} - 2x\frac{df(x)}{dx} + \lambda(\lambda+1)f(x) = 0 \qquad (5.0.24)$$

Show that this solution leads to the **Legendre polynomials** and compute the first few terms.

Proceeding in the usual fashion, we compute derivatives of the power series solution and substitute them into the Legendre equation.

$$(1-x^2)\sum_{j=2}^{\infty} j(j-1)a_j x^{j-2} - 2x\sum_{j=1}^{\infty} j a_j x^{j-1} + \lambda(\lambda+1)\sum_{j=0}^{\infty} a_j x^j = 0 \qquad (5.0.25)$$

We have changed the indices of the sums, as they only contribute non-zero terms starting at their index. We then multiply this out and correct the powers:

$$\sum_{j=0}^{\infty}(j+2)(j+1)a_{j+2}x^j - \sum_{j=0}^{\infty}j(j-1)a_jx^j - 2\sum_{j=0}^{\infty}ja_jx^j + \lambda(\lambda+1)\sum_{j=0}^{\infty}a_jx^j = 0 \quad (5.0.26)$$

Collecting this nicely and realising each coefficient vanishes, we generate the recurrence formula:

$$a_{j+2} = -\frac{(\lambda+j+1)(\lambda-j)}{(j+1)(j+2)}a_j \qquad (5.0.27)$$

Evaluating this for even or odd values of l will generate the Legendre polynomials.

Part II

Lagrangian Mechanics

Lagrangian mechanics was developed by Joseph-Louis Lagrange around a hundred years after Newton's breakthrough works. Lagrange's method is a complete upheaval and reformulation of the Newtonian approach. Central to understanding Lagrangian mechanics is the principle of stationary action, sometimes termed Hamilton's principle.

In this section, you will see that Lagrangian mechanics tackles constrained systems and conserved quantities in a much easier way than Newtonian physics. It does away with the specification of vectors and forces, instead replacing them with generalised coordinates and the Euler-Lagrange equation. The versatility of Lagrange's method is apparent if our problem requires us to change coordinate system; the equations we go on to prove are devoid of any mention of inertial frames or of forces and we see that the dynamics of a system is entirely determined by a single scalar function called the Lagrangian.

We start our discussion by setting some definitions firmly in place and giving the reader an understanding of the terminology of Lagrangian mechanics. Personally, I think the most important part of this chapter is to introduce the generalised coordinates and appreciate that we have a geometrical underpinning to mechanics. The focus of the chapter is ultimately the derivation of the Euler-Lagrange equation. To do this, we can follow two methods: (i) Hamilton's principle and (ii) d'Alembert's principle, both of which will be studied to cover a holistic teaching of the origin of Lagrangian mechanics. In this way, it is hoped the reader will get a better understanding of the finer points of each approach and how they differ. We will consider symmetries and conservation laws, using Noether's theorem and the definition of the Jacobi energy function. References to the appendices are probably best skipped upon first reading, since the appendices contain advanced optional discussions of Lagrangian mechanics and require advanced geometrical material developed in subsequent chapters.

6
Coordinates & Constraints

Analytical mechanics concerns itself with scalar quantities of a dynamic system, namely the potential and kinetic energies of the particle. This is in opposition to Newton's vectorial mechanics method which relies upon defining the positions of the particles in three-dimensional space and the forces acting upon it. This method was first approached by Leibniz in his *Specimen Dynamicum*, where he challenged Newton's view that space and time were not absolute, claiming instead that they were relative, meaning they depend on how you observe the system and, thus set about describing the system by scalars. From this, he planted the seeds of analytical mechanics and even formulated some of the ideas that would appear in Einstein's theory of general relativity. Indeed, Leibniz' contribution to science is often overlooked, since his major impacts were in metaphysics and philosophy but, in fact, arguably one of the most-used rules in differential calculus, the *product rule*, was first developed by Leibniz.

After Leibniz first formulated this alternative approach, we see three contributions to the topic of mechanics before Lagrange generalised their findings in the *Euler-Lagrange* equations. The first of these is Bernoulli's principle of *virtual work* for static systems, and the second is d'Alembert's invaluable extension of this to dynamic systems. The third development came from Maupertuis in 1744 and is called the *principle of stationary action*; and it is the cornerstone of theoretical physics. Finally, *Hamilton's action principle* is the polished-up action theorem that we base our discussion on; it supersedes Maupertui's principle. We later make the distinction between Hamilton's action principle in the Lagrangian framework and in the canonical framework but we need not discern this for now. In order to transition from the Newtonian to the Lagrangian methodology some authors cling to the concepts rooted in the former in order to express the later in a more accessible way. Therefore we will show how Newton's laws tie in to analytical mechanics, using Hamilton's stationary action principle and d'Alembert's virtual work during our discussion. For now, we introduce constraints, generalised coordinates and the various spaces of Lagrangian mechanics.

If a system is **constrained**, it means that it can't do something that it could if it didn't have the restriction placed upon it. If two people are in a room with an open door they can move in and out of the room as they please. If the door is locked, then neither of the people can leave the room; they are constrained to stay inside, and the constraint on the system that dictates their motion is the locked door.

Lagrangian & Hamiltonian Dynamics. Peter Mann, Oxford University Press (2018).
© Peter Mann. DOI: 10.1093/oso/9780198822370.001.0001

The constrained system's equations of motion must satisfy the restrictions; for our example, any equation of motion we may write about the people in the locked room must exclude solutions where they are outside. A particularly useful constraint is the *rigid body* constraint, which, for our purposes, can be expressed as approximating a fixed bond length when calculating rotations, but there are many more. An important distinction between constraints is whether they are **holonomic** or **non-holonomic**. The most general constraint can be written to be a function of the position, velocity and time $\phi(t, q, v)$ (i.e. it is a differential equation). If we can write $v = v(q)$, then the constraint is dependent only on positions and not differentials; in this case, it is holonomic $\phi(t, q)$. Further, if we can write the constraint as a set of constant values $\phi_i = C_i$ that must be satisfied, then it is an **isoperimetric constraint**.

A **degree of freedom** is a variable that can be used to specify the state of a system; it is a specific mode of motion of the system. A free particle can be parameterised by three independent cartesian coordinates x, y, z, and three independent velocity components, giving a total of six degrees of freedom. If we just consider the three position components, then the system's location is defined by three coordinates per particle; for a system of N particles, we have to specify $3N$ coordinates to define their location, or **configuration**.

A constraint on a system can reduce the number of degrees of freedom a system can have. If we consider a particle restricted to two-dimensional rotational motion in the xy-plane, we realise that the system is constrained to two dimensions: the z-axis displacement is zero for all times. Therefore, the system has two degrees of freedom $(i.e. 3 - 1)$. More generally, for $3N$ coordinates that specify the state of a system that has m holonomic constraints imposed on it, there are $3N - m = n$ degrees of freedom. For a two-dimensional rotor, we can fully specify the system using the polar coordinates r and θ or just x and y. The coordinates of a point in configuration space are called the **generalised coordinates** q^i, where $i = 1, \ldots, n$ (we say that the index runs from 1 to n). They can be any physical parameter that describes fully the configuration of a system and there may be fewer of them than the cartesian coordinates; for a diatomic molecule, we might use the separation between the atoms as a coordinate, or the translational motion of the centre of mass, or even angular parameters characterising rotations, and so on. So, for a point p in the n-dimensional configuration space, we would write the coordinates as $q^1, q^2, \ldots, q^n$. We assume that the set of coordinates we choose to use are **linearly independent**, which means there is no equation linking them; this is not true for constrained systems!

A central part of any problem is choosing the generalised coordinates, and the right choice will simplify the problem immensely. For the most part, we imagine switching from cartesian coordinates to spherical coordinates, say, so we would need to find the relation of the form $x(r, \theta, \phi)$, $y(r, \theta, \phi)$ and $z(r, \theta, \phi)$ and then take their time derivatives (for later use). There is nothing stopping us from **inverting** this relation to express the problem in the original coordinates $r(x, y, z)$, $\theta(x, y, z)$ and $\phi(x, y, z)$, so be mindful that, whenever we change coordinates, we can always take the inverse transformation back to the original set, provided it is indeed a linearly independent minimal set.

The space where generalised coordinates live is called **configuration space**. It is a surface within a higher-dimensional space, or **manifold**. The concept of a manifold can be a bit tricky to understand; you could think of a flat piece of paper as being a two-dimensional space within a three-dimensional world. A point in a configuration manifold is characterised by the generalised coordinates $q^1, \ldots, q^n \in \mathbb{R}^n$ of each dimension, in the same way that a point in Euclidean space is determined by $x, y, z \in \mathbb{R}^3$ (see figure 6.1). The whole structure of configuration space is built over a smooth time line that parametrises dynamics, that is to say, a curve γ in the configuration space is a smooth function of time, $\gamma(t)$. You can use the terms 'configuration space' and 'configuration manifold' interchangeably in this discussion and, at this point, it is not expected that you have read the mathematics section on differential geometry.

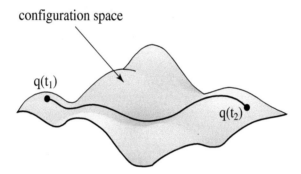

configuration space

$q(t_1)$

$q(t_2)$

Fig. 6.1: A configuration manifold with a trajectory curve connecting two points.

We can see that the constraints on a system will have an intimate relationship with the equations of motion of a system, since they dictate in some way the time evolution of the coordinates. By redefining the coordinates we express the system in, we remove the constraints from the problem: as the forces in that dimension don't do any work, by definition we can't move in that mode, so we just get rid of them via this procedure.

In terms of configuration manifolds, this is removing a dimension from the space, so the constraint equations dictate what part the system can move in. For instance, if a free particle with coordinates (x, y, z) was constrained to the xy-plane, then we would really be reducing the dimensionality of the surface in configuration space in which the motion is free to move. If we want to be a bit more fancy with our mathematical terms, we would say that the system is constrained to an embedded **submanifold**, which is just a lesser-dimensional surface. As the system moves, its coordinates change, so the particle moves its position in the configuration space and, as it does so, it maps out a trajectory between a starting point and an end point. Going back to the example of the people in the locked room, we could say that the total configuration space was the entire building, say, while the actual region they can move is the one room, which is a subspace of the entire space; however the building could also be embedded in the street, which is embedded within a town, and so on. In general, you can aways redefine the manifold to eliminate constraints; situations where you can't are discussed much later on. The people's positions will be described by the coordinates in the room, and their trajectories will lie inside the room for all time, obeying the constraint.

However, to predict motion (to be **deterministic**), it is not enough to know where something is at a given snapshot; we must know the velocity, too! If you took a photograph of a race car, then you wouldn't know how fast it was going and wouldn't be able to tell the precise instant when it was going to finish the race. What this means is that we require a **generalised velocity** $\dot{q}^i$ coordinate too. The problem is that configuration space is the space of generalised positions; it is n dimensional, with a coordinate for each unconstrained mode. So, we require a larger space that is precisely $2n$ dimensional and is spanned by the generalised coordinates and their time derivatives, the generalised velocities. Such a space is called the **velocity phase space** (see figure 6.2). Being technical, we would call it the **tangent bundle** $\mathbb{R}^{2n}$ to a configuration manifold $\mathbb{R}^n$ but that discussion can wait for when we are ready.

The motion of the system is therefore a curve in the velocity phase space, parametrised by the position $q^i(t)$ and velocity $\dot{q}^i(t)$ at each point. Any smooth function defined on this space will be a function of these variables. The velocity phase space is another manifold of exactly twice the dimension of the configuration space and it is the space where **Lagrangian mechanics** lives. An important point for later on is that the generalised velocities are **independent** of the generalised coordinates. Think of a race car again; we can choose what speed it goes without worrying exactly where it is on the track. However, once both position and velocity are chosen a **variation** in the coordinate q^i will induce a variation in the velocity $\dot{q}^i$. This is due to how we lift the curve from configuration space to velocity phase space; we have to ensure that both paths describe the same motion, so they are coupled. Before we look at the dynamics, note that q and v are independent variables; once we solve the equations of motion, the trajectory is exactly $v(t) = \dot{q}(t) = (\frac{d}{dt}q)(t)$.

Lastly, we develop the **generalised acceleration** $\ddot{q}^i(t)$ for each coordinate. From the generalised acceleration, we must integrate twice to obtain the generalised position and velocities, thereby introducing $2n$ independent integration constants, precisely what we need to solve our system of differential equations.

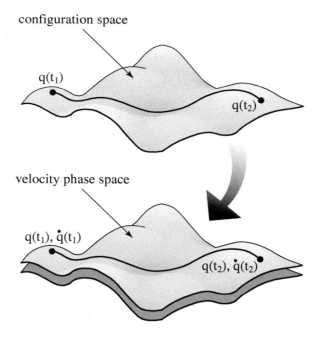

configuration space

$q(t_1)$

$q(t_2)$

velocity phase space

$q(t_1), \dot{q}(t_1)$

$q(t_2), \dot{q}(t_2)$

Fig. 6.2: An n-dimensional configuration manifold and the $2n$-dimensional velocity phase space.

Chapter summary

- A set of **linearly independent generalised coordinates** q^i are introduced, giving us the freedom to parametrise the system in more convenient ways.

- The n-dimensional **configuration manifold** is introduced as the space of q^is, a point in which is like a snapshot of the system.

- The $2n$-dimensional **velocity phase space** is introduced as the space in which we do dynamics. It is parametrised by both the generalised coordinates and the generalised velocity $\dot{q}^i$, which is taken to be the time derivative of the corresponding q^i.

- It is found that a **constraint** on a system reduces the number of **degrees of freedom** by 1, restricting the motion to a subsurface of the total space. For for N particles with $3N$ degrees of freedom and subject to m constraints, the number of generalised coordinates or the number of independent dimensions in configurations space is $3N - m = n$.

Exercise 6.1 We can represent the transition state of the homolysis of a diatomic molecule $AA \rightarrow A + A$ as an effective mass μ on a potential energy surface. The path of minimum potential energy from AA to $A + A$ can projected onto the configuration space $\mathcal{C}$. The generalised coordinates that we will use are defined as (x, s), where x is distance from any point on the surface to the nearest point on the curve, and s is the length along $\mathcal{C}$ to that point on the curve, as shown in figure 6.3. The total energy can be written as the sum of the kinetic and potential terms:

$$E = \frac{p_x^2}{2\mu} + \frac{p_s^2}{2\mu\kappa(x)} + U_1(s) + U_2(x, s) \tag{6.0.1}$$

where $\kappa(x)$ defines the curvature of the curve at the point (x, s), $U_1(s)$ is the potential energy along $\mathcal{C}$ and $U_2(x, s)$ is the potential energy term arising from the deviation from the curve. The **reaction coordinate** describes the bond geometries that minimise the potential from reactants to products and is defined as the curve for which U_2 is minimised $x = x_0(s)$:

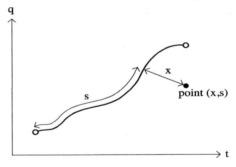

Fig. 6.3: The generalised coordinates for our system.

7

The Stationary Action Principle

In this chapter, we discuss the Euler-Lagrange equation of motion as derived from the **stationary action principle**. This method was developed by Euler and applied to mechanics by Lagrange (figure 7.1). We will first go through the idea behind the method and then go through a long-hand derivation using ordinary calculus before using the notationally different functional derivative to tidy things up. At the end of the chapter, we look at some interesting topics that often cause confusion when first investigating mechanics.

We pose the question that, if we know the configuration of a system at an initial time and we know its configuration at a final time, then what path or trajectory does the particle take on getting from the initial to the final points? This is the fundamental question to ask of a **variational problem**. There are plenty of variational problems in physics and perhaps the most general in this book is Noether's field

Fig. 7.1: Joseph Louis Lagrange. (1736–1813).

theoretic variational problem. What we want to introduce now is the stationary action variational **principle**. The importance of saying *principle* rather than *problem* is that it is still unknown why exactly this particular one works to give us the correct equations of motion. There is no good intuitive reason why we should impose the conditions that we impose, except one- that it works.

We start by constructing two vague yet important quantities to the velocity phase space. The first is a scalar function $\mathscr{L} = \mathscr{L}(\mathbf{q}, \dot{\mathbf{q}}, t)$, which we will use to construct the equations of motion. This function is called the **Lagrangian**. It is simply some function that can be used in the Euler-Lagrange machinery (which we come to in a minute) and will give us the physically correct equations of motion for the system. The Lagrangian of a system may not be unique, so a system may have more than one to describe the same dynamics. In elementary discussions, it is written as the kinetic energy minus the potential energy but expect this to generalise at a later point:

$$\mathscr{L}(\mathbf{q}, \dot{\mathbf{q}}, t) = T(\dot{\mathbf{q}}) - U(\mathbf{q}) \qquad \textbf{Lagrangian function} \qquad (7.0.1)$$

Lagrangian & Hamiltonian Dynamics. Peter Mann, Oxford University Press (2018).
© Peter Mann. DOI: 10.1093/oso/9780198822370.001.0001

The Lagrangian may or may not be an explicit function of time so, at several times, we will ignore the t dependance or bring it in again when required for our discussion.

The second construct is called the **action**, which we denote in this chapter by $A = A(\mathscr{L}) = A[\mathbf{q}]$. The action is a smooth **functional** of the generalised coordinates on the velocity phase space, that is to say, it is a function of the Lagrangian. Note that square brackets indicate *functional of*, and bold symbols are used to indicate all the coordinates:

$$A[\mathbf{q}] = \int_{t_0}^{t_1} \mathscr{L}(\mathbf{q}, \dot{\mathbf{q}}, t)dt \quad \textbf{action functional} \tag{7.0.2}$$

Why it takes the form of a time integral over the Lagrangian is simply that it is constructed in this way: the calculus of variations requires us to have a smooth functional over a scalar function (see figure 7.2). The real question should be why the calculus of variations works at all! We proceed now with the calculus of variations in explicit form, suppressing summations for ease of calculation.

Every possible path between two fixed points in configuration space has its own value of the action; by changing the path slightly, you change the action, too. It is a law of physics that dynamical systems follow a trajectory that makes the action functional **critical** or **stationary** with respect to infinitesimal **first-order variations** in the coordinates with fixed endpoints. This is **Hamilton's principle** and we will see in a second the important features that distinguish it from general variational problems:

$$\delta A[\mathbf{q}, \alpha] = \left.\frac{dA}{d\alpha}\right|_{\alpha=0} = A[\mathbf{q} + \delta\mathbf{q}] - A[\mathbf{q}] = 0 \tag{7.0.3}$$

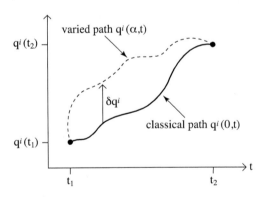

Fig. 7.2: The calculus of variations.

Note they are just different ways we can write the same thing and we will use all where convenient. By **stationary**, we mean that, at every point along the trajectory, the action functional is either a minima or a saddle point just as in calculus; *however*, classical mechanics excludes *maxima*, a topic which is explored by C. G. Gray and E. F. Taylor's (2007) article, When action is not least. The path that does this is known as the **extremal path** of the action A.

If we vary the trajectory infinitesimally, then, by **Fermat's theorem**, the variation between the extremal path and a neighbouring path infinitesimally close to it is zero; this is the definition of an extremal curve. The formal problem we now face is to find the trajectory that makes the action integral stationary, that is, the extremal path. The way to do this is by the calculus of variations. The result is the Euler-Lagrange equation of motion. Here we will follow a simplified version of the calculus of variations and save technicalities for the mathematical background, where notation and procedure are explained in depth.

Paths with the same boundary conditions as the extremal path between two fixed points are called the **variation curves**. We can form a variation curve in terms of the extremal curve by infinitesimally varying the extremal curve slightly. Consider a family of curves that are parameterised by α, namely $q^i(\alpha, t)$; alpha is just a small number, $\eta_i(t)$ is any arbitrary function and t is the independent variable. We can write any path $q^i(\alpha, t)$ in terms of variations to one in which $\alpha = 0$, by modulating α. We impose that this family of functions all have the same boundary conditions; we can see that, in order for all $q^i(\alpha, t)$ to be the same at two certain locations, alpha must be zero, giving us $q^i(0, t)$; that is, the variations $\alpha\eta_i(t)$ must vanish at the **boundary points**:

$$q^i(\alpha, t) = q^i(0, t) + \alpha\eta_i(t) \tag{7.0.4}$$

We can think of $\alpha\eta_i(t)$ as a variation of the shortest path between fixed points, and $q^i(0, t)$ as the desired extremal. The action is a functional of α because, as the perturbation term gets larger, the path changes $A[\mathbf{q}, \alpha]$. An equation describing the family of functions of the time derivative of q is denoted with a dot, but takes the same form:

$$\dot{q}^i(\alpha, t) = \dot{q}^i(0, t) + \alpha\dot{\eta}_i(t) \tag{7.0.5}$$

Start by taking the derivative of the action with respect to alpha, and consider for each variable. The partial with respect to the time term disappears now because time does not depend on alpha (i.e. $dt/d\alpha = 0$):

$$\frac{dA}{d\alpha} = \int_{t_0}^{t_1} \sum_{i=1}^{n} \left(\frac{\partial\mathcal{L}}{\partial q^i}\frac{\partial q^i}{\partial \alpha} + \frac{\partial\mathcal{L}}{\partial \dot{q}^i}\frac{\partial \dot{q}^i}{\partial \alpha} \right) dt \tag{7.0.6}$$

Now look at equations 7.0.4 and 7.0.5, note that we can differentiate them with respect to alpha and substitute them into equation 7.0.6.

$$\frac{d}{d\alpha}q^i(\alpha, t) = \eta_i(t), \quad \text{and} \quad \frac{d}{d\alpha}\dot{q}^i(\alpha, t) = \dot{\eta}_i(t) \tag{7.0.7}$$

such that

$$\frac{dA}{d\alpha} = \int_{t_0}^{t_1} \sum_{i=1}^{n} \left(\frac{\partial\mathcal{L}}{\partial q^i}\eta_i(t) + \frac{\partial\mathcal{L}}{\partial \dot{q}^i}\dot{\eta}_i(t) \right) dt \tag{7.0.8}$$

We can split the integral up into two parts like any integral of this kind, and evaluate the second integral by parts as follows, remembering $\int u\,dv = uv - \int v\,du$:

$$\frac{dA}{d\alpha} = \int_{t_0}^{t_1} \frac{\partial \mathscr{L}}{\partial q^i} \eta_i(t) dt + \int_{t_0}^{t_1} \frac{\partial \mathscr{L}}{\partial \dot{q}^i} \dot{\eta}_i(t) dt \tag{7.0.9}$$

Now consider the second of these integrals:

$$\int_{t_0}^{t_1} \frac{\partial \mathscr{L}}{\partial \dot{q}^i} \frac{d\eta}{dt} dt = \left. \frac{\partial \mathscr{L}}{\partial \dot{q}^i} \eta_i(t) \right|_{t_0}^{t_1} - \int_{t_0}^{t_1} \eta_i(t) \frac{d}{dt}\left(\frac{\partial \mathscr{L}}{\partial \dot{q}}\right) dt \tag{7.0.10}$$

where we have used,

$$u = \frac{\partial \mathscr{L}}{\partial \dot{q}^i}, \qquad du = \frac{d}{dt}\left(\frac{\partial \mathscr{L}}{\partial \dot{q}^i}\right) dt, \qquad dv = \frac{d\eta_i}{dt} dt, \qquad v = \eta_i$$

The boundary term disappears, since by definition we are saying that the perturbation term is zero at the start and final times; that is, for all of the functions we consider, no matter what they are, all have the same boundary conditions:

$$\eta_i(t_0) = \eta_i(t_1) = 0 \tag{7.0.11}$$

So, when we evaluate the limit, it is zero, with no variation at the start or end points, and the blue term disappears entirely:

$$\frac{dA}{d\alpha} = \sum_{i=1}^{n} \int_{t_0}^{t_1} \frac{\partial \mathscr{L}}{\partial q^i} \eta_i(t) dt - \sum_{i=1}^{n} \int_{t_0}^{t_1} \eta_i(t) \frac{d}{dt}\left(\frac{\partial \mathscr{L}}{\partial \dot{q}^i}\right) dt$$

$$= \int_{t_0}^{t_1} \sum_{i=1}^{n} \left[\frac{\partial \mathscr{L}}{\partial q^i} - \frac{d}{dt}\left(\frac{\partial \mathscr{L}}{\partial \dot{q}^i}\right)\right] \eta_i(t) dt \tag{7.0.12}$$

We now impose the condition that, in order to make the action integral stationary, the equation in brackets must vanish for each variation. From equations 7.0.4 and 7.0.5, we can see that, when alpha disappears, the perturbation disappears and the functions above are now just differential equations for the extremal functions of $q^i(t)$ and $\dot{q}^i(t)$. The calculus of variations has not given us the value of $q(0, t)$ but rather a differential equation in terms of it and its time derivative:

$$\int_{t_0}^{t_1} \sum_{i=1}^{n} \left[\frac{\partial \mathscr{L}}{\partial q^i} - \frac{d}{dt}\left(\frac{\partial \mathscr{L}}{\partial \dot{q}^i}\right)\right] \eta_i(t) dt = 0 \tag{7.0.13}$$

Upon inspection of this equation, it can be seen that, since $\eta_i(t)$ can really be any set of independent functions, -well, any with the same boundary conditions -and that we can take the second derivative of. In order to make this equation valid for all choices of $\eta_i(t)$, the term in brackets must exactly equal zero:

$$\frac{\partial \mathscr{L}}{\partial q^i} - \frac{d}{dt}\left(\frac{\partial \mathscr{L}}{\partial \dot{q}^i}\right) = 0, \qquad i = 1, \ldots, n \qquad \textbf{Euler-Lagrange equation} \qquad (7.0.14)$$

This is the **Euler-Lagrange equation of motion**. It is a set of n **second-order** (see end of this section) ODEs in time with $2n$ variables. Note that, for each variable q^i there is a corresponding equation of this form; therefore, we generalise to say that, for any general coordinate, be it $x, y, z, r, \theta, \varphi, \phi, \ldots$ and its time derivative, which is signified by the dot above $\dot{q}^i$, the solution to equation 7.0.14 is the classical trajectory that makes the action integral stationary, which by Hamilton's principle is the physical path followed. Note that, because the Lagrange equation is second order, we can make two *independent* choices of initial conditions but, once they are chosen, subsequent variations to the $q^i s$ will *induce* a variation in the $\dot{q}^i$s. This is why we can treat them as independent variables in the Lagrangian $\mathscr{L}(\mathbf{q}, \dot{\mathbf{q}})$ but treat their variations as inherently linked. The equations of motion for systems we wish to study are found by computing the Lagrangian for that system and using it to solve the Euler-Lagrange equation. It is therefore the specific form of the Lagrangian that determines the equations of motion, -a singe scalar function!

To use the Lagrange equation, we start by defining the Lagrangian for the system in terms of the generalised coordinates and their time derivatives; this is definitely the most difficult step. In general the Lagrangian is well known for simple systems and can usually be transformed without too much work to different coordinate systems. We then use the Lagrange equation to give the corresponding equation of motion for the system; which hopefully, we can then solve. Lagrange's method is far superior for determining conserved quantities and energies, compared to Newton's method, but is entirely equivalent. We will now consider some examples of using Lagrangian mechanics.

Exercise 7.1 To illustrate the Lagrange method, consider the harmonic oscillator. The first step is to construct the Lagrangian; we know that the kinetic energy is $\frac{1}{2}m\dot{x}^2$ and the potential is $\frac{1}{2}kx^2$:

$$\mathscr{L} = T - U = \frac{1}{2}m\dot{x}^2 - \frac{1}{2}kx^2$$

We now compute two derivatives: $\partial\mathscr{L}/\partial\dot{q} = m\dot{x}$ and $\partial\mathscr{L}/\partial q = -kx$. The Euler-Lagrange equation is now $\frac{d}{dt}(m\dot{x}) + kx = 0$ and, if we assume constant mass, gives the familiar equation of motion $m\ddot{x} = -kx$.

Exercise 7.2 We consider now the **central force problem** of two particles. The system is two masses that interact via forces directed along the line that connects them. The first step is to express the cartesian kinetic energy in the polar coordinates $r(t)$ and $\theta(t)$. We know that there is a constraint on the system namely the z displacement is zero for all times; this is independent of the other two dimensions. We also assume that θ is a function of time. Using $x = r\cos\theta$ and $y = r\sin\theta$ and their time derivatives $\dot{x} = \dot{r}\cos\theta - r\dot{\theta}\sin\theta$ and

$\dot{y} = \dot{r}\sin\theta + r\dot{\theta}\cos\theta$, where we have chain ruled the $\cos\theta$ and $\sin\theta$ terms, due to the time dependance of $\theta(t)$.

$$T = \frac{1}{2}m\dot{x}^2 + \frac{1}{2}m\dot{y}^2 \tag{7.0.15}$$

$$\begin{aligned}\dot{x}^2 + \dot{y}^2 &= (\dot{r}\cos\theta - r\dot{\theta}\sin\theta)^2 + (\dot{r}\sin\theta + r\dot{\theta}\cos\theta)^2 \\ &= \dot{\theta}^2 r^2\sin^2\theta - 2\dot{\theta}r\dot{r}\sin\theta\cos\theta + \dot{r}^2\cos^2\theta + \dot{\theta}^2 r^2\cos^2\theta + 2\dot{\theta}r\dot{r}\sin\theta\cos\theta + \dot{r}^2\sin^2\theta \\ &= \dot{\theta}^2 r^2 + \dot{r}^2 \end{aligned} \tag{7.0.16}$$

using $\cos^2\theta + \sin^2\theta = 1$. Now the kinetic energy is given in terms of r and θ:

$$T = \frac{1}{2}m\big(\dot{\theta}^2 r^2 + \dot{r}^2\big) \tag{7.0.17}$$

The potential energy term is assumed to be just a function of r (this is important later) and we can now form the Lagrangian for the system. We then use this term in the Euler-Lagrange equation for each coordinate θ and r:

$$\mathscr{L} = \frac{1}{2}m\big(\dot{\theta}^2 r^2 + \dot{r}^2\big) - U(r) \tag{7.0.18}$$

We note that we have two coordinates, θ and r, and their time derivatives; in this case $\mathscr{L}(\boldsymbol{q},\dot{\boldsymbol{q}}) = \mathscr{L}(q^1,q^2,\dot{q}^1,\dot{q}^2) = \mathscr{L}(\theta,r,\dot{\theta},\dot{r})$, so we will need to compute two sets of Euler-Lagrange equations -one for each coordinate. So in this system i runs from 1 to 2. Firstly, we consider the θ coordinate; substitution of θ for q^i in the Lagrange equation gives us the Lagrange equation for the θ coordinate:

$$\frac{d}{dt}\left(\frac{\partial\mathscr{L}}{\partial\dot{\theta}}\right) - \frac{\partial\mathscr{L}}{\partial\theta} = 0 \tag{7.0.19}$$

We now write in the Lagrangian and perform the partial derivatives:

$$\frac{d}{dt}\big(m\dot{\theta}r^2\big) - 0 = 0 \tag{7.0.20}$$

Interestingly, we note that, because the Lagrangian did not depend on θ but, rather, only its time derivative, the second term is zero. This shows that the time derivative of the quantity in brackets equals zero, which is the criterion for it being a **conserved quantity**! We will keep this example in mind when we discuss **Noether's theorem** a little later on but, for now, all we need realise is that the quantity in brackets is constant for this system so we can give it a symbol L (it is actually angular momentum).

We perform the same procedure again for the r coordinate, noting that the Lagrangian *is* a function of both r and $\dot{r}$:

$$\frac{d}{dt}\left(\frac{\partial\mathscr{L}}{\partial\dot{r}}\right) - \frac{\partial\mathscr{L}}{\partial r} = 0$$

Substituting in the Lagrangian we obtain

$$\frac{d}{dt}(m\dot{r}) - mr\dot{\theta}^2 + \frac{\partial U(r)}{\partial r} = 0 \qquad (7.0.21)$$

We can use L from the first Euler-Lagrange equation to substitute into this and we can also simplify by assuming constant mass:

$$m\ddot{r} = \frac{L^2}{mr^3} - \frac{\partial U(r)}{\partial r} \qquad (7.0.22)$$

Therefore, 7.0.20 and 7.0.22 are our equations of motion for the central force problem; you can see how it wasn't that hard to compute them once we had the expression for the Lagrangian.

Above all, we should prove to ourselves that the Lagrangian formulation is indeed entirely equivalent to Newton's equations of motion. Each trajectory between fixed end points has its own action; we have said that the path the particle actually takes is an extremum of the action meaning, $\delta A = 0$, and, when it does, its Lagrangian will obey the Euler-Lagrange equation.

We might like to observe the equivalence of the two formalisms with a little more generality. Therefore, consider a general trajectory $x(t)$ and a small variation to the path $\delta x(t)$ such that, at both the start and the finish, the variations disappear and so, no matter what path we take we always arrive at the same final point from an initial point.

The action functional can be written as the time integral of the Lagrangian, and the stationary action principle dictates that $\delta A = A[x + \delta x] - A[x] = 0$. In the second step, we are Taylor expanding the potential (more on this later) and ignoring terms in δx^2:

$$\delta A = \int_{t_0}^{t_1} \left[\frac{1}{2}m(\dot{x} + \delta\dot{x})^2 - U(x + \delta x) \right] dt - \int_{t_0}^{t_1} \left[\frac{1}{2}m\dot{x}^2 - U(x) \right] dt$$

$$= \int_{t_0}^{t_1} \left[\frac{1}{2}m(\dot{x}^2 + 2\dot{x}\cdot\delta\dot{x} + \delta\dot{x}^2) - U(x) - \vec{\nabla}U \cdot \delta x \right] dt - \int_{t_0}^{t_1} \left[\frac{1}{2}m\dot{x}^2 - U(x) \right] dt$$

$$= \int_{t_0}^{t_1} \left[m\dot{x}\cdot\delta\dot{x} - \vec{\nabla}U \cdot \delta x \right] dt \qquad (7.0.23)$$

We now realise that it would be handy to collect the δx and $\delta\dot{x}$ terms; but, in order to do, this we have to integrate the first term by parts to get rid of that dot. Therefore, let $u = m\dot{x}$, $du = m\ddot{x}$, $dv = \delta\dot{x}$ and $v = \delta x$:

$$\int_{t_0}^{t_1} m\dot{x}\cdot\delta\dot{x}\,dt = m\dot{x}\cdot\delta x\Big|_{t_1}^{t_2} - \int_{t_1}^{t_2} \delta x \cdot \ddot{x}m \qquad (7.0.24)$$

which we can substitute back in and so collect terms:

$$\delta A = \left[m\dot{x} \cdot \delta x \right]_{t_1}^{t_2} - \int_{t_1}^{t_2} \left[m\ddot{x} + \vec{\nabla} U \right] \cdot \delta x \qquad (7.0.25)$$

The variation terms are zero at the start and end points, so we know that the first term disappears, since we are evaluating the variation at the initial and final times. Substitute the negative sign into the second part and realise that, in order for any arbitrary variation to the extremal path to equal zero, the term in the brackets must equal zero:

$$m\ddot{x} = -\vec{\nabla} U(x) \qquad (7.0.26)$$

Newton's laws of motion are, in fact, describing the extremal of the action. Wow! We will now transition to the use of qs for the general coordinates and reserve xs for vectorial coordinates. There is something really important to note which may seem irrelevant right now but, when we advance our discussions will become crucial: the action functional as presented here is not a functional of the velocity coordinates. It is a functional of the generalised coordinates. A variation in the generalised coordinates automatically leads to a variation in the velocity coordinates since, they are just the time derivatives. We are treating the $\dot{q}$s independently of qs but their variations are inherently coupled, since a variation in the generalised coordinate will induce a variation in the velocity.

As we have seen, the Lagrange equation can be used to give the same result as Newton's equations. Which approach is best to use really does depend on the system. If we have the scalar energy quantities and we know the initial and final points in configuration space, then it would be advantageous to take a Lagrangian approach, especially if there are constraints on the system, or the reference frames are accelerating. One may raise the point, as indeed Poincaré did in the 19th century, that, in order to construct the extremal path, the particle needs to know where its going in order to decide which path to follow in the first place; this is known as the **teleological problem**. This teleological setting of classical mechanics is in opposition to Newton's stepwise approach, which is perhaps more intuitive. A satisfactory resolution is that, in any case, we can construct an action functional whose extremals are the motions of the system, as dictated by the Lagrangian, and both methods give the same result.

Lagrangian mechanics focuses on the motion of the system when it is limited to a particular dimensional surface; it takes the constraints into account by lessening the degrees of freedom of a system which is described instead by generalised coordinates. Newton's approach assigns a force to each constraint and solves the resulting equations; if we would rather analyse the forces on the system and use vectorial coordinates in inertial frames, then Newtonian physics is a fine approach. We will see that **nonholonomic constraints** would be a lot more complex to deal with in the Lagrangian form than **holonomic constraints** because, in the least action principle, we are integrating the Lagrangian and, by definition, non-holonomic constraints are not integrable.

We discuss constraints later when we use **Lagrange multipliers** in an effort to solve a constrained Lagrangian.

Lets us derive the Lagrange equation, using the slightly different notation of **functional derivatives**:

$$\delta A = \delta \int_{t_0}^{t_1} \mathscr{L}(\mathbf{q}, \dot{\mathbf{q}}, t)dt$$

$$= \int_{t_0}^{t_1} \sum_{i=1}^{n} \left[\frac{\partial \mathscr{L}}{\partial q^i} \delta q^i + \frac{\partial \mathscr{L}}{\partial \dot{q}^i} \delta \dot{q}^i + \frac{\partial \mathscr{L}}{\partial t} \delta t \right] dt$$

$$= \int_{t_0}^{t_1} \sum_{i=1}^{n} \left[\frac{\partial \mathscr{L}}{\partial q^i} \delta q^i - \frac{d}{dt} \left(\frac{\partial \mathscr{L}}{\partial \dot{q}^i} \right) \delta q^i \right] dt$$

$$= \int_{t_0}^{t_1} \sum_{i=1}^{n} \left[\frac{\partial \mathscr{L}}{\partial q^i} - \frac{d}{dt} \left(\frac{\partial \mathscr{L}}{\partial \dot{q}} \right) \right] \delta q^i dt$$

where we have integrated the velocity term by parts as before, assumed vanishing variations at the boundary points $\delta q^i(t_0) = \delta q^i(t_1) = 0$, and collected like variations; in the first step, $\delta t = 0$ disappears, since these spatial variations occur at a fixed time[1]. Again, Hamilton's principle states that, for arbitrary independent variations, the variation of the action must be set to zero (made stationary) and hence each term must vanish independently and we generate the Euler-Lagrange equation of motion. With this notation, it is much more straight forward to derive the Lagrange equation from the action principle.

Finally, we may recall that the aim of mechanics was to find the generalised accelerations, right? So, how do we get accelerations from the Lagrange equation of motion? The answer is simple if we consider the Euler-Lagrange equation and expand the time derivative:

$$\frac{d}{dt} \left(\frac{\partial \mathscr{L}(\dot{q}, q)}{\partial \dot{q}^i} \right) - \frac{\partial \mathscr{L}}{\partial q^i} = \left(\frac{\partial^2 \mathscr{L}}{\partial \dot{q}^j \partial \dot{q}^i} \ddot{q}^i + \frac{\partial^2 \mathscr{L}}{\partial \dot{q}^j \partial q^i} \dot{q}^i \right) - \frac{\partial \mathscr{L}}{\partial q^j}$$

$$= A \frac{d^2}{dt^2} q + B \frac{d}{dt} q - C \qquad (7.0.27)$$

Then, with a slight rearrangement we obtain

$$\ddot{q}^i = \left[\frac{\partial^2 \mathscr{L}}{\partial \dot{q}^j \partial \dot{q}^i} \right]^{-1} \left(\frac{\partial \mathscr{L}}{\partial q^j} - \sum_i \frac{\partial^2 \mathscr{L}}{\partial \dot{q}^j \partial q^i} \dot{q}^i \right) \qquad (7.0.28)$$

---

[1]Such variations are called **virtual variations** and they are used to investigate a constant-time slice of velocity phase space. We investigate these in our discussion of d'Alembert's principle in a little while.

This is the second form of the Euler-Lagrange equation we can consider to solve problems with. It is a bit more complicated than using the other form, since we have **second-order** terms and, hence, this is why the Lagrange formalism is built on a second-order equation. Later on, we will question whether the term in square brackets really can be inverted the way it has been here. It is known as the **Hessian condition** and is very important to the transition between the Lagrangian and Hamiltonian formalisms.

7.1   The Inverse Problem

In any general mechanical n-dimensional problem, the equations of motion can be written in the following form:

$$\ddot{q}_k = f_k(t, q^1, \ldots, q^n, \dot{q}, \ldots, q^n), \qquad k = 1, \ldots, n \tag{7.1.1}$$

The Lagrangian for such a system is $\mathscr{L}(t, q^1, \ldots, q^n, \dot{q}^1, \ldots, \dot{q}^n)$ and we know that the equation of motion can be derived from this Lagrangian by using the Euler-Lagrange equation of motion. Considering this, however, from the reverse viewpoint, we observe that a system of coupled second-order ODEs (equations of motion) admit a scalar function (Lagrangian) which obeys the second order Euler-Lagrange equation. The equations of motion must satisfy the **Jacobi last multiplier theorem** (see chapter 35) and the Jacobi multiplier together with the solution to the coupled system of ODEs must satisfy the following condition:

$$\frac{\partial M}{\partial t} + \sum_{k=1}^{n} \frac{\partial(q^{(k)}M)}{\partial q^{(k-1)}} = 0 \tag{7.1.2}$$

Therefore, we now move to express the mechanical Lagrangian system in a way that demonstrates compliance with this result. To do this, we take the time derivative of $\partial\mathscr{L}/\partial\dot{q}^j$ just as before, which is a useful quantity in dynamics that we meet later (see **conjugate momentum** in chapter 11) and enter that into the Euler-Lagrange equation of motion:

$$\frac{d}{dt}\left(\frac{\partial\mathscr{L}}{\partial\dot{q}^j}\right) - \frac{\partial\mathscr{L}}{\partial q^j} = \frac{\partial^2\mathscr{L}}{\partial t\partial\dot{q}^j} + \sum_{k=1}^{n}\left(\dot{q}^k\frac{\partial^2\mathscr{L}}{\partial q^k\partial\dot{q}^j} + f_k\frac{\partial^2\mathscr{L}}{\partial\dot{q}^k\partial\dot{q}^j}\right) - \frac{\partial\mathscr{L}}{\partial q^j} = 0 \tag{7.1.3}$$

with $j = 1, \ldots, n$. Although such an expression may look complicated at first glance, there is nothing other than the mathematics we have used previously, and treating $\partial\mathscr{L}/\partial\dot{q}^i$ as a single object will simplify this greatly. Differentiating this with respect to $\dot{q}^i$ and swapping the order of partial differentiation (see **Clairaut's theorem** in chapter 32) we find:

$$\frac{\partial}{\partial t}\left(\frac{\partial^2 \mathscr{L}}{\partial \dot{q}^i \partial \dot{q}^j}\right) + \sum_{k=1}^{n}\left[\frac{\partial}{\partial q^k}\left(\dot{q}^k \frac{\partial^2 \mathscr{L}}{\partial \dot{q}^i \partial \dot{q}^j}\right) + \frac{\partial}{\partial \dot{q}^k}\left(f_k \frac{\partial^2 \mathscr{L}}{\partial \dot{q}^i \partial \dot{q}^j}\right)\right]$$

$$+ \sum_{k=1}^{n}\left(\frac{\partial f_k}{\partial \dot{q}^i}\frac{\partial^2 \mathscr{L}}{\partial \dot{q}^k \partial q^j} - \frac{\partial f_k}{\partial \dot{q}^k}\frac{\partial^2 \mathscr{L}}{\partial \dot{q}^i \partial \dot{q}^j}\right) + \frac{\partial^2 \mathscr{L}}{\partial q^i \partial \dot{q}^j} - \frac{\partial^2 \mathscr{L}}{\partial \dot{q}^i \partial q^j} = 0$$

If you then swap the $i \to j$ indices and add that result to this one (cancelling a factor of 2 throughout), again remembering that the swapping of order of partial derivatives is allowed, we generate the following expression:

$$\frac{\partial}{\partial t}\left(\frac{\partial^2 \mathscr{L}}{\partial \dot{q}^i \partial \dot{q}^j}\right) + \sum_{k=1}^{n}\left[\frac{\partial}{\partial q^k}\left(\dot{q}^k \frac{\partial^2 \mathscr{L}}{\partial \dot{q}^i \partial \dot{q}^j}\right) + \frac{\partial}{\partial \dot{q}^k}\left(f_k \frac{\partial^2 \mathscr{L}}{\partial \dot{q}^i \partial \dot{q}^j}\right)\right]$$

$$+ \frac{1}{2}\sum_{k=1}^{n}\left[\frac{\partial f_k}{\partial \dot{q}^i}\frac{\partial^2 \mathscr{L}}{\partial \dot{q}^k \partial q^j} + \frac{\partial f_k}{\partial \dot{q}^j}\frac{\partial^2 \mathscr{L}}{\partial \dot{q}^k \partial \dot{q}^i}\right] - \sum_{k=1}^{n}\frac{\partial f_k}{\partial \dot{q}^k}\frac{\partial^2 \mathscr{L}}{\partial \dot{q}^i \partial \dot{q}^j} = 0$$

We now assume that $\partial f_k / \partial \dot{q}^k = 0$ and hence obtain

$$\frac{\partial}{\partial t}\left(\frac{\partial^2 \mathscr{L}}{\partial \dot{q}^i \partial \dot{q}^j}\right) + \sum_{k=1}^{n}\left[\frac{\partial}{\partial q^k}\left(\dot{q}^k \frac{\partial^2 \mathscr{L}}{\partial \dot{q}^i \partial \dot{q}^j}\right) + \frac{\partial}{\partial \dot{q}^k}\left(f_k \frac{\partial^2 \mathscr{L}}{\partial \dot{q}^i \partial \dot{q}^j}\right)\right] = 0 \tag{7.1.4}$$

Identifying the Jacobi last multiplier with the Hessian $M_{ij} = \frac{\partial^2 \mathscr{L}}{\partial \dot{q}^i \partial \dot{q}^j}$, we obtain

$$\frac{\partial M_{ij}}{\partial t} + \sum_{k=1}^{n}\left[\frac{\partial}{\partial q^k}\left(\dot{q}^k M_{ij}\right) + \frac{\partial}{\partial \dot{q}^k}\left(f^k M_{ij}\right)\right] = 0 \tag{7.1.5}$$

Allowing for notational differences, it has been shown that the Euler-Lagrange equation obeys Jacobi's last multiplier theorem. As will be shown in chapter 32, the knowledge of the Jacobi last multipliers can lead to first integrals of the motion and provide a way to fully understand a dynamical system of ODEs.

More generally, the **inverse problem of Lagrangian mechanics** considers whether a system of ODEs $\ddot{q}^i = f^i(q, \dot{q})$ admits a Euler-Lagrange description for a regular Lagrangian function. An equivalent question is under what conditions can a system of second-order differential equations be derived from a variational principle and, in practice, reduces to a matter of finding Lagrangians for an equation of motion. It is the **Douglas theorem** that states there exist **multipliers** $g_{ij}(q, \dot{q})$ such that the following condition holds:

$$g_{ij}(\ddot{q}^j - f^j) = \frac{d}{dt}\left(\frac{\partial \mathscr{L}}{\partial \dot{q}^i}\right) - \frac{\partial \mathscr{L}}{\partial q^i} \tag{7.1.6}$$

Provided that the multipliers satisfy the **Helmholtz conditions**, we expect the multipliers to form a symmetric non-singular matrix g whose entries g_{ij} are symmetric under permutation of indices:

$$\det(g) \neq 0, \qquad g_{ij} = g_{ji}, \qquad \frac{\partial g_{ij}}{\partial \dot{q}^k} = \frac{\partial g_{ik}}{\partial \dot{q}^j} \qquad (7.1.7)$$

And the additional two Helmholtz conditions are

$$\left(\dot{q}^i \frac{\partial}{\partial q^i} + f^i \frac{\partial}{\partial \dot{q}^i} \right) g_{ij} + \frac{1}{2} \frac{\partial f^i}{\partial \dot{q}^j} g_{ik} + \frac{1}{2} \frac{\partial f^j}{\partial \dot{q}^i} g_{kj} = 0, \qquad \text{and} \qquad g_{ik} \Phi^k_j = g_{jk} \Phi^k_i \quad (7.1.8)$$

where

$$\Phi^k_j = \frac{g_{ik}}{2} \left(\dot{q}^i \frac{\partial}{\partial q^i} + f^i \frac{\partial}{\partial \dot{q}^i} \right) \frac{\partial f^k}{\partial \dot{q}^j} - \frac{\partial f^k}{\partial q^j} - \frac{1}{4} \frac{\partial f^l}{\partial \dot{q}^j} \frac{\partial f^k}{\partial \dot{q}^l} \qquad (7.1.9)$$

The Helmholtz conditions provide a method of generating Lagrangians for a physical system described by a second-order equation of motion. These properties arise later in a discussion of velocity-dependent potentials when we investigate the symmetry conditions of potentials. In particular, we use the ruining of the symmetry of the index permutation to show that velocity-dependent forces do not admit potentials.

7.2 Higher-Order Theories & the Ostrogradsky Equation

One may ask, if the Lagrangian had been a function of the higher-order derivatives $q, \dot{q}$ and $\ddot{q}$, would the Euler-Lagrangian be of the same form, following the calculus of variations? Let us consider the following Euler-Lagrange equation for $\mathscr{L}(\mathbf{q}, \dot{\mathbf{q}}, \ddot{\mathbf{q}})$, using the calculus of variations with functional derivative notation:

$$\begin{aligned}
\delta A[\boldsymbol{q}] &= \delta \int_{t_1}^{t_2} \mathscr{L}(\mathbf{q}, \dot{\mathbf{q}}, \ddot{\mathbf{q}}) dt \\
&= \int_{t_1}^{t_2} dt \left[\frac{\partial \mathscr{L}}{\partial q^i} \delta q^i + \frac{\partial \mathscr{L}}{\partial \dot{q}^i} \delta \dot{q}^i + \frac{\partial \mathscr{L}}{\partial \ddot{q}^i} \delta \ddot{q}^i \right] \\
&= \int_{t_1}^{t_2} dt \left[\frac{\partial \mathscr{L}}{\partial q^i} - \frac{d}{dt} \left(\frac{\partial \mathscr{L}}{\partial \dot{q}^i} \right) + \frac{d^2}{dt^2} \left(\frac{\partial \mathscr{L}}{\partial \ddot{q}^i} \right) \right] \delta q^i
\end{aligned}$$

where we have integrated the $\ddot{q}^i$ term by parts twice, the first time obtaining $-\frac{d}{dt} \left(\frac{\partial \mathscr{L}}{\partial \ddot{q}^i} \right) \delta \dot{q}^i$. Imposing the stationary action principle, we obtain

$$\frac{\partial \mathscr{L}}{\partial q^i} - \frac{d}{dt} \left(\frac{\partial \mathscr{L}}{\partial \dot{q}^i} \right) + \frac{d^2}{dt^2} \left(\frac{\partial \mathscr{L}}{\partial \ddot{q}^i} \right) = 0 \qquad (7.2.1)$$

Note it is a fourth-order equation, and extending the calculus of variations requires us to impose another two endpoint conditions: in addition to fixing the positions at the endpoints, we must also fix the velocities, too. In general, the **Ostrogradsky equation** for a Lagrangian dependent of n derivatives $\mathscr{L}(q, \dot{q}, \ddot{q}, \ldots, q^{(n)})$ is a $2n$-th-order equation:

$$\sum_{i=0}^{n}(-1)^i\frac{d^i}{dt^i}\left(\frac{\partial\mathscr{L}}{\partial q^{(i)}}\right)=0 \qquad \textbf{Ostrogradsky equation} \qquad (7.2.2)$$

with $q^{(i)} = d^i q/dt^i$, $i = 0, 1, \ldots, n$. This, however, is not necessary to describe deterministic dynamics since, as a fundamental postulate of the Newtonian formalism, we only require the position and the velocity (or momentum) to predict the evolution of the system. Extending this to the Hamiltonian formalism leads to the famous **Ostrogradsky instability problem**, which is commonly known as Ostrogradsky's ghost, and it is for this reason that most Lagrangians in mechanics are only second order. Outside of mechanics, higher-order Lagrangians are necessary in the EinsteinHilbert action principle, which is the realm of general relativity, but, in engineering, the Ostrogradsky equation is in common use for structural optimisation problems.

7.3 The Second Variation

We can characterise a classical trajectory as a minimum or stationary point by considering the second-order variation to the action. Taylor expand the action in the small variation evaluated on the physical path q_0:

$$A = A\big|_{q_0} + \delta A\big|_{q_0} + \frac{1}{2}\delta^2 A\big|_{q_0} + \cdots \qquad (7.3.1)$$

Hamilton's principle says that the middle term is zero when evaluated on a true classical path. It is the **second variation** $\delta^2 A_0 = \delta A(q_0 + \delta q) - \delta A(q_0)$ that provides the detail on the characteristics of the action, just as in standard calculus. If $\delta^2 A_0 > 0$, the action is a minimum and is positive for all variations. It is important to note that we are varying the physical path between boundary points: we are characterising the **variation field** around the classical trajectory:

$$\mathscr{L}(\boldsymbol{q}, \dot{\boldsymbol{q}}) = \mathscr{L}(\boldsymbol{q}, \dot{\boldsymbol{q}})\big|_{q_0} + \delta\mathscr{L}\big|_{q_0} + \frac{1}{2}\delta^2\mathscr{L}\big|_{q_0} + \cdots$$

Evaluation of these variations using the standard variation procedure yields,

$$= \mathscr{L}(\boldsymbol{q}, \dot{\boldsymbol{q}})\big|_{q_0} + \left[\left(\frac{\partial\mathscr{L}}{\partial q^i}\right)_{q_0}\delta q^i + \left(\frac{\partial\mathscr{L}}{\partial \dot{q}^i}\right)_{q_0}\delta \dot{q}^i\right] + \frac{1}{2}\left[\left(\frac{\partial^2\mathscr{L}}{\partial q^i\partial q^j}\right)_{q_0}\delta q^i\delta q^j\right.$$

$$\left. + \left(\frac{\partial^2\mathscr{L}}{\partial \dot{q}^i\partial \dot{q}^j}\right)_{q_0}\delta \dot{q}^i\delta \dot{q}^j + \left(\frac{\partial^2\mathscr{L}}{\partial q^i\partial \dot{q}^j}\right)_{q_0}\delta q^i\delta \dot{q}^j + \left(\frac{\partial^2\mathscr{L}}{\partial \dot{q}^i\partial q^j}\right)_{q_0}\delta \dot{q}^i\delta q^j\right] + \cdots$$

The classical path q_0 will minimise the action if the second variation is greater than zero for the variation field. If we integrate the mixed variation term by parts, this expression simplifies, as follows:

$$
\begin{aligned}
\delta^2 A &= \int_1^2 \left\{ \left(\frac{\partial^2 \mathscr{L}}{\partial q^i \partial q^j} \right)_{q_0} \delta q^i \delta q^j + \left(\frac{\partial^2 \mathscr{L}}{\partial \dot{q}^i \partial \dot{q}^j} \right)_{q_0} \delta \dot{q}^i \delta \dot{q}^j + 2 \left(\frac{\partial^2 \mathscr{L}}{\partial q^i \partial \dot{q}^j} \right)_{q_0} \delta q^i \delta \dot{q}^j \right\} dt \\
&= \int_1^2 \left\{ \left(\frac{\partial^2 \mathscr{L}}{\partial q^i \partial q^j} \right)_{q_0} \delta q^i \delta q^j + \left(\frac{\partial^2 \mathscr{L}}{\partial \dot{q}^i \partial \dot{q}^j} \right)_{q_0} \delta \dot{q}^i \delta \dot{q}^j - \delta q^i \delta q^j \frac{d}{dt} \left(\frac{\partial^2 \mathscr{L}}{\partial q^i \partial \dot{q}^j} \right) \right\} dt \\
&= \int_1^2 \left\{ \left(\frac{\partial^2 \mathscr{L}}{\partial \dot{q}^i \partial \dot{q}^j} \right)_{q_0} \delta \dot{q}^i \delta \dot{q}^j + \delta q^i \delta q^j \left(\frac{\partial^2 \mathscr{L}}{\partial q^i \partial q^j} - \frac{d}{dt} \left(\frac{\partial^2 \mathscr{L}}{\partial q^i \partial \dot{q}^j} \right) \right)_{q_0} \right\} dt
\end{aligned}
\tag{7.3.2}
$$

We now treat this as if it where a Lagrangian function and find it's Euler-Lagrange equation:

$$
\boxed{ \delta q^j \left[\frac{\partial^2 \mathscr{L}}{\partial q^i \partial q^j} - \frac{d}{dt} \left(\frac{\partial^2 \mathscr{L}}{\partial q^i \partial \dot{q}^j} \right) \right] - \frac{d}{dt} \left(\frac{\partial^2 \mathscr{L}}{\partial \dot{q}^i \partial \dot{q}^j} \delta \dot{q}^i \right) = 0 }
\tag{7.3.3}
$$

This equation is the **Jacobi equation** for the variation field around a curve. If the classical path q_0 is a minimum of the action functional, then the Hessian must be greater than or equal to zero for all time in the interval. This is the **Legendre condition** that is necessary for minimisation. If the Legendre condition is satisfied and the Jacobi equations have only trivial solutions for the variations $\delta q^i = 0$, then the action is a minimum.

7.4   Functions & Functionals

We are aware that we can construct an action *functional* $A[q]$ whose critical points describe the physical trajectory that the system traverses between two fixed end points $q_0^i(t_0)$ and $q_1^i(t_1)$ over a time interval $[t_0, t_1]$ and generate the Lagrange equation of motion by setting first-order variations to zero. The action is a functional of all the paths between the fixed endpoints, in addition to the physical one, so we refer to it as the **off-shell action**:

$$
A[q] = \int_{t_0}^{t_1} \mathscr{L}(q, \dot{q}, t) dt
\tag{7.4.1}
$$

If there is an off-shell action, there must be an **on-shell action**, that particular action that corresponds to the physical path between the two fixed boundary points. It is a *functional* of the classical path only, $A[q_{\text{cl}}]$, but it's a *function* of the **Dirichlet boundary conditions** or the endpoints:

$$
A[q_{\text{cl}}(t)](q_0, t_0, q_1, t_1) = \int_{t_0}^{t_1} \mathscr{L}(q_{\text{cl}}, \dot{q}_{\text{cl}}, t) dt
\tag{7.4.2}
$$

Both action functionals are exactly the same when the off-shell action is evaluated on the particular path that corresponds to its stationary point, which is obvious if they both correspond to the same physical motion[2].

The physical path between two boundary points in configuration space may not be unique and, as we increase the time interval, there could, in fact, be lots of different trajectories all corresponding to physical motion between the two points. These may be due to instantons or gauge symmetries in the system and we investigate an example at the end of this section. If we were to specify **initial conditions**, then there would be one deterministic path only. Specifying initial conditions is not the same as specifying boundary points. Initial conditions would correspond to a point in velocity space, that is to say, we require the position and velocity to describe a unique Laplacian determined path. This subtlety behind the action principle is very important point to appreciate; it is the difference between a Newtonian initial value problem and the Lagrangian boundary point problem.

If we know the action function between boundary points, then we know all possible solutions to the motion. It is the aim of Hamilton-Jacobi theory to tackle this, which we look at in chapter 19. There we will refer to the action with a symbol S, as we have referred to the Hamilton-Jacobi action in section 11.4 as $\mathcal{A}$. This is simply to distinguish between the variables we are treating and shouldn't be fussed over too much. The on-shell action function can be identified with the principal function S and $\mathcal{A}$ in section 11.4 for extremal paths [3].

Exercise 7.3 As an example of how the specification of boundary conditions does not lead to a particular solution being singled out, we consider a rotating rigid body. The Lagrangian for the system is $\mathscr{L} = \frac{1}{2}I\dot{\theta}^2$ with moment of inertia I. The Dirichlet boundary conditions are $\theta(t_i) - \theta_i \in 2\pi n$ and $\theta(t_f) - \theta_f \in 2\pi n$, where $n \in \mathbb{Z}$ is an integer. Clearly, the cyclic boundary conditions lead to an infinite number of solutions to the equation of motion, $\ddot{\theta} = 0$, one rotation on. If, however, we were to specify a particular winding, then we uniquely specify the problem. Using Noether's theorem, (see chapter 11) we know that the angular momentum L is conserved on-shell.

---

[2]When I first encountered this, I spent a lot of time worrying about how we can use a function for a functional variational problem. The answer is, you can't. When expressed as a function of the boundary points, the action has other uses, namely as a solution to the Hamilton-Jacobi problem, but it can't be used in Hamilton's variational principle. They are definitely two distinct mathematical objects that are equal when describing the extremal classical path.

[3]It is perhaps unfortunate that all these quantities are referred to as "action", since they are truly distinct objects. This topic definitely confused me for a great length of time, as I constantly wanted to unite and unify all these quantities that I saw written as the action. I could not understand the relationship (or lack of) between them and until I let this vacuous notion go, I found it difficult to progress. It is not until I encountered this problem in differential geometry, where the Hamilton-Jacobi action is defined as a vertical endomorphism and the functional action as . . . , well, a functional on the space of curves, that I felt much better about the different quantities we call action. My feeling is that it is the shared properties of the solutions of the equations of motion that give rise to the different ways that mechanics can be presented, not that there is one single action manifesting in different ways.

$$L = I \frac{\theta_f - \theta_i}{t_f - t_i} \tag{7.4.3}$$

The on-shell action $A(\theta_i, t_i, \theta_f, t_f)$ is given by,

$$A(\theta_i, t_i, \theta_f, t_f) = \frac{I}{2} \frac{(\theta_f - \theta_i)^2}{t_f - t_i} \tag{7.4.4}$$

This is equivalent to the specification of Dirichlet conditions for the action functional $A[\theta]$:

$$A[\theta] = \int_1^2 \mathscr{L} \, dt \quad \text{with} \quad \theta(1) = \theta_i(t_i), \quad \theta(2) = \theta_f(t_f) \tag{7.4.5}$$

7.5   Boundary Conditions

If the time interval $[t_1, t_2]$ is specified but the initial and final configurations are not, Hamilton's principle *generates* the necessary boundary conditions for the problem. The usual calculus of variations would have the conditions $\delta q^i(t_1) = \delta q^i(t_2) = 0 \; \forall i$, with the extremal path satisfying

$$\delta A = \frac{\partial \mathscr{L}}{\partial \dot{q}^i} \delta q^i \Big|_{t_1}^{t_2} + \int_{t_1}^{t_2} \left(\frac{\partial \mathscr{L}}{\partial q^i} - \frac{d}{dt} \frac{\partial \mathscr{L}}{\partial \dot{q}^i} \right) \delta q^i dt \tag{7.5.1}$$

If the initial and final configurations are not specified, then there are $2n$ boundary conditions of the form

$$\frac{\partial \mathscr{L}}{\partial \dot{q}^i} \Big|_{t_1} = 0, \quad \frac{\partial \mathscr{L}}{\partial \dot{q}^i} \Big|_{t_2} = 0. \tag{7.5.2}$$

For any $\mathscr{L}(q, \dot{q}, t)$, the boundary conditions may be *essential* or *natural* boundary conditions (see figure 7.3). In most cases, this is equivalent to **Dirichlet** or **Neumann** boundary conditions, respectively. Including the boundary terms, the calculus of variations gives us the following result:

$$\delta A = \int_1^2 \left\{ \left[\frac{\partial \mathscr{L}}{\partial q^j} - \frac{d}{dt} \left(\frac{\partial \mathscr{L}}{\partial \dot{q}^j} \right) \right] \delta q^j + \frac{d}{dt} \left[\frac{\partial \mathscr{L}}{\partial \dot{q}^j} \delta q^j \right] \right\} dt \tag{7.5.3}$$

The Dirichlet conditions correspond to setting $q^j(t_0) = q_0^j$ and $q^j(t_1) = q_1^j$ such that we can write an on-shell action function $A(q_0, q_1, t_0, t_1)$. Natural boundary conditions, however, would be anything that forces $\partial \mathscr{L} / \partial \dot{q}^j$ to vanish at t_0 and t_1; most of the time, this is a Neumann condition of $\dot{q}^j = 0$, but that may not always be the case if

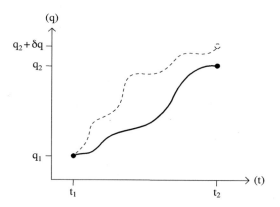

Fig. 7.3: An example of a variational problem that leads to the Hamilton Jacobi equation via the relaxation of imposed boundary conditions. Here the final configuration is not specified, although the time interval and initial configurations are! In order for the action to vanish, in addition to the Euler-Lagrange equations, we require $(\partial \mathscr{L} / \partial \dot{q}^i)_{t_1} = 0$.

the conjugate momenta is not proportional to the generalised velocity. As an example, consider the following action with the addition of a total time derivative:

$$A = \int_{t_1}^{t_2} Ldt \qquad \text{with} \qquad L = \mathscr{L} - \lambda \frac{d}{dt}\left(q^i \frac{\partial \mathscr{L}}{\partial \dot{q}^i}\right) \tag{7.5.4}$$

Application of the functional derivative for vertical variations ($\delta t = 0$) generates the following:

$$\delta A = \int_1^2 \left\{ \left(\frac{\partial \mathscr{L}}{\partial q^i}\delta q^i + \frac{\partial \mathscr{L}}{\partial \dot{q}^i}\delta \dot{q}^i \right) - \lambda \delta \left[\frac{d}{dt}\left(q^i \frac{\partial \mathscr{L}}{\partial \dot{q}^i} \right) \right] \right\} dt$$

$$= \int_1^2 \left\{ \left[\frac{\partial \mathscr{L}}{\partial q^i} - \frac{d}{dt}\left(\frac{\partial \mathscr{L}}{\partial \dot{q}^i} \right) \right]\delta q^i + \frac{d}{dt}\left[\frac{\partial \mathscr{L}}{\partial \dot{q}^i}\delta q^i \right] \right\} dt - \lambda \delta \left(q^i \frac{\partial \mathscr{L}}{\partial \dot{q}^i} \right)\Big|_1^2 \tag{7.5.5}$$

The first term is the usual Euler-Lagrange equation, the second is the boundary term and the final term is the variation from the last part of L; note it was a total time derivative. The middle term is a total time derivative too, allowing us to collect the last two terms together:

$$= \int_1^2 \left[\frac{\partial \mathscr{L}}{\partial q^i} - \frac{d}{dt}\left(\frac{\partial \mathscr{L}}{\partial \dot{q}^i} \right) \right] dt + \left[\frac{\partial \mathscr{L}}{\partial \dot{q}^i}\delta q^i - \lambda \delta \left(q^i \frac{\partial \mathscr{L}}{\partial \dot{q}^i} \right) \right]_1^2$$

$$= \int_1^2 \left[\frac{\partial \mathscr{L}}{\partial q^i} - \frac{d}{dt}\left(\frac{\partial \mathscr{L}}{\partial \dot{q}^i} \right) \right] dt + \left[(1-\lambda)\frac{\partial \mathscr{L}}{\partial \dot{q}^i}\delta q^i - \lambda q^i \delta \left(\frac{\partial \mathscr{L}}{\partial \dot{q}^i} \right) \right]_1^2 \tag{7.5.6}$$

In order that the boundary terms vanish for a second-order ODE, we must specify four boundary conditions; alternatively, we could impose $\lambda = 0$ and set the Dirichlet conditions $q(1) = q(2) = 0$, as a shortcut!

7.6   Variations

There is often a lot of confusion regarding variations in classical mechanics. The calculus of variations is a general machinery that we can insert different variations into. Variations may be at the same time or different times. Consider the configuration of a neighbouring path $q_i'(t)$ to the classical path $q_i(t)$ at the *same time*; we might call this a vertical variation:

$$q'(t) = q(t) + \delta q \tag{7.6.1}$$

We may also consider a variation in time $t' = t + \delta t$. We may then consider the variation of the configuration *in combination* with the variation in time a **non contemporaneous variation**, or a **Noether-type variation**:

$$\delta_0 q(t') = q'(t') - q(t) = q'(t + \delta t) - q(t) \tag{7.6.2}$$

If we Taylor expand $q'(t + \delta t)$ to the first order, we have $q'(t + \delta t) \approx q'(t) + \dot{q}'(t)\delta t$, so

$$\delta_0 q(t') = \delta q + \dot{q}\delta t \tag{7.6.3}$$

Under the variation $\delta_0 A$, we obtain

$$\delta_0 A = \delta_0 \int_1^2 \mathscr{L}(t, q, \dot{q})dt = \delta \int_1^2 \mathscr{L}(t, q, \dot{q})dt + \mathscr{L}\delta t \Big|_1^2 \tag{7.6.4}$$

The first term is Hamilton's principle:

$$\delta \int_1^2 \mathscr{L}(t, q, \dot{q})dt = \frac{\partial \mathscr{L}}{\partial \dot{q}^i}\delta q^i \Big|_{t_1}^{t_2} + \int_{t_1}^{t_2} \left(\frac{\partial \mathscr{L}}{\partial q^i} - \frac{d}{dt}\frac{\partial \mathscr{L}}{\partial \dot{q}^i} \right) \delta q^i dt \tag{7.6.5}$$

We can now let $\delta q = \delta_0 q - \dot{q}\delta t$ and substitute into the action, to obtain

$$\delta_0 A = \frac{\partial \mathscr{L}}{\partial \dot{q}^i}\delta_0 q^i \Big|_1^2 + \left(\mathscr{L} - \frac{\partial \mathscr{L}}{\partial \dot{q}^i}\dot{q}^i \right)\delta t \Big|_1^2 + \int_{t_1}^{t_2} \left(\frac{\partial \mathscr{L}}{\partial q^i} - \frac{d}{dt}\frac{\partial \mathscr{L}}{\partial \dot{q}^i} \right) \delta q^i dt \tag{7.6.6}$$

This is exactly the **Noether condition** we derive in appendix A; later on, we will understand this gives rise to conservation laws! For more information on variations, see chapter 13 and, in particular, exercise 13.2.

7.7   Weierstrass-Erdmann Conditions for Broken Extremals

In practical applications, the continuous extremal paths of the action don't always have continuous gradients $\dot{q}$ on $[t_1, t_2]$. We call these point *corners of discontinuity* in the broken extremal at point(s) $t' \in [t_1, t_2]$. The action for an extremal with a single corner can be written as two integrals:

$$A = \int_1^{t'} \mathscr{L}\, dt + \int_{t'}^2 \mathscr{L}\, dt \tag{7.7.1}$$

With a fixed t' we have to vary these integrals independently and search for imposed conditions on the subsequent free variation at the corner point. The variation at the corner is free in both time and configuration and hence we use the Noether-type variations δ_0 and assume satisfaction of the Euler-Lagrange equations:

$$\delta_0 A_1 = \left.\frac{\partial \mathscr{L}}{\partial \dot{q}^i}\delta_0 q^i\right|_1^{t'} + \left.\left(\mathscr{L} - \frac{\partial \mathscr{L}}{\partial \dot{q}^i}\dot{q}^i\right)\delta t\right|_1^{t'} \tag{7.7.2}$$

$$\delta_0 A_2 = \left.\frac{\partial \mathscr{L}}{\partial \dot{q}^i}\delta_0 q^i\right|_{t'}^2 + \left.\left(\mathscr{L} - \frac{\partial \mathscr{L}}{\partial \dot{q}^i}\dot{q}^i\right)\delta t\right|_{t'}^2 \tag{7.7.3}$$

The vanishing of the action then forces us to obtain the following:

$$\left.\frac{\partial \mathscr{L}}{\partial \dot{q}^i}\right|_{t'_-} = \left.\frac{\partial \mathscr{L}}{\partial \dot{q}^i}\right|_{t'_+} \tag{7.7.4}$$

and

$$\left(\mathscr{L} - \frac{\partial \mathscr{L}}{\partial \dot{q}^i}\dot{q}^i\right)_{t'_-} = \left(\mathscr{L} - \frac{\partial \mathscr{L}}{\partial \dot{q}^i}\dot{q}^i\right)_{t'_+}. \tag{7.7.5}$$

Hence, the functions $\partial \mathscr{L}/\partial \dot{q}$ and $\mathscr{L} - \dot{q}(\partial \mathscr{L}/\partial \dot{q})$ must be continuous at the corner (and hence all along the extremal in $[t_1, t_2]$) where $\dot{q}$ is discontinuous for Hamilton's principle to hold. Due to the discontinuity at t' we had to define two limits: $t'_- = \lim_{\tau \nearrow 0}(t' + \tau)$ and $t'_+ = \lim_{\tau \searrow 0}(t' + \tau)$, from the left and from the right, respectively. Later in the book, we will define these two functions as the generalised momenta and the Jacobi energy function!

7.8   Hamilton-Suslov Principle

Here, we will generalise Hamilton's principle in a subtle way by adding and subtracting a term during the variation.

$$\delta A = \delta \int_1^2 \mathscr{L}(\mathbf{q}, \dot{\mathbf{q}}, t) dt$$

$$= \int_1^2 \left\{ \frac{\partial \mathscr{L}}{\partial q^i} \delta q^i + \frac{\partial \mathscr{L}}{\partial \dot{q}^i} \delta \dot{q}^i \right\} dt$$

$$= \int_1^2 \left\{ \frac{\partial \mathscr{L}}{\partial q^i} \delta q^i + \frac{\partial \mathscr{L}}{\partial \dot{q}^i} \delta \frac{d}{dt} q^i + \frac{\partial \mathscr{L}}{\partial \dot{q}^i} \left(\delta \frac{d}{dt} q^i - \frac{d}{dt} \delta q^i \right) \right\} dt \qquad (7.8.1)$$

where we added and subtracted the following term:

$$+ \frac{\partial \mathscr{L}}{\partial \dot{q}^i} \frac{d}{dt} \delta q^i - \frac{\partial \mathscr{L}}{\partial \dot{q}^i} \frac{d}{dt} \delta q^i. \qquad (7.8.2)$$

With the usual partial integration of the second term, we arrive at the modified **Hamilton-Suslov principle**:

$$\delta A = \int_1^2 dt \left\{ \frac{\partial \mathscr{L}}{\partial q^i} \delta q^i - \frac{d}{dt} \left(\frac{\partial \mathscr{L}}{\partial \dot{q}^i} \right) \delta q^i + \frac{\partial \mathscr{L}}{\partial \dot{q}^i} \left(\delta \frac{d}{dt} q^i - \frac{d}{dt} \delta q^i \right) \right\} + \frac{\partial \mathscr{L}}{\partial \dot{q}^i} \delta q^i \Big|_1^2 \qquad (7.8.3)$$

If the variation commutes with the time derivative, together with the variations at the boundary vanishing, then we regenerate the standard Euler-Lagrange equation. In such a case, the variations are of **Holder type**. We will return to this in section 8.2 in our discussion of **non-holonomic constrained dynamics**, where we see the non-commutation of the δ and d/dt operator under the **Suslov variation**.

Chapter summary

- The dynamical information of a system is found to be encapsulated by a single scalar function called its **Lagrangian**: $\mathscr{L}(\mathbf{q}, \dot{\mathbf{q}}, t)$.
- The equations of motion for a system are found by solving the **Euler-Lagrange equation** for the system's Lagrangian:

$$\frac{d}{dt} \left(\frac{\partial \mathscr{L}}{\partial \dot{q}^i} \right) - \frac{\partial \mathscr{L}}{\partial q^i} = 0$$

  Together with suitable boundary conditions, the Lagrangian is all we need to predict and solve the equations of motion.

- The **principle of stationary action** is introduced and used to derive the Euler-Lagrange equation. Importantly, we discuss the calculus of variations as applied to classical mechanics and note the imposed conditions of **vanishing boundary points** and **independent variations**.

- We show how the Lagrange equation relates to Newton's second law and introduce the **Hessian condition**.

Exercise 7.4 Consider the following Lagrangian for a particle in a central potential in three dimensions:

$$\mathscr{L} = \frac{1}{2}m(\dot{x}^2 + \dot{y}^2 + \dot{z}^2) - U(r)$$

We can express this in spherical coordinates: $(x, y, z) = (r \sin\theta \cos\phi, r \sin\theta \sin\phi, r \cos\theta)$; these will be our generalised coordinates and you will need to compute their time derivatives to write the Lagrangian:

$$\mathscr{L} = \frac{1}{2}m(\dot{r}^2 + r^2\dot{\theta}^2 + r^2\dot{\phi}^2 \sin^2\theta) - U(r) \tag{7.8.4}$$

We now have three Euler-Lagrange equations:

$$\frac{d}{dt}\left(\frac{\partial\mathscr{L}}{\partial\dot{\phi}}\right) - \frac{\partial\mathscr{L}}{\partial\phi}, \qquad \frac{d}{dt}\left(\frac{\partial\mathscr{L}}{\partial\dot{\theta}}\right) - \frac{\partial\mathscr{L}}{\partial\theta} = 0, \qquad \frac{d}{dt}\left(\frac{\partial\mathscr{L}}{\partial\dot{r}}\right) - \frac{\partial\mathscr{L}}{\partial r} = 0 \tag{7.8.5}$$

The first equation in the azimuth coordinate ϕ gives us a conserved quantity:

$$\frac{d}{dt}(mr^2\dot{\phi}\sin^2\theta) = 0 \tag{7.8.6}$$

The θ coordinate gives us

$$\frac{d}{dt}(mr^2\dot{\theta}) = mr^2\dot{\phi}^2 \sin\theta \cos\theta \tag{7.8.7}$$

The radial equation is

$$\frac{d}{dt}(m\dot{r}) = mr(\dot{\theta}^2 + \dot{\phi}^2 \sin^2\theta) - \frac{\partial U}{\partial r} \tag{7.8.8}$$

Exercise 7.5 In this section, we investigate the Lagrangian description of a charged particle in an electric field **E**. The **Lorentz force F** is the combination of electric and magnetic fields that a point charge e of velocity $\dot{\mathbf{r}}$ experiences in an electromagnetic field $\mathbf{E}(\mathbf{r}, t)$ and $\mathbf{B}(\mathbf{r}, t)$:

$$\mathbf{F} = e(\mathbf{E} + \dot{\mathbf{r}} \times \mathbf{B}) = m\ddot{\mathbf{r}} \tag{7.8.9}$$

This is the equation of motion for the system, so you may wonder what else we can actually do by re-formulating it in Lagrangian mechanics? Having a Lagrangian formulation tells us much more than the equations of motion of the system; it gives us energies and conserved quantities and provides an avenue for quantisation (perhaps through the Hamiltonian formalism) and gauge information (see Noether's theorem in chapter 11), so it is well worth investigation.

We can write both fields in terms of their **vector** and **scalar** potentials $\mathbf{A}(\mathbf{r}, t)$ and $\varphi(\mathbf{r}, t)$ respectively:

$$\mathbf{E}(\mathbf{r}, t) = -\vec{\nabla}_r \varphi(\mathbf{r}, t) - \frac{\partial}{\partial t}\mathbf{A}(\mathbf{r}, t), \qquad \mathbf{B}(\mathbf{r}, t) = \vec{\nabla}_r \times \mathbf{A}(\mathbf{r}, t) \tag{7.8.10}$$

The Lagrangian of the charged particle is also written in terms of $\mathbf{A}(\mathbf{r}, t)$ and $\varphi(\mathbf{r}, t)$:

$$\mathscr{L}(\mathbf{r}, \dot{\mathbf{r}}, t) = \frac{1}{2}m\dot{\mathbf{r}} \cdot \dot{\mathbf{r}} + e\dot{\mathbf{r}} \cdot \mathbf{A} - e\varphi \tag{7.8.11}$$

Plugging this in the Lagrange equation should give us the Lorentz force law:

$$\frac{\partial \mathscr{L}}{\partial r^i} = e \sum_j \dot{r}^j \frac{\partial A_j}{\partial r^i} - e \frac{\partial \varphi}{\partial r^i}$$

$$\frac{d}{dt}\left(\frac{\partial \mathscr{L}}{\partial \dot{r}^i}\right) = m\ddot{r}^i + e\frac{dA_i}{dt} = m\ddot{r}^i + e\left[\sum_j \frac{\partial A_i}{\partial r^j}\dot{r}^j + \frac{\partial A_i}{\partial t}\right]$$

Hence,

$$m\ddot{r}^i = e\left[-\frac{\partial \varphi}{\partial r^i} - \frac{\partial A_i}{\partial t}\right] + e\sum_j \dot{r}^j\left[\frac{\partial A_j}{\partial r^i} - \frac{\partial A_i}{\partial r^j}\right] = e(E_i + \dot{r}^i \times B_i) \qquad (7.8.12)$$

Entering the **static expressions** for the scalar and vector potentials for the addition of a second charged particle to the field and substituting into the Lagrangian we obtain

$$\varphi = \frac{1}{4\pi\epsilon_0}\frac{e_2}{|r_1 - r_2|}, \qquad \mathbf{A} = \frac{\mu_0}{4\pi}\frac{e_2\dot{\mathbf{r}}_2}{|r_1 - r_2|} \qquad (7.8.13)$$

giving us,

$$\mathscr{L} = \frac{1}{2}m(\dot{\mathbf{r}} \cdot \dot{\mathbf{r}})_1 + \frac{\mu_0}{4\pi}e_1 e_2 \frac{\dot{\mathbf{r}}_1 \cdot \dot{\mathbf{r}}_2}{|r_1 - r_2|} - \frac{1}{4\pi\epsilon_0}\frac{e_1 e_2}{|r_1 - r_2|} \qquad (7.8.14)$$

If we now extend this for N charged particles subject to an external field and sum over a pairwise interaction, we obtain the **Clausius Lagrangian:s**

$$\mathscr{L} = \frac{1}{2}\sum_{i=1}^{N} m_i(\dot{\mathbf{r}} \cdot \dot{\mathbf{r}})_i - \frac{1}{8\pi\epsilon_0}\sum_{i=1}^{N}\sum_{j\neq i}^{N}\frac{e_i e_j}{r_{ij}}\left(1 - \frac{\dot{\mathbf{r}}_i \cdot \dot{\mathbf{r}}_j}{c^2}\right) \qquad (7.8.15)$$

where we have divided the summation over i, j by 2 due to over-counting and defined $c = 1/\sqrt{\mu_0\epsilon_0}$. The Clausius Lagrangian contains a kinetic energy term for each particle.

Exercise 7.6 In this exercise, we will explore a simple problem in close detail, using the Lagrangian $\mathscr{L}(\dot{q}, t) = \dot{q}^2/4t$ with the boundary points $q(1) = 5$ and $q(2) = 11$. We know that the Lagrange equation will give us the extremal curve; we can then manipulate to find $q(t)$ and impose the two boundary conditions, as follows:

$$A[\mathbf{q}] = \int_1^2 \frac{\dot{q}^2}{4t}dt \qquad (7.8.16)$$

$$\frac{d}{dt}\left(\frac{\partial \mathscr{L}}{\partial \dot{q}}\right) - \frac{\partial \mathscr{L}}{\partial q} = \frac{d}{dt}\left(\frac{\dot{q}}{2t}\right) - 0 = 0 \qquad (7.8.17)$$

Therefore, the term in brackets is equal to a constant a; this can then be integrated to find $q(t)$:

$$\dot{q} = 2t \cdot a \qquad \longrightarrow \qquad \int_1^2 \dot{q}dt = \int_1^2 2t \cdot a\,dt \qquad (7.8.18)$$

We then find $q = at^2 + b$, where b is an integration constant, and substitute in the boundary conditions:

$$5 = a(1)^2 + b, \qquad 11 = a(2)^2 + b \tag{7.8.19}$$

Solving these simultaneous equations gives us $a = 2$ and $b = 3$; so, out of all the possible paths between $q(1)$ and $q(2)$, the extremal is $q(t) = 2t^2 + 3$.

Exercise 7.7 Consider the motion of a particle of mass m under the potential $U(\mathbf{r})$ from a non-inertial frame whose origin coincides with the rotating frame such that $\mathbf{A_R} = 0$ (building on concepts in rotational dynamics in the Newtonian mechanics). The velocity of the particle as measured in the fixed-axis frame $\mathbf{v}$ is written in terms of the inertial frame's quantities by application of equation 3.0.35, so that the transformation operator is $\mathbf{v} = \dot{\mathbf{r}} + \omega \times \mathbf{r}$, thereby giving us the following Lagrangian:

$$\mathscr{L}(\mathbf{r}, \dot{\mathbf{r}}) = \frac{m}{2}\|\dot{\mathbf{r}} + \omega \times \mathbf{r}\|^2 - U(\mathbf{r}) \tag{7.8.20}$$

The bracketed term may be expanded as follows:

$$\begin{aligned}
\|\dot{\mathbf{r}} + \omega \times \mathbf{r}\|^2 &= (\dot{\mathbf{r}} + \omega \times \mathbf{r}) \cdot (\dot{\mathbf{r}} + \omega \times \mathbf{r}) \\
&= \dot{\mathbf{r}} \cdot \dot{\mathbf{r}} + 2\dot{\mathbf{r}} \cdot (\omega \times \mathbf{r}) + (\omega \times \mathbf{r}) \cdot (\omega \times \mathbf{r})
\end{aligned} \tag{7.8.21}$$

The Lagrangian now becomes

$$\mathscr{L}(\mathbf{r}, \dot{\mathbf{r}}) = \frac{m}{2}\left\{\dot{\mathbf{r}} \cdot \dot{\mathbf{r}} + 2\dot{\mathbf{r}} \cdot (\omega \times \mathbf{r}) + (\omega \times \mathbf{r}) \cdot (\omega \times \mathbf{r})\right\} - U(\mathbf{r}) \tag{7.8.22}$$

Consider $d\mathscr{L}(\mathbf{r}, \dot{\mathbf{r}})$:

$$d\mathscr{L}(\mathbf{r}, \dot{\mathbf{r}}) = \frac{m}{2}\left[2\dot{\mathbf{r}} \cdot d\dot{\mathbf{r}} + 2\left[d\dot{\mathbf{r}} \cdot (\omega \times \mathbf{r}) + \dot{\mathbf{r}} \cdot (\omega \times d\mathbf{r})\right] + 2(\omega \times d\mathbf{r}) \cdot (\omega \times \mathbf{r})\right] - \frac{\partial U}{\partial \mathbf{r}} d\mathbf{r} \tag{7.8.23}$$

Now, use a clever vector identity $\mathbf{A} \cdot (\mathbf{B} \times \mathbf{C}) = \mathbf{B} \cdot (\mathbf{C} \times \mathbf{A}) = \mathbf{C} \cdot (\mathbf{A} \times \mathbf{B})$ in two places (the second is a little more obscure to spot):

$$\overbrace{\dot{\mathbf{r}} \cdot (\omega \times d\mathbf{r})}^{\mathbf{A} \cdot (\mathbf{B} \times \mathbf{C})} = \overbrace{d\mathbf{r} \cdot (\dot{\mathbf{r}} \times \omega)}^{\mathbf{C} \cdot (\mathbf{A} \times \mathbf{B})}, \qquad \overbrace{(\omega \times \mathbf{r}) \cdot (\omega \times d\mathbf{r})}^{\mathbf{A} \cdot (\mathbf{B} \times \mathbf{C})} = \overbrace{d\mathbf{r} \cdot \left((\omega \times \mathbf{r}) \times \omega\right)}^{\mathbf{C} \cdot (\mathbf{A} \times \mathbf{B})} \tag{7.8.24}$$

The Euler-Lagrange equation is now constructed as follows:

$$\frac{d}{dt}\left(\frac{\partial \mathscr{L}}{\partial \dot{\mathbf{r}}}\right) = m(\ddot{\mathbf{r}} + \dot{\omega} \times \mathbf{r} + \omega \times \dot{\mathbf{r}}) \tag{7.8.25}$$

$$\frac{\partial \mathscr{L}}{\partial \mathbf{r}} = -m[\omega \times \dot{\mathbf{r}} + \omega \times (\omega \times \mathbf{r})] - \vec{\nabla} U(\mathbf{r}) \tag{7.8.26}$$

All together, the Euler-Lagrange equations is

$$m\ddot{\mathbf{r}} = -m[\dot{\omega} \times \mathbf{r} + 2\omega \times \dot{\mathbf{r}} + \omega \times (\omega \times \mathbf{r})] - \vec{\nabla} U(\mathbf{r}) \tag{7.8.27}$$

This is exactly the same result we would expect from the Newtonian analysis measuring from a fixed frame (f); to re-connect with the earlier notation, write $\vec{F} = m\ddot{\mathbf{r}}$ and $m\hat{a}_f = -\vec{\nabla} U(\mathbf{r})$.

We don't often talk about reference frames in analytical mechanics because our analysis is a little more abstract than in Newtonian physics; however, there are two cases where we may want to discuss this. The first case is the one we highlighted above, where the Lagrangian is written from the start to include all the necessary equipment for a non-inertial frame. This is somewhat unusual but completely allowed and is useful in the rotating frames of particle accelerators and rotational-vibrational motion. The second case is when we actively use coordinates that are non-inertial, that is, they are at rest with the fixed frame (f) rather than the inertial frame of the particle (p) as would be the standard approach.

The most elementary case would be to construct a Lagrangian in a frame of reference with coordinates at rest in that frame. We then observe the Lagrangian from another frame of reference. If the Lagrangian is at rest in this second frame, then it produces the same Euler-Lagrange equations. If, however, we observe the Lagrangian and it is not at rest with the other frame, then the two Lagrangians are not equal. This is a manifestation of the fictitious forces that appear in the second frame. Both Lagrangians must describe the same event, however. Therefore, we conclude that, as the Lagrangian transforms between different coordinate frames, it does so in such a way as to account for the correct dynamics in that frame, (e.g. the transforming between an inertial and non-inertial frame would automatically generate the terms required for the fictitious forces in the equations of motion). In this way, a Lagrangian is not unique for a system. The equations of motion will not look the same but we expect the following condition to hold irrespective:

$$\frac{d}{dt}\left(\frac{\partial \mathscr{L}}{\partial \dot{q}^i}\right) - \frac{\partial \mathscr{L}}{\partial q^i} = 0$$

This is precisely the Euler-Lagrange equation and, therefore, we haven't brought anything new by discussing non-inertial frames. Analytical mechanics has more powerful concepts which we discuss later in this section in two separate topics, **point transformations** (in chapter 9) and **gauge invariance** (in chapter 11). This does not mean that we can't write Lagrangians for relativistic systems! Consider the following set of Lagrangians for (1) a relativistic particle under a conservative potential, (2) a relativistic oscillator and (3) a relativistic particle in an electromagnetic field:

$$\mathscr{L} = -mc^2\sqrt{1 - \frac{v^2}{c^2}} - V(\mathbf{r}), \tag{7.8.28}$$

$$\mathscr{L} = -mc^2\sqrt{1 - \frac{v^2}{c^2}} - \frac{1}{2}k\mathbf{r}^2, \tag{7.8.29}$$

$$\mathscr{L} = -mc^2\sqrt{1 - \frac{v^2}{c^2}} - e\varphi + e\dot{\mathbf{r}} \cdot \mathbf{A} \tag{7.8.30}$$

Exercise 7.8 Constructing a Lagrangian is probably the most difficult step to any problem in mechanics and, in most elementary examples, we can take it to just be the kinetic energy minus the potential energy. When a Lagrangian is not known, then either clever guesswork based on physical observation is required or backtracking from other known quantities of the system is necessary. In any case, Lagrangians depending on lower-order derivative terms are favoured, since they require fewer initial conditions to be specified. The Lagrangian for a particle under a conservative potential can be constructed as,

$$\mathscr{L} = \frac{1}{2}m\dot{q}^2 - U(q)$$

and, for a harmonic oscillator,

$$\mathscr{L} = \frac{1}{2}m\dot{q}^2 - \frac{1}{2}kx^2$$

It is important to express the kinetic energy in terms of the velocity, not the momentum, so writing $p^2/2m$ would not be appropriate. Sometimes we can also construct Lagrangians by thinking of the entire system as being composed of **non-interacting subsystems**. In this case, the Lagrangian for the entire system is the sum of Lagrangians of each of it's subsystems. We can then model their interaction through a potential term.

To understand this, imagine a two-electron system orbiting a stationary nucleus. The nucleus is assumed to be a coulombic potential of charge Q interacting favourably with two electrons of charge q at distances $|r_1|$ and $|r_2|$ away. The two electrons interact repulsively; the interacting term dependant on their separation is $|r_1 - r_2|$:

$$\begin{aligned}
\mathscr{L} &= \mathscr{L}_{e_1^-} + \mathscr{L}_{e_2^-} + \mathscr{L}_{\text{int}} \\
&= \left(\frac{1}{2}m_e|\dot{r}_1^2| + \frac{Q \cdot q}{4\pi\epsilon_0|r_1|} \right) + \left(\frac{1}{2}m_e|\dot{r}_2^2| + \frac{Q \cdot q}{4\pi\epsilon_0|r_2|} \right) - \frac{q \cdot q}{4\pi\epsilon_0|r_1 - r_2|}
\end{aligned} \tag{7.8.31}$$

Note the sign convention for repulsive and attractive potential terms, and then their subsequent reversal using equation 7.0.1. Even this rather complicated Lagrangian, however, is not quite enough to be entirely useful; a generalisation for charged particles is the **Clausius Lagrangian** which we look at later since it requires knowledge of **electrodynamics**.

Hopefully, this demonstrates how you can construct a Lagrangian for most problems and that it might not always look too pretty, but the principle is the same; it is simply a scalar function that we need in order to obtain the equations of motion. Try to construct Lagrangians for toy problems and solve their equations of motion. In each case, consider the variables you will use and the degrees of freedom they can move in.

Exercise 7.9 Proteins are giant biological molecules that are fundamental to cellular processes and metabolism. They are composed of amino acid chains that fold into minimum-energy structures which are often helical in parts connected by strands that afford the flexibility to adopt complex topologies (see figure 7.4). The resulting structure is held together by intermolecular forces (hydrogen bonds) that arrange themselves such that non-polar regions have minimal contact to polar areas and vice versa, as well as reducing the overall torsion and stress. The study of **protein folding** from a structural biology standpoint offers insight into diseases caused by misfolding events such as occur in Alzheimer's and other neurological illness.

Given that the adopted structure follows an energy minimum, we can use the action principle to minimise an energy functional and therefore predict protein morphology. To construct the functional, we use the **Frenet-Serret frame** that describes a curve $\gamma(s)$ in $\mathbb{R}^3$ within an orthogonal basis of tangent, $\vec{T}$, normal, $\vec{N}$, and binormal, $\vec{B}$, vectors. The tangent is simply a unit vector to the curve in the direction of motion. The unit normal is the derivative of $\vec{T}$ with respect to the arclength of the curve, s, divided by its length, and the binormal vector is the cross product of $\vec{T}$ and $\vec{N}$. The Frenet-Serret equations describe how these vectors change as we move along the curve in terms of each other:

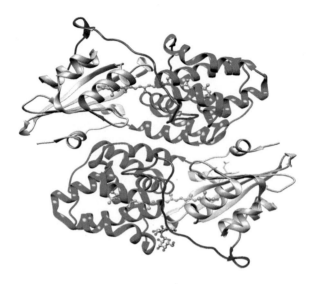

Fig. 7.4: A ribbon diagram of an orange carotenoid protein, this illustrates the helical nature of the protein structures interspaced with strands that allow flexibility in the quaternary structure. Image kindly provided by Cheryl Kerfeld from http://www.kerfeldlab.org/images.html.

$$\frac{d\vec{T}}{ds} = \kappa\vec{N}, \qquad \frac{d\vec{N}}{ds} - \kappa\vec{T} + \tau\vec{B}, \qquad \frac{d\vec{B}}{ds} = -\tau\vec{N} \qquad (7.8.32)$$

where κ is the **curvature** (failure of the curve to be a straight line) and τ is the **torsion** of the curve (failure of the curve to be planar).

We begin by analogising the protein backbone to the curve $\gamma(s) = (x(s), y(s), z(s))$ and set up an energy functional $\mathcal{E}(\kappa, \tau, \dot{\kappa}, \dot{\tau})$ that depends in the most general case on κ and τ and their derivatives with respect to the arc length $\dot{\kappa}$ and $\dot{\tau}$. The calculus of variations then considers a variation to the curve in each of the basis directions, although the $\vec{T}$ variation does not encode new information about the curve (only reparametrisations) and can be ignored:

$$\tilde{\gamma}(s) = \gamma(s) + \sum_{i=1} \epsilon_i \eta_i(s)\vec{e}_i, \qquad i = 1, 2, 3 \qquad (7.8.33)$$

where the parameters ϵ_i are small variations, η_i are smooth functions that vanish on the boundary and $\vec{e}_i$ are the basis vectors $\vec{T}, \vec{N}$ and $\vec{B}$. Within this frame, the curvature and torsion are well known formulas:

$$\kappa = \frac{|\dot{\gamma} \times \ddot{\gamma}|}{|\dot{\gamma}|^3}, \qquad \tau = \frac{\det(\dot{\gamma}, \ddot{\gamma}, \dddot{\gamma})}{|\dot{\gamma} \times \ddot{\gamma}|^2} \tag{7.8.34}$$

where dots indicate d/ds. The equations of motion can then be found by extremising $\mathcal{E}$ in the usual (although complicated) way:

$$\delta S[\gamma(s)] = \delta \int \mathcal{E}(\kappa, \tau, \dot{\kappa}, \dot{\tau}) |\tilde{\gamma}'| ds \tag{7.8.35}$$

The details of this calculation can be found in full in the article by Thamwattana *et al.* (2008), where the authors derive the equations of motion for the normal and binormal directions before studying energy densities dependent on only some of the curve parameters $\mathcal{E}(\kappa), \mathcal{E}(\kappa, \tau)$ and $\mathcal{E}(\kappa, \dot{\kappa})$. In the simplest case of $\kappa(s)$, the Euler-Lagrange equations are shown below for normal and binormal directions, respectively:

$$\frac{d^2}{ds^2}\frac{\partial \mathcal{E}}{\partial \kappa} + (\kappa^2 - \tau^2)\frac{\partial \mathcal{E}}{\partial \kappa} - \mathcal{E}\kappa = 0 \tag{7.8.36}$$

$$2\frac{d}{ds}\frac{\partial \mathcal{E}}{\partial \kappa}\tau + \frac{\partial \mathcal{E}}{\partial \kappa}\dot{\tau} = 0 \tag{7.8.37}$$

Solutions to these equations of motion that give rise to helix structures can be used to model proteins and their foldings.

Exercise 7.10 The action functionals considered in this chapter have all been **local** functionals, meaning that they depend only on the generalised coordinates *at a given point in time*. **Non-local actions** conversely, depend on more than one point in time. Vanilla classical mechanics is a local theory; however, non-locality is apparent in some field theories (here, the action functional will depend on more than one point on a space-time manifold). The same is true of actions that depend on an infinite number of derivatives (e.g. time derivatives of q^i), this is because a Taylor series can approximate a small shift $q(t + \delta t)$ exactly using an infinite series of derivatives. Local actions can be viewed as special cases of non-local action functionals, and it is always beneficial to attempt to reduce a non-local theory to a combination of local terms where possible.

As an exercise, find the Euler-Lagrange equation for the following non-local Lagrangian:

$$\mathscr{L} = \frac{m}{2}\dot{q}^2(t) + q(t)\int_0^t f(t - t')q(t')dt' \tag{7.8.38}$$

The action integral is then:

$$A[q] = \int_1^2 \mathscr{L}\,dt \tag{7.8.39}$$

With the variation $\delta A = A[q + \delta q] - A[q]$. As with any functional derivative, simply replace $x(s) \to x(s) + \delta x(s)$ for every coordinate or velocity:

$$q(s) \to q(s) + \delta q(s), \qquad \dot{q}(s) \to \dot{q}(s) + \delta\dot{q}(s) = \dot{q}(s) + \frac{d}{dt}\delta q(s) \tag{7.8.40}$$

to obtain

$$A[q + \delta q] = \int_1^2 dt \left\{ \frac{m}{2} \left(\dot{q}(t) + \frac{d}{dt} \delta q(t) \right)^2 + (q(t) + \delta q(t)) \int_0^t f(t - t')(q(t') + \delta q(t')) dt' \right\}$$

$$= \int_1^2 dt \left\{ \frac{m}{2} \left(\dot{q}(t)^2 + \left(\frac{d}{dt} \delta q(t) \right)^2 + 2\dot{q} \frac{d}{dt} \delta q(t) \right) + \int_0^t f(t - t') [q(t)q(t') \right.$$

$$\left. \delta q(t)q(t') + q(t)\delta q(t') + \delta q(t)\delta q(t') \right] dt' \Bigg\} \tag{7.8.41}$$

We now cancel terms in $\mathcal{O}(\delta^2)$, and integrate by parts:

$$\int_1^2 \dot{q}(t) \frac{d}{dt} \delta q(t) = \dot{q}(t)\delta q(t) \Big|_1^2 - \int_1^2 \ddot{q}(t)\delta q(t) dt \tag{7.8.42}$$

Inserting this and collecting variations, we obtain

$$A[q + \delta q] = \int_1^2 dt \left\{ \left(-\ddot{q}(t)m + \int_0^t dt' f(t - t')q(t') \right) \delta q(t) + \int_0^t dt' f(t - t')q(t)\delta q(t') \right.$$

$$\left. + \frac{m}{2} \dot{q}(t)^2 + \int_0^t dt' f(t - t')q(t)q(t') \right\} \tag{7.8.43}$$

The variational term in $\delta q(t')$ can be rewritten by changing $t \longleftrightarrow t'$ and changing the integral limits:

$$\int_1^2 dt \int_0^t dt' f(t - t')q(t)\delta q(t') = \int_1^2 dt' \int_0^{t'} dt \delta q(t)q(t')f(t' - t) \tag{7.8.44}$$

$$= \int_1^2 dt \int_t^2 dt' \delta q(t)q(t')f(t' - t) \tag{7.8.45}$$

With this, the full variation now becomes:

$$\delta A = \int_1^2 dt \delta q(t) \left(-m\ddot{q}(t) + \int_1^2 dt' f(|t - t'|)q(t') \right) \tag{7.8.46}$$

and the Euler-Lagrange equation is:

$$m\ddot{q}(t) = \int_1^2 dt' f(|t - t'|)q(t') \tag{7.8.47}$$

We continue this discussion in exercise 16.8.

8

Constrained Lagrangian Mechanics

Many problems in mechanics involve **constraints** on the possible dynamics. When there is a constraint, the generalised coordinates are no longer all independent, which was a requirement for the application of the calculus of variations, since the variations were assumed to be arbitrary. Therefore, it puts chapter 7 in jeopardy. In this case there will be an equation linking m of the coordinates which is called the **constraint equation**. There will be one constraint equation for each constraint so, if we have m constraints, there will be m constraint equations, reducing the number of generalised coordinates to $n - m$.

8.1 Holonomic Constraints

In the case of **holonomic constraints**, we can write the constraint equation as some function of the generalised coordinates and time (recall our discussion from the start of part II). The index j runs from 1 to m, so we have m copies of equation 8.1.1:

$$\phi_j(\boldsymbol{q}, t) = \phi_j(q^1, \dots, q^n, t) = 0 \qquad \textbf{constraint equation} \qquad (8.1.1)$$

A relation like this allows us to write one generalised coordinate in terms of the others, meaning that they are no longer independent but coupled. We could use it to reduce the number of coordinates until we obtain a linearly independent minimal set that describes a **constraint surface** within configuration space, and then set up and solve $n - m$ Lagrange equations.

There is another way we could approach this that involves keeping hold of all of the generalised coordinates; it is called the method of undetermined **Lagrange multipliers**. We would find this method suitable if the constraint equation was difficult to solve or we wanted to know details about the constraint itself; eliminating wouldn't be all too useful, then! We will now consider the calculus of variations with two coordinates and a single holonomic constraint equation linking them:

$$\phi(q^1, q^2, t) = 0 \qquad (8.1.2)$$

We can form equations for given trajectories in terms extremal paths and perturbations as we are familiar with, just extending the idea slightly to a two-dimensional holonomic constraint:

Lagrangian & Hamiltonian Dynamics. Peter Mann, Oxford University Press (2018).
© Peter Mann. DOI: 10.1093/oso/9780198822370.001.0001

$$q^1(\alpha, t) = q^1(0, t) + \alpha \eta_1(t), \qquad \text{and} \qquad q^2(\alpha, t) = q^2(0, t) + \alpha \eta_2(t) \qquad (8.1.3)$$

The variation to the action can be written as follows, returning to chapter 7 for assistance:

$$\frac{dA}{d\alpha} = \int_{t_0}^{t_1} \left\{ \left[\frac{\partial \mathscr{L}}{\partial q^1} - \frac{d}{dt}\left(\frac{\partial \mathscr{L}}{\partial \dot{q}^1} \right) \right] \eta_1(t) + \left[\frac{\partial \mathscr{L}}{\partial q^2} - \frac{d}{dt}\left(\frac{\partial \mathscr{L}}{\partial \dot{q}^2} \right) \right] \eta_2(t) \right\} dt = 0 \qquad (8.1.4)$$

Now, this is the part where we deviate from before. However, because the constraints are a function of two coordinates, the variation functions $\eta_1(t)$ and $\eta_2(t)$ are no longer *independently vanishing*; instead only their *sum* vanishes. This does not mean that the coefficients in the square brackets have to be equal to zero; hence, it does not mean that each coordinate will satisfy the Lagrange equation.

To proceed, we take the total derivative of the constraint function with respect to alpha, the coefficient of the perturbation term. This tells us how the constraint varies for different paths on the constraint surface. *By definition*, we know that any *physical* trajectory must satisfy the constraint at all times, so it must remain on the constraint surface and hence any perturbed path will not change the constraint function. Therefore, we set the derivative to zero, and substituting in the expressions for q^1 and q^2 in equation 8.1.3, we can write the first perturbation in terms of the other:

$$\frac{d\phi}{d\alpha} = \frac{\partial \phi}{\partial q^1}\frac{\partial q^1}{\partial \alpha} + \frac{\partial \phi}{\partial q^2}\frac{\partial q^2}{\partial \alpha} = \frac{\partial \phi}{\partial q^1}\eta_1 + \frac{\partial \phi}{\partial q^2}\eta_2 = 0 \qquad (8.1.5)$$

giving

$$\eta_2 = -\frac{\partial \phi / \partial q^1}{\partial \phi / \partial q^2}\eta_1 \qquad (8.1.6)$$

which can be entered into equation 8.1.4:

$$\frac{dA}{d\alpha} = \int_{t_0}^{t_1} \left\{ \left[\frac{\partial \mathscr{L}}{\partial q^1} - \frac{d}{dt}\left(\frac{\partial \mathscr{L}}{\partial \dot{q}^1} \right) \right] - \left[\frac{\partial \mathscr{L}}{\partial q^2} - \frac{d}{dt}\left(\frac{\partial \mathscr{L}}{\partial \dot{q}^2} \right) \right]\left(\frac{\partial \phi / \partial q^1}{\partial \phi / \partial q^2} \right) \right\} \eta_1 dt = 0 \quad (8.1.7)$$

Since $\eta_1(t)$ is an independent variation, we can now say that the term in the curly brackets must equal zero if any choice of arbitrary variational function must vanish:

$$\left[\frac{\partial \mathscr{L}}{\partial q^1} - \frac{d}{dt}\left(\frac{\partial \mathscr{L}}{\partial \dot{q}^1} \right) \right]\left(\frac{1}{\partial \phi / \partial q^1} \right) = \left[\frac{\partial \mathscr{L}}{\partial q^2} - \frac{d}{dt}\left(\frac{\partial \mathscr{L}}{\partial \dot{q}^2} \right) \right]\left(\frac{1}{\partial \phi / \partial q^2} \right) \qquad (8.1.8)$$

Note, however, that both sides take the same form for their respective coordinate and are functions of only one generalised coordinate and time; therefore, to maintain equality, they must both be equal to a function of time only: $\lambda(t)$:

$$\left[\frac{\partial \mathscr{L}}{\partial q^i} - \frac{d}{dt}\left(\frac{\partial \mathscr{L}}{\partial \dot{q}^i} \right) \right]\left(\frac{1}{\partial \phi / \partial q^i} \right) = \lambda(t)$$

With slight rearrangement,

$$\frac{\partial \mathscr{L}}{\partial q^i} - \frac{d}{dt}\left(\frac{\partial \mathscr{L}}{\partial \dot{q}^i}\right) - \lambda(t)\frac{\partial \phi}{\partial q^i} = 0, \qquad i = 1, 2 \tag{8.1.9}$$

We now have three equations, including the constraint equation of 8.1.2, that can be solved for the generalised coordinates and the undetermined multiplier $\lambda(t)$. If we peak at section 32 on Lagrange multipliers, we realise that we are actually finding the stationary point of the action that lies on the constraint surface. The importance of this technique is the decoupling of the coordinates to give independent variations, such that we can generate Lagrange equations in terms of a single coordinate and its time derivative, with no mixed terms. If we have m holonomic constraint equations, we

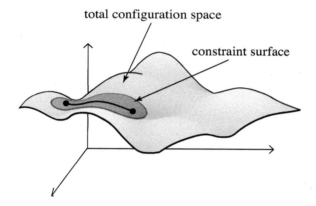

Fig. 8.1: A constraint submanifold in a configuration space.

can extend the result of the undetermined multiplier method: (while the proof in this notation is an extension of the above, it is easier to show using functional derivatives, which we do later):

$$\frac{\partial \mathscr{L}}{\partial q^i} - \frac{d}{dt}\left(\frac{\partial \mathscr{L}}{\partial \dot{q}^i}\right) - \sum_{j=1}^{m} \lambda_j(t)\frac{\partial \phi_j}{\partial q^i} = 0 \tag{8.1.10}$$

where $i = 1, \ldots, n$ and $j = 1, \ldots, m$. We now have $n+m$ equations to solve for n generalised coordinates and m undetermined multipliers, along with the constraint equations. Each one of the m functions represents an $(n-1)$-dimensional **hypersurface**[1] within the total configuration space, upon which that particular constraint is

---

[1]As we discuss later in more detail, a hypersurface is a surface of dimension $n-1$ within an n-dimensional space.

satisfied, as in figure 8.1. The point where all these surfaces intersect each other is a $(n-m)$-dimensional surface, where all of the constraints are simultaneously satisfied[2]. In order for the motion of the system or the curve in configuration space to satisfy the constraints, any motion must be confined to this surface and not venture off it for all time.

For a system whose coordinates are defined such that they account for the constraints, we do not require any of the Lagrange multipliers and we can simply write the familiar Lagrange equation that we have used so far. Trying to draw such a constrained velocity phase space is tricky, since we must imagine a lesser dimensional space within a larger one; however, the figure indicates that the brown patch is the constraint surface within the total phase space. The physically allowed dynamics is restricted to this submanifold.

We will now repeat the previous work using the functional derivative notation to derive the constrained Lagrange equation. Notice that, if we look at 8.1.10, it tidies up nicely if we define a new Lagrangian function L:

$$ L := \mathscr{L} - \sum_{j=1}^{m} \lambda_j \phi_j \qquad \text{then} \qquad \frac{d}{dt}\left(\frac{\partial L}{\partial \dot{q}^i}\right) - \frac{\partial L}{\partial q^i} = 0 \qquad (8.1.11)$$

The stationary action principle for this Lagrangian will give us the constrained Euler-Lagrange equation:

$$ \delta A[\boldsymbol{q}] = \delta \int_{t_0}^{t_1} L\, dt$$

$$ = \int_{t_0}^{t_1} \left[\delta \mathscr{L}(\boldsymbol{q}, \dot{\boldsymbol{q}}, t) - \delta \sum_{j=1}^{m} \left(\lambda_j(t)\phi_j(\boldsymbol{q}, t)\right)\right] dt$$

$$ = \int_{t_0}^{t_1} \left[\frac{\partial \mathscr{L}}{\partial q^i}\delta q^i + \frac{\partial \mathscr{L}}{\partial \dot{q}^i}\delta \dot{q}^i + \frac{\partial \mathscr{L}}{\partial t}\delta t - \sum_{j=1}^{m} \lambda_j \left[\frac{\partial \phi_j}{\partial q^i}\delta q^i + \frac{\partial \phi_j}{\partial t}\delta t\right] + \sum_{j=1}^{m} \phi_j \frac{\partial \lambda_j}{\partial t}\delta t\right] dt$$

$$ = \int_{t_0}^{t_1} \left[\frac{\partial \mathscr{L}}{\partial q^i} - \frac{d}{dt}\left(\frac{\partial \mathscr{L}}{\partial \dot{q}^i}\right) - \sum_{j=1}^{m} \lambda_j \frac{\partial \phi_j}{\partial q^i}\right] \delta q^i\, dt$$

where we have used the product rule in the second step $\delta AB = \delta(A)B + A\delta(B)$. At this point, we realise that $\delta t = 0$ for **virtual variations** (as highlighted in d'Alembert's principle a little later) and we integrated the blue term by parts, with the resulting term evaluated at the boundary disappearing by the postulates of Hamilton's principle:

$$ \int_{t_0}^{t_1} \frac{\partial \mathscr{L}}{\partial \dot{q}^i}\delta \dot{q}^i\, dt = \frac{\partial \mathscr{L}}{\partial \dot{q}^i}\delta q^i \bigg|_{t_0}^{t_1} - \int_{t_0}^{t_1} \frac{d}{dt}\left(\frac{\partial \mathscr{L}}{\partial \dot{q}^i}\right)\delta q^i\, dt$$

[2]Trying to imagine it is a bit fruitless, since we can really only think about, at most, three dimensions.

We now impose the stationary action principle $\delta A = 0$ and assume independent variations, meaning that the term in square brackets vanishes for each $i = 1, \ldots, n$, giving us the Euler-Lagrange equation of motion for constrained systems, as above.

8.2 Non-Holonomic Constraints

We might now ask ourselves, what if the constraint had consisted of velocity-dependent (i.e. non-holonomic), differential equation relations? We are aware that d'Alembert's principle/Gibbs-Appell equations (see chapter 13) will, in the case of linear or affine non-holonomic constraints, yield the correct equations of motion. In these formulations, we project the curve of the free motion onto the constrained subbundle of the tangent bundle, using a Lagrange multiplier or generalised force. We will now proceed to perform the calculus of variations with constraints $\phi_k(q^1, \ldots, q^n, \dot{q}^1, \ldots, \dot{q}^n, t)$, with $k = 1, \ldots, m$. We consider λ as a variable and assume that the constraints are applied before making the action stationary; in other words, the variations are compatible with the constraint surface:

$$
\delta A = \delta \int_1^2 (\mathscr{L}(t, q, \dot{q}) + \lambda_k \phi_k(t, q, \dot{q}) dt
$$

$$
= \int_1^2 \left(\frac{\partial \mathscr{L}}{\partial q^i} \delta q^i + \frac{\partial \mathscr{L}}{\partial \dot{q}^i} \delta \dot{q}^i + \lambda_k \left(\frac{\partial \phi_k}{\partial q^i} \delta q^i + \frac{\partial \phi_k}{\partial \dot{q}^i} \delta \dot{q}^i \right) + \phi \delta \lambda_k \right) dt \qquad (8.2.1)
$$

We then integrate by parts and *assume* that the variation commutes with the time derivative:

$$
\delta A = \left(\frac{\partial \mathscr{L}}{\partial \dot{q}^i} + \lambda_k \frac{\partial \phi_k}{\partial \dot{q}^i} \right) \delta q^i \Big|_1^2 + \int_1^2 \left[\left(\frac{\partial \mathscr{L}}{\partial q^i} - \frac{d}{dt} \left(\frac{\partial \mathscr{L}}{\partial \dot{q}^i} \right) + \lambda_k \frac{\partial \phi_k}{\partial q^i} \right. \right. \qquad (8.2.2)
$$

$$
\left. \left. - \dot{\lambda}_k \frac{\partial \phi_k}{\partial \dot{q}^i} - \lambda_k \frac{d}{dt} \left(\frac{\partial \phi_k}{\partial \dot{q}^i} \right) \right) \delta q^i + \phi_k \delta \lambda_k \right] dt \qquad (8.2.3)
$$

when the coefficients of the variations in q and λ vanish, the constraints are satisfied. Due to this, we can choose the Lagrange multipliers such that m components of δq^i are zero in $[t_1, t_2]$. The rest of the components are then independent and we can write the integral portion as the Euler-Lagrange equation with a modified Lagrangian $L = \mathscr{L} + \lambda_k \phi_k$ together with the m constraint equations $\phi_k = 0$, $k = 1, \ldots, m$:

$$
\frac{d}{dt} \left(\frac{\partial \mathscr{L}}{\partial \dot{q}^i} \right) - \frac{\partial \mathscr{L}}{\partial q^i} = -\lambda_k \left[\frac{d}{dt} \left(\frac{\phi_k}{\partial \dot{q}^i} \right) - \frac{\partial \phi_k}{\partial q^i} \right] - \dot{\lambda} \frac{\partial \phi_k}{\partial \dot{q}^i}, \qquad i = 1, \ldots, n \qquad (8.2.4)
$$

These are known as the **vakonomic equations**, after V. Koslov in 1983, and are equations derived by a variational principle that searches for the extremals of the curves which satisfy the constraints. We then have $n+m$ equations for finding the classical trajectory parametrised by n independent configuration space variables and m Lagrange multipliers. The action then becomes equal to the boundary terms, which, in general, are known variations $\delta q(t_1) = \delta q(t_2) = 0$ and hence the action vanishes:

$$\delta A = \left(\frac{\partial \mathscr{L}}{\partial \dot{q}^i} + \lambda_k \frac{\partial \phi_k}{\partial \dot{q}^i} \right) \delta q^i \bigg|_1^2 = 0. \tag{8.2.5}$$

An interesting detail now arises: *these equations of motion do not coincide with those found by application of d'Alembert's principle.* In other words, the trajectory of a non-holonomic system is not the extremal of the action integral in 8.2.4. The precise reason for this is not well understood in the literature; however, both formulations coincide when restricted to the holonomic case. We write here for brevity the central Lagrange equation as derived later in the chapter on d'Alembert's principle (see chapter 13):

$$\frac{d}{dt} \left(\frac{\partial \mathscr{L}}{\partial \dot{q}^j} \right) - \frac{\partial \mathscr{L}}{\partial q^j} = \lambda \left(\frac{\partial \phi_k}{\partial \dot{q}^i} \right) \tag{8.2.6}$$

It is a strange feature that vakonomic mechanics and d'Alembert's principle both contain similar ingredients: a Lagrangian and a set of constraints. The two equations are very similar, only differing by the addition of the Euler-Lagrange equation for the constraint. Of course, at the time of writing, this issue is not resolved; however, we can get some insight by considering three factors: the commutativity of the delta and the time derivative; the treatment of λ as a variable on which to perform variations; and the nature of the variations themselves.

Firstly, we note in d'Alembert's principle that the the variations do not always commute with the time derivatives (we outline this later in chapter 13 where we discuss the principle of virtual work):

$$\delta \frac{dq_i}{dt} - \frac{d}{dt} \delta q_i \neq 0 \tag{8.2.7}$$

This was something that we assumed above to complete our derivation. It becomes tempting therefore to consider the **Suslov** form of the vakonomic equations, which we introduced in the chapter 7:

$$\delta A = \int_1^2 dt \left\{ \frac{\partial L}{\partial q^i} \delta q^i - \frac{d}{dt} \left(\frac{\partial L}{\partial \dot{q}^i} \right) \delta q^i + \frac{\partial L}{\partial \dot{q}^i} \left(\delta \frac{d}{dt} q^i - \frac{d}{dt} \delta q^i \right) \right\} + \frac{\partial L}{\partial \dot{q}^i} \delta q^i \bigg|_1^2 + \int_1^2 dt \frac{\partial L}{\partial \lambda_k} \delta \lambda_k$$

with $L(q, \dot{q}, \lambda, t) = \mathscr{L}(q, \dot{q}, t) - \lambda(t)\phi(q, \dot{q}, t)$. For holonomic constraints, the variation and the delta commute and the middle term vanishes. Second, and perhaps somewhat related, is the nature of the variations themselves. In d'Alembert's principle, we consider variations to the free system and project the result onto the constraint surface using the constraint force, as we depict in figure 8.2. In vakonomic mechanics, we applied the constraint equations at the same time as we found the equations of motion, thereby

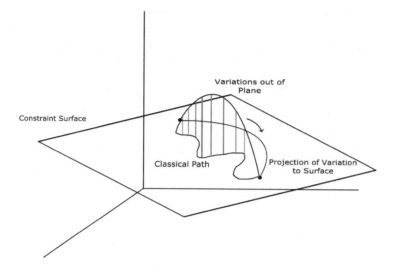

Fig. 8.2: The variations in d'Alembert's principle are free to leave the constraint surface before being projected down. In the vakonomic variational principle, the variations are confined to the surface straight away.

finding the extremal of the admissible curves from the start. Finally, the status of the Lagrange multipliers as variables of the system in the vakonomic approach is quite different from their status in the d'Alembert strategy, where they can be eliminated from the picture using the constraint equations!

We will return to the discussion of constrained dynamics in the Hamiltonian setting with the *Dirac theory* for constrained Hamiltonian systems. For further reading, please see (Arnold *et al.*s (1996), *Mathematical Aspects of Classical and Celestial Mechanics*).

Chapter summary

- Constraints make the generalised coordinates no longer independent since they are linked through a **constraint equation**. Coordinate independence was one of the criteria we required when we derived the Euler-Lagrange equation. There is one constraint equation for each imposed constraint.

- In this chapter, we affirmed our ideas about how constraint equations reduce the dimensionality of the system and confine motion to an $n - m$ dimensional **constraint surface**.

- We extend the Euler-Lagrange equation of motion to account for the constraints by using undetermined Lagrange multipliers:

$$\frac{\partial \mathscr{L}}{\partial q^i} - \frac{d}{dt}\left(\frac{\partial \mathscr{L}}{\partial \dot{q}^i}\right) - \sum_{j=1}^{m} \lambda_j \frac{\partial \phi_j}{\partial q^i} = 0$$

We can derive this from the stationary action principle by using a modified Lagrangian that accounts for holonomic constraint equations:

$$L = \mathscr{L} - \lambda_j \phi_j$$

- Non-holonomic mehanics was studied using a variational principle that fails to yield the correct equations of motion given by d'Alembert's principle.

Exercise 8.1 Consider a hydrogen atom consisting of proton orbited by an electron at fixed radius such that the electron is constrained to move on the surface of a sphere about the nucleus. We will be using spherical polar coordinates (r, θ, ϕ) to describe this system. The constraint is the fixed radius $r = l$, where l is an arbitrary constant indicating the fixed length. There are two approaches: the reduction of the generalised coordinates to a minimal set, and the Lagrange multiplier method.

The reduction method recognises that $r = l$ so that $\delta r = 0$; therefore, our two generalised coordinates are θ and ϕ. The Lagrangian for the system is written below; consult chapter 9 on point transformations to derive this result:

$$\mathscr{L}(l, \theta, \phi) = \frac{1}{2}ml^2\left(\dot{\theta}^2 + \dot{\phi}^2 \sin^2\theta\right) - U(l, \theta, \phi) \tag{8.2.8}$$

The Euler-Lagrange equations for each coordinate are given by

$$\frac{d}{dt}\left(\frac{\partial \mathscr{L}}{\partial \dot{\theta}}\right) - \frac{\partial \mathscr{L}}{\partial \theta} = 0 = ml^2\ddot{\theta} - ml^2\dot{\phi}^2 \sin\theta\cos\theta + \frac{\partial U}{\partial \theta} \tag{8.2.9}$$

$$\frac{d}{dt}\left(\frac{\partial \mathscr{L}}{\partial \dot{\phi}}\right) - \frac{\partial \mathscr{L}}{\partial \phi} = 0 = \frac{d}{dt}\left(ml^2\dot{\phi}\sin^2\theta\right) + \frac{\partial U}{\partial \phi} \tag{8.2.10}$$

Notice that there are only two equations of motion. The Lagrange multiplier method uses all of the three spherical coordinates (r, θ, ϕ) and each has a Euler-Lagrange equation. In addition, we develop the constraint equation

$$\varphi = r - l = 0 \tag{8.2.11}$$

The Lagrangian is now

$$L(r, \theta, \phi, \lambda) = \frac{1}{2}m(\dot{r}^2 + r^2\dot{\theta}^2 + r^2\dot{\phi}^2 \sin^2 \theta) - U(r, \theta, \phi) - \lambda(r - l) \tag{8.2.12}$$

We now have four equations to solve:

$$\frac{\partial \mathscr{L}}{\partial r} - \frac{d}{dt}\left(\frac{\partial \mathscr{L}}{\partial \dot{r}}\right) - \lambda\frac{\partial \varphi}{\partial r} = 0, \tag{8.2.13}$$

$$\frac{\partial \mathscr{L}}{\partial \theta} - \frac{d}{dt}\left(\frac{\partial \mathscr{L}}{\partial \dot{\theta}}\right) - \lambda\frac{\partial \varphi}{\partial \theta} = 0, \tag{8.2.14}$$

$$\frac{\partial \mathscr{L}}{\partial \phi} - \frac{d}{dt}\left(\frac{\partial \mathscr{L}}{\partial \dot{\phi}}\right) - \lambda\frac{\partial \varphi}{\partial \phi} = 0, \tag{8.2.15}$$

$$\varphi = 0$$

We note that $\partial\varphi/\partial r = 1$ such that the equations of motion proceed as follows:

$$mr\dot{\theta}^2 + mr\dot{\phi}\sin^2\theta - \frac{\partial U}{\partial r} - m\ddot{r} - \lambda = 0 \tag{8.2.16}$$

$$mr^2\dot{\phi}^2\cos\theta\sin\theta - \frac{\partial U}{\partial \theta} - \frac{d}{dt}(mr^2\dot{\theta}) = 0 \tag{8.2.17}$$

$$-\frac{\partial U}{\partial \phi} - \frac{d}{dt}(mr^2\dot{\phi}\sin^2\theta) = 0 \tag{8.2.18}$$

We now solve the constraint equation for r; we find this to be $r = l$ and substitute this result in to the equations of motion; noticing that $\dot{l} = \ddot{l} = 0$:

$$ml\dot{\theta}^2 + ml\dot{\phi}^2\sin^2\theta - \frac{\partial U}{\partial r}\Big|_{r=l} - \lambda = 0 \tag{8.2.19}$$

$$ml^2\dot{\phi}^2\cos\theta\sin\theta - \frac{\partial U}{\partial \theta}\Big|_{r=l} - ml^2\ddot{\theta} = 0 \tag{8.2.20}$$

$$-\frac{\partial U}{\partial \phi}\Big|_{r=l} - \frac{d}{dt}(ml^2\dot{\phi}\sin^2\theta) = 0 \tag{8.2.21}$$

We can see that exactly the same equations of motion have been generated for θ and ϕ, and the radial equation is an expression for the undetermined multiplier. Therefore, both methods are equivalent. The approach you choose will depend on the situation at hand; although not in this case, sometimes the Lagrange multiplier method can be an easier avenue. This example provides background reading for the example in chapter 21 on constrained Hamiltonian dynamics.

Exercise 8.2 Biological membranes are well studied from a biochemical and structural biology standpoint. These self-assembled structures allow organisms to compartmentalise regions and set up complex molecular processes and chemical reactions.

The membranes of a cell form a bilayer made from phospholipids with hydrophilic heads and hydrophobic tails made from chains of fatty acids (see figure 8.3). The composition of the phospholipids varies and the presence of unsaturated C-C bonds leads to different tail lengths and kinks and therefore densities. Non-polar molecules tend to aggregate in the lipid region, while polar molecules in aqueous media are attracted to the charged phosphate heads. On the exterior of the membrane, there is often a coating of carbohydrates (glycolipids, if attached to phosphate anions, or glycoproteins, if attached to proteins) for cellular recognition and adhesion. Dotted throughout the membrane are various proteins that allow ion transportation and gradient flows. Proteins are complex molecules made from amino acid chains folded into organised shapes and structures. They are responsible for most of the exciting molecular events within the cell. On the interior, there is a cytoskeleton made from microtubules; this facilitates vesicle transport within the cell. The cytoskeleton connects to membrane proteins and receptors and links cell structures together.

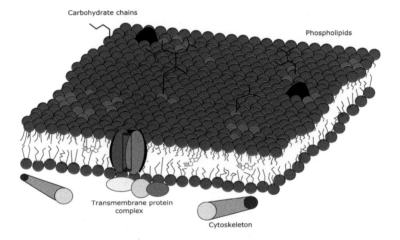

Carbohydrate chains

Phospholipids

Transmembrane protein
complex

Cytoskeleton

Fig. 8.3: A section of a cell membrane.

Membranes can maintain their structural integrity even when subject to enormous deformations, which might include shearing, bending compressions or rarefactions of the layers. The resulting membrane structure can be modelled as a two-dimensional surface embedded in $\mathbb{R}^3$. The shape adopted is assumed to minimise the energy of the membrane and hence we create an energy functional $G[\boldsymbol{x}]$ which is a function of the surface $\boldsymbol{x} = (x^1, x^2)$. The cell must maintain its volume and surface area and hence the constrained Euler-Lagrange equations provide the equilibrium membrane shape. Such studies have led to the elucidation of protein crowding and lipid rafts within membrane structures, by what is known as the **Canham-Helfrich-Evans theory**:

$$G[\boldsymbol{x}] = \frac{1}{2} \int dx^1 x^2 \left\{ \kappa (\Delta f)^2 + \sigma (\vec{\nabla} f)^2 \right\} \tag{8.2.22}$$

Here G is the energy functional, f is the function characterising the surface, $dx^1 dx^2$ is an area element of the surface, σ is the surface tension of the membrane and κ is the bending modulus. Such an action leads to a fourth-order linear PDE called the **shape equation**. This theory can by applied to different surfaces to investigate their equilibrium properties. It could also be applied to surface science to investigate monolayer morphologies over step sites, for instance.

Exercise 8.3 Show that the equations of motion for the following constrained system can be put in **Sturm-Liouville form**. Hence, show the equivalence of the Lagrange multiplier in the variational problem and the eigenvalue of the corresponding Sturm-Liouville problem:

$$\delta A[q] = \delta \int_1^2 (\mathscr{L} - \lambda \phi) dt \qquad (8.2.23)$$

with $\mathscr{L} = a\dot{q}^2 + bq^2$ and $\phi = cq^2$, $a, b, c > 0 \in \mathbb{R}$:

$$-\frac{d}{dt}(a\dot{q}) + bq = \lambda cq \qquad (8.2.24)$$

This indicates that we can convert an appropriate variational problem into a corresponding Sturm-Liouville problem, and vice versa with an integrating factor. We can then estimate the lowest eigenvalue using the **Rayleigh-Ritz** procedure with a suitable trial function. The logic can be reversed and we can take a real second-order linear ODE equation of motion, put it into Sturm-Liouville form and generate its variational principle! The benefit of this is the vast mathematical might behind the Sturm-Liouville problem for finding eigenfunctions and eigenvalues.

9

Point Transformations in Lagrangian Mechanics

Sometimes when solving problems, it is useful to change coordinates in velocity phase space to better suit and simplify the system at hand; this is a requirement of any physical theory. This change is often motivated by some physicality of the system we have observed experimentally or may highlight new conserved quantities that we might have overlooked using the old description. In the Newtonian formalism, it was a bit of a hassle to change coordinates, and the equations of motion will look quite different.

The Lagrange equation has so far not required us to specify the form of the generalised coordinates, so we expect it to be invariant with respect to a change in coordinates for physical paths; this is one of the reasons it is so powerful. The aim of this chapter is therefore to prove the following relation holds for some change of coordinates $(q) \to (Q)$:

$$\frac{d}{dt}\left(\frac{\partial \mathscr{L}_{\text{old}}}{\partial \dot{q}^i}\right) - \frac{\partial \mathscr{L}_{\text{old}}}{\partial q^i} \stackrel{?}{=} \frac{d}{dt}\left(\frac{\partial \mathscr{L}_{\text{new}}}{\partial \dot{Q}^i}\right) - \frac{\partial \mathscr{L}_{\text{new}}}{\partial Q^i} = 0 \qquad (9.0.1)$$

where $\mathscr{L}_{\text{old}}(q, \dot{q}, t) = \mathscr{L}_{\text{new}}(Q, \dot{Q}, t) = \mathscr{L}_{\text{old}}(q, (Q), \dot{q}(Q, \dot{Q}), t)$. This means that the new Lagrangian is nothing but the old one in disguise, expressed in the new coordinates; it transforms like a **scalar** under point transformations. This traces back to the geometry of Lagrangian mechanics and we will show much later how the Lagrangian (Hamiltonian) determines the structure of the velocity (momentum) phase space. This is not to say, however, that the look of the expression may not be completely different, just that the answer they give will be the same.

We now set about proving this invariance. The old coordinates are a function of the new coordinates $q^i = q^i(Q, t)$, with the index running from 1 to n, just as the new coordinates are a function of the old coordinates $Q^i = Q^i(q, t)$. Importantly, the transformation does not depend on the velocities, so $\partial q/\partial \dot{Q} = \partial Q/\partial \dot{q} = 0$ and we start by computing some partial derivatives:

$$\frac{\partial \mathscr{L}}{\partial \dot{Q}^i} = \sum_j \left[\frac{\partial \mathscr{L}}{\partial q^j}\frac{\partial q^j}{\partial \dot{Q}^i} + \frac{\partial \mathscr{L}}{\partial \dot{q}^j}\frac{\partial \dot{q}^j}{\partial \dot{Q}^i}\right], \qquad \frac{\partial \mathscr{L}}{\partial Q^i} = \sum_j \left[\frac{\partial \mathscr{L}}{\partial q^j}\frac{\partial q^j}{\partial Q^i} + \frac{\partial \mathscr{L}}{\partial \dot{q}^j}\frac{\partial \dot{q}^j}{\partial Q^i}\right] \qquad (9.0.2)$$

Lagrangian & Hamiltonian Dynamics. Peter Mann, Oxford University Press (2018).
© Peter Mann. DOI: 10.1093/oso/9780198822370.001.0001

Exploring the first of these terms, we can see that $\partial q/\partial \dot{Q} = 0$, but we can work on the $\dot{q}$ term too:

$$\dot{q}^i = \sum_k \frac{\partial q^i}{\partial Q^k}\dot{Q}^k + \frac{\partial q^i}{\partial t}, \qquad \text{therefore} \qquad \frac{\partial \dot{q}^i}{\partial \dot{Q}^j} = \frac{\partial q^i}{\partial Q^j}$$

This equation is used in the discussion of d'Alembert's principle a little later on. Writing the Lagrange equation out, we obtain

$$\frac{d}{dt}\left(\frac{\partial \mathscr{L}}{\partial \dot{Q}^i}\right) - \frac{\partial \mathscr{L}}{\partial Q^i} = \frac{d}{dt}\left[\sum_j \frac{\partial \mathscr{L}}{\partial \dot{q}^j}\frac{\partial q^j}{\partial Q^i}\right] - \sum_j \left[\frac{\partial \mathscr{L}}{\partial q^j}\frac{\partial q^j}{\partial Q^i} + \frac{\partial \mathscr{L}}{\partial \dot{q}^j}\frac{\partial \dot{q}^j}{\partial Q^i}\right]$$

$$= \sum_j \left\{\left[\frac{\partial \mathscr{L}}{\partial \dot{q}^j}\frac{d}{dt}\left(\frac{\partial q^j}{\partial Q^i}\right) + \frac{d}{dt}\left(\frac{\partial \mathscr{L}}{\partial \dot{q}^j}\right)\frac{\partial q^j}{\partial Q^i}\right] - \frac{\partial \mathscr{L}}{\partial q^j}\frac{\partial q^j}{\partial Q^i} - \frac{\partial \mathscr{L}}{\partial \dot{q}^j}\frac{\partial \dot{q}^j}{\partial Q^i}\right\}$$

$$= \sum_j \left\{\frac{\partial q^j}{\partial Q^i}\left[\frac{d}{dt}\left(\frac{\partial \mathscr{L}}{\partial \dot{q}^j}\right) - \left(\frac{\partial \mathscr{L}}{\partial q^j}\right)\right] + \frac{\partial \mathscr{L}}{\partial \dot{q}^j}\left[\frac{d}{dt}\left(\frac{\partial q^j}{\partial Q^i}\right) - \frac{\partial \dot{q}^j}{\partial Q^i}\right]\right\}$$

The dark blue term is the Lagrange equation for the old coordinate, which we know was zero, leaving us with,

$$\frac{d}{dt}\left(\frac{\partial \mathscr{L}}{\partial \dot{Q}^i}\right) - \frac{\partial \mathscr{L}}{\partial Q^i} = \sum_j \left\{\frac{\partial \mathscr{L}}{\partial \dot{q}^j}\left[\frac{d}{dt}\left(\frac{\partial q^j}{\partial Q^i}\right) - \frac{\partial \dot{q}^j}{\partial Q^i}\right]\right\} \qquad (9.0.3)$$

If we change the order of differentiation of the time derivative on the right-hand side, we see that the square bracket is exactly zero and the entire expression is equal to zero (you would be right to be weary of swapping the order of total derivatives like this in general; however, since, in this case, the total derivative is equal to the partial, we are free to do so; convince yourself using the above relations for the transformation, or cheat and look in appendix C):

$$\frac{d}{dt}\left(\frac{\partial \mathscr{L}}{\partial \dot{Q}^i}\right) - \frac{\partial \mathscr{L}}{\partial Q^i} = 0 \qquad (9.0.4)$$

This type of coordinate transformation is called a **point transformation** and, by definition, it only changes the coordinates, so we say that the Euler-Lagrange equation is **covariant**. This is the power of the Lagrange formalism; we may choose our description to suit ourselves. This is intuitive from the stationary action principle, since we didn't require a specific choice of coordinates to derive it or it's stationary points!

The coordinate transformation is assumed to be invertible, since both sets of coordinates are, by assumption, a linearly independent minimal set. However, we impose the condition that, to be invertible from one set to the other, the following matrix must have a non-zero determinant:

$$\det \left[\boldsymbol{D}q(Q) \right] = \frac{\partial (q^1, q^2, \ldots, q^n)}{\partial (Q^1, Q^2, \ldots, Q^n)}$$

$$= \det \begin{pmatrix} \partial q^1 / \partial Q^1 & \partial q^1 / \partial Q^2 & \ldots & \partial q^1 / \partial Q^n \\ \partial q^2 / \partial Q^1 & \partial q^2 / \partial Q^2 & \ldots & \ldots \\ \vdots & \vdots & \ddots & \vdots \\ \partial q^n / \partial Q^1 & \partial q^n / \partial Q^2 & \ldots & \partial q^n / \partial Q^n \end{pmatrix} \neq 0$$

This matrix is the **Jacobian** $\boldsymbol{D}q(Q)$ of the transformation and, when the **Jacobian determinant** is non-zero, we can invert the expression freely, as per the **implicit function theorem**. The inverse transformation will also have its own Jacobian $\boldsymbol{D}Q(q)$. This Jacobian is such that it is the inverse matrix of the "forward" Jacobian, therefore satisfying,

$$(\boldsymbol{D}Q(q))(\boldsymbol{D}q(Q)) = \mathbb{I} \qquad \text{where} \qquad (\boldsymbol{D}Q(q)) = (\boldsymbol{D}q(Q))^{-1} \qquad (9.0.5)$$

This is obvious from the chain rule as applied to the determinants of the square matrices:

$$\frac{\partial (q^1, q^2, \ldots, q^n)}{\partial (Q^1, Q^2, \ldots, Q^n)} \frac{\partial (Q^1, Q^2, \ldots, Q^n)}{\partial (q^1, q^2, \ldots, q^n)} = \frac{\partial (Q^1, Q^2, \ldots, Q^n)}{\partial (q^1, q^2, \ldots, q^n)} \frac{\partial (q^1, q^2, \ldots, q^n)}{\partial (Q^1, Q^2, \ldots, Q^n)}$$

$$= \mathbb{I}$$

Therefore, we expect the coordinate transformation to be smooth in the forward direction, and invertible and smooth in the reverse:

$$\det \left[\boldsymbol{D}q(Q) \right] \neq 0 \qquad \text{and} \qquad \det \left[\boldsymbol{D}Q(q) \right] \neq 0 \qquad (9.0.6)$$

If the Lagrangian transforms such as these conditions hold, then we say it transforms as a *scalar*. This means that $\mathscr{L}(\boldsymbol{q}, \dot{\boldsymbol{q}}, t) = \mathscr{L}(\boldsymbol{Q}, \dot{\boldsymbol{Q}}, t)$, provided that both Lagrangians satisfy their own Euler-Lagrange equation:

$$\frac{d}{dt} \left(\frac{\partial \mathscr{L}(\boldsymbol{Q}, \dot{\boldsymbol{Q}}, t)}{\partial \dot{Q}^i} \right) - \frac{\partial \mathscr{L}(\boldsymbol{Q}, \dot{\boldsymbol{Q}}, t)}{\partial Q^i} = 0$$

$$\frac{d}{dt} \left(\frac{\partial \mathscr{L}(\boldsymbol{q}, \dot{\boldsymbol{q}}, t)}{\partial \dot{q}^i} \right) - \frac{\partial \mathscr{L}(\boldsymbol{q}, \dot{\boldsymbol{q}}, t)}{\partial q^i} = 0 \qquad (9.0.7)$$

This is why we don't bother with inertial frames in Lagrangian mechanics: the universality of the Euler-Lagrange equation with the correct Lagrangian will reproduce the results.

In analytical mechanics, we may also consider a change of **parameter** $t \to t+\tau = \beta$, in addition to coordinate changes. If we can define variables $\beta(t)$, $q(\beta)$ and $\dot{q}(\beta)$, where $\dot{q}(\beta) = dq/d\beta$, then the Euler-Lagrange equation will still hold and calculations are repeated, except with t replaced by β (i.e. the solutions are invariant if $q_i(\beta) = q_i(\beta(t))$). In this case, we say that the action is **parametrisation invariance** and we say that $t \to \beta$ transforms as a scalar. It is not a general property of mechanics, however, that theories are invariant under a change of parameter; and we will investigate an example of this at the end of chapter 11 a little later.

Chapter summary

- Coordinate transformations are a necessity for any practitioner of classical mechanics. We show that the Lagrange equation of motion is **invariant** under a special class of coordinate transformations called **point transformations**.
- In general, performing a point transformation will make the form of the Lagrangian and the equations of motion look vastly different, but it is just disguise and both will give the same result, just expressed differently.

Exercise 9.1 Transforming a Lagrangian between cartesian and spherical polar coordinates is a very useful operation but can be rather tiresome due to the lengths of algebra required. Here we show part of the working and quote the result to the reader, who is encouraged to fill in the intermediary step with several sheets of paper! We start as follows:

$$x = r \sin\theta \cos\phi \qquad y = r \sin\theta \sin\phi \qquad z = r \cos\theta \tag{9.0.8}$$

The time derivative of each of these is computed using product and chain rules:

$$\dot{x} = \dot{r} \sin\theta \cos\phi + \dot{\theta} r \cos\theta \cos\phi - \dot{\phi} r \sin\theta \sin\phi \tag{9.0.9}$$

$$\dot{y} = \dot{r} \sin\theta \sin\phi + r\dot{\theta} \cos\theta \sin\phi + r\dot{\phi} \sin\theta \cos\phi \tag{9.0.10}$$

$$\dot{z} = \dot{r} \cos\theta - r\dot{\theta} \sin\theta \tag{9.0.11}$$

The kinetic energy is then formed by summing the squares of these relations in the following formula:

$$T = \frac{1}{2}m(\dot{x}^2 + \dot{y}^2 + \dot{z}^2) \tag{9.0.12}$$

After a lot of basic algebra and cancelation of cross terms, and so on, we find the transformed kinetic energy to be

$$T = \frac{1}{2}m(\dot{r}^2 + r^2\dot{\theta}^2 + r^2\dot{\phi}^2 \sin^2\theta) \tag{9.0.13}$$

The potential can be written $U(x, y, z) \to U(r, \theta, \phi)$.

Exercise 9.2 Consider the transformation from Cartesian to cylindrical coordinates:

$$x = \rho \cos \phi \qquad y = \rho \sin \phi \qquad z = z \tag{9.0.14}$$

The time derivatives are constructed as follows:

$$\dot{x} = \dot{\rho} \cos \phi - \rho \dot{\phi} \sin \phi \qquad \dot{y} = \dot{\rho} \sin \phi + \rho \dot{\phi} \cos \phi \qquad \dot{z} = \dot{z} \tag{9.0.15}$$

The Lagrangian is then formed by entering these expressions into the kinetic energy:

$$\mathscr{L}(x, y, z, \dot{x}, \dot{y}, \dot{z}) = \frac{1}{2} m(\dot{x}^2 + \dot{y}^2 + \dot{z}^2) - U(x, y, z) = \frac{1}{2} m(\dot{\rho}^2 + \rho^2 \dot{\phi}^2 \dot{z}^2) - U(\rho, \phi, z) \tag{9.0.16}$$

Exercise 9.3 We will consider an example of performing a point transformation of a Lagrangian from cartesian to polar coordinates, $(x, y) \rightarrow (r, \theta)$. Let the Lagrangian be

$$\mathscr{L}(x, y, \dot{x}, \dot{y}) = \frac{1}{2}\dot{x}^2 + \frac{1}{2}\dot{y}^2 + \frac{1}{(x^2 + y^2)^{3/2}} \tag{9.0.17}$$

The equations of motion are found to be

$$\frac{d}{dt}\left(\frac{\partial \mathscr{L}}{\partial \dot{x}}\right) - \frac{\partial \mathscr{L}}{\partial x} = \ddot{x} + \frac{3x}{(x^2 + y^2)^{5/2}} = 0, \qquad \frac{d}{dt}\left(\frac{\partial \mathscr{L}}{\partial \dot{y}}\right) - \frac{\partial \mathscr{L}}{\partial y} = \ddot{y} + \frac{3y}{(x^2 + y^2)^{5/2}} = 0$$

To perform the differentiation $\partial \mathscr{L}/\partial x$, let $u = x^2 + y^2$ and chain rule. Let's now change to polar coordinates $x = r \cos \theta$ and $y = r \sin \theta$, with $\dot{x} = \dot{r} \cos \theta - r\dot{\theta} \sin \theta$ and $\dot{y} = \dot{r} \sin \theta + r\dot{\theta} \cos \theta$:

$$\mathscr{L}(r, \theta, \dot{r}, \dot{\theta}) = \frac{1}{2}\dot{r}^2 + \frac{1}{2}\dot{\theta}^2 r^2 + \frac{1}{r^3} \tag{9.0.18}$$

The equations of motion are now much more simple in polar form:

$$\frac{d}{dt}\left(\frac{\partial \mathscr{L}}{\partial \dot{r}}\right) - \frac{\partial \mathscr{L}}{\partial r} = \ddot{r} - r\dot{\theta}^2 + 3r^{-4} = 0, \qquad \frac{d}{dt}\left(\frac{\partial \mathscr{L}}{\partial \dot{\theta}}\right) - \frac{\partial \mathscr{L}}{\partial \theta} = \frac{d}{dt}(r^2\dot{\theta}) - 0 = 0 \tag{9.0.19}$$

Exercise 9.4 In the chapter 7 we used the electromagnetic Lagrangian in the Euler-Lagrange equation of motion to derive the Lorentz force law. In this exercise, we see if application of Hamilton's principle and the calculus of variations does, in fact, yield the Lagrange equation of motion. You could view this as a coordinate transformation $(q) \rightarrow (r)$, so we are investigating if we generate from the calculus of variations the Lagrange equation from a transformed Lagrangian. If we obtain a positive result, we expect the same equation of motion in both cases, which would show us that the Lagrangian method is consistent, without referring to qs and $\dot{q}$s:

$$\delta A[\boldsymbol{q}] = \delta \int_{t_0}^{t_1} \left[\frac{1}{2}m\dot{\boldsymbol{r}} \cdot \dot{\boldsymbol{r}} + e\dot{\boldsymbol{r}} \cdot \boldsymbol{A} - e\varphi\right] dt$$

$$= \int_{t_0}^{t_1} \left[m\dot{\boldsymbol{r}} \cdot \delta\dot{\boldsymbol{r}} + e\delta\dot{\boldsymbol{r}} \cdot \boldsymbol{A} + e\dot{\boldsymbol{r}} \cdot \frac{\partial \boldsymbol{A}}{r^i}\delta r^i - e\frac{\partial \varphi}{\partial r^i}\delta r^i\right] dt$$

Now integrate all the variations in $\delta\dot{\boldsymbol{r}}$ by parts:

$$= \int_{t_0}^{t_1} \left[-m\ddot{\boldsymbol{r}} \cdot \delta\boldsymbol{r} - e\delta\boldsymbol{r} \cdot \dot{\boldsymbol{A}} + e\dot{\boldsymbol{r}} \cdot \frac{\partial \boldsymbol{A}}{\partial r^i} \delta r^i - e \frac{\partial \varphi}{\partial r^i} \delta r^i \right] dt$$

$$= \int_{t_0}^{t_1} \left[-m\ddot{r}^i - e\frac{dA_i}{dt} + e\dot{\boldsymbol{r}} \cdot \frac{\partial \boldsymbol{A}}{\partial r^i} - e \frac{\partial \varphi}{\partial r^i} \right] \delta r^i dt$$

$$= \int_{t_0}^{t_1} \left[-m\ddot{r}^i - e\sum_j \frac{\partial A_i}{\partial r^j} \dot{r}^j - e\frac{\partial A_i}{\partial t} + e\sum_j \dot{r}^j \frac{\partial A_j}{\partial r^i} - e \frac{\partial \varphi}{\partial r^i} \right] \delta r^i dt$$

$$= \int_{t_0}^{t_1} \left[-m\ddot{r}^i - e\left(-\frac{\partial \varphi}{\partial r^i} - \frac{\partial A_i}{\partial t} \right) + e\sum_j \dot{r}^j \left(\frac{\partial A_j}{\partial r^i} - \frac{\partial A_i}{\partial r^j} \right) \right] \delta r^i dt \qquad (9.0.20)$$

If we then impose Hamilton's principle $\delta A = 0$ and enter the expressions for vector and scalar potentials, then we arrive at the Lorentz force law:

$$m\ddot{r}^i = e(E_i + \dot{r}^i \times B_i) \qquad (9.0.21)$$

This again highlights that the Lagrange equation of motion holds, no matter what coordinates we choose to express the Lagrangian, the stationary action principle and the calculus of variations will yield the correct results.

Exercise 9.5 Consider the calculus of variations for coordinate transformation $q(\boldsymbol{Q}, t)$ $\dot{q}(\boldsymbol{Q}, \dot{\boldsymbol{Q}}, t)$ and inverse transformations $Q(\boldsymbol{q}, t)$ $\dot{Q}(\boldsymbol{q}, \dot{\boldsymbol{q}}, t)$:

$$\delta A = \delta \int_{t_0}^{t_1} \mathscr{L}(q(\boldsymbol{Q}, t), \dot{q}(\boldsymbol{Q}, \dot{\boldsymbol{Q}}, t)) dt$$

$$= \int_{t_0}^{t_1} \left(\frac{\partial \mathscr{L}}{\partial Q^k} \frac{\partial Q^k}{\partial q^i} \delta q^i + \frac{\partial \mathscr{L}}{\partial \dot{Q}^k} \frac{\partial \dot{Q}^k}{\partial q^i} \delta q^i + \frac{\partial \mathscr{L}}{\partial \dot{Q}^k} \frac{\partial \dot{Q}^k}{\partial \dot{q}^i} \delta \dot{q}^i + \frac{\partial \mathscr{L}}{\partial Q^k} \frac{\partial Q^k}{\partial \dot{q}^i} \delta \dot{q}^i \right) dt \qquad (9.0.22)$$

Using results from appendix C, this becomes

$$= \int_{t_0}^{t_1} \left(\frac{\partial \mathscr{L}}{\partial Q^k} \frac{\partial Q^k}{\partial q^i} \delta q^i + \frac{\partial \mathscr{L}}{\partial \dot{Q}^k} \frac{d}{dt} \left(\frac{\partial Q^k}{\partial q^i} \right) \delta q^i + \frac{\partial \mathscr{L}}{\partial \dot{Q}^k} \frac{\partial Q^k}{\partial q^i} \delta \dot{q}^i \right) dt$$

$$= \int_{t_0}^{t_1} \left(\frac{\partial \mathscr{L}}{\partial Q^k} \frac{\partial Q^k}{\partial q^i} + \frac{\partial \mathscr{L}}{\partial \dot{Q}^k} \frac{d}{dt} \left(\frac{\partial Q^k}{\partial q^i} \right) - \frac{d}{dt} \left(\frac{\partial \mathscr{L}}{\partial \dot{Q}^k} \frac{\partial Q^k}{\partial q^i} \right) \right) \delta q^i dt \qquad (9.0.23)$$

The last two terms can be collected using the product rule to generate a collected expression.

$$= \int_{t_0}^{t_1} \left(\frac{\partial \mathscr{L}}{\partial Q^k} - \frac{d}{dt} \left(\frac{\partial \mathscr{L}}{\partial \dot{Q}^k} \right) \right) \frac{\partial Q^k}{\partial q^i} \delta q^i dt \qquad (9.0.24)$$

Hence, by Hamilton's principle, we see that the two Lagrange equations are related by the Jacobian of the transformation:

$$\left(\frac{\partial \mathscr{L}}{\partial Q^k} - \frac{d}{dt} \left(\frac{\partial \mathscr{L}}{\partial \dot{Q}^k} \right) \right) \frac{\partial Q^k}{\partial q^i} = \frac{\partial \mathscr{L}}{\partial q^i} - \frac{d}{dt} \left(\frac{\partial \mathscr{L}}{\partial \dot{q}^i} \right) = 0 \qquad (9.0.25)$$

10
The Jacobi Energy Function

In this chapter we consider how the Lagrange formalism treats the energy of the system. Our discussion will lead us nicely to conservation laws and symmetries, which is the focus of chapter 11.

Earlier, we seemed to disregard **explicit time dependance** in the Lagrangian $\mathcal{L}(\boldsymbol{q}, \dot{\boldsymbol{q}})$ or $\mathcal{L}(\boldsymbol{q}, \dot{\boldsymbol{q}}, t)$. But what exactly is explicit time dependence? Surely, the Lagrangian is always going to change with time, since both $q^i(t)$ and $\dot{q}^i(t)$ are functions of time?

To illustrate explicit time dependance, let's follow an analogy and imagine a playground slide. As we slide down it, our Lagrangian will change, since our position $q^i(t)$ and velocity $\dot{q}^i(t)$ coordinates are changing. If the slide was indoors and perfectly maintained, then, no matter what time of the day/month/year we go back to it, we always get the same experience when we slide on it. If the slide was outdoors and *not* well maintained, it will go all rusty and weathered and, when you slide down it, the exact experience will be different. Your experience will depend exactly when you slide on it and hence so will your Lagrangian. When the slide is inside, the slide's Lagrangian is not an explicit function of time, even though it did change with time through the coordinate's time dependance. When the slide is outside, the slide's Lagrangian is an explicit function of time, since it matters exactly when we choose to slide. The indoor Lagrangian is **time translationally invariant** (it doesn't change for a time translation $t' = t + \delta t$); however both Lagrangians are also **implicit functions of time**, since they both change as we slide down through the coordinates and velocities.

With this idea in mind, let's take the total time derivative of the Lagrangian $\mathcal{L}(\boldsymbol{q}, \dot{\boldsymbol{q}}, t)$:

$$\frac{d\mathcal{L}}{dt} = \sum_i \frac{\partial \mathcal{L}}{\partial q^i} \dot{q}^i + \sum_i \frac{\partial \mathcal{L}}{\partial \dot{q}^i} \ddot{q}^i + \frac{\partial \mathcal{L}}{\partial t} \tag{10.0.1}$$

if we assume that the Euler-Lagrange equation is satisfied, we can substitute for the first partial derivative term; from there, we use the reverse product rule $\frac{d}{dt}(AB) = \dot{A}B + A\dot{B}$:

Lagrangian & Hamiltonian Dynamics. Peter Mann, Oxford University Press (2018).
© Peter Mann. DOI: 10.1093/oso/9780198822370.001.0001

$$\frac{d\mathcal{L}}{dt} = \sum_i \left[\frac{d}{dt}\left(\frac{\partial\mathcal{L}}{\partial\dot{q}^i}\right)\dot{q}^i + \frac{\partial\mathcal{L}}{\partial\dot{q}^i}\ddot{q}^i \right] + \frac{\partial\mathcal{L}}{\partial t}$$

$$= \sum_i \frac{d}{dt}\left(\frac{\partial\mathcal{L}}{\partial\dot{q}^i}\dot{q}^i\right) + \frac{\partial\mathcal{L}}{\partial t} \tag{10.0.2}$$

Collecting the terms in d/dt, we obtain

$$\frac{d}{dt}\left[\sum_i \left(\frac{\partial\mathcal{L}}{\partial\dot{q}^i}\dot{q}^i\right) - \mathcal{L}\right] = \frac{d}{dt}J = -\frac{\partial\mathcal{L}}{\partial t} \tag{10.0.3}$$

We call the quantity J the **Jacobi energy function**[1] (we must note that J is a **conserved quantity** when the Lagrangian is not an explicit function of time):

$$J(q,\dot{q},t) = \sum_i \frac{\partial\mathcal{L}}{\partial\dot{q}^i}\dot{q}^i - \mathcal{L}(q,\dot{q},t) \qquad \text{\textbf{Jacobi energy function}} \tag{10.0.4}$$

So, how do we know that J is an energy quantity? Okay, so now we make a statement: J is the total energy of the system if the potential energy is not a function of the general velocity coordinate and if the cartesian coordinate is not a function of time:

$$\frac{\partial x_i}{\partial t} = 0, \qquad U = U(q) \tag{10.0.5}$$

When this is true, we can rewrite J as follows:

$$J = \sum_i \dot{q}^i \frac{\partial T}{\partial\dot{q}^i} - (T - U) \tag{10.0.6}$$

We know that we can write the transformation from cartesian coordinates to generalised coordinates as $x_i = x_i(q,t)$, and the inverse transformation as $q^i = q^i(x,t)$. We also know the familiar expression for kinetic energy, into which we can change variables by the chain rule:

$$v_i = \sum_j \frac{\partial x_i}{\partial q^j}\dot{q}^j + \frac{\partial x_i}{\partial t} \tag{10.0.7}$$

The kinetic energy is now

$$T = \sum_i \frac{1}{2}m_i {v_i}^2 = \sum_i \frac{1}{2}m_i \left(\sum_j \frac{\partial x_i}{\partial q_j}\dot{q}_j + \frac{\partial x_i}{\partial t}\right) \cdot \left(\sum_k \frac{\partial x_i}{\partial q_k}\dot{q}_k + \frac{\partial x_i}{\partial t}\right) \tag{10.0.8}$$

---

[1]Traditionally, this energy function is given the symbol h, to associate it with the Hamiltonian which we come to later in chapter 14. I have called it J in order to attribute it to Jacobi, whom sometimes gets overlooked in the discussion of analytical mechanics yet whose vast contribution is evident all throughout the subject.

Multiply out the brackets and tidy up some terms:

$$T = \sum_i \frac{1}{2} m_i \sum_j \sum_k \left(\frac{\partial x_i}{\partial q^j} \cdot \frac{\partial x_i}{\partial q^k} \right) \dot{q}^j \dot{q}^k$$

$$+ \sum_i m_i \sum_j \left(\frac{\partial x_i}{\partial q^j} \cdot \frac{\partial x_i}{\partial t} \right) \dot{q}^j + \frac{1}{2} \sum_i m_i \left(\frac{\partial x_i}{\partial t} \right)^2 \qquad (10.0.9)$$

If we apply the condition of equation 10.0.5 then we lose the last two terms. Physically, this means the transformation from cartesian to generalised coordinates is time invariant and is the same no matter when you do it; it is not an explicit function of time.

$$T = \sum_i \frac{1}{2} m_i \sum_{j,k} \left(\frac{\partial x_i}{\partial q^j} \cdot \frac{\partial x_i}{\partial q^k} \right) \dot{q}^j \dot{q}^k \qquad (10.0.10)$$

Now that we have developed an expression for kinetic energy that is **homogeneously quadratic** in velocity, we can start to work out J in equation 10.0.6 by taking the partial derivative, but, again, it need not be with respect to the same coordinate, so we might say $\partial T / \partial \dot{q}^l$. However the only situation when doing this gives us a non-zero answer is when q^j, q^k or both equal q^l. To account for this, we rewrite the problem, where, just for ease of calculation, we have removed the sum over i, the mass and a factor of a half, because we know they are just constants for this problem, and reduced it to the important bits:

$$\frac{\partial T}{\partial \dot{q}^l} = \frac{\partial}{\partial \dot{q}^l} \left[\sum_{j,k} \left(\frac{\partial x_i}{\partial q^j} \cdot \frac{\partial x_i}{\partial q^k} \right) \dot{q}^j \dot{q}^k \right] \qquad (10.0.11)$$

Next, take the two situations where we get non-zero terms in the sum, namely where $j = l$ and $k = l$:

$$\frac{\partial T}{\partial \dot{q}^l} = \frac{\partial}{\partial \dot{q}^l} \left(\sum_{j,l} \left[\frac{\partial x_i}{\partial q^j} \cdot \frac{\partial x_i}{\partial q^l} \right] \dot{q}^j \dot{q}^l + \sum_{l,k} \left[\frac{\partial x_i}{\partial q^l} \cdot \frac{\partial x_i}{\partial q^k} \right] \dot{q}^l \dot{q}^k \right) \qquad (10.0.12)$$

Now perform the differentiation:

$$\frac{\partial T}{\partial \dot{q}^l} = \left(\sum_{j,l} \left[\frac{\partial x_i}{\partial q^j} \cdot \frac{\partial x_i}{\partial q^l} \right] \dot{q}^l + \sum_{l,k} \left[\frac{\partial x_i}{\partial q^l} \cdot \frac{\partial x_i}{\partial q^k} \right] \dot{q}^k \right) \qquad (10.0.13)$$

All we need to do is to put the constants back in to finalise this; but, before we do this, we can further multiply by the lth general velocity and sum over l, as equation 10.0.3

tells us to do. This leaves us with two kinetic energy terms, once we put in the terms we left out:

$$\sum_l \dot{q}^l \frac{\partial T}{\partial \dot{q}^l} = \sum_{l,j} \sum_i \frac{1}{2} m_i \left[\frac{\partial x_i}{\partial q^j} \cdot \frac{\partial x_i}{\partial q^l} \right] \dot{q}^j \dot{q}^l + \sum_{l,k} \sum_i \frac{1}{2} m_i \left[\frac{\partial x_i}{\partial q^l} \cdot \frac{\partial x_i}{\partial q^k} \right] \dot{q}^l \dot{q}^k \qquad (10.0.14)$$

Note we can always write notationally $jk = kj$, so we can simplify this expression to

$$\sum_l \dot{q}^l \frac{\partial T}{\partial \dot{q}^l} = 2 \sum_{j,k} \sum_i \frac{1}{2} m_i \left[\frac{\partial x_i}{\partial q^j} \cdot \frac{\partial x_i}{q^k} \right] \dot{q}^j \dot{q}^k = 2T \qquad (10.0.15)$$

Plugging this all back into the Jacobi integral, we obtain

$$J = \sum_l \dot{q}^l \frac{\partial T}{\partial \dot{q}^l} - (T - U) = 2T - T + U \qquad (10.0.16)$$

which is the **total energy** $T + U$. So, to recap:

- The Jacobi energy function is the total energy if the potential is not a function of the generalised velocities $U(\boldsymbol{q})$ and the coordinate transformation is **scleronomic** (i.e. $\partial x_i / \partial t = 0$).
- The Jacobi energy function is conserved if the Lagrangian does not have explicit time dependance $\partial \mathscr{L} / \partial t = 0$.

What does this mean if there is no time dependance in converting the cartesian coordinates to the generalised counterparts? It means we have a **non-inertial** generalised coordinate system: if we specify a generalised coordinate at one time and then perform an experiment, the coordinates value is not the same at some other time that we might choose to start the same experiment. This is different from saying that the Lagrangian is time dependent, which means that the experiment is different at different times. The two conditions are expressed mathematically as follows:

$$\frac{\partial x(\boldsymbol{q}, t)}{\partial t} \neq 0 \qquad \text{and} \qquad \frac{\partial \mathscr{L}(\boldsymbol{q}, \dot{\boldsymbol{q}}, t)}{\partial t} \neq 0 \qquad (10.0.17)$$

For a one-dimensional system with a simple Lagrangian, we can convert the cartesian coordinate x into a generalised coordinate q if we take (q) with the inverse function being $q(x)$; then the cartesian velocity is given simply by the chain rule:

$$\frac{dx}{dt} = \frac{dx}{dq} \frac{dq}{dt} \qquad (10.0.18)$$

When we construct the Lagrangian, we obtain

$$\mathscr{L} = \frac{1}{2} m \left(\frac{dx}{dq} \right)^2 \dot{q}^2 - U(q) \qquad (10.0.19)$$

We have shown that the first term of the Jacobi energy integral equals $2T$; let's just explicitly show this for one generalised coordinate:

$$\frac{\partial T}{\partial \dot{q}} \dot{q} = m \left(\frac{\partial x}{\partial q} \right)^2 \dot{q}^2 \qquad (10.0.20)$$

which we recognise as $2T$ from before; with this substitution, J now becomes the total energy of the system, $2T - T + U = T + U$.

What about when the coordinate transformation is dependent of time, such that $x = x(q,t)$ and the inverse $q = q(x,t)$? First, we work out an expression for the cartesian velocity by taking the time derivative of x; we then substitute that result into the Lagrangian and expand the brackets before finding the expression[2] for J:

$$\dot{x} = \left(\frac{\partial x}{\partial q} \right) \dot{q} + \frac{\partial x}{\partial t} \qquad (10.0.21)$$

$$\mathscr{L} = \underbrace{\frac{1}{2} m \left[\left(\frac{\partial x}{\partial q} \right)^2 \dot{q}^2 + \left(\frac{\partial x}{\partial t} \right)^2 + 2 \left(\frac{\partial x}{\partial q} \right) \left(\frac{\partial x}{\partial t} \right) \dot{q} \right]}_{T} - U(x(q,t)) \qquad (10.0.22)$$

$$\frac{\partial T}{\partial \dot{q}} = m \left[\left(\frac{\partial x}{\partial q} \right)^2 \dot{q} + \left(\frac{\partial x}{\partial q} \right) \left(\frac{\partial x}{\partial t} \right) \right] \qquad (10.0.23)$$

$$\frac{\partial T}{\partial \dot{q}} \dot{q} = m \left[\left(\frac{\partial x}{\partial q} \right)^2 \dot{q}^2 + \left(\frac{\partial x}{\partial q} \right) \left(\frac{\partial x}{\partial t} \right) \dot{q} \right] \qquad (10.0.24)$$

$$\frac{\partial T}{\partial \dot{q}} \dot{q} - \mathscr{L} = \frac{1}{2} m \left(\frac{\partial x}{\partial q} \right)^2 \dot{q}^2 - \frac{1}{2} m \left(\frac{\partial x}{\partial t} \right)^2 + U(x(q,t)) \qquad (10.0.25)$$

We can now express J as the total energy *minus* some other terms. From this, it is quite clear that, in this case, the Jacobi energy function *is not the total energy of the system*; rather, it is some lesser energy of the system and may or may not be conserved:

$$J = T + U - m \left[\left(\frac{\partial x}{\partial q} \right) \left(\frac{\partial x}{\partial t} \right) \dot{q} + \left(\frac{\partial x}{\partial t} \right)^2 \right] \qquad (10.0.26)$$

The bottom line is, if the entire system is portrayed in the Lagrangian, then J will be the total energy $T + V$; if the Lagrangian does not represent a closed system with no external forces acting upon it, then it will not give the total energy and we can redefine our ideas about what we include as being part of the system until it is closed and comprises only internal forces.

---

[2] Again, as we have in all previous discussions, we are assuming the potential is velocity independent so that the partial derivative of the Lagrangian with respect to the general velocity is basically the same as saying the partial derivative of the kinetic energy with respect to the general velocity, ignoring the potential term (in case you think there is a discrepancy between the expression for J and what we actually calculate).

Chapter summary

- We derive the **Jacobi energy function** $J(q, \dot{q}, t)$ and claim that, under certain circumstances, it is the **total energy** of the system:

$$J(q, \dot{q}, t) = \sum_i \frac{\partial \mathscr{L}}{\partial \dot{q}^i} \dot{q}^i - \mathscr{L}(q, \dot{q}, t)$$

- We explore how it is a **conserved quantity** of the system when the Lagrangian is not an **explicit function of time**.

Exercise 10.1 Can we say that the Jacobi energy function for the harmonic oscillator Lagrangian is the total energy? We start with the following equations:

$$\mathscr{L} = \frac{1}{2}m\dot{q}^2 - \frac{1}{2}kq^2, \qquad J = \frac{1}{2}m\dot{q}^2 + \frac{1}{2}kq^2$$

Next, compute the partial derivative of the Lagrangian with respect to the generalised velocity and then multiply by $\dot{q}$ and subtract the Lagrangian:

$$J = m\dot{q}^2 - \left(\frac{1}{2}m\dot{q}^2 - \frac{1}{2}kq^2\right) = \frac{1}{2}m\dot{q}^2 + \frac{1}{2}kq^2$$

Exercise 10.2 Use the Lagrangian for the **central force problem** (see chapter 7, exercise 7.2) to compute the Jacobi energy function for the system:

$$\mathscr{L} = \frac{1}{2}mr^2\dot{\theta}^2 + \frac{1}{2}m\dot{r}^2 - U(r), \qquad J = \frac{1}{2}m\dot{r}^2 + \frac{1}{2}mr^2\dot{\theta}^2 + U(r)$$

Is this energy conserved?

The central force problem is slightly more complex, since we have two coordinates, but the process is exactly the same:

$$J = \frac{\partial \mathscr{L}}{\partial \dot{r}}\dot{r} + \frac{\partial \mathscr{L}}{\partial \dot{\theta}}\dot{\theta} - \mathscr{L} = m\dot{r}^2 + mr^2\dot{\theta}^2 - \frac{1}{2}mr^2\dot{\theta}^2 - \frac{1}{2}m\dot{r}^2 + U(r)$$

Since the Lagrangian doesn't contain explicit time dependance, we know J will be conserved; the proof can be found directly from the equation of motion (equation 7.0.22).

$$m\ddot{r} = \frac{L^2}{mr^3} - \frac{\partial U(r)}{\partial r}$$

If we integrate the first term on the right with respect to r but then differentiate the result, we can collect terms in d/dr:

$$m\ddot{r} = -\frac{d}{dr}\left(\frac{L^2}{2mr^2} + U(r)\right)$$

Multiplying both sides by $\dot{r} = dr/dt$ and collecting, we obtain

$$\frac{d}{dt}\left(\frac{1}{2}m\dot{r}^2\right) = -\frac{d}{dt}\left(\frac{L^2}{2mr^2} + U(r)\right)$$

Where we have canceled the dr from top and bottom of the right and (assuming constant mass) used $\frac{1}{2}m\frac{d}{dt}(\dot{r}r) = m\ddot{r}\dot{r}$.

$$\frac{d}{dt}\left(\frac{1}{2}m\dot{r}^2 + \frac{1}{2}\frac{L^2}{mr^2} + U(r)\right) = 0$$

Substitution of the angular momentum $L = m\dot{\theta}r^2$ yields

$$\frac{d}{dt}J = 0$$

which is the condition for a conserved quantity; hence, J is conserved, as we expected.

Exercise 10.3 Find the Jacobi function for a particle of charge e and mass m moving in a static electric field $\boldsymbol{E}$; does it represent the total energy? We start as follows:

$$\mathscr{L} = \frac{1}{2}m\dot{\boldsymbol{r}} \cdot \dot{\boldsymbol{r}} + e\dot{\boldsymbol{r}} \cdot \boldsymbol{A}(r) - e\varphi$$

We find that

$$J = \frac{\partial \mathscr{L}}{\partial \boldsymbol{r}}\dot{\boldsymbol{r}} - \mathscr{L}$$

$$= m\dot{\boldsymbol{r}} \cdot \dot{\boldsymbol{r}} + e\dot{\boldsymbol{r}} \cdot \boldsymbol{A} - \left[\frac{1}{2}m\dot{\boldsymbol{r}} \cdot \dot{\boldsymbol{r}} + e\dot{\boldsymbol{r}} \cdot \boldsymbol{A} - e\varphi\right]$$

$$= \frac{1}{2}m\dot{\boldsymbol{r}} \cdot \dot{\boldsymbol{r}} + e\varphi$$

In this case, J represents the total energy but may not be conserved.

Exercise 10.4 Consider a particle in an inertial frame (denoted (p)) of reference that is observed from a non-inertial (denoted (f)) frame where both coordinate axes coincide. Drawing on previous work, we let the velocity of the particle in the non-inertial frame be denoted by $\boldsymbol{v}$ and, using the transformation operator defined in equation 3.0.35 we can construct this from the quantities in the inertial frame $\dot{\boldsymbol{r}}, \boldsymbol{r}$, as follows:

$$\boldsymbol{v} = \dot{\boldsymbol{r}} + \boldsymbol{\omega} \times \boldsymbol{r} \tag{10.0.27}$$

where $\boldsymbol{\omega}$ is the angular velocity of the inertial frame. Further, we can rearrange the expression to obtain $\dot{\boldsymbol{r}}$ in terms of $\boldsymbol{v}$ by using $\dot{\boldsymbol{r}} = \boldsymbol{v} - (\boldsymbol{\omega} \times \boldsymbol{r})$. The Lagrangian for such a system can be written as,

$$\mathscr{L} = \frac{1}{2}mv^2 - U(\boldsymbol{r}) \qquad (10.0.28)$$

The Jacobi energy function as measured from the non-inertial frame $J_{(f)}$ takes its usual formulation

$$J_{(f)}(\boldsymbol{r}, \dot{\boldsymbol{r}}) = \frac{\partial \mathscr{L}}{\partial \dot{\boldsymbol{r}}} \cdot \dot{\boldsymbol{r}} - \mathscr{L}(\boldsymbol{r}, \dot{\boldsymbol{r}}) \qquad (10.0.29)$$

We know from previous work that

$$\frac{\partial \mathscr{L}}{\partial \dot{\boldsymbol{r}}} = m(\dot{\boldsymbol{r}} + \boldsymbol{\omega} \times \boldsymbol{r}) = m\boldsymbol{v} \qquad (10.0.30)$$

Therefore, we can construct the energy function as

$$J_{(f)}(\boldsymbol{r}, \dot{\boldsymbol{r}}) = m\boldsymbol{v} \cdot (\boldsymbol{v} - \boldsymbol{\omega} \times \boldsymbol{r}) - (\frac{1}{2}mv^2 - U(\boldsymbol{r}))$$

$$= \frac{1}{2}mv^2 + U(\boldsymbol{r}) - m\boldsymbol{v} \cdot (\boldsymbol{\omega} \times \boldsymbol{r}) \qquad (10.0.31)$$

The first two terms constitute the total energy in the inertial frame: let's denote this energy as $J_{(p)}$. The second term can be manipulated by using the properties of the triple product under cyclic permutations, in particular $\boldsymbol{A} \cdot (\boldsymbol{B} \times \boldsymbol{C}) = -\boldsymbol{B} \cdot (\boldsymbol{A} \times \boldsymbol{C})$. Note, however, that the angular momentum of the inertial frame equals $\boldsymbol{L} = (\boldsymbol{r} \times m\boldsymbol{v})$ and that $\boldsymbol{A} \times \boldsymbol{B} = -\boldsymbol{B} \times \boldsymbol{A}$; we thus arrive at

$$J_{(f)}(\boldsymbol{r}, \dot{\boldsymbol{r}}) = J_{(p)} - \boldsymbol{\omega} \cdot \boldsymbol{L} \qquad (10.0.32)$$

Therefore, the energy in the non-inertial frame is less than the energy in the inertial frame by $\boldsymbol{\omega} \cdot \boldsymbol{L}$. This term is largest when the two vectors are aligned using $\boldsymbol{A} \cdot \boldsymbol{B} = \|\boldsymbol{A}\|\|\boldsymbol{B}\| \cos\theta$; therefore, the energy of the system in the rotating frame is lowest when the particle's angular momentum aligns with the rotation axis. This effect is observed in nuclear chemistry and rotational-vibrational spectroscopy, as we discuss later when we consider the linear triatomic molecule (see chapter 12 exercise 12.3).

11

Symmetries & Lagrangian-Hamilton-Jacobi Theory

In this chapter, we discuss conservation laws in Lagrangian mechanics. Newtonian physics has no real mechanism by which we can identify conserved quantities, and this is a real shortfall, since often these quantities turn out to be of the utmost importance to the system, as we draw out later. In chapter 10 we proved the conservation of the Jacobi energy function by rearranging the Euler-Lagrange equation and then manipulating it a bit. We will see here that certain conservation laws are just particular examples of a more fundamental theory called **Noether's theorem**, after Amalie "*Emmy*" Noether, who first discovered it in 1918. In this way, analytical mechanics has a recipe for obtaining the conserved quantities of a system, one which also paves the way for symmetries in quantum field theory and general relativity.

11.1 Noether's Theorem

Noether's theorem is very simple and can be summarised in one sentence: for every **continuous symmetry** of the system, there is a **conserved quantity**. We will now spend some time investigating this and, of course, proving it for the general case.

We start by defining the **conjugate momentum** p_i for a coordinate q^i as the partial derivative of the Lagrangian $\mathscr{L}(\boldsymbol{q}, \dot{\boldsymbol{q}}, t)$ with respect to the generalised velocity of that coordinate:

$$p_i := \frac{\partial \mathscr{L}}{\partial \dot{q}^i} \qquad \textbf{conjugate momentum} \qquad (11.1.1)$$

It may seem a bit odd to call momentum 'conjugate' but you can see that, for $\mathscr{L} = \frac{1}{2}m\dot{q}^2$, this quantity is none other than the standard linear momentum, hence its name. Note that the indices are written as subscripts, since they are elements of the cotangent bundle, the dual space to the tangent bundle (ignore this for now). You have already met this before in the Lagrange equation of motion (see equation 7.0.14) which we might like to rewrite as

$$\frac{d}{dt}(p_i) - \frac{\partial \mathscr{L}}{\partial q^i} = 0 \qquad (11.1.2)$$

Lagrangian & Hamiltonian Dynamics. Peter Mann, Oxford University Press (2018).
© Peter Mann. DOI: 10.1093/oso/9780198822370.001.0001

Thinking closely about this expression, we have the makings of a conservation law. If the Lagrangian is independent (not a function of) a given coordinate q^i, then its conjugate momentum is **conserved** (i.e. it's a constant of the motion). Generalised coordinates for which this is true are called **cyclic** or **ignorable** coordinates (note that the name has nothing to do with circular motion).

Exercise 11.1 Using the above definitions, show that $\frac{d}{dt}(p\dot{q}) = \frac{d}{dt}\mathscr{L}$. (Hint: use the product rule.)

The implicit function theorem tells us that a function can be inverted, provided that the following Jacobian determinant is non-zero:

$$\det\left(\frac{\partial p_i}{\partial \dot{q}^j}\right) \neq 0 \qquad \textbf{Hessian condition}$$

When this condition holds, we may invert the relationship to express $\dot{q}^j$ as a function of (q, p, t). Substituting the definition of the conjugate momentum into this, we realise that we have the **Hessian condition** we discussed at the end of chapter 7. We will pick this back up in the discussion of the Dirac-Bergmann theory of constrained Hamiltonian dynamics, in chapter 21.

As we have seen, the Lagrangian dictates the dynamical equations of the system and, if it is independent of a given coordinate, then changing that coordinate a little isn't going to change the Lagrangian. We say that the coordinate is a **symmetry** of the Lagrangian. In this case, a symmetry of a function is a transformation of its variables that leaves the function exactly the same as before. Therefore, let us summarise:

> **Noether's theorem**: For every continuous symmetry of the Lagrangian, there is a conserved momentum conjugate to the cyclic coordinate.

We distinguish between **discrete** and **continuous** symmetries in Noether's theorem. A discrete symmetry is like rotating a trigonal bipyramidal molecule about the axial bonds (it has a C_3 axis) or an octahedral molecule about one of its C_4 axes; the point is, you can only do it in jumps. Meanwhile, a continuous symmetry is a smooth process; it is like rotating a CO_2 molecule about the internuclear axis. In physics, continuous symmetries are often a lot more interesting mathematically than discrete ones are and, importantly, Noether's theorem deals with continuous symmetries only; while this may seem to limit its power, make no mistake- Noether's theorem is of fundamental importance to theoretical physics!

So, a continuous symmetry is a smooth **transformation** that leaves the Lagrangian and, hence, the action invariant. You can think of it as a *mapping* from the solutions into the solutions of equations of motion and, as such, you may think of these transformations as a **Lie algebra** whose Lie group are the **generators** of the transformations.

However, I think we are getting ahead of ourselves a little bit. We will return to cyclic coordinates and conserved momenta when we discuss the Poisson bracket in chapter 17, but we will return to Noether's theorem when we discuss **field theory** in part IV. It is in the language of fields that Noether's theorem is best described and understood. There, we can talk about **proper, improper, quasi-, global** and **local** symmetries and that there are in fact two Noether theorems, conveniently called the **first** and **second Noether theorems**, respectively.

In literature and amongst physicists and mathematicians, Noether's theorem is widely regarded as one of the most important results to come out of Lagrangian mechanics. Indeed, Emmy Noether first derived her theory in the Lagrange formalism but it is equally applicable to Hamiltonian mechanics and beyond, including quantum mechanics, stochastic dynamics, particle physics (the origin of the term **strangeness** for quarks) and general relativity. So, although presented here in the language of the Lagrange formalism, the connection between symmetries and conservation laws is much deeper. The strong connection between symmetries and conservation laws is an idea that predates Noether, perhaps even to Galileo, and certainly to Lagrange, Jacobi, Hamilton, and so on, so you might wonder exactly why Noether gets the credit. However, it was Emmy Noether who proved the general relation between symmetries and conservation laws; a version of her work can be found in appendix A and in the section covering classical fields later on (see chapter 28. Not only this; she distinguished between two types of variation (global and local). Her work was held in the highest regard by Einstein, Weyl, Klein and Wigner, all giants in modern theoretical physics.

To prove Noether's theorem, we can start from the variation of the action functional $A[\boldsymbol{q}]$; invariance of the action is similar to invariance of the Lagrangian but easier to work with. In fact, we often term *symmetries* as invariance of the action but reserve the term *quasisymmetries* for invariance of the Lagrangian function[1]! We say Noether's theorem is derived from a **variational problem**, similar to the stationary action **principle** but *without* the vanishing boundary restrictions. Noether's theorem searches for those cases where the first-order variation in the action vanishes ($\delta A = 0$) without imposing the conditions of Hamilton's principle.

If we assume that $q^i(t)$ is a symmetry of the action, then we denote a smooth transformation by $q'(t) = q(t) + \delta q(t)$, where δq is an infinitesimal variation (continuous). Since Noether's variational problem is truly very general, we are also free to vary both the dependant (q^i) and independent variables (t) (i.e. for this case *only* the variations in time are allowed $\delta t \neq 0$ but we will leave this until appendix A). Thus we obtain

---

[1]We may transform the coordinates $q(t) \to q'(t)$ and/or the time $t \to t'$ in a transformation and, correspondingly, we can apply it to the action, equations of motion, the Lagrangian or even the solutions of the equation of motion! If the given object is invariant under the transformation, then that object has a symmetry. Two special cases are quasisymmetries: where the transformed object is the Lagrangian, and the interesting scenario where the solution coordinate $q(t)$ doesn't share a symmetry that the equations of motion have, which we call a **spontaneous symmetry breaking**.

$$\delta A = \left. \frac{\partial \mathscr{L}}{\partial \dot{q}^i} \delta q^i \right|_{t_0}^{t_1} + \int_{t_0}^{t_1} dt \left[\frac{\partial \mathscr{L}}{\partial q^i} - \frac{d}{dt} \left(\frac{\partial \mathscr{L}}{\partial \dot{q}^i} \right) \right] \delta q^i \qquad (11.1.3)$$

When evaluated on extremal paths, the Euler-Lagrange equation of motion vanishes and we are left with the term evaluated at the boundaries and the time variation:

$$\delta A[\boldsymbol{q}] = \frac{\partial \mathscr{L}}{\partial \dot{q}^i}(t_1)\delta q^i(t_1) - \frac{\partial \mathscr{L}}{\partial \dot{q}^i}(t_0)\delta q^i(t_0) = 0 \qquad (11.1.4)$$

This implies that both terms are equal to each other, but we didn't require the boundary variations to be zero and, since t_0 and t_1 are completely arbitrary, they must be constant in time:

$$\frac{\partial \mathscr{L}}{\partial \dot{q}^i}(t_1)\delta q^i(t_1) = \frac{\partial \mathscr{L}}{\partial \dot{q}^i}(t_0)\delta q^i(t_0) \qquad (11.1.5)$$

Hence,

$$\boxed{\frac{d}{dt}\left(\frac{\partial \mathscr{L}}{\partial \dot{q}^i} \delta q^i \right) = 0} \qquad (11.1.6)$$

This is Noether's theorem for the variation δq^i. Don't think, however, that $\delta A = 0$ in any old case; it is only so in those cases, where there is a conserved quantity: although we assumed that the variation in the action vanished, we see it is only the case when there is a conserved quantity.

There is a large body of modern research dedicated to **reduction techniques** whose general aim is to use cyclic coordinates and conserved momenta to reduce the dimensionality of the problem at hand. We discuss one such technique a little later called **Routhian reduction** (see chapter 16 section 16.2) but there are plenty more! We mentioned that there are symmetries not derived from Noether's theorem. One set of such symmetries is a result of the **Hojman-Harleston-Lutzky theorem**, which generates conserved quantities from non-Noetherian symmetries! For a review of these non-Noether symmetries, consult the work by Mimura and Nono (1983).

So far, we have been very vague about the nature of the transformation and the conserved quantity. We have seen plenty of previous examples where application of the Euler-Lagrange equation gave us a conserved quantity but we have not yet discussed what they actually are.

When discussing the Jacobi energy function J, we mentioned that, if the Lagrangian was time translationally invariant, then the energy was conserved. Time translational invariance leads to the conservation of energy, since time is viewed as a cyclic coordinate. But there are other examples: *translational invariance* leads to conservation of *linear momenta*, *rotational invariance* leads to conservation of *angular momenta*, and so on. What this is actually telling us is that the equations of classical mechanics don't, in general, pick favoured positions in space, time or direction so that is why these quantities are usually conserved.

The nature of the variation is another subtle area of Noether's theorem. When first learning about symmetries, it is often easiest to think of δq^i as a little boost in coordinates or as an infinitesimal rotation or a little time translation. In all these cases, we are (or at least I am) imagining δq^i as a **constant**, which is fine. However, why not let it be a **function** instead? Symmetries where the variations are functions (of q and or t) are called **local** symmetries because their value changes from point to point and their treatment is generally more complex. We will discuss this later, when we cover field theory in chapter 28. Please see appendix A for a proper derivation of Noether's theorem and some examples of using the Noether condition, including sone that will tie in with our discussion of field theory.

Exercise 11.2 Noether's theorem states that the invariance of a Lagrangian under an infinitesimal transformation leads to a conserved quantity. There is no parallel to this in Newtonian mechanics and, as a prime example we will consider the conservation of angular momentum. If we recall, we spent quite a while in chapter 3 proving that the total angular momentum is equal to the total torque of the external forces:

$$\frac{d}{dt}\vec{L}_{\text{total}} = \frac{d}{dt}\sum_i (\boldsymbol{r}_i \times \boldsymbol{p}_i) = \vec{\tau}_{\text{total}}^{(e)} \tag{11.1.7}$$

This was quite a contrived derivation and certainly not immediately obvious. Noether's theorem changes this. Imagine a system whose coordinate origin is entered on an axis of rotation.

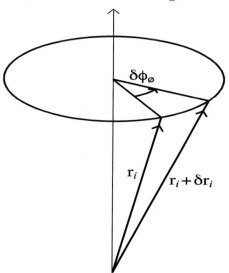

Fig. 11.1: The rotation of the system.

If we rotate a system by an angle $\delta\phi$ about an axis, where $\boldsymbol{\delta\phi}$ is a vector pointing along the axis of rotation, the magnitude is $\delta\phi$, according to figure 11.1. As we rotate the system, the ith particle will be displaced by δr_i and the velocity will change by δv_i where r_i and v_i are the initial positions and velocities. If the Lagrangian $\mathscr{L}(r,v)$ is invariant under this infinitesimal rotation, then $\delta\mathscr{L}=0$. This is enough to prove the conservation of angular momentum. We start by taking the variation of the Lagrangian:

$$\delta\mathscr{L}=\sum_i\left(\frac{\partial\mathscr{L}}{\partial r_i}\cdot\delta r_i+\frac{\partial\mathscr{L}}{\partial v_i}\cdot\delta v_i\right) \tag{11.1.8}$$

We use the Lagrange equation to replace the first term with the time derivative of the conjugate momentum:

$$\delta\mathscr{L}=\sum_i\left(\dot{p}_i\cdot\delta r_i+p_i\cdot\delta v_i\right) \tag{11.1.9}$$

This is the same equation 11.1.6. We now let $\delta r_i=\delta\phi\times r_i$ and $\delta v_i=\delta\phi\times v_i$:

$$\delta\mathscr{L}=\sum_i\left(\dot{p}_i\cdot(\delta\phi\times r)+p_i\cdot(\delta\phi\times v_i)\right)$$
$$=\sum_i\delta\phi\cdot(r_i\times\dot{p}_i+\dot{r}_i\times p_i)$$
$$=\delta\phi\cdot\sum_i\frac{d}{dt}(r_i\times p_i) \tag{11.1.10}$$

We know that $\delta\phi$ is non-zero; therefore, if the Lagrangian is invariant the time derivative vanishes:

$$\frac{d}{dt}\sum_i(r_i\times p_i)=0 \tag{11.1.11}$$

Therefore, we prove that the **isotropy of space** leads to the conservation of angular momentum. This was a much more intuitive derivation than talking about torques and forces.

11.2 Gauge Theory

From our elementary theory, we know that the kinetic and potential energies are not unique functions; rather, they depend on the choice of generalised coordinates used to describe them and, as a result, the Lagrangian and the Jacobi energy function are not unique, either. In reality, lots of different Lagrangians can be used to describe the same physical system; as we will see in this section, the **total time derivative** of some arbitrary function can be added to any Lagrangian without changing the physics of what it describes; this is because it disappears when we calculate stationary points in the action. This means that a single physical phenomena can be described by an undetermined number of Lagrangians that yield the same Euler-Lagrange equations. This non-uniqueness is called **gauge freedom**.

Let us consider some arbitrary function of configuration space variables $\varphi = \varphi(q^1, q^2, \ldots, q^n, t)$ and its time derivative,

$$\frac{d\varphi}{dt} = \sum_i \frac{\partial \varphi}{\partial q^i} \frac{\partial q^i}{\partial t} + \frac{\partial \varphi}{\partial t}$$

Then consider the derivative of this with respect to $\dot{q}^i$:

$$\frac{\partial}{\partial \dot{q}^i} \frac{d\varphi}{dt} = \frac{\partial \varphi}{\partial q^i}$$

Then, again take the time derivative and utilise the above two relations to obtain

$$\frac{d}{dt} \frac{\partial}{\partial \dot{q}^i} \left(\frac{d\varphi}{dt} \right) = \frac{d}{dt} \left(\frac{\partial \varphi}{\partial q^i} \right) = \frac{\partial}{\partial q^i} \left(\frac{d\varphi}{dt} \right) \tag{11.2.1}$$

but note carefully how we swap the order of differentiation around in the last term. It does not change the result and we can therefore define a quantity that satisfies the Lagrange equation, namely $d\varphi/dt$:

$$\frac{d}{dt} \frac{\partial}{\partial \dot{q}^i} \left(\frac{d\varphi}{dt} \right) - \frac{\partial}{\partial q^i} \left(\frac{d\varphi}{dt} \right) = 0 \tag{11.2.2}$$

This forces us to reconsider the definition of the Lagrangian! If we can go around willy-nilly adding the total time derivatives of functions on the end of the Lagrangian *and still satisfy the same equations of motion*, then the Lagrangian itself will always be uncertain up to a total time derivative of some function $d\varphi/dt$. Therefore, the Lagrangian is not unique for a physical system and we can re-write it as

$$L = \mathscr{L} + \frac{d\varphi}{dt} \tag{11.2.3}$$

So, we have seen that an arbitrary total derivative will satisfy the Lagrange equation, but how do we know that we can indeed express $\dot{\varphi}$ in equation 11.2.2 with the Lagrange equation? In other words, how do we know that we can add this and still satisfy the same equations of motion? The answer lies in the action principle. If we define the action in terms of the new Lagrangian we obtain

$$A' = \int_{t_1}^{t_2} \mathscr{L}(\boldsymbol{q}, \dot{\boldsymbol{q}}, t) dt + \int_{t_1}^{t_2} \frac{d\varphi}{dt} dt$$

$$= \int_{t_1}^{t_2} \mathscr{L}(\boldsymbol{q}, \dot{\boldsymbol{q}}, t) dt + \int_{t_1}^{t_2} d\varphi$$

$$= A + \varphi(\boldsymbol{q}(t_2), t_2) - \varphi(\boldsymbol{q}(t_1), t_1)$$

The only contribution from the function is at the end points and therefore the function does not contribute to the allowed variations of the action, since the variation of a constant is zero. We call such terms **surface terms**. As such, A' is only stationary when A is stationary, and no new criteria. The critical points of both actions are equivalent! This means that the dynamics of the system is invariant with respect to a **gauge transformation** of the Lagrangian. The gauge refers to the redundant degrees of freedom that the system has left over after choosing the generalised coordinates. Another way to express this is by saying that the action is invariant up to surface terms. In some cases, we require additional boundary conditions to admit the same Euler-Lagrange equation.

A gauge transformation is not a coordinate transformation, instead, it changes the Lagrangian itself for a given set of coordinates by bringing in forgotten dimensions. Even though two different Lagrangians can lead to the same equations of motion, their conjugate momenta will be different and, as a result, they will give *different* equations of motion in Hamilton's formalism. This is another reason why gauge transformations are different from point transformations. We remark that, in certain special cases, an equation of motion may posses several Lagrangians that are not related via gauge transformations. These coincidental examples are sometimes important and sometimes just trivialities of mathematics.

11.3   Isotropic Symmetries

In this section, we consider the conserved quantities that arise outside of Noether's theorem. These symmetries were developed by D. Currie, E. Saletan, S. Hojman and M. Lutzky in the 1980s and 1990s and generalise the set of integrals we can obtain from equivalent non-singular Lagrangians. The theory is based around **s-equivalent Lagrangians**, which are defined by the coincidence of the solutions to their equations of motion. For two non-singular s-equivalent Lagrangians $\mathscr{L}$ and L, we may write

$$\frac{d}{dt}\left(\frac{\partial \mathscr{L}}{\partial \dot{q}^i}\right) - \frac{\partial \mathscr{L}}{\partial q^i} = \Lambda_i^j \left[\frac{d}{dt}\left(\frac{\partial L}{\partial \dot{q}^i}\right) - \frac{\partial L}{\partial q^i}\right] \tag{11.3.1}$$

The theorem states that the trace of all integer powers of Λ are constants of the motion. To show that this relation holds, we write the Euler-Lagrange equations for both Lagrangians:

$$\frac{\partial^2 \mathscr{L}}{\partial \dot{q}^i \partial \dot{q}^j}\ddot{q}^j + \frac{\partial^2 \mathscr{L}}{\partial \dot{q}^i \partial \dot{q}^j}\dot{q}^j - \frac{\partial \mathscr{L}}{\partial q^i} = 0 \tag{11.3.2}$$

$$\frac{\partial^2 L}{\partial \dot{q}^i \partial \dot{q}^j}\ddot{q}^j + \frac{\partial^2 L}{\partial \dot{q}^i \partial \dot{q}^j}\dot{q}^j - \frac{\partial \mathscr{L}}{\partial q^i} = 0 \tag{11.3.3}$$

We then define the Hessian matrix elements W_{ij}, $\overline{W}_{ij}$ and their inverse elements U^{ij}, $\overline{U}^{ij}$:

$$W_{ij} = \frac{\partial^2 \mathscr{L}}{\partial \dot{q}^i \partial \dot{q}^j}, \qquad \overline{W}_{ij} = \frac{\partial^2 L}{\partial \dot{q}^i \partial \dot{q}^j} \tag{11.3.4}$$

We then rearrange equation 11.3.2 for $\ddot{q}^j$ and pop this into equation 11.3.3.

$$W_{ij}\overline{U}^{ji}\left(\frac{\partial L}{\partial q^i} - \frac{\partial^2 L}{\partial \dot{q}^j \partial q^i}\dot{q}^i\right) = \frac{\partial \mathscr{L}}{\partial q^i} - \frac{\partial^2 \mathscr{L}}{\partial \dot{q}^i \partial q^j}\dot{q}^j \tag{11.3.5}$$

We can add $\ddot{q}\partial_{\dot{q}\dot{q}}L = \ddot{q}\Lambda\partial_{\dot{q}\dot{q}}\mathscr{L}$ to arrive at the result shown in equation 11.3.1 and identify Λ_i^j:

$$\Lambda_i^j = W_{ij}\overline{U}^{ji} \tag{11.3.6}$$

Showing that the time derivative of Λ_i^j vanishes is an involved process and can readily be found in D. G. Currie and E. J. Saletan's (1966) book, *q-Equivalent Particle Hamiltonians. I. The Classical One-Dimensional Case* (1966).

11.4   Caratheodory-Hamilton-Jacobi theory

An important theory within classical mechanics is the **Hamilton-Jacobi theory** and it is usually discussed within the context of canonical/Hamiltonian dynamics. Here we take the unorthodox step of formulating the Hamilton-Jacobi equation from **Carathéodory's equivalent Lagrangian method** and use it to solve the dynamics of two simple systems, assuming that the Lagrangians are regular and hence that the Hessian condition is satisfied.

We are by now well aware that we can formulate from a Lagrangian function $\mathscr{L}(q, \dot{q}, t)$ another one related by a total time derivative of some function such that the action integrals share critical points. Let $L(q, \dot{q}, t)$ be such a Lagrangian, allowing us to write,

$$L(q, \dot{q}, t) = \mathscr{L}(q, \dot{q}, t) - \frac{d\mathcal{A}(q, t)}{dt} \tag{11.4.1}$$

If we choose $\mathcal{A}(q, t)$ such that L is made stationary, then the corresponding values of this function, q and $\dot{q}$, are exactly the ones that solve the variational problem. Hence, we are solving the usual variational problem in a rather sneaky way! We look for combinations of $\mathcal{A}(q, t)$, q^i and $\dot{q}^i$ such that the following conditions hold:

$$L(q, \dot{q}, t)\big|_{\dot{q}^i = \alpha^i} = 0, \quad \text{and} \quad \frac{\partial L}{\partial \dot{q}^i}\bigg|_{\dot{q}^i = \alpha^i} = 0 \tag{11.4.2}$$

From the first equality, we generate

$$\frac{\partial \mathcal{A}(q, t)}{\partial t} = \mathscr{L}(q, \dot{q}, t) - \frac{\partial \mathcal{A}(q, t)}{\partial q^i}\dot{q}^i \tag{11.4.3}$$

The right-hand side is slightly more tricky, since we have to product rule the differential operator and the argument d/dt and $\mathcal{A}$:

$$\frac{\partial L}{\partial \dot{q}^i} = \frac{\partial \mathscr{L}}{\partial \dot{q}^i} - \frac{\partial}{\partial \dot{q}^i}\left(\frac{d\mathcal{A}(q,t)}{dt}\right)$$

$$= \frac{\partial \mathscr{L}}{\partial \dot{q}^i} - \frac{\partial \dot{\mathcal{A}}}{\partial \dot{q}^i} - \frac{d}{dt}\left(\frac{\partial \mathcal{A}}{\partial \dot{q}^i}\right)$$

$$= \frac{\partial \mathscr{L}}{\partial \dot{q}^i} - \frac{\partial \mathcal{A}}{\partial q^i} - \frac{d}{dt}\left(\frac{\partial \mathcal{A}}{\partial \dot{q}^i}\right) \qquad (11.4.4)$$

In the final step, we have "cancelled the dot", as shown in appendix C; in addition, we realise that $\mathcal{A} \neq \mathcal{A}(\dot{q})$ so that $\partial \mathcal{A}/\partial \dot{q}^i = 0$ and hence, upon evaluation of the expression at $\dot{q}^i = \alpha^i$, we obtain $\partial \mathcal{A}/\partial q^i = \partial \mathscr{L}/\partial \dot{q}^i$:

$$\frac{\partial \mathcal{A}}{\partial t} + J\left(q, \frac{\partial \mathcal{A}}{\partial q^i}, t\right) = 0 \quad \text{with} \quad \frac{\partial \mathcal{A}}{\partial q^i} = \frac{\partial \mathscr{L}}{\partial \dot{q}^i}, \qquad J = \frac{\partial \mathcal{A}}{\partial q^i}\dot{q} - \mathscr{L} \qquad (11.4.5)$$

The two equations provide the fundamental basis of the **Lagrangian Hamilton-Jacobi** approach and the focus is now to find the function $\mathcal{A}(q,t)$ that satisfies these equations and which should contain within it all solutions to the dynamics of the problem. The general approach is to find the Lagrangian, find $\partial \mathcal{A}/\partial q^i$ and then write the "Lagrangian-Hamilton-Jacobi" equation; this will be introduced in the Hamiltonian formalism in chapter 19.

Chapter summary

- In this chapter, we introduced the **conjugate momentum** $p_i = \partial \mathscr{L}/\partial \dot{q}^i$ for a coordinate q^i.

- The strong connection between **symmetries** and **conservation laws** is explored through **Noether's variational problem**. While Hamilton's variational principle forces the variations to vanish at the boundary points, Noether's considers broader variations and investigates what general conditions are required to make first-order variation in the action vanish.

- The proof of Noether's theorem is a little involved and so is left until appendix A.

- The Lagrangian was shown to be unique only up to a total time derivative of a function. It is shown that this extra term vanishes on-shell and leads to the same equations of motion. This is known as **gauge invariance**.

- The existence of the Hamilton-Jacobi problem in a Lagrangian setting was investigated and used to solve particle dynamics in a different way from that of the traditional Euler-Lagrange approach.

Exercise 11.3 As we know, the Lagrangian for the harmonic oscillator is a function of $\dot{q}$ and q, as shown below:

$$\mathscr{L}_{HO} = \frac{1}{2}m\dot{q}^2 - \frac{1}{2}m\omega^2 q^2$$

This yields the familiar Euler-Lagrange equation of motion, $\ddot{q} = -\omega^2 q$. If we now modify this Lagrangian by adding $\dot{q}$, we can see that the Lagrange equations of motion are invariant:

$$L = \mathscr{L}_{HO} + \dot{q}$$

The Euler-Lagrange equations are linear and we can write $EL[L] = EL[\mathscr{L}_{HO}] + EL[\dot{q}]$ and hope that the last term will disappear:

$$EL[\dot{q}] = \frac{d}{dt}\left(\frac{\partial\dot{q}}{\partial\dot{q}}\right) - \frac{\partial\dot{q}}{\partial q} = 0$$

Since the generalised velocity is independent of the generalised coordinate, the last term does in fact vanish, meaning that both Lagrangians correspond to the same equation of motion, giving us gauge freedom. We expect this, since $\dot{q}$ is a total derivative.

Exercise 11.4 The Lagrangian for a particle of mass m free to move in one dimension is simply the kinetic energy $\mathscr{L}(\dot{q}) = \frac{1}{2}m\dot{q}^2$. Does the following Lagrangian also describe the free particle?

$$\mathscr{L}(q,\dot{q}) = \frac{1}{2}m\dot{q}^2 - q\dot{q} \qquad (11.4.6)$$

On the surface, we might say that it is a different Lagrangian; however, we can express $q\dot{q}$ as a total time derivative $q\dot{q} = \frac{1}{2}\frac{d}{dt}q^2$. Therefore, both Lagrangians correspond to the same equations of motion; compute them!

Exercise 11.5 In this example, we consider the gauge invariance of the Lagrangian, admitting the same equations of motion only when additional boundary conditions are imposed. Consider

$$L = \mathscr{L}(q,\dot{q}) + \frac{d}{dt}\left(\sum_i q^i\frac{\partial\mathscr{L}}{\partial\dot{q}^i}\right) \qquad (11.4.7)$$

Inputting this Lagrangian into the stationary action principle and then cancelling the dt/dt factor gives us, in addition to the Euler-Lagrange equation for $\mathscr{L}$, the following exact derivative:

$$EL[L] = EL[\mathscr{L}] + \delta\left[\sum_i q^i\frac{\partial\mathscr{L}}{\partial\dot{q}^i}\right]_1^2 \qquad (11.4.8)$$

The ultimate objective is to prove the vanishing of the second term. To do this, we expand the variation term:

$$\delta\left[\sum_i q^i\frac{\partial\mathscr{L}}{\partial\dot{q}^i}\right]_1^2 = \sum_i\left[\delta q^i\frac{\partial\mathscr{L}}{\partial\dot{q}^i} + q^i\delta\left(\frac{\partial\mathscr{L}}{\partial\dot{q}^i}\right)\right]_1^2 = \sum_i\delta q^i\frac{\partial\mathscr{L}}{\partial\dot{q}^i}\Big|_1^2 + \sum_{ij}q^i\left[\frac{\partial^2\mathscr{L}}{\partial q^i\partial\dot{q}^j}\delta q^j + \frac{\partial^2\mathscr{L}}{\partial\dot{q}^i\partial\dot{q}^j}\delta\dot{q}^j\right]_1^2$$

In order for this term to vanish we have to impose the additional boundary condition that $\delta \dot{q} = 0$ when evaluated at $2, 1$. The alternative is to assume $q^i = 0 \; \forall \; i$, which isn't very profitable.

Exercise 11.6 Consider the following Lagrangian in spherical polar coordinates and rewrite it in terms of the conjugate momentum for each coordinate:

$$\mathscr{L}(r, \theta, \phi) = \frac{1}{2} m(\dot{r}^2 + r^2 \dot{\theta}^2 + r^2 \dot{\phi}^2 \sin^2 \theta) - U(r, \theta, \phi) \qquad (11.4.9)$$

The first step is to compute the conjugate momentum for each coordinate:

$$p_r = \frac{\partial \mathscr{L}}{\partial \dot{r}} = m\dot{r}, \qquad p_\theta = \frac{\partial \mathscr{L}}{\partial \dot{\theta}} = mr^2 \dot{\theta}, \qquad p_\phi = \frac{\partial \mathscr{L}}{\partial \dot{\phi}} = mr^2 \dot{\phi} \sin^2 \theta \qquad (11.4.10)$$

These are then used to construct the following expression for the Lagrangian:

$$\mathscr{L}(r, \theta, \phi, p_r, p_\theta, p_\phi) = \frac{p_r^2}{2m} + \frac{1}{2I} \left(p_\theta^2 + \frac{p_\phi^2}{\sin^2 \theta} \right) - U(r, \theta, \phi) \qquad (11.4.11)$$

where the moment of inertia is taken as $I = mr^2$. This function is now a function of the coordinates and their conjugate momenta. Does it give rise to the correct equations of motion?

Exercise 11.7 Consider two Lagrangians that are not related by a total time derivative (gauge transformation) but nevertheless lead to the same Euler-Lagrange equations. It should not surprise us that two different Lagrangians can lead to the same equations of motion, since we can always multiply $\mathscr{L}$ by a constant $\lambda \neq 1$ to obtain $\mathscr{L}' = \lambda \mathscr{L}$, which generates the same Euler-Lagrange equations. However, the difference between the two Lagrangians is not a total time derivative $\mathscr{L}' - \mathscr{L} = (\lambda - 1)\mathscr{L}$.

Consider an isotropic oscillator that can oscillate independently in two degrees of freedom such that there is no coupling between the oscillations; we say it is **separable**. The following two Lagrangians both lead to the same equations of motion, where $\omega^2 = k/m$:

$$\mathscr{L}_A = \frac{1}{2} [\dot{q}_1^2 + \dot{q}_2^2 - \omega^2(q_1^2 + q_2^2)] \qquad \text{and} \qquad \mathscr{L}_B = \dot{q}_1 \dot{q}_2 - \omega^2 q_1 q_2 \qquad (11.4.12)$$

The equations of motion are in both cases found to be

$$\ddot{q}_1 + \omega^2 q_1 = 0 \qquad \text{and} \qquad \ddot{q}_2 + \omega^2 q_2 = 0 \qquad (11.4.13)$$

If we perform a point transformation to polar coordinates $(q_1, q_2) \rightarrow (r, \theta)$, we can see from the Lagrangian that the conjugate momentum to θ is a conserved quantity:

$$\mathscr{L}_A(q_1, q_2) = \frac{1}{2} m(\dot{r}^2 + r^2 \dot{\theta}^2) - \frac{1}{2} kr^2, \qquad \Rightarrow \qquad \frac{d}{dt}(p_\theta) = \frac{d}{dt}(mr^2 \dot{\theta}) = 0 \qquad (11.4.14)$$

This is exactly the angular momentum $\vec{L}$ for the system $\vec{L} = q_1 \dot{q}_2 - q_2 \dot{q}_1$. An inconsistency now arises when we ask, to what does this symmetry correspond? If we use $\mathscr{L}_A$, we will obtain a different symmetry than if we use $\mathscr{L}_B$. A more trivial example is the system $\mathscr{L}_A = q^2\dot{q}^4$ and $\mathscr{L}_B = q^3\dot{q}^6$; for which it can be shown that both Lagrangians lead to the same equations of motion by careful application of the product rule $q\dot{q}^2[\dot{q}^2 + 2q\ddot{q}] = 0$.

As an exercise, consider the equations of motion of the following Lagrangian functions:

$$\mathscr{L}_A = \frac{1}{2}(\dot{q} + q\omega\tan\omega t)^2, \qquad \mathscr{L}_B = \frac{1}{q^2}(q\cos t - \dot{q}\sin t)\ln\left(\frac{q\cos t - \dot{q}\sin t}{q\sin t + \dot{q}\cos t}\right),$$

Exercise 11.8 If the action is invariant under an infinitesimal time transformation $t \to t\lambda$ such that at the boundary points $\lambda = 0$, then the energy function J vanishes identically. This is an example of a **parametrisation invariance**. This cannot be a general result, since we are well aware that, in general, $J \neq 0$. Assuming that the transformation is infinitesimal, we can linearise it to $\delta q^i = q^i + \dot{q}^i\lambda$ and $\delta\dot{q}^i = \dot{q}^i + \frac{d}{dt}(\dot{q}\lambda)$, where $\delta t = t + \lambda$:

$$\delta A = \int_1^2 dt\left[\frac{\partial\mathscr{L}}{\partial q^i}(q^i + \dot{q}^i\lambda) + \frac{\partial\mathscr{L}}{\partial\dot{q}^i}\left(\dot{q}^i + \frac{d}{dt}(\dot{q}^i\lambda)\right) + \frac{\partial\mathscr{L}}{\partial t}(t + \lambda)\right] \tag{11.4.15}$$

$$- \int_1^2 dt\left[\frac{\partial\mathscr{L}}{\partial q^i}q^i + \frac{\partial\mathscr{L}}{\partial\dot{q}^i}\dot{q}^i + \frac{\partial\mathscr{L}}{\partial t}t\right]$$

Where $\delta A = A[q + \delta q] - A[q]$,

$$= \int_1^2\left[\left(\frac{\partial\mathscr{L}}{\partial q^i}\dot{q}^i + \frac{\partial\mathscr{L}}{\partial\dot{q}^i}\ddot{q}^i + \frac{\partial\mathscr{L}}{\partial t}\right)\lambda + \left(\frac{\partial\mathscr{L}}{\partial\dot{q}^i}\dot{q}^i\right)\frac{d\lambda}{dt}\right]$$

$$= \int_1^2 dt\left[\frac{d\mathscr{L}}{dt}\lambda + \left(\frac{\partial\mathscr{L}}{\partial\dot{q}^i}\dot{q}^i\right)\frac{d\lambda}{dt}\right] \tag{11.4.16}$$

Then consider the product rule $\frac{d}{dt}(\mathscr{L}\lambda) = \mathscr{L}\frac{d\lambda}{dt} + \lambda\frac{d\mathscr{L}}{dt}$,

$$= \mathscr{L}\lambda\Big|_1^2 + \int_1^2 dt\left(\frac{\partial\mathscr{L}}{\partial\dot{q}^i}\dot{q}^i - \mathscr{L}\right)\frac{d\lambda}{dt} \tag{11.4.17}$$

The first term is evaluated on the boundary where we set $\lambda = 0$. By imposition of the invariance for any arbitrary transformation, we necessarily require that $J = 0$. Such symmetries are called **non-dynamical symmetries** and are vastly different from the conservation laws that we are familiar with, where $J = $ constant. The symmetries we are familiar with arise due to invariance of the action under the motion of the system and lead to conserved Noether currents (for continuous symmetries). Non-dynamical symmetries, in contrast, do not assume that the equations of motion have been satisfied and give rise to **identities** (in the above example, $J = 0$) holding for all trajectories, not just physical ones. Invariance of the action $\delta A = 0$ under reparameterisation $t \to \beta$ implies that there are solutions to the equations of motion other than $q^i(t)$ since $w^i(t) = q^i(\beta(t))$ that lead to zero-energy functions, implying that $q^i(t)$ are 0-forms on the timeline $\mathbb{R}$. Non-dynamical symmetries are therefore associated with redundant descriptions of the system, that is, extra degrees of freedom, and are therefore a special type of gauge theory.

Exercise 11.9 It is a property of coordinate reparameterisation that involves derivatives that the equations of motion change order. Consider the behaviour of the Lagrangian $\mathcal{L} = \frac{1}{2}\dot{q}^2$ under a reparameterisation $q = Q + \epsilon\dot{Q}$. Show that the equations of motion are first order in q and second order in the new variables.

Exercise 11.10 When the Lagrangian depends on higher-order time derivatives of the generalised coordinate $\mathcal{L}(q, \dot{q}, \ddot{q}, \ldots, q^{(n)})$, Hamilton's variational principle gives us the Ostrogradsky equation. The definition of conjugate momenta also changes to reflect this higher-order dependence. The **Ostrogradsky momentum** is given by,

$$p_i = \sum_{j=i}^{n} (-1)^{j-i} \left(\frac{d}{dt}\right)^{j-i} \frac{\partial \mathcal{L}}{\partial q^{(j)}} \tag{11.4.18}$$

Show that for $\mathcal{L}(q, \dot{q}, \ddot{q})$, and hence $i = 0, 1, 2$, that the following Ostrogradsky momenta are obtained:

$$p_1 = \frac{\partial \mathcal{L}}{\partial \dot{q}} - \frac{d}{dt}\left(\frac{\partial \mathcal{L}}{\partial \ddot{q}}\right) \quad \text{and} \quad p_2 = \frac{\partial \mathcal{L}}{\partial \ddot{q}} \tag{11.4.19}$$

Exercise 11.11 Consider a Lagrangian depending on a single variable $q(t)$, and on its time derivatives to nth order:

$$A[q] = \int_{t_1}^{t_2} \mathcal{L}\left(q, \dot{q}, \ddot{q}, \ldots, \frac{d^n q}{dt^n}, t\right) dt \tag{11.4.20}$$

Using a Noether variation $q(t) \to q'(t')$, such that there is a symmetry in the action $A[q] = A'[q']$, derive the Noether condition.

The first step is to insert the variation into the action integral:

$$A'[q'] = \int_{t_1}^{t_2} \mathcal{L}\left(q + \delta_0 q, \dot{q} + \frac{d}{dt}(\delta_0 q), \ldots, q^{(n)} + \frac{d^n}{dt^n}(\delta_0 q), t\right) + \mathcal{L}\delta t \Big|_{t_1}^{t_2} \tag{11.4.21}$$

The quantity $\delta A = A'[q'] - A[q]$ becomes:

$$\delta A = \sum_{i=0}^{n} \int_{t_1}^{t_2} \left\{ \frac{d^i(\delta_0 q)}{dt^i} \frac{\partial \mathcal{L}}{\partial q^{(i)}} \right\} dt + \mathcal{L}\delta t \Big|_{t_1}^{t_2} \tag{11.4.22}$$

Integration of this expression by parts results in:

$$\delta A = \int_{t_1}^{t_2} \delta q \sum_{i=0}^{n} (-1)^i \frac{d^i}{dt^i}\left(\frac{\partial \mathcal{L}}{\partial q^{(i)}}\right) dt + \left[\mathcal{L}\delta t - \sum_{i=1}^{n} \sum_{j=1}^{i} (-1)^j \frac{d^{i-j}(\delta_0 q)}{dt^{i-j}} \frac{d^{j-1}}{dt^{j-1}}\left(\frac{\partial \mathcal{L}}{\partial q^{(i)}}\right)\right]_{t_1}^{t_2}$$

In the case that $n = 2$, use this result to show that the Noether condition is given by

$$\delta A = \int_{t_1}^{t_2} \left\{ \frac{\partial \mathcal{L}}{\partial q} - \frac{d}{dt}\left(\frac{\partial \mathcal{L}}{\partial \dot{q}}\right) + \frac{d^2}{dt^2}\left(\frac{\partial \mathcal{L}}{\partial \ddot{q}}\right) \right\} \delta_0 q \, dt + \left\{ \frac{\partial \mathcal{L}}{\partial \dot{q}} - \frac{d}{dt}\left(\frac{\partial \mathcal{L}}{\partial \ddot{q}}\right) \right\} \delta_0 q + \frac{\partial \mathcal{L}}{\partial \ddot{q}} \delta_0 \dot{q}$$

Exercise 11.12 Show that the two-dimensional harmonic oscillator Lagrangian:

$$\mathscr{L} = \frac{m}{2}(\dot{x}^2 + \dot{y}^2) - \frac{k}{2}(x^2 + y^2) \qquad (11.4.23)$$

exhibits rotational invariance which leads to the following conserved angular momentum:

$$L_z = m(x\dot{y} - y\dot{x}) \qquad (11.4.24)$$

12
Near-Equilibrium Oscillations

In this chapter, we explore small **perturbations** (small nudges or tiny shifts) to an equilibrium point in configuration space. We consider only leading-order terms contributing to the Taylor expansion of the Lagrangian and which turn out to be **harmonic**. Therefore, near-equilibrium dynamics is governed by **oscillatory** behaviour; hence, the importance of the harmonic oscillator to general theoretical physics. In this chapter, we require a little bit of linear algebra when dealing with matrices, as well as an understanding of differential equations, but that shouldn't hold us back from the physics of what is going on.

The first thing to appreciate is, what is meant by an **equilibrium point** on a configuration manifold? We assume that the potential is conservative and smooth up to second order for the following discussion. A point in configuration space is an *equilibrium point* q_0^i if the following condition holds:

$$-\frac{\partial U}{\partial q^i}\bigg|_{q_0^i} = 0 \qquad \forall i \tag{12.0.1}$$

That is to say, the point is a critical point of the potential energy, or the **generalised force** vanishes. The equilibrium point is **stable** if it is a local minima, **unstable** if it is a local maxima and **static** if $\dot{q}_0^i = 0$ (see figure 12.1).

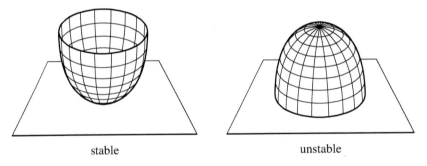

stable   unstable

Fig. 12.1: Stable and unstable equilibrium points.

Lagrangian & Hamiltonian Dynamics. Peter Mann, Oxford University Press (2018).
© Peter Mann. DOI: 10.1093/oso/9780198822370.001.0001

If we perturb a static, stable potential slightly, then we can use a Taylor expansion to characterise the small variations in the **near-equilibrium potential**. Any constant terms or total derivatives are not important in this stability analysis. We need a small quantity to expand about, so we choose the difference between the origin of the equilibrium configuration and a small variation to that: $\eta_i = q^i - q_0^i$ and $\dot{\eta}_i = \dot{q}^i - \dot{q}_0^i = \dot{q}^i$. We thus obtain

$$
U(\boldsymbol{q}) = U(q_0^i) + \sum_i \frac{\partial U}{\partial q^i}\bigg|_0 \eta_i + \sum_{i,j} \frac{1}{2}\left(\frac{\partial^2 U}{\partial q^i \partial q^j}\right)\bigg|_0 \eta_i \eta_j
$$

$$
+ \sum_{i,j,k} \frac{1}{6}\left(\frac{\partial^3 U}{\partial q^i \partial q^j \partial q^k}\right)\bigg|_0 \eta_i \eta_j \eta_k + \cdots \tag{12.0.2}
$$

The first term is set to zero, since we only consider potential differences, not absolutes, and the second term is zero when evaluated at the equilibrium point, meaning that the third term is the leading term in the expansion. This term is **homogeneously quadratic** in the displacement vectors η_i and, for this reason, termination of the series with this term gives a **harmonic approximation**.

We move now to the kinetic energy, building on our previous definition:

$$
T = \sum_h \frac{1}{2} m_i \sum_{i,j} \left(\frac{\partial x_h}{\partial q^i} \cdot \frac{\partial x_h}{\partial q^j}\right) \dot{q}^i \dot{q}^j = \frac{1}{2} M_{ij} \dot{\eta}_i \dot{\eta}_j
$$

where M_{ij} is the **mass matrix**. The mass matrix is a symmetric matrix, meaning $M_{ij} = M_{ji}$, and can be thought of as an array of all the things related to the mass (the masses themselves, and all the things that influence their value). If we want to consider the most general of systems, then M_{ij} can be a function of the generalised coordinates themselves and hence we must Taylor expand about the variation from the equilibrium point, just as for the potential:

$$
M_{ij}(\boldsymbol{q}) = M_{ij}(\boldsymbol{q_0}) + \sum_k \left(\frac{\partial M_{ij}}{\partial q^k}\right)\bigg|_0 \eta_k + \sum_{k,l} \frac{1}{2}\left(\frac{\partial^2 M_{ij}}{\partial q^k \partial q^l}\right)\bigg|_0 \eta_k \eta_l + \cdots \tag{12.0.3}
$$

However, the zeroth order term in this expansion will make the kinetic energy quadratic in the displacement and so will be the only term retained if we truncate the perturbation at second order as above. The Lagrangian is then,

$$
\mathcal{L} = \frac{1}{2} \sum_{i,j} (M_{ij} \dot{\eta}_i \dot{\eta}_j - V_{ij} \eta_i \eta_j) \tag{12.0.4}
$$

with

$$M_{ij} = M_{ij}(\boldsymbol{q}_0), \qquad V_{ij} = \left(\frac{\partial^2 U}{\partial q^i \partial q^j}\right)\Bigg|_0 \qquad (12.0.5)$$

which is the same result as if we had perturbed the Lagrangian directly. We can also write this in matrix form; since the expansions were evaluated at the equilibrium, they are **constant** matrices, meaning all their entries are constants. In addition, they are real and symmetric, meaning $\boldsymbol{V}^\dagger = \boldsymbol{V}$ and $\boldsymbol{M}^\dagger = \boldsymbol{M}$ are both **Hermitian**. Providing that $\boldsymbol{M}$ non-singular $\det(\boldsymbol{M}) \neq 0$, we can rotate the displacement coordinates $\eta'^i = R^i_j \eta^j$ by some orthogonal matrix $\boldsymbol{R}$ to diagonalise $\boldsymbol{M}$ and can even scale the coordinates so all the masses $m_1, \ldots, m_n$ are equal ± 1:

$$\mathscr{L} = \frac{1}{2}\dot{\boldsymbol{\eta}}^t \cdot \boldsymbol{M} \cdot \dot{\boldsymbol{\eta}} - \frac{1}{2}\boldsymbol{\eta}^t \cdot \boldsymbol{V} \cdot \boldsymbol{\eta}, \quad \text{with} \quad \boldsymbol{V} = \begin{pmatrix} V_{11} & V_{21} & \cdots & V_{n1} \\ V_{21} & V_{22} & \cdots & \vdots \\ \vdots & \vdots & \ddots & \vdots \\ V_{n1} & \cdots & \cdots & V_{nn} \end{pmatrix}, \quad V_{ij} = \frac{\partial U^2}{\partial q^i \partial q^j}$$

with $\boldsymbol{\eta}^t$ being the transpose of the column vector $\boldsymbol{\eta}$, and similarly for $\dot{\boldsymbol{\eta}}^t$ with $\dot{\boldsymbol{\eta}}$. The Lagrange equation is now used to write the equations of motion for the system, with the index running from 1 to n:

$$\sum_j (M_{ij}\ddot{\eta}_j + V_{ij}\eta_j) = 0, \qquad \text{or} \qquad \boldsymbol{M} \cdot \ddot{\boldsymbol{\eta}} + \boldsymbol{V} \cdot \boldsymbol{\eta} = 0 \qquad (12.0.6)$$

These are a set of **linearised**, homogeneous, **coupled** ODEs. They are described as *linear* because they are linear in the displacements η and their derivatives $\ddot{\eta}$ and they are termed *coupled* because the behaviour of each variable influences another. It is helpful to think of an extended array of atoms oscillating about their equilibrium position; when one atom is displaced, it will displace it's neighbour, since there is an interaction between them. In other words, they are coupled.

Mathematically, there exists a special type of solution for these equations of motion; such solutions are called **normal modes**. The most general solution to the problem is then a linear combination of normal modes. A normal mode assumes that all the solutions oscillate at the same frequency ω and phase φ and can be written,

$$\eta_j = a_j \cos(\omega t - \varphi), \qquad \text{or} \qquad \boldsymbol{\eta} = \boldsymbol{a} \cos(\omega t - \varphi) \qquad (12.0.7)$$

where $\boldsymbol{a} = (a_1, a_2, \ldots, a_n)$ is an amplitude column vector of the mode; while all the amplitudes may be different they are assumed to be *constant* in time. Entering this solution, we generate a system of simultaneous **linear algebraic equations** in the coordinate amplitudes:

$$\sum_j (V_{ij}a_j - \omega^2 M_{ij}a_j) = 0, \qquad \text{or} \qquad (\boldsymbol{V} - \omega^2 \boldsymbol{M}) \cdot \boldsymbol{a} = 0 \tag{12.0.8}$$

This equation is of the form $\sum A_{ij} \cdot a_j = 0$; if A_{ij} has an inverse, then we can multiply both sides to get $\boldsymbol{A}^{-1}\boldsymbol{A}\boldsymbol{a} = \boldsymbol{a} = 0$, the **trivial solution**. The trivial solution is where all the amplitudes of oscillation are zero for any frequency $\boldsymbol{a} = 0$, $\forall\, \omega$; but, in this case, $\boldsymbol{\eta} = \boldsymbol{a}\cos(\omega t - \varphi) = 0$, meaning we didn't displace in the first place. The condition for a non-trivial solution is the vanishing of the **determinant** of the linear algebraic equations, which we know means that the inverse of the matrix doesn't exist:

$$\begin{vmatrix} V_{11}-\omega^2 M_{11} & V_{12}-\omega^2 M_{12} & V_{13}-\omega^2 M_{13} & \dots & V_{1n}-\omega^2 M_{1n} \\ V_{21}-\omega^2 M_{21} & V_{22}-\omega^2 M_{22} & V_{23}-\omega^2 M_{23} & \dots & \dots \\ V_{31}-\omega^2 M_{31} & V_{32}-\omega^2 M_{32} & V_{33}-\omega^2 M_{33} & \dots & \dots \\ \dots & \dots & \dots & \dots & \dots \\ V_{n1}-\omega^2 M_{n1} & \dots & \dots & \dots & V_{nn}-\omega^2 M_{nn} \end{vmatrix} = 0 \tag{12.0.9}$$

or,

$$\det(V_{ik} - \omega^2 M_{ik}) = 0 \tag{12.0.10}$$

When this is true, we have an nth-degree polynomial in ω^2; this is called the **characteristic equation** and can be solved for n real, positive roots $\omega_1^2, \omega_2^2, \dots, \omega_n^2$ and entered into the above algebraic equations to solve for each amplitude a_r^j and each **characteristic frequency** ω_r. We generate an **eigenvalue equation** for the operator $\boldsymbol{M}^{-1}\boldsymbol{V}$. Although both $\boldsymbol{M}$ and $\boldsymbol{V}$ are symmetric matrices, $\boldsymbol{M}^{-1}\boldsymbol{V}$ need not be symmetric (but don't worry about that for now):

$$(M)^{-1}V\boldsymbol{a} = \omega^2\boldsymbol{a} \tag{12.0.11}$$

The aim of this equation is to solve for each characteristic frequencies ω_i and amplitude vector $\boldsymbol{a}$. If some of the roots are repeated in the polynomial, then there will be more than one mode corresponding to that characteristic; we say that **degenerate modes** share the same frequency, but there are always exactly n modes. The degeneracy of modes is important in the splitting of atomic energy levels, as a degenerate energy level can separate into its component modes under the influence of an external electromagnetic field.

For each characteristic frequency ω_r, the Lagrange equation of motion is

$$\sum_j (V_{ij} - \omega_r^2 M_{ij})a_{jr} = 0, \qquad \text{or} \qquad (\boldsymbol{V} - \omega_r^2 \boldsymbol{M})\boldsymbol{a}_r = 0 \tag{12.0.12}$$

The general solution to the small oscillations problem is then a linear superposition of normal modes:

$$\eta_r = \sum_r a_r \cos(\omega_r t - \varphi_r) \qquad (12.0.13)$$

Substituting the definition of η, we generate the solution to the harmonic oscillation:

$$q^i = q_0^i + \sum_{r=1}^{n} a_{jr} \cos(\omega_r t - \varphi_r) \qquad (12.0.14)$$

Exercise 12.1 Lets familiarise all this with an example. Consider a particle oscillating in three dimensions and with different spring constants in each direction. Using cartesian coordinates, the Lagrangian is

$$\mathscr{L} = \frac{1}{2}m(\dot{x}^2 + \dot{y}^2 + \dot{z}^2) - \frac{1}{2}(k_1 x^2 + k_2 y^2 + k_3 z^2) \qquad (12.0.15)$$

Since there are no crossed terms between the different coordinates, we expect the mass and potential matrices to look pretty nice and, indeed, we find that they are both diagonal:

$$V = \begin{pmatrix} k_1 & 0 & 0 \\ 0 & k_2 & 0 \\ 0 & 0 & k_3 \end{pmatrix}, \qquad M = \begin{pmatrix} m & 0 & 0 \\ 0 & m & 0 \\ 0 & 0 & m \end{pmatrix} \qquad (12.0.16)$$

Constructing the array of linear algebraic equations and taking its determinant we obtain

$$\det \begin{pmatrix} k_1 - m\omega^2 & 0 & 0 \\ 0 & k_2 - m\omega^2 & 0 \\ 0 & 0 & k_3 - m\omega^2 \end{pmatrix} = (k_1 - m\omega^2)(k_2 - m\omega^2)(k_3 - m\omega^2) = 0 \quad (12.0.17)$$

from which we obtain three roots:

$$\omega_i = \sqrt{\frac{k_i}{m}} \qquad i = 1, 2, 3 \qquad (12.0.18)$$

Exercise 12.2 In solid state chemistry, we are familiar with small molecules adsorbing onto the surface of materials (see figure 12.2); often we use spectroscopies to identify what species are present. However, the act of binding to the surface changes the frequency of the characteristic oscillations that fingerprint a molecule. To illustrate this phenomena, we consider a uniform metal surface with a homonuclear diatomic molecule adsorbed along the internuclear axis:

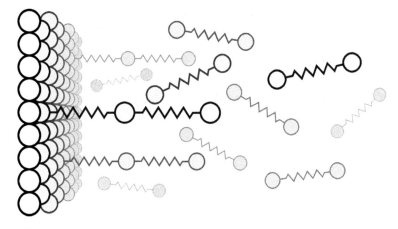

Fig. 12.2: Gas molecules adsorbing onto the surface of a solid.

We model both the surface-substrate interaction and the molecular bond by using springs, which, for convenience, we will assume are identical. The coordinates for the system are the displacements of the masses from their equilibrium positions, or **relative coordinates**. If we take q_0 as the equilibrium configuration of the coupled oscillator, then, at any moment, the instantaneous position of a mass is $q(t)$. The deviation of this from equilibrium is then $q(t) - q_0$, which we call $\eta(t)$. We turn this into our generalised coordinate. The potential for the interaction consists of two terms. The first is the interaction of the surface and the adjoining atom. We assume that the surface interaction falls with distance such that we can ignore interactions of the second atom and the surface. The second term in the potential is the interaction between the atoms of the diatomic molecule. We assume that this is a function of the distance between the atoms and we can write $q_{21} = q_2 - q_1$, which, in relative coordinates, is $q_{21} = \eta_2 - \eta_1$. Since we have a harmonic oscillator, both these appear as squared terms $-\frac{1}{2}kx^2$, by Hooke's law. The Lagrangian for this system is given below and we see how we have constructed a Lagrangian from subsystems!

$$\mathscr{L} = \frac{1}{2}m(\dot{\eta}_1^2 + \dot{\eta}_2^2) - \frac{1}{2}k\left[\eta_1^2 + (\eta_2 - \eta_1)^2\right] \tag{12.0.19}$$

We now construct the Euler-Lagrange equation for each coordinate η_1 and η_2, respectively:

$$\frac{d}{dt}\left(\frac{\partial\mathscr{L}}{\partial\dot{\eta}_1}\right) - \frac{\partial\mathscr{L}}{\partial\eta_1} = 0 \qquad \longrightarrow \qquad m\ddot{\eta}_1 + k(2\eta_1 - \eta_2) = 0 \tag{12.0.20}$$

where we have chain ruled the potential term by substituting $u = (\eta_2 - \eta_1)$; we then obtian

$$\frac{\partial}{\partial\eta_1}\left[-\frac{1}{2}k(\eta_1^2 + (\eta_2 - \eta_1)^2)\right] = -\frac{1}{2}k\left[2\eta_1 + \frac{\partial(u^2)}{\partial u}\frac{\partial u}{\partial\eta_1}\right]$$

$$= -\frac{1}{2}k\left[2\eta_1 + 2u\cdot(-1)\right]$$

$$= -k\left[2\eta_1 - \eta_2\right]$$

Proceeding similarly with the second coordinate we find that

$$\frac{d}{dt}\left(\frac{\partial \mathscr{L}}{\partial \dot{\eta}_2}\right) - \frac{\partial \mathscr{L}}{\partial \eta_2} = 0 \qquad \longrightarrow \qquad m\ddot{\eta}_2 + k(\eta_2 - \eta_1) = 0 \qquad (12.0.21)$$

We are left with two coupled differential equations of motion in the displacements. Using the solution $\eta_i = a_i \sin(\omega t)$ for both coordinates and subsequent cancelation of $\sin(\omega t)$ from both sides leaves us with,

$$-a_1 m\omega^2 + k(2a_1 - a_2) = 0 \qquad (12.0.22)$$
$$-a_2 m\omega^2 + k(a_2 - a_1) = 0 \qquad (12.0.23)$$

It is convenient to rewrite these as

$$(2k - \omega^2 m)a_1 - ka_2 = 0 \qquad (12.0.24)$$
$$(k - \omega^2 m)a_2 - ka_1 = 0 \qquad (12.0.25)$$

This can also be written as a the product of a 2×2 matrix with a column vector. The first matrix is formed by considering the first column as the coefficients of a_1, and the second column as the coefficients of a_2, while the rows are for the two equations, respectively:

$$\begin{pmatrix} 2k - \omega^2 m & -k \\ -k & k - \omega^2 m \end{pmatrix} \begin{pmatrix} a_1 \\ a_2 \end{pmatrix} = \begin{pmatrix} 0 \\ 0 \end{pmatrix} \qquad (12.0.26)$$

The determinant of the linear algebraic equations is then found and solved for the frequency:

$$\det \begin{pmatrix} 2k - \omega^2 m & -k \\ -k & k - \omega^2 m \end{pmatrix} = k^2 - 3\omega^2 km + \omega^4 m^2 = 0 \qquad (12.0.27)$$

This equation can be factored and each of the resulting two brackets can be set to zero to give equations of the forms $x^2 + xy - y^2 = 0$ and $x^2 - xy - y^2 = 0$. These quadratic equations have a well-known solution called the **golden ratio**:

$$k^2 - 3\omega^2 km + \omega^2 m^2 = [\omega^2 - \omega\sqrt{k/m} - (k/m)][\omega^2 + \omega\sqrt{k/m} - (k/m)] = 0 \qquad (12.0.28)$$

The oscillatory frequencies are then,

$$\omega_\pm = \frac{k}{m}\sqrt{\frac{3 \pm \sqrt{5}}{2}} \qquad (12.0.29)$$

Therefore, the two frequencies of the adsorbed molecule are modified versions of the frequencies of a free diatomic molecule: one at a higher energy and one at a lower energy; they correspond to in-phase and out-of-phase vibrations. To improve this model, try repeating the calculation using springs with different spring constants, different masses and molecules with more atoms in them. Experimentally the stretching frequencies of individually adsorbed molecules can be measured using a technique known as scanning tunnelling spectroscopy. Such experiments are

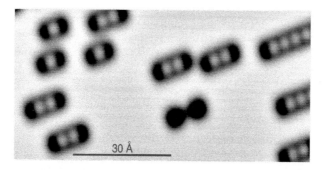

Fig. 12.3: Scanning tunnelling microscopy image showing carbon monoxide molecules on a Cu(110) surface. Image courtesy of K. Mann, Schaub group, University of St Andrews. For further information see chapter 24.1.

routinely performed under ultra-high vacuum and at very low temperatures (around 4 K), as they require extreme mechanical stability. The frequencies depend on the crystal structure of the metal surface and the attachment geometry of the adsorbed molecule.

12.1 Normal Modes

Mathematical Background 1

Lets investigate the properties of the eigenvalue problem for small oscillations. Multiplying both sides by the conjugate transpose or **adjoint** of the amplitude vector $\boldsymbol{a}^\dagger$, and, for ease of notation, let $\omega^2 = \lambda$:

$$\boldsymbol{a}^\dagger \boldsymbol{V} \boldsymbol{a} = \lambda^* \boldsymbol{a}^\dagger \boldsymbol{M} \boldsymbol{a} \tag{12.1.1}$$

But, since V and M are real and symmetric (Hermitian),

$$(\boldsymbol{a}^\dagger \boldsymbol{V} \boldsymbol{a})^t = \boldsymbol{a}^\dagger \boldsymbol{V}^\dagger \boldsymbol{a} = \boldsymbol{a}^\dagger \boldsymbol{V} \boldsymbol{a} \tag{12.1.2}$$

$$(\boldsymbol{M}^\dagger \boldsymbol{V} \boldsymbol{M})^t = \boldsymbol{M}^\dagger \boldsymbol{V}^\dagger \boldsymbol{M} = \boldsymbol{M}^\dagger \boldsymbol{V} \boldsymbol{M} \tag{12.1.3}$$

Then the **Rayleigh quotient**, 12.1.4, tells us that the frequencies are finite and positive, meaning that the matrices are **positive semi-definite** with all positive or zero eigenvalues. Physically, this is because we are expanding the potential from a minimum and the presence of negative eigenvalues would imply that there was a direction of lower potential:

$$\lambda = \frac{\boldsymbol{a}^\dagger \boldsymbol{V} \boldsymbol{a}}{\boldsymbol{a}^\dagger \boldsymbol{M} \boldsymbol{a}} \tag{12.1.4}$$

> We are perturbing around a minimum potential and we only have positive kinetic energies, so ω^2 is a real eigenvalue. The Rayleigh quotient is used in analysis to estimate the eigenvalue problem. Therefore, we have used the properties of $\boldsymbol{V}$ and $\boldsymbol{M}$ to show that we have only positive eigenvalues.

Generally, η_r is not a periodic function of time; however, since it can be written as the sum of harmonic functions, there will exist a set of coordinates ζ that are periodic in one of the frequencies and therefore drastically simplify the problem. We call these special coordinates **normal coordinates**. In mathematical language we are **diagonalising** both $\boldsymbol{M}$ and $\boldsymbol{V}$ at the same time by a single coordinate transformation. Diagonal matrices make finding their eigenvalues and eigenvectors a piece of cake and, since we are considering such a general problem of perturbing an equilibrium point, it is exciting to think that such a nice solution exists.

We define the coordinate transformation from $\eta \to \zeta$ and its inverse as follows:

$$\boldsymbol{\eta} = \mathcal{Q} \cdot \boldsymbol{\zeta}, \qquad \text{and} \qquad \boldsymbol{\zeta} = \mathcal{Q}^{-1} \cdot \boldsymbol{\eta} \tag{12.1.5}$$

where $\mathcal{Q}$ is a time-independent **transformation matrix** whose columns are linearly independent. The Lagrangian is now written

$$\mathcal{L} = \frac{1}{2}(\mathcal{Q}\dot{\boldsymbol{\zeta}})^t \cdot \boldsymbol{M} \cdot (\mathcal{Q}\dot{\boldsymbol{\zeta}}) - \frac{1}{2}(\mathcal{Q}\boldsymbol{\zeta})^t \cdot \boldsymbol{V} \cdot (\mathcal{Q}\boldsymbol{\zeta})$$

$$= \frac{1}{2}\dot{\boldsymbol{\zeta}}^t(\mathcal{Q}^t \boldsymbol{M} \mathcal{Q}) \cdot \dot{\boldsymbol{\zeta}} - \frac{1}{2}\boldsymbol{\zeta}^t \cdot (\mathcal{Q}^t \boldsymbol{V} \mathcal{Q}) \cdot \boldsymbol{\zeta} \tag{12.1.6}$$

In these coordinates, the amplitude vectors of different normal modes are said to be **mutually orthogonal** to each other. This means that they satisfy the following **orthonormality relations**:

$$\boldsymbol{a}_i^t \cdot \boldsymbol{M} \cdot \boldsymbol{a}_j = \delta_{ij} \tag{12.1.7}$$

The particular transformation matrix that we are interested in finding is the one whose columns are the amplitude vectors of the n modes of the system:

$$\mathcal{Q} = \begin{pmatrix} \boldsymbol{a_1} \ \boldsymbol{a_2} \ \dots \ \boldsymbol{a_n} \end{pmatrix} \tag{12.1.8}$$

Since we impose the orthonormality relations between the amplitude column vectors of each mode, the columns of the transformation matrix are **linearly independent**. In this way, $\mathcal{Q}^t \cdot \boldsymbol{M} \cdot \mathcal{Q} = \mathbb{I}$, which is the **identity matrix**, since all off-diagonal elements are zero by the orthonormality relations.

Turning to similarly treat the top line of the Rayleigh quotient:

$$\mathcal{Q}^t \cdot \boldsymbol{V} \cdot \mathcal{Q} = \begin{pmatrix} \boldsymbol{a}_1^t \\ \boldsymbol{a}_2^t \\ \vdots \\ \boldsymbol{a}_n^t \end{pmatrix} \cdot V \cdot \begin{pmatrix} \boldsymbol{a}_1 \ \boldsymbol{a}_2 \ \dots \ \boldsymbol{a}_n \end{pmatrix} \tag{12.1.9}$$

Consider the ijth element; we can use the eigenvalue equation $\boldsymbol{a}_k V = \omega^2 M \boldsymbol{a}_k$ to express each element in terms of the diagonalised mass matrix:

$$\boldsymbol{a}_i^t \cdot \boldsymbol{V} \cdot \boldsymbol{a}_j = \boldsymbol{a}_i^t \cdot (\omega_j^2 \boldsymbol{M} \cdot \boldsymbol{a}_j) = \omega_j^2 (\boldsymbol{a}_i^t \boldsymbol{M} \boldsymbol{a}_j) = \omega_j^2 \delta_{ij} \tag{12.1.10}$$

Therefore, all off-diagonal elements of the top line vanish, leaving us with a diagonalised matrix of characteristic frequencies:

$$\mathcal{Q}^t \cdot \boldsymbol{V} \cdot \mathcal{Q} = \boldsymbol{\Lambda}^2, \qquad \text{with} \qquad \boldsymbol{\Lambda} = \begin{pmatrix} \omega_1 & 0 & \cdots & 0 \\ 0 & \omega^2 & \cdots & 0 \\ \vdots & \vdots & \ddots & \vdots \\ 0 & 0 & \cdots & \omega_n \end{pmatrix} \tag{12.1.11}$$

Having diagonalised both the top and the bottom of the Rayleigh quotient, the Lagrangian becomes

$$\mathcal{L} = \frac{1}{2} \sum_{i=1}^{n} (\dot{\zeta}_i^2 - \omega_i^2 \zeta_i^2) \tag{12.1.12}$$

Hence, the equations of motion for the near-equilibrium harmonic oscillations comprise a system of **decoupled** second-order linear ODEs which requires $2n$ initial conditions to solve:

$$\ddot{\zeta}_i + \omega_i^2 \zeta_i = 0, \qquad \text{or} \qquad \ddot{\boldsymbol{\zeta}} + \boldsymbol{\Lambda}^2 \cdot \boldsymbol{\zeta} = 0 \tag{12.1.13}$$

Each normal coordinate is a periodic function in only *one* characteristic frequency, which we now call a **normal frequency** with any oscillatory motion being described by a linear superposition of amplitude-weighted normal modes oscillating at their own unique frequency. The energy of the normal modes is constructed in the usual fashion from the Lagrangian:

$$J = \dot{\zeta}_i^2 - \frac{1}{2}(\dot{\zeta}_i - \omega_i^2 \zeta_i^2) = \frac{1}{2}\dot{\zeta}^2 + \frac{1}{2}\omega_i^2 \zeta_i^2 \tag{12.1.14}$$

Normal mode theory may seem a little unnecessary since we start by converting the *known solution* into something easier. It is, however, very useful in practical problems.

Chapter summary

- In this chapter, we defined **equilibrium points** on a configuration manifold to be regions where the potential energy is critical and showed how small perturbations can be **linearised** about any stable equilibrium point to exhibit oscillatory motion.

- We expanded the Lagrangian to leading-order terms **homogeneously quadratic** in its coordinates to generate a system of **coupled** differential equations, a basic solution which is a called a **mode**. The most general solution is then seen to be a linear combination of modes, and the problem is reduced to an **eigenvalue equation**.

- We then investigated **normal coordinates** and showed how we can **decouple** the equations of motion by clever choice of coordinates.

Exercise 12.3 In this example, we compute classical modes of molecular oscillations by assuming that the nuclei can be treated as classical particles; we will consider a free, symmetric **linear triatomic molecule** ABA. The importance of the *free* is that it is allowed to translate freely with constant velocity.

In any introductory chemistry class, we are taught how to count the vibrations of a polyatomic molecule. For N atoms, there are $3N$ degrees of freedom, three of which correspond to the translation of the centre of mass. If the molecule is non-linear, we have three orthogonal rotational axes of the rigid molecule but only two orthogonal axes for linear molecules, since rotating about the internuclear axis doesn't actually change anything.

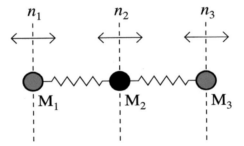

Fig. 12.4: The linear triatomic molecule.

Considering the linear case, we therefore have $3N - 5$ degrees of freedom left and the only thing they can be is **vibrations**. We can split these vibrations up by looking down the internuclear axis and saying which vibrations are **longitudinal** and which ones are **transverse**. The nuclei have N degrees of freedom down the internuclear axis; one of these is free translation, leaving $N - 1$ longitudinal modes. Therefore, of the total number of vibrations, we can say there are $(3N - 5) - (N - 1) = (2N - 4)$ transverse oscillations for a linear polyatomic molecule.

We can experimentally observe vibrations in infra-red or Raman spectroscopy; in both cases, the vibrations are both photon absorption processes (the former to an **excited mode** and the latter to a **virtual mode** (sometimes called a scattering experiment)); anyway, back to oscillations!

We assume that the bonds of the linear triatomic molecule are well approximated by identical springs with equilibrium separation b (bond length), an approximation that decays with increasing excitation from the grounds state or for large displacements from equilibrium. It is also assumed that the terminal masses are the same but different from the middle mass, so, in essence, we are considering the vibrations of CO_2 or CS_2.

The first step is to define our generalised coordinates to write the Lagrangian and we choose the **relative coordinate** $\eta_i = q^i - q_0$, which is the difference between instantaneous and equilibrium positions for nucleus i. The potential will depend on the *distance* between adjacent nuclei; and this is expected to change over the course of a molecular vibration.

For two adjacent nuclei i and j, we write the difference between them as $q_{ij} = q_i - q_j$. Writing this in terms of the relative coordinates this becomes $\eta_i - \eta_j$; this is the separation between the nuclei, re-normalised to the equilibrium configuration. That is to say, when at equilibrium, this expression vanishes and is equal to the bond length b. The potential is written

$$U = \frac{1}{2}k(\eta_2 - \eta_1)^2 + \frac{1}{2}k(\eta_3 - \eta_2)^2 \tag{12.1.15}$$

where η_2 is identified with the central atom. The Lagrangian is then

$$\begin{aligned}\mathscr{L} &= \frac{1}{2}m_1(\dot{\eta}_1^2 + \lambda\dot{\eta}_2^2 + \dot{\eta}_3^2) - \frac{1}{2}k(\eta_2 - \eta_1)^2 - \frac{1}{2}k(\eta_3 - \eta_2)^2 \\ &= \frac{1}{2}m_1(\dot{\eta}_1^2 + \lambda\dot{\eta}_2^2 + \dot{\eta}_3^2) - \frac{1}{2}k(\eta_1^2 + 2\eta_2^2 + \eta_3^2 - 2\eta_1\eta_2 - 2\eta_2\eta_3)\end{aligned} \tag{12.1.16}$$

with $\lambda = m_2/m_1$. We now solve the Lagrange equation for each generalised coordinate η_i and rearrange to give

$$0 = m_1\ddot{\eta}_1 + k\eta_1 - k\eta_2 \tag{12.1.17}$$
$$0 = \lambda m_1\ddot{\eta}_2 - k\eta_1 + 2k\eta_2 - k\eta_3 \tag{12.1.18}$$
$$0 = m_1\ddot{\eta}_3 - k\eta_2 + k\eta_3 \tag{12.1.19}$$

We can form M and V matrices out of these equations. The mass matrix has columns of each equation and row of the coordinates, elements being non-zero where there is a mass term. The potential matrix formed from the coefficients of the three coordinates is the same result as was obtained from differentiating the original coordinates with the potential expression from earlier. Notice that the mass matrix is already diagonalised but the potential matrix is not, suggesting that our coordinates are not the normal coordinates of the system:

$$M = \begin{pmatrix} m_1 & 0 & 0 \\ 0 & \lambda m_1 & 0 \\ 0 & 0 & m_1 \end{pmatrix}, \qquad V = \begin{pmatrix} k & -k & 0 \\ -k & 2k & -k \\ 0 & -k & k \end{pmatrix} \tag{12.1.20}$$

Recall the eigenvalue equation $(V_{ij} - \omega^2 M_{ij})a = 0$ only has a non-trivial solution if the secular determinant vanishes:

$$\det(\boldsymbol{V} - \omega^2 \boldsymbol{M}) = \begin{pmatrix} k - m_1\omega^2 & -k & 0 \\ -k & 2k - m_2\omega^2 & -k \\ 0 & -k & k - m_1\omega \end{pmatrix} \quad (12.1.21)$$

Expanding the determinant and employing a little algebra we obtain

$$
\begin{aligned}
0 &= k - \omega^2 m_1 (2k - \omega^2 m_2)(k - \omega^2 m_1) - (-k \cdot -k) - (-k)(-k)(k - \omega^2 m_1) - (-k)(0) + 0 \\
&= (k - \omega^2 m_1)^2 (2k - \omega^2 m_2) - 2k^2 (k - \omega^2 m_1) \\
&= (k - \omega^2 m_1)[(k - \omega^2 m_1)(2k - \omega^2 m_2) - 2k^2] \\
&= (k - \omega^2 m_1)[2k^2 - 2k^2 - k\omega^2 m_2 - 2k\omega^2 m_1 + \omega^2 m_1 m_2] \\
&= (\omega^2)(k - \omega^2 m_1)[-k(m_2 + 2m_1) + \omega^2 m_2 m_1]
\end{aligned}
\quad (12.1.22)
$$

This has three roots, which can be seen by setting each bracket to zero and solving for the frequency:

$$\omega_1 = 0, \qquad \omega_2 = \sqrt{\frac{k}{m_1}}, \qquad \omega_3 = \sqrt{\frac{km_T}{m_2 m_1}} \quad (12.1.23)$$

Notice that $2m_1 + m_2 = m_T$, the total mass. We interpret these frequencies as follows. The first result, ω_1, is a zero-valued frequency and it corresponds to the free translation of the entire molecule through a laboratory axis. Nothing is displaced relative to it's internal equilibrium, so there is no restoring force and no oscillatory frequency; $\ddot{\eta} + (0)\eta = 0$ becomes $\ddot{\eta} = 0$. The second result is the **symmetric mode** of the molecule; it only contains the terminal atom's mass indicating that the centre atom is stationary. The third result, ω_3, is an **asymmetric mode**; it involves both types of atoms but, crucially, the centre of mass is still at rest! This means that, while the two terminal atoms move in one direction, the centre mass moves in the opposite direction, to compensate. The most general oscillation will be a linear combination of these modes, weighted by their amplitudes, which are determined by the initial conditions. Each normal mode is characterised by each atom passing through its equilibrium position at the same time as the others; the general oscillation might not look like this.

You will recall that we can count the number of modes we expect from a molecule, using the simple rules outlined above, and, for CO_2 there should be four vibrational modes; we already have two (since ω_1 actually turned out to be one of the translations). We can also say that there are two longitudinal modes, which we have already found, leaving two transverse modes. We expect these two transverse oscillations to comprise the displacement of the carbon atom from the internuclear axis to form a v-shaped state and, since the equilibrium state is symmetric, we expect these two modes to be **degenerate** (no special preference of x or y axes) with the same vibrational frequency. If they are superimposed upon one another, they can be either be in phase or out of phase; the in-phase motion is another true vibration, while the out-of-phase motion causes the nuclei to perform rotational motion about the internuclear axis. This is known as rotational-vibrational coupling, or **Coriolis coupling** and leads to a loss of symmetry momentarily along the internuclear axis. With this sudden loss of symmetry, there is now a moment of inertia and hence angular momentum about the axis, leading to rotation. This coupling has well-known effects in physical chemistry and is well studied.

At this point we could conclude our example content with knowledge of the frequency of the modes, but there is still the matter of the amplitudes to determine and, to do this, we return to the eigenvalue problem. Since we have three eigenfrequencies, we will introduce a new subscript: $r = 1, 2, 3$:

$$(\boldsymbol{V} - \omega_r^2 \boldsymbol{M})\boldsymbol{a}_r = \begin{pmatrix} k - m_1\omega_r^2 & -k & 0 \\ -k & 2k - m_2\omega_r^2 & -k \\ 0 & -k & k - m_1\omega_r^2 \end{pmatrix} \begin{pmatrix} a_{1r} \\ a_{2r} \\ a_{3r} \end{pmatrix} = 0 \qquad (12.1.24)$$

Multiplying out this $3x3$ matrix by the column vector gives us

$$\begin{pmatrix} (k - m_1\omega_r^2)a_{1r} - ka_{2r} \\ -ka_{1r} + (2k - m_2\omega_r^2)a_{2r} - ka_{3r} \\ -ka_{2r} + (k - m_1\omega_r^2)a_{3r} \end{pmatrix} = 0 \qquad (12.1.25)$$

So, let's turn to the first mode, $\omega_1^2 = 0$:

$$\omega_1 = 0 \rightarrow \begin{cases} a_{11}k - a_{21}k = 0 \\ -a_{11}k + a_{21}2k - a_{31}k = 0 \\ -a_{21}k + a_{31}k = 0 \end{cases} \qquad (12.1.26)$$

If we look at the top and bottom two equations, we see, $a_{11} = a_{21} = a_{31}$, which trivialises the middle equation. For the first mode, we can therefore write

$$\boldsymbol{a}_1 = a_{11} \begin{pmatrix} 1 \\ 1 \\ 1 \end{pmatrix} \qquad (12.1.27)$$

We now need to normalise this result to align with the diagonalisation of the mass matrix so they can satisfy the orthonormality relations:

$$\boldsymbol{a}_1^t \boldsymbol{M} \boldsymbol{a}_1 = a_{11}^2 \begin{pmatrix} 1 & 1 & 1 \end{pmatrix} \begin{pmatrix} m_1 & 0 & 0 \\ 0 & m_2 & 0 \\ 0 & 0 & m_1 \end{pmatrix} \begin{pmatrix} 1 \\ 1 \\ 1 \end{pmatrix} = 1 \qquad (12.1.28)$$

Multiplying this all out using the rules of matrix multiplication, we obtain $a_{11}^2(m_2 + 2m_1) = 1$, which is the first of our normalisation conditions, allowing us to write a normalised eigenvector for ω_1:

$$\boldsymbol{a}_1 = \frac{1}{\sqrt{m_2 + 2m_1}} \begin{pmatrix} 1 \\ 1 \\ 1 \end{pmatrix} \qquad (12.1.29)$$

Repeating the same procedure for the second mode ω_2 shows us that $a_{22} = 1$ and $a_{12} = -a_{32}$, which is then normalised to give

$$\boldsymbol{a}_2 = \frac{1}{\sqrt{2m_1}} \begin{pmatrix} 1 \\ 0 \\ -1 \end{pmatrix} \qquad (12.1.30)$$

The third frequency, ω_3, proceeds in exactly the same way but it is helpful to simplify the diagonal matrix elements before trying to multiply them all out. It will lead to the relations

$a_{13} = -\frac{m_2}{2m_1}a_{23}$ and $a_{13} = a_{33}$. The normalised eigenvector for the third frequency is then found to be

$$\boldsymbol{a}_3 = \frac{1}{\sqrt{2m_1(1 + 2m_1/m_2)}} \begin{pmatrix} 1 \\ -2m_1/m_2 \\ 1 \end{pmatrix} \tag{12.1.31}$$

As we expect, $\boldsymbol{a}_1$ has three equivalent components corresponding to uniform translation. The second eigenvector corresponds to a symmetric stretch where the middle mass (say, the carbon of CO_2) is at rest (hence the zero) and the two oxygens move in opposite directions symmetrically. The final mode shows that the two oxygens move in the same direction, opposite to the carbon's motion; the $-2m_1/m_2$ factor is telling us that the centre of mass is at rest or travelling at constant velocity.

Normal mode theory can be applied to non-linear polyatomic molecules in much the same way, however, the reduction in molecular symmetry leads to enormous complication of the characteristic polynomial and difficulty in finding the frequencies. Another type of problem is to consider an infinite chain of oscillators such as a one-dimensional crystal; this is treated when we discuss field theory. The reader is referred to R. Lavin and J. Pozo's (2011) article, The study and teaching of the vibrational modes of the triatomic molecule, for analysis of a non-linear triatomic molecule.

Exercise 12.4 Consider a particle of mass m subject to a restoring force $\vec{F} = -kx$ and a frictional force $\vec{F} = -\alpha\dot{x}$. Write the Lagrangian for the system and show that the equation of motion is given by

$$m\frac{d^2x}{dt^2} = -kx - \alpha\frac{dx}{dt} \tag{12.1.32}$$

Solve this for the general case to find

$$x(t) = a_1 e^{\gamma_1 t} + a_2 e^{\gamma_3 t} \tag{12.1.33}$$

Identify both γ_1 and γ_2 and analyse this solution for three regimes.

Exercise 12.5 Atomic force microscopy is used to image surface topologies at an atomic scale and to estimate sample-probe forces as a function of distance. An atomic force microscopy experiment consists of tracing an oscillating microcantilever with a silicone wafer or a silicone nitride (Si_3N_4) tip across a surface (see figure 12.5). As the tip scans over the surface it interacts with adsorbed molecules and surface structures (such as step edges) to create *tip-sample interactions*. This greatly alters the oscillation modes of the cantilever as it bends under the tip-surface separation-dependent force.

In order to pick these changes up, a laser beam is focused on the tip, and the reflected rays are sensed by a photodetector. The current generated is then used to rebuild an image of the material surface.

There are three modes of operation: *contact mode* (used in ambient conditions), where the tip contacts the adsorbed fluid on the surface, and the force is kept constant via a feedback mechanism; *tapping mode*, where the tip has contact with the surface during its oscillation, so the contact damps the motion; and *non-contact mode* (used in vacuum conditions), where the tip is held above the sample.

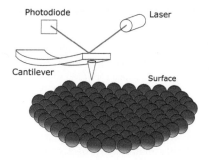

Fig. 12.5: The atomic force microscopy experimental set-up.

Classical models of the tip dynamics can provide insight into artefact resolution, image interpretation and mode analysis. Therefore, we will model the oscillating tip by a second-order non-linear ODE as if it where a damped, driven harmonic oscillator:

$$m\ddot{q} + \frac{m\omega_0}{Q}\dot{q} + kq = F_{ts}(q) + F(t) \tag{12.1.34}$$

where q is the tip deflection normal to the surface and m, Q, k and ω_0 are the cantilever mass, the quality factor for the mode, the force constant and resonant frequency, respectively. The driving or *excitation force* is given by $F(t)$ and its frequency by ω. The tip-surface interaction, F_{ts}, is a combination of short-range repulsive factors, longer-range attraction factors, adhesion and contact forces, electrostatics, magnetostatics and capillary action terms.

In the absence of any tip-surface interactions, the driven oscillator can be solved using a Fourier series; recognising that the equation of motion can be viewed as a linear operator on $q(t)$, we realise that the solution will be a superposition of harmonic oscillations:

$$q(t) = \frac{1}{2\pi}\int_{-\infty}^{\infty}\tilde{q}(\omega)e^{i\omega t}d\omega, \qquad F(t) = \frac{1}{2\pi}\int_{-\infty}^{\infty}\tilde{F}(\omega)e^{i\omega t}d\omega \tag{12.1.35}$$

Insertion of this into 12.1.34 creates an algebraic equation:

$$\left\{\frac{d^2}{dt^2} + \lambda\frac{d}{dt} + \omega_0^2\right\}\int_{-\infty}^{\infty}\tilde{q}(\omega)e^{i\omega t}d\omega = \frac{1}{m}\int_{-\infty}^{\infty}\tilde{F}(\omega)e^{i\omega t}d\omega$$

$$\int_{-\infty}^{\infty}\left(\omega_0^2 - \omega^2 + i\lambda\omega\right)\tilde{q}(\omega)e^{i\omega t}d\omega = \frac{1}{m}\int_{-\infty}^{\infty}\tilde{F}(\omega)e^{i\omega t}d\omega \tag{12.1.36}$$

We now simply take the Fourier transform of both sides and arrange for $\tilde{q}(\omega)$. To find $q(t)$, we then just need to take the inverse transform again:

$$q(t) = \frac{1}{2\pi}\int_{-\infty}^{\infty}\frac{1}{m}\frac{\tilde{F}(\omega)}{\omega_0^2 - \omega^2 + i\lambda\omega}e^{i\omega t}d\omega \tag{12.1.37}$$

In practical application, the form of the driving force will be known and the system can be analysed for its modes and resonances by solving the resulting contour integral.

Depending on the operational mode, the system is highly non-linear, as the vibrational amplitudes of the tips are larger than the tip-sample interaction times. Therefore, the theory of small oscillations fails to meet such chaotic motions and, instead, researchers turn to **contact mechanics**. The cantilever is then best described using the **Euler-Bernoulli beam theory**, the Lagrangian density (see chapter 25) of which is given by

$$\mathscr{L} = \frac{1}{2} \int_0^l \left\{ \mu \left(\frac{\partial q}{\partial t} \right)^2 - \mathcal{Y} \left(\frac{\partial^2 q}{\partial x^2} \right)^2 \right\} dx \qquad (12.1.38)$$

where q is the vertical displacement, and $x \in [0, l]$, where l is the cantilever length and μ and $\mathcal{Y}$ are the mass density and elasticity modulus, respectively:

$$\delta A = \int_{t_1}^{t_2} \int_0^l \left\{ \mu \left(\frac{\partial q}{\partial t} \right) \delta \left(\frac{\partial q}{\partial t} \right) - \mathcal{Y} \left(\frac{\partial^2 q}{\partial x^2} \right) \delta \left(\frac{\partial^2 q}{\partial x^2} \right) \right\} dx dt \qquad (12.1.39)$$

We commute the variation with the partial derivatives and integrate both terms by parts:

$$\int_{t_1}^{t_2} \int_0^l \mu \frac{\partial q}{\partial t} \frac{\partial}{\partial t} (\delta q) dx dt = \int_0^l \mu \frac{\partial q}{\partial t} \delta q \Big|_{t_1}^{t_2} dx - \int_{t_1}^{t_2} \int_0^l \mu \frac{\partial^2 q}{\partial t^2} (\delta q) dx dt \qquad (12.1.40)$$

$$\int_{t_1}^{t_2} \int_0^l \mathcal{Y} \left(\frac{\partial^2 q}{\partial x^2} \right) \frac{\partial^2}{\partial x^2} (\delta q) dx dt = \int_{t_1}^{t_2} \left\{ \mathcal{Y} \frac{\partial^2 q}{\partial x^2} \delta \left(\frac{\partial q}{\partial x} \right) - \mathcal{Y} \frac{\partial^3 q}{\partial x^3} \delta q \right\} \Big|_0^l dt$$
$$+ \int_{t_1}^{t_2} \int_0^l \mathcal{Y} \frac{\partial^4 q}{\partial x^4} \delta q \, dx dt \qquad (12.1.41)$$

Therefore, we obtain,

$$\delta A = \int_0^l \mu \frac{\partial q}{\partial t} \delta q \Big|_{t_1}^{t_2} dx + \mathcal{Y} \int_{t_1}^{t_2} \left\{ \frac{\partial^2 q}{\partial x^2} \delta \left(\frac{\partial q}{\partial x} \right) - \frac{\partial^3 q}{\partial x^3} \delta q \right\} \Big|_0^l dt \qquad (12.1.42)$$

$$- \int_{t_1}^{t_2} \int_0^l \left\{ \mu \frac{\partial^2 q}{\partial t^2} + \mathcal{Y} \frac{\partial^4}{\partial x^4} q \right\} \delta q \, dx dt = 0 \qquad (12.1.43)$$

The time boundary conditions $\delta(x, t_1) = \delta q(x, t_2) = 0$ mean the first term vanishes while the second term is determined by the beam boundary conditions. In this case, we assume that the start of the beam is clamped while the other end is free. In this case, the boundary conditions are $q(0, t) = \partial q(0, t)/\partial x = \partial^2 q(l, t)/\partial x^2 = \partial^3 q(l, t)/\partial x^3 = 0$. The Euler-Lagrange equation is then given by

$$\mu \frac{\partial^2 q}{\partial t^2} + \mathcal{Y} \frac{\partial^4 q}{\partial x^4} = 0 \qquad (12.1.44)$$

This is quite a difficult equation to solve.

13

Virtual Work & d'Alembert's Principle

Fig. 13.1: Jean d'Alembert (1717–1783).

In this chapter we will investigate d'Alembert's principle and other formulations of classical mechanics (for a portrait of d'Alembert see figure 13.1). We will discover exactly what context Lagrange's mechanics is valid in and investigate any assumptions we make in an exercise at the end of the chapter. Most textbooks will cover this material prior to discussing Lagrangian mechanics, as it is halfway between Newton's paradigm and analytical mechanics. The final type of mechanics we look at is the Gibbs-Appell formulation. These equations are perhaps the most general in mechanics as they incorporate non-holonomic constraints.

We begin by revisiting the generalised coordinates briefly and defining **virtual displacement**. Suppose we have a system of N particles described by a set of $3N$ cartesian coordinates. We subject the system to m constraints (meaning the accelerations are not independent), leaving $3N - m = n$ generalised coordinates defined by the relation

$$x_i = x_i(q^1, q^2, \ldots, q^n), \qquad q^i = q^i(x_1, x_2, \ldots, x_n) \tag{13.0.1}$$

We can find the inverse transformation if the Jacobian determinant is non-zero. Referring to chapter 9, we state two useful relations which will come in handy in a moment:

$$\frac{\partial x_i}{\partial q_j} = \frac{\partial \dot{x}_i}{\partial \dot{q}^j}, \qquad \frac{d}{dt}\frac{\partial x_i}{\partial q^j} = \frac{\partial \dot{x}_i}{\partial q^j} \tag{13.0.2}$$

The proof of the second is straightforward enough:

$$\frac{d}{dt}\left(\frac{\partial x_i}{\partial q^j}\right) = \sum_k \frac{\partial}{\partial q^k}\left(\frac{\partial x_i}{\partial q^j}\right)\dot{q}^k + \frac{\partial}{\partial t}\left(\frac{\partial x_i}{\partial q^j}\right)$$

Lagrangian & Hamiltonian Dynamics. Peter Mann, Oxford University Press (2018).
© Peter Mann. DOI: 10.1093/oso/9780198822370.001.0001

But, recalling that the order of partial derivatives is irrelevant,

$$\frac{\partial}{\partial q^j}\left(\frac{dx_i}{dt}\right) = \frac{\partial}{\partial q^j}\left(\sum_k \frac{\partial x_i}{\partial q^k}\dot{q}^k + \frac{\partial x_i}{\partial t}\right) = \sum_k \frac{\partial}{\partial q^j}\left(\frac{\partial x_i}{\partial q^k}\right)\dot{q}^k + \frac{\partial}{\partial q^j}\left(\frac{\partial x_i}{\partial t}\right)$$

A **virtual displacement** δx_i is an infinitesimal displacement in a given coordinate x_i; it is carried out at a given instant $dt = 0$ and is consistent with the forces and constraints. Therefore, the difference between the differential and the virtual displacement as follows:

$$dx_i = \sum_{j=1} \frac{\partial x_i}{\partial q^j}dq^j + \frac{\partial x_i}{\partial t}dt, \quad \text{while} \quad \delta x_i = \sum_{j=1} \frac{\partial x_i}{\partial q^j}\delta q^j \qquad (13.0.3)$$

What if we now return to the Newtonian formalism with generalised coordinates? Can we define a generalised force for instance? Consider the **virtual work** δW of an applied force $F^{(a)}$, that is, the work done by a system as it moves through virtual displacements δx_i:

$$\delta W = \sum_{i=1}^{3N} F_i^{(a)}\delta x_i = \sum_{i=1}^{3N} F_i^{(a)}\left(\sum_{j=1}^{3N} \frac{\partial x_i}{\partial q^j}\delta q^j\right) = \sum_{j=1}^{3N}\left(\sum_{i=1}^{3N} F_i^{(a)}\frac{\partial x_i}{\partial q^j}\right)\delta q^j \qquad (13.0.4)$$

If we recall that work is force dot distance, then we define a **generalised force** $\mathcal{F}_j$ as

$$\delta W = \sum_{j=1}^{3N} \mathcal{F}_j\delta q^j, \quad \text{with} \quad \mathcal{F}_j := \sum_{i=1}^{3N} F_i^{(a)}\frac{\partial x_i}{\partial q^j} \qquad (13.0.5)$$

Suppose that the system is in **static equilibrium**, that is, that the net force on each of the particles is zero ($\vec{F}_i = 0$) so there are no unbalanced forces (see chapter 12). An arbitrary virtual displacement in the coordinates that is consistent with the constraints is written $\vec{F}_i \cdot \delta\vec{x}_i$, which is still zero; summing them is just summing zeroes, so we can write that the virtual work is zero:

$$\delta W = \sum_i \vec{F}_i \cdot \delta x_i = 0 \qquad (13.0.6)$$

We split $\vec{F}_i = F_i^{(a)} + F_i^{(c)}$ into **applied forces** $F_i^{(a)}$ and **constraint forces** $F_i^{(c)}$, the forces keeping the system on the constraint surface and, by definition, the applied force is the total force minus the constraint force; we then obtain

$$\delta W^{(a)} + \delta W^{(c)} = \sum_i F_i^{(a)} \cdot \delta x_i + \sum_i F_i^{(c)} \cdot \delta x_i = 0 \qquad (13.0.7)$$

However, if the displacements are to be *consistent* with the constraints, then the net virtual work by $F_i^{(c)}$ is zero, the constraint forces are perpendicular to the constraint surface and the sum over all dot products vanishes[1]: $\sum F_i^{(c)} \cdot \delta x_i = 0$. This is, in fact, by definition, since we are assuming that the system is constrained to lie on the surface of satisfied constraint; if there was a net force from the constraints, then they would accelerate the system in some way, but this is not the nature of a constraining force. We thus obtain

$$\sum_i F_i^{(a)} \cdot \delta x_i = 0 \quad \textbf{virtual work} \tag{13.0.8}$$

This is the **principle of virtual work**. Since the variations can be linked via the constraint equations, we cannot say that the applied forces vanish independently. If, however, we say there is a linearly independent set of coordinates q^j, then we can write the virtual work as above and claim that the generalised force vanishes: $\mathcal{F}_i = 0$.

What if we now consider a dynamical system that has a net force? By Newton's second law, $\vec{F}_i = \dot{p}_i$. Therefore, as in the static system, we can build a formula around $\vec{F}_i - \dot{p}_i = 0$; in other words, the system is in equilibrium under the force $f_i = \vec{F}_i - \dot{p}_i$. Take virtual variations consistent with constraints and sum over zero elements:

$$\sum_i (\vec{F}_i - \dot{p}_i) \cdot \delta x_i = 0 \tag{13.0.9}$$

We then separate the constraint forces out:

$$\sum_i (\vec{F}_i^{(a)} - \dot{p}_i) \cdot \delta x_i + \sum_i F_i^{(c)} \cdot \delta x_i = 0 \tag{13.0.10}$$

Assuming the constraint forces do no work, the second term vanishes: this is the most important step. We thus obtain

$$\sum_i (\vec{F}_i^{(a)} - \dot{p}_i) \cdot \delta x_i = -\sum_i F_i^{(c)} \cdot \delta x_i = 0 \quad \textbf{d'Alembert's principle} \tag{13.0.11}$$

This is **d'Alembert's principle**. It states that the sum of the differences between the applied forces acting on the system and the rate of change of momenta of the system along any virtual displacement δx_i that respects the constraint forces is zero. It is celebrated because it removes the constraints from the picture. We should now split that up into two parts:

$$\sum_i F_i^{(a)} \cdot \delta x_i, \quad \text{and} \quad \sum_i \dot{p}_i \cdot \delta x_i \tag{13.0.12}$$

---

[1] We could be stricter and require that individual dot products vanish ($F^{(c)} \cdot \delta x_i = 0$) but it is more restrictive than only their sum vanishing.

Assuming that there exists a suitable set of generalised coordinates, we know how to deal with the first term, so let's focus on the second term. What follows is a long-winded series of intricate manipulations:

$$\sum_i \dot{p}_i \cdot \delta x_i = \sum_i m_i \ddot{x}_i \cdot \delta x_i$$

$$= \sum_i m_i \ddot{x}_i \cdot \left(\sum_j \frac{\partial x_i}{\partial q^j} \delta q^j \right) = \sum_i \sum_j \left(m_i \ddot{x}_i \cdot \frac{\partial x_i}{\partial q^j} \right) \delta q^j \qquad (13.0.13)$$

Consider the product rule $\frac{d}{dt}(A \cdot B) = \dot{A} \cdot B + A \cdot \dot{B}$ and then re-arrange for $\dot{A} \cdot B = \frac{d}{dt}(A \cdot B) - A \cdot \dot{B}$:

$$m_i \ddot{x}_i \cdot \frac{\partial x_i}{\partial q^j} = \frac{d}{dt}\left(m_i \dot{x}_i \cdot \frac{\partial x_i}{\partial q^j} \right) - m_i \dot{x}_i \cdot \frac{d}{dt}\left(\frac{\partial x_i}{\partial q^j} \right) \qquad (13.0.14)$$

The $\partial x_i / \partial q^j$ term is a familiar relation from the start of this chapter, as is the time derivative on the right. Expression 13.0.13 now becomes,

$$\sum_i \dot{p}_i \cdot \delta x_i = \sum_{i,j} \left[\frac{d}{dt}\left(m_i \dot{x}_i \cdot \frac{\partial \dot{x}_i}{\partial \dot{q}^j} \right) - m_i \dot{x}_i \cdot \frac{\partial \dot{x}_i}{\partial q^j} \right] \delta q^j \qquad (13.0.15)$$

In the final step, we consider two derivatives of the kinetic energy, one with respect to q^j and the other to $\dot{q}^j$:

$$\frac{\partial T}{\partial q^j} = \frac{\partial}{\partial q^j}\left(\sum_i \frac{1}{2} m_i \dot{x}_i^2 \right) = \sum_i m_i \dot{x}_i \cdot \frac{\partial \dot{x}_i}{\partial q^j}, \quad \frac{\partial T}{\partial \dot{q}^j} = \frac{\partial}{\partial \dot{q}^j}\left(\sum_i \frac{1}{2} m_i \dot{x}_i^2 \right) = \sum_i m_i \dot{x}_i \cdot \frac{\partial \dot{x}_i}{\partial \dot{q}^j}$$

These two expressions are exactly the interior of the brackets in the expression we are working on:

$$\sum_i \dot{p}_i \cdot \delta x_i = \sum_j \left[\frac{d}{dt}\left(\frac{\partial T}{\partial \dot{q}^j} \right) - \frac{\partial T}{\partial q^j} \right] \delta q^j \qquad (13.0.16)$$

Returning to d'Alembert's equation we find that

$$\sum_i (\vec{F}_i^{(a)} - \dot{p}_i) \cdot \delta x_i = \sum_j \left[\mathcal{F}_j - \frac{d}{dt}\left(\frac{\partial T}{\partial \dot{q}^j} \right) + \frac{\partial T}{\partial q^j} \right] \delta q^j = 0 \qquad (13.0.17)$$

Since the generalised coordinates are independent, then each bracket under the summation will vanish independently so we obtain

$$\boxed{ \mathcal{F}_j = \frac{d}{dt}\left(\frac{\partial T}{\partial \dot{q}^j} \right) - \frac{\partial T}{\partial q^j} \qquad \textbf{Central Lagrange equation} } \qquad (13.0.18)$$

with $j = 1, \ldots, n$. This is an alternative way in which to express d'Alembert's principle[2]. We can now think of the principle of virtual work as an action principle of sorts. As a particle moves between two points it could take a variety of trajectories, each with its own work. The principle of virtual work states that the physical trajectory is the path for which the virtual work (of the forces) is zero for all virtual displacements from equilibrium. The bracketed (forces) can be either $F_i^{(a)}$, in the static case or, $f_i = \vec{F}_i - \dot{p}_i$, in the dynamic case. Either way, we are saying that first-order virtual variations in the work vanish for the classical path.

So, how does the central Lagrange equation relate to the Euler-Lagrange equation? There are three particulars that we can consider for the generalised force and they all depended on the nature of the applied force $F_i^{(a)}$. The first is when $F_i^{(a)} = -\vec{\nabla}_i U(x)$, that is, when the applied force is derived from the gradient of a scalar potential. In this case,

$$\mathcal{F}_i = \sum_i F_i^{(a)} \cdot \frac{\partial x_i}{\partial q^j} = -\sum_i \frac{\partial U}{\partial x_i} \cdot \frac{\partial x_i}{\partial q^j} = -\frac{\partial U}{\partial q^j}$$

in which case we can add $-\partial U / \partial \dot{q}^j$ for neatness, since it is zero:

$$\frac{d}{dt}\left(\frac{\partial(T-U)}{\partial \dot{q}^j}\right) - \frac{\partial(T-U)}{\partial q^j} = \frac{d}{dt}\left(\frac{\partial \mathcal{L}}{\partial \dot{q}^j}\right) - \frac{\partial \mathcal{L}}{\partial q^j} = 0 \tag{13.0.19}$$

This is the familiar Euler-Lagrange equation of motion and it applies to conservative forces only. The next case assumes the applied force is velocity dependant: $F_i^{(a)} = -\frac{\partial U}{\partial q^j} + \frac{d}{dt}\left(\frac{\partial U}{\partial \dot{q}^j}\right)$. However, we know that a total time derivative of a function can be added to a Lagrangian without changing the equations of motion, so we regenerate the first case.

The general case is where the applied force cannot be derived from a scalar potential, known as the **Lagrange-d'Alembert equation**:

$$\frac{d}{dt}\left(\frac{\partial \mathcal{L}}{\partial \dot{q}^j}\right) - \frac{\partial \mathcal{L}}{\partial q^j} = \mathcal{F}_j, \quad \textbf{or} \quad \int_1^2 (\delta \mathcal{L} + \mathcal{F}\delta q)dt = 0 \tag{13.0.20}$$

where $\mathcal{L}$ contains those forces that are derivable from a scalar function, and $\mathcal{F}_j$ concerns the other forces. This equation is fully equivalent to Newton's second law and may be regarded as the most general equation in dynamics. The non-conservative forces can be identified with the constraint forces; and to show this consider the second law $\dot{p}_i = F_i^{(a)} + F_i^{(c)}$ and the generalised force. Note, however, that, since we are including the constraint forces, our generalised coordinates will not be the minimal set, since they

---

[2]To be technical, we should really reserve "Euler-Lagrange equation" for the result of Hamilton's principle and call this result the "Lagrange equation"; however, it is of little importance in reality.

must be inherently coupled by a constraint equation and we vary without respecting the constraints specifically:

$$\mathcal{F}_j = \sum_i \dot{p}_i \cdot \frac{\partial x_i}{\partial q^j} = \sum_i F_i^{(a)} \cdot \frac{\partial x_i}{\partial q^j} + \sum_i F_i^{(c)} \cdot \frac{\partial x_i}{\partial q^j} = -\frac{\partial U}{\partial q^j} + F_i^{(c)} \qquad (13.0.21)$$

Simply by plugging this into the Lagrange-d'Alembert equation, we obtain

$$\mathcal{F}_j = \frac{d}{dt}\left(\frac{\partial T}{\partial \dot{q}^j}\right) - \frac{\partial T}{\partial q^j} = -\frac{\partial U}{\partial q^j} + F_i^{(c)} \qquad (13.0.22)$$

Therefore, identifying the scalar potentials and kinetic energies with a Lagrangian function, we may write

$$\frac{d}{dt}\left(\frac{\partial (T-U)}{\partial \dot{q}^j}\right) - \frac{\partial (T-U)}{\partial q^j} = F_i^{(c)} \qquad (13.0.23)$$

Comparing this work with our previous discussion of the Lagrange multiplier method (see chapter 8), we identify the constraint forces with the constraint functions in the following way:

$$F_i^{(c)} = \sum_k \lambda_k \frac{\partial \phi_k}{\partial q^j} \qquad (13.0.24)$$

The most important step in d'Alembert's principle, however, is the vanishing of the constraint forces on the constraint surface. Note that there are some types of constraints which are complicated to treat and so, in those cases, we cannot exactly follow this procedure. We have assumed that the constraint can be defined such that $\delta\phi(\boldsymbol{q}, t) = 0$; then

$$\delta W^{(c)} = \sum_j \mathcal{F}_j^{(c)} \delta q^j = \sum_j \sum_k \lambda_k \frac{\partial \phi_k}{\partial q^j} \delta q^j = \sum_k \lambda_k \delta \phi_k = 0 \qquad (13.0.25)$$

Those that can be defined in this way are called **holonomic constraints**; those that cannot are called **non-holonomic** (see chapter 8). Non-holonomic systems have constraints on their velocity. Non-holonomic mechanics requires a more general framework and, in general, does not admit a variational principle approach. There is a wealth of theory exploring non-holonomic mechanics, including most notably **vakonomic mechanics** as we discussed earlier in chapter 8, where we tried to derive a variational principle for non-holonomic systems. Here, we will investigate non-holonomic systems in d'Alembert's principle.

The variation in the configuration δq^j induces a variation in the constraint $\delta\phi_k$, which we Taylor expand to first order and set to vanish:

$$\delta\phi_k = \phi_k(\boldsymbol{q} + \delta\boldsymbol{q}, \dot{\boldsymbol{q}} + \delta\dot{\boldsymbol{q}}, t) - \phi_k(\boldsymbol{q}, t)$$

$$= \left(\frac{\partial \phi_k}{\partial q^i}\right)\delta q^i + \left(\frac{\partial \phi_k}{\partial \dot{q}^i}\right)\delta\dot{q}^i \qquad (13.0.26)$$

In order that $\delta\phi_k = 0$, the configuration variations δq^i must be orthogonal to the gradients $\partial\phi_k/\partial\dot{q}^i$, which, in turn, are normal to the surface:

$$\left(\frac{\partial\phi}{\partial\dot{q}}\right)\cdot\delta q = 0 \tag{13.0.27}$$

Hence, using d'Alembert's principle of workless constraints, we have an expression for the generalised force in non-holonomic systems:

$$\delta W^{(c)} = \left[\mathcal{F}^{(c)} - \lambda\left(\frac{\partial\phi}{\partial\dot{q}}\right)\right]\cdot\delta q = 0 \tag{13.0.28}$$

Of course, δq is arbitrary, and the Lagrange-d'Alembert equation becomes

$$\frac{d}{dt}\left(\frac{\partial\mathscr{L}}{\partial\dot{q}^j}\right) - \frac{\partial\mathscr{L}}{\partial q^j} = \sum_k \lambda_k \frac{\partial\phi_k}{\partial\dot{q}^j}. \tag{13.0.29}$$

These equations of motion are the correct equations for non-holonomic systems (when the constraints are linear r affine I should add). Contrast these to those found in chapter 8 using the vakonomic variational approach! We note that the variations here were created without reference to the constraints but were then restricted to satisfying the constraint equation via the constraint force and λ.

13.1   Gauss's Least Constraint & Jourdain's Principle

Well before Hamilton's action principle, Johann Carl Friedrich Gauss in 1829 reformulated d'Alembert's principle into a minimisation procedure. In analogy to a least squares formula, Gauss's principle states that, at every moment the motion of a system of particles whose positions and velocities at time t are known, maximises its similarity to the free movement or obtains the **least constraint**. Constraint, Z, is measured by the sum of the products of mass and the square difference between the constrained acceleration and the acceleration of the free unconstrained motion (not satisfying the constraint):

$$Z := \sum_{i=1}^{N} \frac{1}{2m_i}(\vec{F}_i^{(a)} - \dot{p}_i)^2 \tag{13.1.1}$$

This formula is readily found from d'Alembert's formula; we then consider Taylor expansions of the actual coordinates $x_i(t+\delta t)$ and the coordinates of the unconstrained free motion $x_i^{(a)}(t+\delta t)$ to second order:

$$x_i(t+\delta t) = x_i(t) + \dot{x}_i(t)\delta t + \frac{1}{2}\ddot{x}_i(t)\delta t^2 \tag{13.1.2}$$

$$x_i^{(a)}(t + \delta t) = x_i(t) + \dot{x}_i(t)\delta t + \frac{1}{2}\ddot{x}_i^{(a)}(t)\delta t^2 \tag{13.1.3}$$

An important point to realise is that the positions and velocities at time t are known and hence the contribution to the variation at time $t + \delta t$ occurs from the variation in the accelerations:

$$\delta x_i(t + \delta t) = \frac{1}{2}\delta\left(\frac{\dot{p}_i}{m_i}\right)\delta t^2, \qquad \delta x_i^{(a)}(t + \delta t) = \frac{1}{2}\delta\left(\frac{\vec{F}_i^{(a)}(t)}{m_i}\right)\delta t^2 \tag{13.1.4}$$

Considering that $\delta x_i^{(a)}(t + \delta t) = 0$ (as the unconstrained motions at $t + \delta t$ are all completely specified by the information given at t (e.g $F^{(a)} = F^{(a)}(t, q^1, \ldots, q^n, \dot{q}^1, \ldots, \dot{q}^n)$)) we can write $\delta x_i(t + \delta t) = \delta x_i(t + \delta t) - \delta x_i^{(a)}(t + \delta t)$. Upon entering these variations into d'Alembert's principle, we obtain Gauss's formula:

$$\sum_{i=1}^{N}(\vec{F}_i^{(a)} - \dot{p}_i) \cdot \delta x_i = \sum_{i=1}^{N}\frac{1}{2m_i}(\vec{F}_i^{(a)} - \dot{p}_i) \cdot \delta(\vec{F}_i^{(a)} - \dot{p}_i) = 0 \tag{13.1.5}$$

If there are no constraints on the system, then Z is zero exactly; otherwise, it is a true minimum. In this way, Z is the difference between the force predicted by the free motion equations and the real value the system exhibits when the constraints are satisfied. The variation of the constraint δZ is then given by,

$$\sum_{i=1}^{N}(\vec{a}_i^{(a)} - \vec{a}_i) \cdot \delta\vec{a}_i = 0 \tag{13.1.6}$$

where we have fixed the free motion $\vec{a}_i^{(a)}$ and kept variations in the actual acceleration. The virtual variation in acceleration $\delta\vec{a}_i$ must be compatible with the constraints $\delta t = \delta x_i = \delta\dot{x}_i = 0$ and $\delta\ddot{x}_i \neq 0$.

A special case of Gauss's principle is **Hertz's least curvature**, which states that the trajectory the system follows under the constraints is the one that is the most straight (the geodesic) compared to any other possible path by minimising the curvature κ. It is valid when there are no applied forces and assumes the masses of the particles are identical.

Similar to both d'Alembert's and Gauss's principles, **Jourdain's principle** (published in 1909), another variational principle of mechanics, expresses the variations in the velocities such that time and position are fixed ($\delta t = \delta x = 0$):

$$\sum_{i=1}^{N}(\vec{F}_i^{(a)} - \dot{p}_i) \cdot \delta\dot{x}_i = 0 \tag{13.1.7}$$

So, we realise that we have three equivalent variational principles, each with different variations and each with their own benefits and drawbacks in practical calculations. To summarise, we list the equations of d'Alembert, Jourdain and Gauss, complete with their constraint-compatible variations:

d'Alembert $\qquad \sum_{i=1}^{N} (\vec{F}_i^{(a)} - m_i \ddot{x}_i) \cdot \delta_0 x_i = 0, \qquad \delta_0 t = 0 \qquad\qquad$ (13.1.8)

Jourdain $\qquad \sum_{i=1}^{N} (\vec{F}_i^{(a)} - m_i \ddot{x}_i) \cdot \delta_1 \dot{x}_i = 0, \qquad \delta_1 t = \delta_1 x_i = 0 \qquad$ (13.1.9)

Gauss $\qquad \sum_{i=1}^{N} (\vec{F}_i^{(a)} - m_i \ddot{x}_i) \cdot \delta_2 \ddot{x}_i = 0, \qquad \delta_2 t = \delta_2 x_i = \delta_2 \dot{x}_i = 0 \quad$ (13.1.10)

We have introduced the notation δ_i with $i = 0, 1, 2$ to represent the order of the variation. In each case, we see that the net forces on the particles $\vec{F}^{(c)}$ are orthogonal to the permissible variations $\delta \vec{\gamma}_i$ in acceleration (Gauss), velocity (Jourdain) or position (d'Alembert) and hence their dot product vanishes. This is generalised into the **Mangeron-Deleanu principle**, where the variations vanish up to a penultimate chosen order following the trend observed above. It is hence clear now why $F^{(a)} = F^{(a)}(t, x_1, \ldots, x_N, \dot{x}_1, \ldots, \dot{x}_N)$ was completely specified in Gauss's principle, as these quantities don't vary. We thus obtain

$$\sum_{i=1}^{N} \vec{F}^{(c)} \cdot \delta \vec{\gamma}_i = 0 \qquad\qquad (13.1.11)$$

The practical use of these equations is very limited, as the higher order terms are often tricky to compute.

In 1961, Kane redeveloped d'Alembert's principle into yet another formulation of classical mechanics. These equations are seldom found in the physics literature yet are a common tool in an engineering handbook due to their apparent computerization! There was much controversy on whether Kane's formulation was indeed novel and, over the ensuing 30 years, many have attacked Kane's equations as a simple reformulation of Jourdain's principle. Their argument lies in the expression of the variations used by Jourdain. Consider the time derivative of x_i:

$$\dot{x}_i = \sum_k \frac{\partial x_i}{\partial q^k} \dot{q}^k + \frac{\partial x_i}{\partial t} \qquad\qquad (13.1.12)$$

Then, by adding a dot, Jourdain's variation is $\delta_1 \dot{x}_i = (\partial \dot{x}_i / \partial \dot{q}^k) \delta_1 \dot{q}^k$. Since $\delta_1 \dot{q}^k$ is arbitrary, we may write Jordain's principle equivalently as Kane's equations, a system of $k = 1, \ldots, n$ scalar equations that describe the dynamics:

$$\sum_{i=1}^{N} \vec{F}_i^{(a)} \cdot \frac{\partial \dot{x}_i}{\partial \dot{q}^k} - \sum_{i=1}^{N} \dot{p}_i \cdot \frac{\partial \dot{x}_i}{\partial \dot{q}^k} = 0 \qquad (13.1.13)$$

The basic method of Kane is to write Newton's laws and then multiply by **partial velocities**, which to you and me are simply the partial derivatives of the position coordinates and the generalised coordinates $\partial \dot{x}_i / \partial \dot{q}^k$. It is actually this step that will eliminate the constraint forces and hence is considered the centre point of Kane's approach.

13.2 The Gibbs-Appell Equations

In this final section, we close this chapter by looking at the **Gibbs-Appell equations of motion**, which were developed independently in 1879 and 1899 by Josiah Gibbs and Paul Appell. The formulation is equivalent to that of d'Alembert but instead utilises generalised coordinates. As with the other principles of mechanics discussed in this chapter, it is more general than Hamilton's principle of stationary action, due to its ability to handle constraints of a difficult nature. We will first derive the equations of motion before considering a brief example at the end of this section. Considering the differential of the coordinates from the start of this chapter, we may write $x_i(q^1, \ldots, q^n)$ for a system of N particles $i = 1, \ldots, N$:

$$\delta x_i = \sum_{j=1}^{n} \frac{\partial x_i}{\partial q^j} \delta q^j \qquad (13.2.1)$$

The work done by the coordinates is δW:

$$\delta W = \sum_{j=1}^{n} \mathcal{F}_j \delta q^j = \sum_{i=1}^{N} \vec{F}_i \cdot \delta x_i = \sum_{i=1}^{N} m_i \vec{a}_i \cdot \delta x_i \qquad (13.2.2)$$

We then simply substitute the expression for the variation and swap the order of the summations:

$$\delta W = \sum_{j=1}^{n} \delta q^j \sum_{i=1}^{N} m_i \vec{a}_i \cdot \left(\frac{\partial x_i}{\partial q^j} \right) \qquad (13.2.3)$$

We therefore identify the generalised force with,

$$\mathcal{F}_j = \sum_{i=1}^{N} m_i \vec{a}_i \cdot \left(\frac{\partial x_i}{\partial q^j} \right) = \sum_{i=1}^{N} m_i \vec{a}_i \cdot \left(\frac{\partial \vec{a}_i}{\partial \alpha^j} \right). \qquad (13.2.4)$$

Upon realising that cancelling the dot (from appendix C) also means we can add the dot to the top and bottom, we can write $\partial x_i / \partial q^j = \partial \ddot{x}_i / \partial \ddot{q}^j$ and we let $\ddot{q}^j = \alpha^j$. By the introduction of the **Gibbs function** $G(q, \dot{q}, \ddot{q}, t)$ as the mass-weighted sum of the square accelerations, we arrive at the Gibbs-Appell equations of motion:

$$\mathcal{F}_j = \frac{\partial G}{\partial \alpha^j} \qquad \textbf{Gibbs-Appell equations} \qquad (13.2.5)$$

The system of $(3N - m)$ equations are perhaps the most general in classical mechanics, as they describe the dynamics of holonomically and non-holonomically constrained systems. The Gibbs-Appell equations see most of their application in engineering and robotics. The equations may also be derived straight from the central Lagrange equation:

$$\left[\frac{d}{dt} \left(\frac{\partial T}{\partial \dot{q}^j} \right) - \frac{\partial T}{\partial q^j} - \mathcal{F}_j \right] \delta q^j = 0 \qquad (13.2.6)$$

If we let $T = (1/2) \sum m_i (\dot{x} \cdot \dot{x})$ and evaluate the partial and time derivatives, we end up with

$$\left[\sum_{i=1}^{N} \left(m_i \ddot{x}_i \cdot \frac{\partial \dot{x}_i}{\partial \dot{q}^j} + m_i \dot{x}_i \cdot \frac{d}{dt} \left(\frac{\partial \dot{x}_i}{\partial \dot{q}^j} \right) - m_i \dot{x}_i \cdot \frac{\partial \dot{x}_i}{\partial q^j} \right) - \mathcal{F}_j \right] \delta q^j = 0 \qquad (13.2.7)$$

From equation 13.2.19, we see that $d/dt(\partial_{\dot{q}} \dot{x}) = \partial_q \dot{x}$, as well as adding a dot top and bottom to the first term; we thus obtain

$$\sum_{i=1}^{N} \left(m_i \ddot{x}_i \cdot \frac{\partial \ddot{x}_i}{\partial \ddot{q}^j} - \mathcal{F}_j \right) \delta q^j = \left(\frac{\partial}{\partial \ddot{q}^j} \sum_{i=1}^{N} \frac{1}{2} m_i (\ddot{x}_i \cdot \ddot{x}_i) - \mathcal{F}_j \right) \delta q^j$$

$$= \left(\frac{\partial G}{\partial \ddot{q}^j} - \mathcal{F}_j \right) \delta q^j \qquad (13.2.8)$$

With independent variations, we arrive at the Gibbs-Appell equations. A third way to derive the Gibbs-Appell equations is to multiply the brackets of Gauss's least constraint formulation and retain only terms which contain the true accelerations. The mixed term turns into the generalised force multiplied by the generalised acceleration:

$$Z = \frac{1}{2} \sum_{i=1}^{N} m_i \ddot{x}_i^2 - \sum_{k=1}^{m} \mathcal{F}_k \ddot{q}^k \qquad (13.2.9)$$

When entered into the Gauss principle, this generates the Gibbs-Appell equations. Gibbs is well known in thermodynamics due to the Gibbs free energy; however, he developed key results in the unification of classical statistical mechanics and thermodynamics, as well as in ergodic theory.

Chapter summary

- The **principle of virtual work** for statics was developed into **d'Alembert's principle** for dynamical systems:

$$\sum_i (\vec{F}_i^{(a)} - \dot{p}_i) \cdot \delta x_i = 0$$

It is a convenient way to eliminate constraint forces in the Newtonian formalism.

- It was shown that d'Alembert's principle can be re-expressed as the **central Lagrange equation**, which is more general than the Euler-Lagrange equation since it incorporates non-conservative forces:

$$\mathcal{F}_j = \frac{d}{dt}\left(\frac{\partial T}{\partial \dot{q}^j}\right) - \frac{\partial T}{\partial q^j}$$

- The principles of d'Alembert, Gauss and Jourdain were compared and the Gibbs-Appell equations were derived.

Exercise 13.1 In going from d'Alembert's principle to the Lagrange equation, we have to make some assumptions along the way. Although we have already shown the equivalence of the central Lagrange equation and that of virtual work, here we will show explicitly the equivalence of d'Alembert's principle to Hamilton's action principle and count the assumptions we make along the way. Consider this a review of Lagrangian mechanics from more general equations. Before we start, we realise that the coordinates x_i are independent. We begin by integrating equation 13.0.11:

$$\int_1^2 \sum_i (\vec{F}_i^{(a)} - \dot{p}_i) \cdot \delta x_i dt = 0 \tag{13.2.10}$$

We make the *assumption* that $\vec{F}_i^{(a)} \cdot \delta x_i = -\delta U(x)$, that is, the external forces are conservative. Next, we consider the total derivative $\dot{p}_i = m_i \dot{v}_i$:

$$\int_1^2 m_i \frac{d}{dt}(v_i \cdot \delta x_i) dt = \int_1^2 m_i \frac{dv_i}{dt} \cdot \delta x_i dt + \int_1^2 m_i v_i \cdot \frac{d}{dt}(\delta x_i) dt \tag{13.2.11}$$

We can see that the second term is present in d'Alembert's principle. The term on the left, being a total derivative, can be ignored provided we *assume* that the variations vanish on the boundary. This is actually not restrictive; as both d'Alembert's principle and Hamilton's principle account for the dynamics of the system that is specified at t_1 and t_2, the variational path should coincide with the true path at these points. The last term on the right of 13.2.11 can be rewritten by *assuming* that the variation δ commutes with the time derivative:

$$\int_1^2 m_i v \cdot \frac{d}{dt}(\delta x_i) dt = \int_1^2 m_i v_i \cdot (\delta v_i) dt = \delta \int_1^2 \left(\frac{1}{2} m_i v_i \cdot v_i\right) dt \tag{13.2.12}$$

The final step is to pop the sum back in and rewrite d'Alembert's equation:

$$\delta \int_1^2 \sum_i \left(\frac{1}{2} m_i v_i \cdot v_i\right) dt - \delta \int_1^2 V \, dt = 0 \tag{13.2.13}$$

It is clear, therefore, that d'Alembert's principle is more general! We can investigate the assumption that the variation commutes with the time derivative; by evaluating the quantities $\dot{x}_i$, δx_i, $\delta \dot{x}_i$ and $d/dt(\delta x_i)$, given $x_i = x_i(t, q^1, \ldots, q^n)$ with $i = 1, \ldots, N$:

$$\dot{x}_i = \sum_j^n \frac{\partial x_i}{\partial q^j} \dot{q}^j + \frac{\partial x_i}{\partial t} \tag{13.2.14}$$

$$\delta x_i = \sum_j^n \frac{\partial x_i}{\partial q^j} \delta q^j, \qquad \delta t = 0 \tag{13.2.15}$$

$$\delta \dot{x}_i = \sum_{j,k}^n \frac{\partial^2 x_i}{\partial q^j \partial q^k} \dot{q}^j \delta q^k + \sum_i^n \frac{\partial x_i}{\partial q^j} \delta \dot{q}^j + \sum_i^n \frac{\partial^2 x_i}{\partial t \partial q^j} \delta q^j \tag{13.2.16}$$

$$\delta_1 \dot{x}_i = \sum_i^n \frac{\partial x_i}{\partial q^j} \delta_1 \dot{q}^j \tag{13.2.17}$$

$$\frac{d}{dt}(\delta x_i) = \sum_{j,k}^n \frac{\partial^2 x_i}{\partial q^j \partial q^k} \dot{q}^j \delta q^k + \sum_j^n \frac{\partial x_i}{\partial q^j} \frac{d}{dt}(\delta q^j) + \sum_j^n \frac{\partial^2 x_i}{\partial q^j \partial t} \delta q^j \tag{13.2.18}$$

$$\ddot{x}_i = \frac{\partial x_i}{\partial q^j} \ddot{q}^j + \frac{\partial^2 x_i}{\partial q^j \partial q^k} \dot{q}^j \dot{q}^j + 2 \frac{\partial^2 x_i}{\partial q^j \partial t} \dot{q}^j + \frac{\partial^2 x_i}{\partial t^2} \tag{13.2.19}$$

We then find that the velocity of the variation and the variation of the velocity are not always equivalent(!):

$$\delta \dot{x}_i - \frac{d}{dt}(\delta x_i) = \sum_j^n \frac{\partial x_i}{\partial q^j} \left(\delta \dot{q}^j - \frac{d}{dt}(\delta q^j)\right) \tag{13.2.20}$$

We see that, while d'Alembert's principle does not assume commutation, the *definition* of the variation in Hamilton's principle assures us that the two commute. The variation between the extremal path q and a neighbouring path q' is $\delta q = q' - q$. It follows that $\delta \dot{q} = \dot{q}' - \dot{q}$ and, by the linearity of the derivative operator, $d/dt(\delta q)$. It is hence commutative by construction.

The final assumption we investigate is that the force must be conservative. A time-dependent force cannot be conservative, since the time and coordinates are not independent of each other. The associated work is then an explicit function of time and thus the total work done may vary, depending on the parametrisation. We may generalise this discussion somewhat by defining the force to be a function of the nth time derivative of $x_i(t)$ with some constant of proportionality greater than zero:

$$F(x) = -k \frac{d^n x}{dt^n}, \qquad U(x^{(n)}) = (-1)^{n/2} \frac{k}{2} \left(\frac{d^{n/2} x}{dt^{n/2}}\right)^2 \tag{13.2.21}$$

where n is a non-negative integer. Three important cases are when $n = 0, 1, 2$ (i.e. the force is dependent on the position, velocity or acceleration). In the case of $n = 0$, we generate Hooke's theory: $F = -kx$, $U(x) = 1/2(kx^2)$. In the case $n = 1$, we generate frictional forces proportional to the velocity:$F = -k\dot{x}$; in addition we see there is no associated potential term. When $n = 2$ and the force is proportional to the acceleration:$F = -k\ddot{x}$, $U = -1/2(k\dot{x} \cdot \dot{x})$. For $n = 3$, we derive the Abraham-Lorentz force $F = -k\dddot{x}$, with $k < 0$. We see the trend here: when n is odd, we cannot find an associated potential function since, in the general case, we do not allow for fractional derivatives.

We conclude by saying that velocity dependent *potentials* do admit an action principle, as we have seen using the Lorentz force, the centrifugal force and the Coriolis forces for rigid bodies. This can also be understood by using Helmholtz conditions in Douglas' theorem on the existence of a variational principle. In particular, we have that the entries of the non-singular matrix g_{ij} are symmetric under permutation of the indices in an action formulation:

$$\frac{\delta}{\delta x^i(t)} \frac{\delta}{\delta x^j(t')} = \frac{\delta}{\delta x^j(t')} \frac{\delta}{\delta x^i(t)} \tag{13.2.22}$$

If the force was equal to the negative functional derivative of the potential part of the action, then we would generate a **Maxwell relation** for the force:

$$F_i(t) = -\frac{\delta}{\delta x^i(t)} \left(\int U dt \right), \qquad \frac{\delta F_i}{\delta x^j} - \frac{\delta F_j}{\delta x^i} = 0 \tag{13.2.23}$$

Performing this analysis for $n = 1$ in the above force-potential system, we find that the Maxwell relation is not satisfied and $n = 1$ forces do not admit velocity-dependent potentials!

$$\frac{\delta F_i(t)}{\delta x^j(t')} = -k \frac{d}{dt} \frac{\delta x^i(t)}{\delta x^j(t')} = -k\delta^{ij} \frac{d}{dt} \delta(t - t') \tag{13.2.24}$$

This is not true in even n cases which are symmetric under permutation. With the Helmholtz conditions not satisfied, the Lagrangian formalism does not hold without modification. If there exists a velocity-dependent potential, we may classify the force as conservative e.g the Lorentz force $U = e(\phi - \vec{v} \cdot \vec{A})$, and $\vec{F} = e(\vec{E} + \vec{v} \times \vec{B})$.

Exercise 13.2 We know that d'Alembert's principle is more general than Hamilton's action principle, due to the previous exercise. Therefore, we may wonder if Noether's theorem holds in d'Alembert's mechanics and the generalisation to time variations! Building off earlier work (see chapter 11 and appendix A), we have the following relations for Noether-type variations:

$$\delta q^k = q'^k(t) - q^k(t), \qquad \delta_0 q^k = q'^k(t + \delta t) - q^k(t) \tag{13.2.25}$$

By Taylor expansion, we may write the variation $\delta_0(*) = \delta(*) + d/dt(*)\delta t$ as follows:

$$\delta_0 q^k = \delta q^k + \dot{q}^k \delta t, \qquad \delta_0 \dot{q}^k = \delta \dot{q}^k + \ddot{q}^k \delta t. \tag{13.2.26}$$

We then have the following two relations by taking the time derivative of the above relations and assuming the commutation of the variation δ and the time derivative (Holder type, see section 7.8). We also find by considering $\delta_0 \dot{q}^k$ and $d/dt(\delta_0 q^k)$, that the Noether-type variation is not commutative with the time derivative:

$$\frac{d}{dt}(\delta_0 q^k) = \delta_0 \dot{q}^k + \dot{q}^k \frac{d}{dt}\delta t, \qquad \frac{d}{dt}(\delta q^k) = \delta_0 \dot{q}^k - \ddot{q}^k \delta t. \qquad (13.2.27)$$

The next step is to insert the variations $\delta q^k = \delta_0 q^k - \dot{q}^k \delta t$ and $d/dt(\delta q^k) = \delta_0 \dot{q}^k - \ddot{q}^k \delta t$ into Lagrange-d'Alembert's equation which we have rewritten using the product rule for the time derivative:

$$\frac{d}{dt}\left\{ \frac{\partial \mathscr{L}}{\partial \dot{q}^k} \delta q^k \right\} - \frac{\partial \mathscr{L}}{\partial \dot{q}^k} \frac{d}{dt}\delta q^k - \frac{\partial \mathscr{L}}{\partial q^k} \delta q^k - \mathcal{F}_k \delta q^k = 0 \qquad (13.2.28)$$

With the addition and subtraction of $(\partial \mathscr{L}/\partial t)\delta t$ and the use of the product rule, we obtain

$$\frac{d}{dt}\left\{ \frac{\partial \mathscr{L}}{\partial \dot{q}^k}(\delta_0 q^k - \dot{q}^k \delta t) + \mathscr{L}\delta t \right\} = \delta_0 \mathscr{L} + \mathscr{L}\frac{d}{dt}\delta t + \mathcal{F}_k(\delta_0 q^k - \dot{q}^k \delta t) \qquad (13.2.29)$$

where we have defined the Noether-type variation of a function as

$$\delta_0 \mathscr{L} = \frac{\partial \mathscr{L}}{\partial q^k} \delta_0 q^k + \frac{\partial \mathscr{L}}{\partial \dot{q}^k} \delta_0 \dot{q}^k + \frac{\partial \mathscr{L}}{\partial t} \delta t. \qquad (13.2.30)$$

If the right-hand side vanishes, then the term in brackets is conserved, therefore showing Noether's theorem generalises to d'Alembert's principle. Extensions of the current form of this derivation to Suslov-type variations should be made!

It should also be clear that we can derive this result from Jourdain's principle too if we consider that $\delta_0^1 q^k = \delta^1 t = 0$ (the superscript 1 indicating Jourdainian variations, and the subscript 0 indicating Noether-type variations). Popping these variations into the above equations (retaining $d/dt(\delta_0^1 q^k) \neq 0$ and $d/dt(\delta^1 t) \neq 0$ terms) yields the same Noether condition as anticipated.

Exercise 13.3 Here we will consider a brief illustration of the Gibbs-Appell equations to derive the equations of motion of a system of **rigid body** particles connected by rigid rods. The general process involves determining the degrees of freedom of the system and evaluating the Gibbs function with $\ddot{\boldsymbol{r}}_i = \ddot{\boldsymbol{r}}_i(\ddot{q}_1, \ldots, \ddot{q}_n)$. The rigid body system has three translational degrees of freedom and three rotational degrees and the total angular momentum is conserved. Building on our previous work in chapter 3, and referring to figure 11.1, we find that the work done δW on rotating the body by an angle $\delta\boldsymbol{\phi}$ is

$$\delta W = \int_1^2 \vec{F} \cdot \delta \boldsymbol{r} = \int_1^2 \vec{F} \cdot (\delta \boldsymbol{\phi} \times \boldsymbol{r}) = \int_1^2 \boldsymbol{r} \times \vec{F} \cdot \delta \boldsymbol{\phi} = \int_1^2 \vec{\tau} \cdot \delta \boldsymbol{\phi}. \qquad (13.2.31)$$

Hence, we identify the torque vector $\vec{\tau}$ as the generalised force in the rotation. Working from the identity $\delta \boldsymbol{r} = \delta \boldsymbol{\phi} \times \boldsymbol{r}$, we have that the velocity of the particle is given by the cross product of the angular velocity $\boldsymbol{\omega}$ and the position $\dot{\boldsymbol{r}} = \boldsymbol{\omega} \times \boldsymbol{r}$. It follows that the acceleration of the ith particle is the time derivative of $\dot{\boldsymbol{r}}_i$:

$$\ddot{\boldsymbol{r}}_i = \left(\frac{d\boldsymbol{\omega}}{dt} \times \boldsymbol{r}_i \right) + \boldsymbol{\omega} \times \dot{\boldsymbol{r}}_i \qquad (13.2.32)$$

The Gibbs function is then constructed by evaluating $(\ddot{\boldsymbol{r}}_i \cdot \ddot{\boldsymbol{r}}_i)$. Note, in the second step, we have canceled the $(\boldsymbol{\omega} \times \dot{\boldsymbol{r}}_i)^2$ as it does not depend on $\boldsymbol{\alpha}_i = d\boldsymbol{\omega}/dt$:

$$
\begin{aligned}
G &= \frac{1}{2} \sum_{i=1}^{N} m_i (\ddot{\boldsymbol{r}}_i \cdot \ddot{\boldsymbol{r}}_i) \\
&= \frac{1}{2} \sum_{i=1}^{N} m_i \left\{ (\boldsymbol{\alpha}_i \times \boldsymbol{r}_i)^2 + 2(\boldsymbol{\alpha}_i \times \boldsymbol{r}_i) \cdot (\boldsymbol{\omega} \times \dot{\boldsymbol{r}}_i) \right\} \\
&= \frac{1}{2} \sum_{i=1}^{N} m_i \left\{ \boldsymbol{\alpha}_i \cdot (\boldsymbol{r}_i \times (\boldsymbol{\alpha} \times \boldsymbol{r}_i)) + 2\boldsymbol{a}_i \cdot (\boldsymbol{r} \times (\boldsymbol{\omega} \times \dot{\boldsymbol{r}}_i)) \right\} \\
&= \frac{1}{2} \sum_{i=1}^{N} m_i \left\{ \boldsymbol{\alpha}_i \cdot (\boldsymbol{r}_i \times (\boldsymbol{\alpha} \times \boldsymbol{r}_i)) + 2\boldsymbol{a}_i \cdot (\boldsymbol{\omega} \times (\boldsymbol{r}_i \times (\boldsymbol{\omega} \times \boldsymbol{r}_i))) \right\}
\end{aligned}
\tag{13.2.33}
$$

We have used the identities $\boldsymbol{a} \cdot (\boldsymbol{b} \times \boldsymbol{c}) = \boldsymbol{b} \cdot (\boldsymbol{c} \times \boldsymbol{a})$ and $\boldsymbol{a} \times (\boldsymbol{b} \times \boldsymbol{c}) = \boldsymbol{b}(\boldsymbol{a} \cdot \boldsymbol{c}) - \boldsymbol{c}(\boldsymbol{a} \cdot \boldsymbol{b})$, which we entitle the *dot* and *triplet relations*, respectively, for later use. In the last step, we realised that $\boldsymbol{r} \times (\boldsymbol{\omega} \times (\boldsymbol{\omega} \times \boldsymbol{r})) = -(\boldsymbol{\omega} \times \boldsymbol{r})(\boldsymbol{r} \cdot \boldsymbol{\omega})$, since $\boldsymbol{\omega} \times \boldsymbol{r}$ is perpendicular to both $\boldsymbol{\omega}$ and $\boldsymbol{r}$ and the first term vanishes in the triplet relation. We then used the fact that this quantity is the same as $\boldsymbol{\omega} \times (\boldsymbol{r} \times (\boldsymbol{\omega} \times \boldsymbol{r}))$ using the triplet relation with $\boldsymbol{a} = \boldsymbol{\omega}$, $\boldsymbol{b} = \boldsymbol{r}$ and $\boldsymbol{c} = (\boldsymbol{\omega} \times \boldsymbol{r})$. We then need to perform the same analysis again on both $\boldsymbol{r} \times (\boldsymbol{\alpha} \times \boldsymbol{r})$ and $\boldsymbol{r} \times (\boldsymbol{\omega} \times \boldsymbol{r})$ using the triplet relation. Hence, the Gibbs function is found to be,

$$
G = \frac{1}{2} \boldsymbol{\alpha}_i \cdot \boldsymbol{I} \cdot \boldsymbol{\alpha}_i + \boldsymbol{\alpha}_i \cdot (\boldsymbol{\omega} \times (\boldsymbol{I} \cdot \boldsymbol{\omega}))
\tag{13.2.34}
$$

where we have introduced the **inertia tensor** for the system. We thus obtain

$$
\sum_{i=1}^{N} \boldsymbol{r}_i \times m_i(\boldsymbol{X} \times \boldsymbol{r}_i) = \boldsymbol{X} \sum_{i=1}^{N} m_i r_i^2 = \boldsymbol{XI}
\tag{13.2.35}
$$

The derivative of the Gibbs function with respect to α, complete with the generalised force, gives the equations of motion which are otherwise known as the **Euler equations** for the rigid body:

$$
\boldsymbol{I} \cdot \boldsymbol{\alpha} + \boldsymbol{\omega} \times (\boldsymbol{I} \cdot \boldsymbol{\omega}) = \boldsymbol{\tau}
\tag{13.2.36}
$$

Confirm that this result is also obtained with Newton's approach and Lagrange's by checking our work previously in this chapter! The practicality of the Gibbs-Appell method is very dependent on the situation it is applied to. In this case, the computation of G was not too involved, but, in general, it is quite a procedure!

Exercise 13.4 In the derivation of d'Alembert's principle it is vital that the constraint forces do no work. For a system of particles, let us here examine the decomposition of the total force acting on the ith particle as

$$
\vec{F}_i = F_i^{(a)} + F_i^{(c)} + \sum_j F_{ji}
\tag{13.2.37}
$$

where $\vec{F}_i$ is the total force, $F_i^{(a)}$ is the applied force, $F_i^{(c)}$ is the constraint force and F_{ji} is the force on the ith particle due to another particle j summed over all particles in the system.

We now consider the virtual work δW done under a virtual displacement δr_i:

$$\delta W = \sum_{i=1}^{N}(F_i^{(a)} + F_i^{(c)}) \cdot \delta r_i + \sum_{i,j \neq i} F_{ji} \cdot \delta r_i \qquad (13.2.38)$$

This is different form our earlier decomposition (of just the applied and constraint forces in equation 13.0.7) due to the double sum. Here, we will show how Newton's third law *and* the assumption that the inter-particle distances $|\vec{r}_{ij}|$ are fixed (e.g are rigid), leads to the vanishing of the double sum, in agreement with previous results in this chapter.

Newton's third law states that the forces between two particles i and j are equal and opposite $\vec{F}_{ij} = -\vec{F}_{ji}$. In addition, we may demand that the inter-particle forces act in a straight line between the two particles. This indicates that $\vec{F}_{ij}$ is parallel to the difference in position vector, $\vec{r}_{ij} = \vec{r}_j - \vec{r}_i$. Given these results we have the condition that the variation of the constrained distances vanish:

$$\delta|\vec{r}_{ij}|^2 = 2\vec{r}_{ij} \cdot (\delta\vec{r}_j - \delta\vec{r}_i) = 0 \qquad (13.2.39)$$

and hence

$$\vec{F}_{ij} \cdot (\delta\vec{r}_j - \delta\vec{r}_i) = 0 \qquad (13.2.40)$$

Now, the double sum vanishes as

$$\sum_{i \neq j} \vec{F}_{ij} \cdot \delta\vec{r}_j - \sum_{i \neq j} \vec{F}_{ij} \cdot \delta\vec{r}_i = \sum_{i \neq j} \vec{F}_{ij} \cdot \delta\vec{r}_j - \sum_{i \neq j} \vec{F}_{ji} \cdot \delta\vec{r}_j \qquad (13.2.41)$$

with a change of index, and

$$2\sum_{i \neq j} \vec{F}_{ij} \cdot \delta\vec{r}_j = 0 \qquad (13.2.42)$$

Therefore, we can legitimately ignore the internal forces F_{ij} when computing the work done.

Part III

Canonical Mechanics

In the previous two sections, we developed the framework of classical mechanics. We come now to the subject of Hamiltonian mechanics as developed by the Irish mathematician Sir William Rowan Hamilton in 1833. Hamiltonian mechanics reformulates mechanics into a momentum rather than the velocity phase space approach. We will see in due course a switch of focus from qs and q̇s to qs and ps, where the conjugate momentum to the general coordinate plays a central role rather than some quantity we derive as an after thought. You may wonder after seeing the power of the Lagrange equation why we should trouble ourselves with another formulation but, as we will see, Hamiltonian mechanics constitutes a mathematically elegant structure of nature and lays the foundation of modern theoretical physics.

Although primarily about Hamiltonian mechanics this, section is far more reaching and will hopefully solidify our ideas about phase space before defining the Poisson formalism of mechanics, Hamilton-Jacobi theory and the Liouville equation of motion, which is extremely useful to molecular mechanics: hence the title "Canonical Mechanics" and not "Hamiltonian Mechanics".

We will cover canonical coordinate transformations and consider associated topics such as action-angle coordinates, integrability and resonant tori in phase space. Constrained dynamics in the Hamiltonian formalism will be considered; here the importance of the Hessian condition to transitioning between Lagrangian and Hamiltonian descriptions plays a vital role. Geometry in mechanics is formulated in both time-dependent and autonomous settings. Canonical perturbation theory and systems with no analytical solutions are explored too, as well as the generalisation of the Hamilton-Nambu formalism.

For the first few chapters, we take a very slow path through the elementary foundations of Hamiltonian dynamics, exploring every question a reader may have regarding the transition from the Lagrangian formalism. However, the style of our journey changes once we reach the midpoint of the Hamilton-Jacobi theory, after which we will know enough about the necessary mathematics and mechanics to be a little more eloquent in our discussion. Such topics after this turning point can be regarded as advanced and may be skipped upon first reading the canonical formalism. The next jump occurs when we discuss geometrical mechanics, due to the formality required inherently by the material. This topic may be skipped to continue to perturbation theory and classical field theory without too much detriment; however, in reality, it is the most enlightening part of the book. Okay, let's get started on what I find the most fascinating area of classical physics!

14

The Hamiltonian & Phase Space

Hamilton's equations can be derived in several ways and, in a fashion similar to that used in previous chapters, we will follow two pathways to arrive at the same result, with this approach giving us insight into the motivation for forming these equations. Both our pathways start from the **Hamiltonian**, the exposition of which is the focus of this chapter. This is done via a **Legendre transform** of the Lagrangian with respect to the generalised velocity coordinates. We will then derive **Hamilton's equations**, firstly via a stationary action principle before using a more traditional approach. Of course, both approaches are entirely equivalent and give us the same equations! The importance of deriving the same result in several ways is to show you that, in physics, there are often several mathematical avenues we can go down and that each time we learn that approaching a problem with, say, the calculus of variations can be entirely as valid as using a differential equation approach.

In order to motivate Hamilton's equations, we must first consider phase space and Lagrange's equations briefly. In Lagrange's formalism, we picked $3N - m$ *independent* generalised coordinates that describes a $3N - m$ dimensional configuration space; then we chose a generalised velocity to go with each of these coordinates and defined the velocity space as being $2(3N - m)$ dimensional, consisting of both q and $\dot{q}$. In this velocity space, we had a function called a Lagrangian, to which we then applied the stationary action principle and found the Lagrange equations for each variable, to arrive at the equations of motion for the system. Alternatively, we could consider the full-dimensional configurational space (rather than some submanifold which the constraint surfaces have been reduced from), and use the full equation where the Lagrange multipliers are present to again derive equations of motion for the system. This is Lagrangian mechanics.

With the generalised coordinate approach, we have $3N - m = n$, second-order ODEs for the equations of motion of a system. There are n independent generalised coordinates and n independent generalised velocities. This means we need to know $2n$ variables at a particular time to describe a point in velocity space: n coordinates and n velocities. The upshot is we need to know $2n$ things about the system in order to have **Laplacian determinacy** over it (defined later). Physically, we can say that it is not enough to know where something is alone; rather, we must specify its time evolution

Lagrangian & Hamiltonian Dynamics. Peter Mann, Oxford University Press (2018).
© Peter Mann. DOI: 10.1093/oso/9780198822370.001.0001

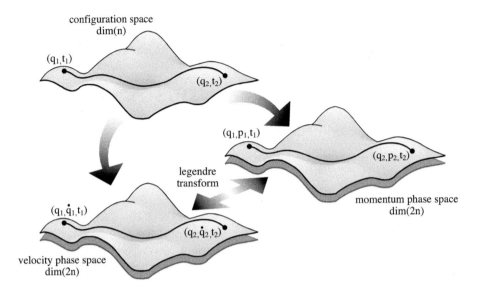

Fig. 14.1: The relationship between configuration, velocity and momentum phase space.

too via a velocity coordinate in order to map out a trajectory. Configuration space is n dimensional, and a curve in it describes the position evolution of the coordinates. Velocity space is $2n$ dimensional; a curve in which is described by position and velocity vectors. Trajectories through configuration space can cross the same point and, as they do, they will have identical generalised coordinates yet can have *different* velocity coordinates.

The Hamiltonian approach is based on **momentum phase space**, not velocity space. A point in phase space is parametrised by $p_j(t)$ and $q^j(t)$, it has twice the dimensionality of the configuration space, since, for every coordinate, we specify a momentum, too. The generalised coordinates q^j and the conjugate momentum p_j are treated equally, and a point in phase space instantly specifies where something is *and* where it's going next; it intrinsically describes the state of the system and its infinitesimal evolution to the next state. Previously, when we talked about Newtonian mechanics, we posited that *knowing the position and momentum allows us to describe fully the state of a system*; well, Hamiltonian mechanics really highlights this: "a point in phase space is a (pure) state of the system" or "a (pure) state is a point in phase space". Velocity space and phase space are essentially the same except that the dynamical description in the former is $(q, \dot{q})$ while in the latter it is (q, p). We use the Legendre transform (below) to convert between velocity and momentum phase spaces: it can be viewed as a mapping between the two spaces (see figure 14.1).

A point in phase space is unique to a given trajectory, since momentum is built into that point and trajectories do not cross over each other, by definition. That is not to say we cannot have a circular loop associated with periodic motion, where the trajectory retraces itself; in fact, this is what the harmonic oscillator does.

Our aim is to find a function that describes this unique evolution in phase space in terms of the generalised coordinates and their conjugate momenta. In our toolbox of mathematical tricks, we know that the **Legendre transform** sounds like a tempting method for changing variables and, since we want to get rid of $\dot{q}^j$, we perform the transform with respect to all of the velocity variables in the Lagrangian. (At this point, you may find it useful to review the Jacobi energy function before proceeding.)

Okay, with this in mind, let's move on to what we initially set out to do, that is, write out a function that is dependent on q^j, p_j and possibly t and which will encode a phase space trajectory rather than a velocity space trajectory like the Lagrangian did. We work from the Jacobi energy function, which, with mathematical subtly, can be converted into a function of the conjugate momenta *instead* of the generalised velocities. To do this, we must invert the relation between the conjugate momenta and the velocities; theoretically this is rather simple but, in practice, it may be a rather difficult step to carry out. Mathematically, this means we rearrange the following expression for $\dot{q}$ once we have a given expression for the Lagrangian, such that we have $\dot{q}^i = \dot{q}^i(\boldsymbol{q}, \boldsymbol{p}, t)$:

$$p_i := \frac{\partial \mathscr{L}}{\partial \dot{q}^i} \qquad \textbf{conjugate momentum} \tag{14.0.1}$$

Exercise 14.1 Let's pause and see what this inversion means for two specific simple Lagrangians. Firstly, if the Lagrangian is of the form $\mathscr{L} = \frac{1}{2}m\dot{x}^2 - U(x)$, then the conjugate momentum is given by $p = m\dot{x}$, which, once inverted (rearranged), gives us $\dot{x} = p/m$.

For a Lagrangian of the form $\mathscr{L} = \frac{1}{2}m(\dot{r}^2 + r^2\dot{\theta}^2) - U(r)$, the conjugate momentum of the θ coordinate is given by $p_{\dot{\theta}} = \partial \mathscr{L}/\partial \dot{\theta}$, which we calculate and invert to give us $\dot{\theta} = p/mr^2$. Note, in both cases, the generalised velocity is a function of the momentum and the coordinate itself.

We now substitute all of the $\dot{q}$s in the Jacobi energy function for these inverted expressions. Note that, although we write $\dot{q}^i$ below, you can see how it can be written purely in terms of p, q and t. Now, J goes from being a function of $q, \dot{q}, t$ to being a function of q, p, t and, when this transition occurs, we call it the **Hamiltonian** H instead of J:

$$H(\boldsymbol{p}, \boldsymbol{q}, t) = \sum_{j=1}^{n} p_i \dot{q}^i - \mathscr{L}(\boldsymbol{q}, \dot{\boldsymbol{q}}, t) \tag{14.0.2}$$

$$H(p,q,t) = p_i \dot{q}^i(p,q,t) - \mathscr{L}(q, \dot{q}(p,q,t), t) \qquad \textbf{Hamiltonian function} \qquad (14.0.3)$$

Both of the $\dot{q}^i$s are the inverted expressions for the velocities in terms of the momenta and the coordinates; here we have also used **Einstein's summation convention** for the sum over all the coordinates. Mathematically, p, q without subscript, reads $(p, q) = (p_1, p_2, \ldots, p_n, q^1, q^2, \ldots, q^n)$; it has been coloured to highlight this point.

The Hamiltonian and the energy function are essentially the same, except that $J(q, \dot{q}, t)$ is a function of the generalised velocity whilst the Hamiltonian is a function of the conjugate momenta instead. One is based on velocity space while the other is based on the phase space. They both describe the same energy, so they will give the same value at the end, but the crucial point is *how* they both describe it[1].

A useful yet elementary example of the difference between J and H can be high-lighted as follows:

$$J(x, \dot{x}) = \frac{1}{2}m\dot{x}^2 + U(x) \qquad \text{and} \qquad H(x,p) = \frac{p^2}{2m} + U(x)$$

Exercise 14.2 Let's have a quick example to see how this works for a familiar Lagrangian. Consider the central force problem from chapter 7, exercise 7.2, where $\mathscr{L} = \frac{1}{2}m(\dot{r}^2 + r^2\dot{\theta}^2) - U(r)$. Take two conjugate momenta; one with respect to $\dot{r}$ and the other with respect to $\dot{\theta}$; these are both accounted for under the summation sign in the definition of the Hamiltonian:

$$H = \sum_i p_i \dot{q}^i - \mathscr{L}$$
$$= m\dot{r}\left(\frac{p_r}{m}\right) + mr^2\dot{\theta}\left(\frac{p_\theta}{mr^2}\right) - \frac{1}{2}m\left(\frac{p_r}{m}\right)^2 - \frac{1}{2}mr^2\left(\frac{p_\theta}{mr^2}\right)^2 + U(r)$$
$$= \frac{p_r^2}{2m} + \frac{1}{2}\frac{p_\theta^2}{mr^2} + U(r)$$

which is the total energy of the central force problem.

At this point, we have established the form of the Hamiltonian and, just as J was born out of the Lagrange equation, we can say that H actually has equivalent equations which together form the Hamiltonian formalism of mechanics.

Although we are now well equipped to go forth and derive Hamilton's equations, for the sake of being a little pedantic, we could really emphasise the Legendre transform aspect of what we have just done; if you feel you don't need this, then just skip ahead to chapter 16.

[1]When I first encountered J and H, I spent a lot of time thinking why their forms are so similar, but turning the question around helped me understand this: *why should they be different if they describe the same quantity, namely the energy, but are just functions of different variables?*

We discuss in detail the Legendre transform in chapter 33, so if necessary, you may wish to have a quick peek there. The point we really want to go over is the many-variable Legendre transform. We should observe an explicit proof that H is indeed the Legendre transform of L with respect to the $\dot{q}$s. One point to note is that, since we are converting the $\dot{q}$s to ps we can ignore any other variables because they simply don't matter a jot to this process. Thus, we will be writing the Lagrangian as if it only depended on the $\dot{q}$'s and ignoring any passive variables such as q or t dependancies. Further to this, we are going to consider a Lagrangian that is a function of two velocity coordinates, not just one. We can then write its total derivative in a standard fashion. Let $\mathscr{L} = \mathscr{L}(\dot{q}^1, \dot{q}^2)$; we then obtain

$$
d\mathscr{L} = \frac{\partial L}{\partial \dot{q}^1}d\dot{q}^1 + \frac{\partial \mathscr{L}}{\partial \dot{q}^2}d\dot{q}^2
$$
$$
= p_1 d\dot{q}^1 + p_2 d\dot{q}^2 \tag{14.0.4}
$$

here, p_1 and p_2 are functions of $\dot{q}^1$ and $\dot{q}^2$, respectively. Suppose now we want to write a function such that it inverts the dependence we have just defined; namely, instead of p being a function of $\dot{q}$, we want to get $\dot{q}$ as a function of p. Let's denote the function that does this as H, such that $H = H(p_1, p_2)$. This function has the effect of inverting the relationship between $\dot{q}^i$ and p_i

$$
\frac{\partial H}{\partial p_j} = \dot{q}^j \tag{14.0.5}
$$

Now, $\dot{q}^1$ and $\dot{q}^2$ are functions of p_1 and p_2, allowing us to write

$$
dH = \frac{\partial H}{\partial p_1}dp_1 + \frac{\partial H}{\partial p_2}dp_2
$$
$$
= \dot{q}^1 dp_1 + \dot{q}^2 dp_2 \tag{14.0.6}
$$

We now have two functions that are linked via their variables. It would be rather nifty if we could express them together into a single equation. To start this off, find the sum of $\mathscr{L} + H$, collecting like terms and simplifying with some reverse product ruling:

$$
d(\mathscr{L} + H) = (p_1 d\dot{q}^1 + p_2 d\dot{q}^2) + (\dot{q}^1 dp_1 + \dot{q}^2 dp_2)
$$
$$
= (p_1 d\dot{q}^1 + \dot{q}^1 dp_1) + (p_2 d\dot{q}^2 + \dot{q}^2 dp_2)
$$
$$
= d(p_1\dot{q}^1 + p_2\dot{q}^2) \tag{14.0.7}
$$

Cleverly, yet beautifully simply, we can now say that

$$
H = (\mathscr{L} + H) - \mathscr{L} \tag{14.0.8}
$$

$$H = p_1\dot{q}^1 + p_2\dot{q}^2 - \mathscr{L} \tag{14.0.9}$$

In order to be explicit about what is a function of what, let's write in the implied relations:

$$H(p_1, p_2) = p_1 \cdot \dot{q}^1(p_1, p_2) + p_2 \cdot \dot{q}^2(p_1, p_2) - \mathscr{L}(\dot{q}^1(p_1, p_2), \dot{q}^2(p_1, p_2)) \tag{14.0.10}$$

If we now generalise this so that $\mathscr{L}$ was originally a function of n velocity variables rather than just two, we can write

$$H(p_1, p_2, \ldots, p_n) = \sum_{i=1}^{n}(p_i\dot{q}^i) - \mathscr{L}(\dot{q}^1, \dot{q}^2, \ldots, \dot{q}^n) \tag{14.0.11}$$

And there we have the Hamiltonian into which we can easily substitute all of our passive variables that we ignored before to regenerate the full expression. Note all the generalised velocities in $\mathscr{L}$ are themselves functions of $p_1, \ldots, p_n$. As a final remark, I like to think of the Hamiltonian as $H(q, p, t) = J(q, \dot{q}(q, p,), t)$. The Hamiltonian is the **total energy** of a system, with conditions similar to whether the Jacobi function was the total energy or not. The equality of H and J depends on whether the equations of motion are satisfied or not; the terms **on-shell** and **off-shell** were coined by Feynman, meaning satisfied and not satisfied respectively. This equality of J and H will not concern us until we discuss constrained dynamics.

Chapter summary

- The **conjugate momentum** was defined to be $p_i = \frac{\partial \mathscr{L}}{\partial \dot{q}^i}$.
- The **Hamiltonian** energy function H was shown to be the direct analogue of the Jacobi energy function J except that the $\dot{q}$'s have been **inverted** to the p_is. This means that, while J is a function of $(q, \dot{q})$, the Hamiltonian is a function of (q, p). This is best highlighted using the following simplistic analogy:

$$J(\boldsymbol{q}, \dot{\boldsymbol{q}}) = \frac{1}{2}m\dot{q}^2 + U(\boldsymbol{q}) \quad \text{and} \quad H(\boldsymbol{q}, \boldsymbol{p}) = \frac{p^2}{2m} + U(\boldsymbol{q})$$

- The **Legendre transform** was then explored for two degrees of freedom.

Exercise 14.3 Find the Hamiltonian for a charged particle in an electric field, given

$$\mathscr{L} = \frac{1}{2}m\dot{\boldsymbol{r}} \cdot \dot{\boldsymbol{r}} + e\dot{\boldsymbol{r}} \cdot \boldsymbol{A} - e\varphi$$

Compute the generalised momentum, treating $\dot{r}$ as the generalised coordinate. We begin by finding the momentum:

$$p = \frac{\partial \mathscr{L}}{\partial \dot{r}} = m\dot{r} + eA$$

The Hamiltonian is then given by $H = p\dot{q} - \mathscr{L}$ so

$$
\begin{aligned}
H &= (m\dot{r} + eA)\dot{r} - (\frac{1}{2}m\dot{r}^2 + e\dot{r}\cdot A - e\varphi)\\
&= \frac{1}{2}m\dot{r}^2 + e\varphi\\
&= \frac{(p - eA)^2}{2m} + e\varphi
\end{aligned}
$$

We require the last step because the middle line of working contains a velocity term; simply invert the generalised momentum to generate the expression for the velocity and substitute it in. Despite having an odd-looking conjugate momentum we are pleased to see that the Hamiltonian is the sum of kinetic and potential energies.

15

Hamilton's Principle in Phase Space

Okay, now we have an expression for the Hamiltonian, which is clearly analogous to the Jacobi function. Both describe the energy of a system, yet J, is a velocity space function, while H, is defined on the momentum space. Let's now derive **Hamilton's equations**; for a portrait of Hamilton, see figure 15.1. There are several entirely equivalent ways to do this, and, since we derived the Lagrange equation from the stationary action principle, it is fitting to do the same with Hamilton's equations.

Fig. 15.1: William Hamilton (1805–1865).

If you recall, in chapter 7, we started by defining the action as the time integral of the Lagrangian; we then varied this whilst keeping the boundary conditions fixed and, insisted that the variation in the action had to obey $\delta A = 0$, for the correct path. Straight from our definition of H above, we can write a new expression for the Lagrangian:

$$\mathscr{L} = \sum_i p_i \dot{q}^i - H \tag{15.0.1}$$

Consequently, we substitute this into the action integral, which becomes equation 15.0.3; this is known as **Hilbert's invariant integral**, and the term under the integral sign is known as the **integral invariant of Poincaré-Cartan**:

$$A[\boldsymbol{q}] = \int_{t_1}^{t_2} \mathscr{L} dt \tag{15.0.2}$$

$$S[\boldsymbol{q}, \boldsymbol{p}] = \int_{t_1}^{t_2} (p_j \dot{q}^j - H(\boldsymbol{q}, \boldsymbol{p})) dt \tag{15.0.3}$$

When we derived the Euler-Lagrange equation, we varied q^j, and hence simultaneously varied the velocity coordinate, since it is the time derivative of q^j. We treat the qs and ps as independent variables. The Lagrangian $\mathscr{L}' = \mathscr{L}'(q, p, \dot{q}, \dot{p}, t)$, is defined on the $4n + 1$-dimensional surface (not the phase space itself, but the tangent bundle

Lagrangian & Hamiltonian Dynamics. Peter Mann, Oxford University Press (2018).
© Peter Mann. DOI: 10.1093/oso/9780198822370.001.0001

to the phase space, don't worry about this for now, we will only understand it when we discuss manifolds properly, see chapter 22). Strikingly, we are calling it $\mathscr{L}'$, not $\mathscr{L}$! This is because $\mathscr{L}'$ and $\mathscr{L}$ are defined in different spaces, hence, we must denote them differently; $\mathscr{L}'$ is sometimes referred to as the **phase space Lagrangian**.

For *arbitrary paths*, the configuration action principle and the phase space action principle are distinct from one another. The phase space Lagrangian and the familiar Lagrangian are equivalent *only* for the extremal path, since in this case, there will exist relations among the coordinates $(q, p, \dot{q}, \dot{p})$. Another point worth considering, is how we should view a Lagrangian. It is just some function that allows us to generate the equations of motion, through the mathematical machinery of the calculus of variations. In general, $\mathscr{L}'$ and $\mathscr{L}$ are not equivalent objects until the equations of motion are satisfied.

We have to define the action functional as S, rather than A. This is because $S = S[q, p]$, whereas $A = A[q]$. Hence, we require a phase space action, which we call the **Hamiltonian action**. All extrema of actions corresponding to the same mechanical system are the same, but off-shell, they don't have to share equality *a priori*. We could have labelled them with the same symbol, but this highlights their subtle, yet important, different functional dependence!

Just as we did with the Lagrange action in chapter 7, we now apply the principle of stationary action to 15.0.3. We require that a variation from the true path must vanish $\delta S = 0$; and, in a similar fashion to chapter 7, we can decompose the perturbed paths into the extremal path plus a small variational term to both p and q. Thus, we obtain:

$$\delta S = \frac{\partial S}{\partial \alpha} d\alpha = \delta \int_{t_1}^{t_2} \mathscr{L}' dt \tag{15.0.4}$$

$$q^j = q_0^j + \alpha \eta_j \tag{15.0.5}$$

$$p_j = p_{0j} + \alpha \eta_j' \tag{15.0.6}$$

Taking the functional derivative of the action with respect to the perturbation alpha, and using the chain rule, we have:

$$\frac{\partial S}{\partial \alpha} d\alpha = \int_{t_1}^{t_2} \left\{ \sum_{j=1}^{n} \left[\frac{\partial \mathscr{L}'}{\partial q^j} \frac{\partial q^j}{\partial \alpha} d\alpha + \frac{\partial \mathscr{L}'}{\partial \dot{q}^j} \frac{\partial \dot{q}^j}{\partial \alpha} d\alpha \right] \right.$$
$$\left. + \sum_{j=1}^{n} \left[\frac{\partial \mathscr{L}'}{\partial p_j} \frac{\partial p_j}{\partial \alpha} d\alpha + \frac{\partial \mathscr{L}'}{\partial \dot{p}_j} \frac{\partial \dot{p}_j}{\partial \alpha} d\alpha \right] \right\} dt \tag{15.0.7}$$

integrating the second term in the first bracket by parts. The first term that results from this treatment will then vanish, since we assume that there is no perturbation in the qs at the end points, so the term $\partial q / \partial \alpha = 0$, when evaluated on the boundary, t_1 and t_2:

$$\int_{t_1}^{t_2} \frac{\partial \mathscr{L}'}{\partial \dot{q}^j} \frac{\partial \dot{q}^j}{\partial \alpha} dt = \frac{\partial \mathscr{L}'}{\partial \dot{q}^j} \frac{\partial q^j}{\partial \alpha} \bigg|_{t_1}^{t_2} - \int_{t_1}^{t_2} \frac{\partial q^j}{\partial \alpha} \frac{d}{dt} \left(\frac{\partial \mathscr{L}'}{\partial \dot{q}^j} \right) dt \qquad (15.0.8)$$

We can substitute this result back into equation 15.0.7, having cancelled the first part of equation 15.0.8:

$$\frac{\partial S}{\partial \alpha} d\alpha = \int_{t_1}^{t_2} \left\{ \sum_{j=1}^{n} \left[\frac{\partial \mathscr{L}'}{\partial q^j} \frac{\partial q^j}{\partial \alpha} d\alpha - \frac{\partial q^j}{\partial \alpha} \frac{d}{dt} \left(\frac{\partial \mathscr{L}'}{\partial \dot{q}^j} \right) d\alpha \right] \right.$$
$$\left. + \sum_{j=1}^{n} \left[\frac{\partial \mathscr{L}'}{\partial p_j} \frac{\partial p_j}{\partial \alpha} d\alpha + \frac{\partial \mathscr{L}'}{\partial \dot{p}_j} \frac{\partial \dot{p}_j}{\partial \alpha} d\alpha \right] \right\} dt \qquad (15.0.9)$$

Next, we do exactly the same thing with the momentum term, but, watch what happens in a subtle move:

$$\int_{t_1}^{t_2} \frac{\partial \mathscr{L}'}{\partial \dot{p}_j} \frac{\partial \dot{p}_j}{\partial \alpha} dt = \frac{\partial \mathscr{L}'}{\partial \dot{p}_j} \frac{\partial p_j}{\partial \alpha} \bigg|_{t_1}^{t_2} - \int_{t_1}^{t_2} \frac{\partial p_j}{\partial \alpha} \frac{d}{dt} \left(\frac{\partial \mathscr{L}'}{\partial \dot{p}_j} \right) dt \qquad (15.0.10)$$

Something rather important happens here, just as above, we remove the first term, but for a different reason. The collective first term in equation 15.0.10 is zero, this is because the phase space Lagrangian (using the explicit form on the left-hand-side of equation 15.0.1) is independent of the $\dot{p}$s, such that, $\partial \mathscr{L}' / \partial \dot{p} = 0$, therefore we are not requiring that the perturbations in the momentum vanish at the boundary points, and hence place no restriction on the momentum's variation. There is nothing stopping us placing this restriction upon allowed paths, but it is not necessary (for the current definition of $\mathscr{L}'$) and, will only serve to restrict the possible paths. It is a general trend, that integrals over the momentum coordinates are unbounded integrals, and the consequences are observed in statistical mechanics (see chapter 20). Inserting the result of equation 15.0.10, back into equation 15.0.9, we obtain:

$$\frac{\partial S}{\partial \alpha} d\alpha = \int_{t_1}^{t_2} \left\{ \sum_{j=1}^{n} \left[\frac{\partial \mathscr{L}'}{\partial q^j} \frac{\partial q^j}{\partial \alpha} d\alpha - \frac{\partial q^j}{\partial \alpha} \frac{d}{dt} \left(\frac{\partial \mathscr{L}'}{\partial \dot{q}^j} \right) d\alpha \right] \right.$$
$$\left. + \sum_{j=1}^{n} \left[\frac{\partial \mathscr{L}'}{\partial p_j} \frac{\partial p_j}{\partial \alpha} d\alpha - \frac{\partial p_j}{\partial \alpha} \frac{d}{dt} \left(\frac{\partial \mathscr{L}'}{\partial \dot{p}_j} \right) d\alpha \right] \right\} dt \qquad (15.0.11)$$

Let's make this neater, by calling $\delta q^j = (\partial q^j / \partial \alpha) d\alpha$ and $\delta p_j = (\partial p_j / \partial \alpha) d\alpha$, to obtain:

$$\frac{\partial S}{\partial \alpha} d\alpha = \int_{t_1}^{t_2} \left\{ \sum_{j=1}^{n} \left[\frac{\partial \mathscr{L}'}{\partial q^j} - \frac{d}{dt} \left(\frac{\partial \mathscr{L}'}{\partial \dot{q}^j} \right) \right] \delta q_j + \sum_{j=1}^{n} \left[\frac{\partial \mathscr{L}'}{\partial p_j} - \frac{d}{dt} \left(\frac{\partial \mathscr{L}'}{\partial \dot{p}_j} \right) \right] \delta p_j \right\} dt \quad (15.0.12)$$

Imposing that the true physical path is the one that makes the action integral stationary, we can set $(\partial S / \partial \alpha) d\alpha = 0$. Now, the importance of treating p and q on equal footing emerges; the variations must vanish independently of one another. This gives rise to *two equations* rather than one:

$$\frac{d}{dt}\left(\frac{\partial \mathscr{L}'}{\partial \dot{q}^j}\right) - \frac{\partial \mathscr{L}'}{\partial q^j} = 0 \tag{15.0.13}$$

$$\frac{d}{dt}\left(\frac{\partial \mathscr{L}'}{\partial \dot{p}_j}\right) - \frac{\partial \mathscr{L}'}{\partial p_j} = 0 \tag{15.0.14}$$

Why two equations? Because, the variations in the qs and the ps are truly independent. On-shell, in the velocity space, the velocity coordinate is the time derivative of the position; we treat them as independent variables in general, since we can choose initial conditions independently. On-shell q and $\dot{q}$ are inherently linked, so a variation in coordinates *induces* a variation in the velocities. We will explore this in chapter 22 in more detail; there we will not assume the equivalence of the velocity coordinate v^i, and the time derivative of the generalised coordinate $\dot{q}^i$, instead we will *derive* their equivalence from first principles.

To finish the derivation, enter the expression for the phase space Lagrangian $\mathscr{L}'(\boldsymbol{q}, \dot{\boldsymbol{q}}, \boldsymbol{p})$, to obtain:

$$\frac{\partial \mathscr{L}'}{\partial \dot{q}^j} = p_j \qquad \text{and} \qquad \frac{\partial \mathscr{L}'}{\partial q^j} = -\frac{\partial H}{\partial q^j}$$

$$\tag{15.0.15}$$

$$\frac{\partial \mathscr{L}'}{\partial \dot{p}_j} = 0 \qquad \text{and} \qquad \frac{\partial \mathscr{L}'}{\partial p_j} = \dot{q}^j - \frac{\partial H}{\partial p_j} = 0$$

Using equations 15.0.15 in 15.0.13 and 15.0.14, gives us **Hamilton's equations of motion**. Hamilton's equations are a set of $2n$ coupled ODEs:

$$dp_j + \frac{\partial H}{\partial q^j}dt = 0, \qquad dq^j - \frac{\partial H}{\partial p_j}dt = 0 \qquad \textbf{Hamilton's equations} \tag{15.0.16}$$

If you are struggling to remember which one has the minus sign, there is an easy way to remember; let the Hamiltonian be $T(\dot{q}) + U(q)$. You can now use Newton's second law; the rate of change of momentum is equal to the negative gradient of the potential, thus $\dot{p} = -\partial H/\partial q$. Note that we get *two first-order* ODEs, rather than *one second-order* Lagrange equation. This is due to the qs and ps being truly independent variables, while q and $\dot{q}$ are linked. However, determinism requires that there be $2n$ known-quantities of the system.

Chapter summary

- The stationary action principle was explored in the momentum phase space. It was used to derive **Hamilton's equations of motion**:

$$\dot{p}_i = -\frac{\partial H}{\partial q^i}, \qquad \dot{q}^i = \frac{\partial H}{\partial p_i}$$

  These are a set of $2n$ first-order ODEs.

- Two technical differences between the Lagrangian and Hamiltonian formulations were investigated; the **phase space Lagrangian** and the **phase space action**.

Exercise 15.1 Using the functional derivative notation, derive Hamilton's equations of motion starting from the Hilbert integral invariant. (see chapter 21.)

Exercise 15.2 Using the calculus of variations, show that the following total time derivative $-\frac{d}{dt}(qp)$, can be added to the phase space Lagrangian, without changing the equations of motion. (see chapter 18.)

16

Hamilton's Equations & Routhian Reduction

In the previous two chapters we have investigated a phase space function called the Hamiltonian (see chapter 14) and derived Hamilton's equations of motion using a phase space action principle (see chapter 15). We are now going to derive Hamilton's equations, by using the traditional approach of the Legendre transform. We don't require any knowledge of the action principle to do this; only the definition of the Hamiltonian function. This contrasts the approach in chapter 15, which appears to rely on the Lagrangian formulation quite heavily, and shows that Hamiltonian mechanics is quite independent from the method of Lagrange.

To begin, we consider a general small change in the Hamiltonian $H(\boldsymbol{p}, \boldsymbol{q}, t)$. We are now going to consider the Hamiltonian as being explicitly dependent on time t; it will lead to a third important relation in addition to Hamilton's equations that, although not an equation of motion, is still useful to discuss. We can start by taking the total differential of the left-hand side of the Hamiltonian. Recalling $H(\boldsymbol{q}, \boldsymbol{p}, t) = p\dot{q} - \mathscr{L}(q, \dot{q}, t)$, we obtain

$$dH = \sum_{j=1}^{n} \left[\frac{\partial H}{\partial q^j} dq^j + \frac{\partial H}{\partial p_j} dp_j \right] + \frac{\partial H}{\partial t} dt \qquad (16.0.1)$$

The total differential of the right-hand side is

$$dH = \sum_{j=1}^{n} \left[\dot{q}^j dp_j + p_j d\dot{q}^j - \frac{\partial \mathscr{L}}{\partial \dot{q}^j} d\dot{q}^j - \frac{\partial \mathscr{L}}{\partial q^j} dq^j \right] - \frac{\partial \mathscr{L}}{\partial t} dt \qquad (16.0.2)$$

The two blue middle terms cancel each other out, leaving us with 16.0.3, which we now match up with 16.0.1, collecting terms equivalent in dq^j and dp_j:

$$dH = \sum_{j=1}^{n} \left[\dot{q}^j dp_j - \frac{\partial \mathscr{L}}{\partial q^j} dq^j \right] - \frac{\partial \mathscr{L}}{\partial t} dt \qquad (16.0.3)$$

Equating equations 16.0.1 and 16.0.3 we obtain

Lagrangian & Hamiltonian Dynamics. Peter Mann, Oxford University Press (2018).
© Peter Mann. DOI: 10.1093/oso/9780198822370.001.0001

$$\dot{q}^j \, dp_j - \frac{\partial \mathscr{L}}{\partial q^j} dq^j - \frac{\partial \mathscr{L}}{\partial t} dt = \frac{\partial H}{\partial q^j} dq^j + \frac{\partial H}{\partial p_j} dp_j + \frac{\partial H}{\partial t} dt \qquad (16.0.4)$$

Instantly, we can see that we can take out the two blue terms:

$$\boxed{\frac{\partial H}{\partial t} = -\frac{\partial \mathscr{L}}{\partial t}} \qquad (16.0.5)$$

$$-\frac{\partial \mathscr{L}}{\partial q^j} dq^j - \frac{\partial H}{\partial q^j} dq^j = \frac{\partial H}{\partial p_j} dp_j - \dot{q}^j \, dp_j \qquad (16.0.6)$$

The only way this relation will hold for any *arbitrary independent* variations is if both sides are exactly zero:

$$-\frac{\partial \mathscr{L}}{\partial q^j} dq^j - \frac{\partial H}{\partial q^j} dq^j = 0 \qquad (16.0.7)$$

$$\frac{\partial H}{\partial p_j} dp_j - \dot{q}^j \, dp_j = 0 \qquad (16.0.8)$$

Take out the variations and use the Lagrange equation to rewrite the first side of equation 16.0.7 as $\partial \mathscr{L}/\partial q^j - \dot{p}_j$ and we have Hamilton's equations;

$$\boxed{\dot{p}_j + \frac{\partial H}{\partial q^j} = 0} \qquad (16.0.9)$$

$$\boxed{\dot{q}^j - \frac{\partial H}{\partial p_j} = 0} \qquad (16.0.10)$$

Importantly, this derivation highlights the need for independent variations in q and p and we again up with two first-order equations for each dimension of phase space. If the general coordinates of the system are subject to constraints, then we should redefine them as to eliminate any constraints and then perform the Legendre transform on the independent, unconstrained Lagrangian. There is an alternative way to approach constrained problems in Hamiltonian mechanics; this was first investigated by Dirac, and is the subject of chapter 21. In basically every other situation, we define the generalised coordinates such that the constraints are eliminated just as we did in chapter 8. In mathematics, surfaces of one dimension less are called **hypersurfaces**. Therefore, each constraint generates a hypersurface in the total phase space; the region in which these surfaces overlap is the region the system is allowed to explore.

Exercise 16.1 As a quick example of using Hamilton's equations, we can use our trusty harmonic oscillator in one dimension. We know by other means that this system has a potential of $\frac{1}{2}kx^2$ and therefore a Lagrangian of $\mathscr{L} = \frac{1}{2}m\dot{x}^2 - \frac{1}{2}kx^2$. Work out the conjugate momentum and invert it to get $\dot{x}$ as $\dot{x} = \dot{x}(p)$. Doing this we get:

$$H = p\dot{x} - \mathscr{L} = p\dot{x} - \left(\frac{1}{2}m\dot{x}^2 - \frac{1}{2}kx^2\right) = p\left(\frac{p}{m}\right) - \frac{1}{2}m\left(\frac{p}{m}\right)^2 + \frac{1}{2}kx^2 = \frac{p^2}{2m} + \frac{kx^2}{2}$$

Now use Hamilton's equations to obtain

$$\dot{x} = \frac{\partial H}{\partial p} = \frac{p}{m} \qquad \text{and} \qquad \dot{p} = -\frac{\partial H}{\partial x} = -kx$$

which reproduces the Newtonian result. This isn't really easier than using Newton's equations and certainly isn't easier than using the Lagrange equation of motion, since you need the Lagrangian anyway to calculate H, but it highlights that they are all equivalent methods, which is what we said right at the start. The real use of Hamiltonian mechanics lies in what theory we can derive as a result of the underlying geometrical mechanics, for instance action-angle variables, symplecticity, Poisson brackets, Liouville's theorem, and so on.

There are at least another two derivations we can follow to arrive at Hamilton's equations; all of these have their own intuitive benefits, so I urge you to find some different ways. The ones we have used in this chapter and in chapter 15 highlight the important points for any discussion we will need for now. We will derive Hamilton's equations from the Hamilton-Jacobi equation and using the geometrical framework a little later on in chapters 19 22 respectively.

16.1 Phase Space Conservation Laws

In the Newtonian formalism, conservation laws were rather difficult to tease out through long, contrived derivations (see for instance the conservation of angular momentum in chapter 3). The Lagrangian formalism revolutionised the way we look at conservation laws and associated symmetries of the system to each conserved property (see chapter 11). The Hamiltonian formalism perhaps simplifies the finding of conservation laws even further.

If H doesn't depend explicitly on time, then the Hamiltonian is a conserved quantity. As we know, the very definition of a conserved quantity is that it is constant in time; it doesn't change as time evolves. Therefore, taking the time derivative of the Hamiltonian should highlight this point, again chain ruling in coordinate dependance:

$$\frac{dH}{dt} = \frac{\partial H}{\partial q^j}\dot{q}^j + \frac{\partial H}{\partial p_j}\dot{p}_j + \frac{\partial H}{\partial t} = -\dot{p}_j\dot{q}^j + \dot{q}^j\dot{p}_j + \frac{\partial H}{\partial t} \tag{16.1.1}$$

This, of course, shows that if the partial time derivative of the Hamiltonian is zero, then so too is the total time derivative and, indeed, H is a constant of the motion, we can tie this in with the Jacobi energy function (see chapter 10) by considering the blue term above in equation 16.0.5.

From Hamilton's equation in 16.0.9, we can identify conserved conjugate momenta in an instant, since the whole theory is built around p. If a coordinate doesn't appear in the Lagrangian, then performing a Legendre transform isn't going to suddenly make it appear; it will be absent in H, too. From the outset, we can show that the partial derivative of the Hamiltonian with respect to this absent coordinate is zero and thus the time derivative of the conserved quantity is zero, from the definition of 16.0.9.

There is a symmetry in Hamiltonian mechanics that links the ps and the qs with conservation laws. If you take a look at both of Hamilton's equations in equations 16.0.9 and 16.0.10, you can easily see that, if the Hamiltonian is independent of a coordinate (canonical momentum): then its canonical momentum's (corresponding coordinate's) time derivative is zero: it is conserved. In this way, you can think of Hamilton's equations as conservation laws for q and p. Symmetries in Hamiltonian mechanics lead to a direct simplification of the dimensionality of the system: if a coordinate is cyclic, then it is absent from the Hamiltonian and its momentum is constant, so our problem is now reduced to $2n - 2$ dimensions. Further, the remaining coordinates do not depend on that generalised coordinate.

The Hamiltonian may or may not be conserved, depending on whether the Lagrangian is an explicit function of time. The Hamiltonian may or may not be the total energy, depending on whether the coordinate transformation is inertial, that is, it is independent of time such $\partial x_i / \partial t = 0$, so that we only keep the term homogeneously quadratic in $\dot{q}^j$ below, just as we did with the Jacobi function J (see chapter 10). Hence, we have built up our formalism around a sort of **generalised energy**, an approach which is much clearer than obtaining energy from the Lagrange equations:

$$T = \frac{1}{2}\sum_{i=1}^{3N} m_i \sum_j \sum_k \left(\frac{\partial x_i}{\partial q^j} \cdot \frac{\partial x_i}{\partial q^k}\right)\dot{q}^j\dot{q}^k + \sum_{i=1}^{3N} m_i \sum_j \left(\frac{\partial x_i}{\partial q^j} \cdot \frac{\partial x_i}{\partial t}\right)\dot{q}^j + \frac{1}{2}\sum_{i=1}^{3N} m_i \left(\frac{\partial x_i}{\partial t}\right)^2$$

$$T = \sum_{jk} a^{jk}\dot{q}^j\dot{q}^k + \sum_j b^j\dot{q}^k + c$$

where,

$$a^{jk} = \frac{1}{2}\sum_{i=1}^{3N} m_i\left(\frac{\partial x_i}{\partial q^j} \cdot \frac{\partial x_i}{\partial q^k}\right), \qquad b^j = \sum_{i=1}^{3N} m_i\left(\frac{\partial x_i}{\partial q^j} \cdot \frac{\partial x_i}{\partial t}\right), \qquad c = \frac{1}{2}\sum_{i=1}^{3N} m_i\left(\frac{\partial x_i}{\partial t}\right)^2$$

We only keep the first term (**homogeneously quadratic** in the $\dot{q}$s) if $x_i = x_i(q^1, q^2, \ldots, q^n)$, and hence $\partial x_i / \partial t = 0$.

It is often remarked that Hamiltonian mechanics uses the "natural coordinates" of a system, namely those of phase space as opposed to those of the velocity space. One may ask, which is the more fundamental theory, Lagrangian theory or Hamiltonian theory? From a historical viewpoint, it must be Newton's theory, since that gave rise to all that we know now. From a physics viewpoint, analytical mechanics is seen as a more fundamental branch than Newtonian mechanics but there is no particular distinction between the Lagrangian, the Hamiltonian or indeed the Poisson and Liouville formalisms we study later; a better question is, *what formalism highlights the physics I want to describe?* You may argue that, in order to derive Hamilton's equations, we require a Lagrangian and the idea of a conjugate momentum. Well, this is sort of true, but only because of the order in which they were developed and we could indeed develop Hamilton's mechanics independently of any knowledge of the Lagrange formalism. Some may argue that the geometrical formalism is the most general formalism of mechanics (probably true), but it isn't very helpful for find the equation of motion of a harmonic oscillator!

16.2 Routhian Mechanics

Here, we will investigate the **Routhian formalism** of classical mechanics; this was named after Edward Routh, who developed it in the late twentieth century. The Routhian take on mechanics is somewhere in-between Lagrange's and Hamilton's and is concerned with exploiting the conserved quantities of a system for ease of computation and the identification of symmetries. Cyclic coordinates do not appear in the Lagrangian expression; namely, $\partial \mathscr{L}/\partial q^i = 0$ if q^i is a **first integral**[1]. If the nth general coordinate q^n is cyclic *but* its time derivative *is* present in the Lagrangian, then we write

$$\mathscr{L} = \mathscr{L}(q^1, \ldots, q^{n-1}, \dot{q}^1, \ldots, \dot{q}^n, t) \tag{16.2.1}$$

with no dependance on q^n. The Lagrange equation for this coordinate is

$$\frac{d}{dt}\left(\frac{\partial \mathscr{L}}{\partial \dot{q}^n}\right) = 0 \tag{16.2.2}$$

The conjugate momentum is an **integral of the motion** (conserved quantity). The Lagrangian is a function of n generalised velocities but $(n-1)$ generalised coordinates. Now that we have this idea, let's make it more general by extending the number of coordinates that are cyclic. Let's say the first k general coordinates are present in the Lagrangian, $q^1 \ldots, q^k$, but $k+1$ to n are cyclic, where n is the total number of coordinates. The generalised velocities in the range of $k+1$ to n are now subjected to

---

[1]We recall that a *constant of motion* is some function $I(q,p,t)$, that is defined on the phase space, over some interval of time, such that, for an extremal path of the action, this quantity does not depend on time. In other words, let $I(q,p,t) : M \times [t_1, t_2] \to \mathbb{R}$, then $\forall\, t \in [t_1, t_2]$ the evaluation of I upon the classical path $(q,p) = \gamma(t)$ does not depend on time $\forall\, (q,p) = \gamma(t)$ in the set of solutions to the system. We define an *integral of the motion* or a *first integral* to be a constant of the motion $I(q,p)$ that is not an explicit function of time $I(q,p) \neq I(q,p,t)$.

a Legendre transformation to their momenta. We now define a scalar function called the **Routhian** that satisfies

$$R = \sum_{l=k+1}^{n} p_l \dot{q}_l - \mathscr{L} \tag{16.2.3}$$

with $R(q^1, \ldots, q^k, \dot{q}^1, \ldots, \dot{q}^k, p_{k+1}, \ldots, p_n, t)$ and $p = \partial \mathscr{L}/\partial \dot{q}$ while the sum extends over $k+1, \ldots, n$; otherwise, if it extended from 1 to n, we would have the Hamiltonian. The Routhian is a function of the integrals of motion of the system, we have removed the general velocities corresponding to the cyclic coordinates by solving their Lagrange equations and writing in the integrals of motion they correspond to. The Routhian is the Legendre conjugate to the Lagrangian and its form is a general result from Legendre transform theory. To move on, we take the total derivative of the Routhian function on both sides:

$$dR = \sum_{i=1}^{k} \left(\frac{\partial R}{\partial q^i} dq^i + \frac{\partial R}{\partial \dot{q}^i} d\dot{q}^i \right) + \sum_{l=k+1}^{n} \left(\frac{\partial R}{\partial p_l} dp_l \right) + \frac{\partial R}{\partial t} dt \tag{16.2.4}$$

$$dR = \sum_{l=k+1}^{n} \left(\dot{q}^l dp_l + p_l d\dot{q}^l - \frac{\partial \mathscr{L}}{\partial \dot{q}^l} d\dot{q}^l \right) - \sum_{i=1}^{k} \left(\frac{\partial \mathscr{L}}{\partial q^i} dq^i + \frac{\partial \mathscr{L}}{\partial \dot{q}^i} d\dot{q}^i \right) - \frac{\partial \mathscr{L}}{\partial t} dt \tag{16.2.5}$$

If we equate both sides and compare terms in like variations and then assume that we vary independently, we obtain **Routh's differential equations of motion**. They comprise a set of coupled ODEs. Here, $i = 1, \ldots, k$ and $l = k + 1, \ldots, n$:

$$\frac{\partial R}{\partial q^i} = -\frac{\partial \mathscr{L}}{\partial q^i}, \qquad \frac{\partial R}{\partial \dot{q}^i} = -\frac{\partial \mathscr{L}}{\partial \dot{q}^i}, \qquad \text{together} \qquad \frac{d}{dt}\left(\frac{\partial R}{\partial \dot{q}^i} \right) - \frac{\partial R}{\partial q^i} = 0 \tag{16.2.6}$$

$$\frac{\partial R}{\partial t} = -\frac{\partial \mathscr{L}}{\partial t} \tag{16.2.7}$$

$$\dot{p}_l = -\frac{\partial R}{\partial q^i}, \qquad \frac{\partial R}{\partial p_l} = \dot{q}^l, \qquad l = k+1, \ldots, n \tag{16.2.8}$$

Look closely at the first set of equations and substitute the two blue terms into the Euler-Lagrange equations. Therefore, the first k coordinates satisfy the Lagrange equation. The vanishing term above can be used to construct Hamilton's equations for the remaining coordinates, which generates the relation for $\dot{p}_l$. Routhian reduction by symmetry is a popular theme in methods to solve problems in Hamiltonian mechanics and is part of a wider theme of **symplectic reduction**.

Exercise 16.2 Building on the central force problem from exercise 7.2, compute Hamilton's equations of motion for each coordinate. We begin by constructing the Hamiltonian:

$$H = \frac{p_r^2}{2m} + \frac{1}{2}\frac{p_\theta^2}{mr^2} + U(r) \tag{16.2.9}$$

The coordinate equations are then:

$$\dot{r} = \frac{\partial H}{\partial p_r} = \frac{p_r}{m}, \qquad \dot{\theta} = \frac{\partial H}{\partial p_\theta} = \frac{p_\theta}{mr^2} \tag{16.2.10}$$

while the momentum equations,

$$\dot{p}_r = -\frac{\partial H}{\partial r} = -\left[-\frac{\partial p_\theta^2}{mr^3} + \frac{\partial U}{\partial r} \right], \qquad \dot{p}_\theta = -\frac{\partial H}{\partial \theta} = 0 \tag{16.2.11}$$

Recalling $p_r = m\dot{r}$, we generate the equations of motion we are familiar with. In a way, the Hamiltonian formalism really gives us all the information at once: the equations of motion, the conserved quantities (look at $\dot{p}_\theta$) and the total energy of the system (conservative).

Exercise 16.3 Consider Hamilton's equations for a charged particle in an electromagnetic field with $H = \frac{1}{2m}(\boldsymbol{p} - e\boldsymbol{A})^2 + e\varphi$ and $\boldsymbol{p} = m\dot{\boldsymbol{r}} + e\boldsymbol{A}$. First, compute a momentum $\dot{\boldsymbol{p}}$ component:

$$\dot{p}_i = m\ddot{r}_i + e\dot{\boldsymbol{A}} = m\ddot{r}_i + e\left(\frac{\partial \boldsymbol{A}}{\partial r_j}\dot{r}_j + \frac{\partial \boldsymbol{A}}{\partial t} \right) \tag{16.2.12}$$

Then, Hamilton's equations are given by,

$$\dot{p}_i = -\frac{\partial H}{\partial r_i} = \frac{1}{m}(\boldsymbol{p} - e\boldsymbol{A}) \cdot e\frac{\partial \boldsymbol{A}}{\partial r_i} - e\frac{\partial \varphi}{\partial r_i} \tag{16.2.13}$$

$$\dot{r}_i = \frac{\partial H}{\partial p_i} = \frac{1}{m}(p_i - eA_i) \cdot 1 \tag{16.2.14}$$

where we have chain ruled in both cases, using $(p - eA)$ as the substituted variable; the second equation merely confirms our inverted expression for the velocity from earlier problems. That marks the end of the concepts from Hamiltonian mechanics we use, but next we will show that the expressions we have are equivalent to the Lorentz force law, as expected. We now have two expressions for $\dot{p}_i$; we can equate these and collect terms:

$$\frac{1}{m}\sum_j (p_j - eA_j) \cdot e\frac{\partial A_j}{r_i} - e\frac{\partial \varphi}{\partial r_i} = m\ddot{r}_i + e\left(\sum_j \frac{\partial A_i}{\partial r_j}\dot{r}_j + \frac{\partial A_i}{\partial t} \right) \tag{16.2.15}$$

where we have replaced the vector notation for a sum over indices of the component expressions. The next step is to re-arrange for $m\ddot{r}_i$ and enter expressions for the electric and magnetic fields:

$$m\ddot{r}_i = \dot{r}_j e\left(\frac{\partial A_j}{\partial r_i} - \frac{\partial A_i}{\partial r_j}\right) - e\left(\frac{\partial A_i}{\partial t} + \frac{\partial \varphi}{\partial r_i}\right) \qquad (16.2.16)$$

We now use a vector identity on the first term on the right-hand side: $\dot{\boldsymbol{r}} \times (\vec{\nabla} \times \boldsymbol{A}) = \vec{\nabla}(\dot{\boldsymbol{r}} \cdot \boldsymbol{A}) - (\dot{\boldsymbol{r}} \cdot \vec{\nabla})\boldsymbol{A}$. We thus obtain

$$m\ddot{\boldsymbol{r}} = e(\dot{\boldsymbol{r}} \times (\vec{\nabla} \times \boldsymbol{A}) + \boldsymbol{E}) \qquad (16.2.17)$$

which is the Lorentz force law. This example not only demonstrates how to use Hamilton's equations to an applied problem but also shows that proceeding via a scalar energy function yields the same results as Newton's force and vector methods do.

Exercise 16.4 From chapter 12, systems near equilibrium can be linearised using the theory of small oscillations. Consider a Lagrangian $\mathcal{L}(\boldsymbol{q}, \dot{\boldsymbol{q}})$ with $\boldsymbol{q} = (q^1, \ldots, q^n)$. We can perform the familiar stability analysis at a point $\boldsymbol{q_0}$ by considering a small perturbation term $\eta^i = q^i - q^i_0$ $\forall i$ and expanding the Lagrangian to second order about this perturbation:

$$\mathcal{L}(\eta, \dot{\eta}) = \frac{1}{2}M_{ij}\dot{\eta}^i\dot{\eta}^j + \frac{1}{2}Y_{ij}\eta^i\dot{\eta}^j - \frac{1}{2}V_{ij}\eta^i\eta^j + f_i\eta^i \qquad (16.2.18)$$

Here, $\boldsymbol{M}, \boldsymbol{Y}$ and $\boldsymbol{V}$ are real matrices; note that $\boldsymbol{Y}$ is antisymmetric ($\boldsymbol{Y} = -\boldsymbol{Y}^t$) while $\boldsymbol{M}$ and $\boldsymbol{V}$ are symmetric.

Find the canonical momentum, the Hamiltonian and hence Hamilton's equations of motion for this Lagrangian.

The canonical momentum is given by

$$p_i = \frac{\partial \mathcal{L}}{\partial \dot{\eta}^i} = M_{ij}\dot{\eta}^j - \frac{1}{2}Y_{ij}\eta^j \qquad (16.2.19)$$

The Hamiltonian is then given by

$$H(\boldsymbol{q}, \boldsymbol{p}) = \frac{1}{2}\boldsymbol{p}^t\boldsymbol{M}^{-1}\boldsymbol{p} + \frac{1}{2}\boldsymbol{p}^t\boldsymbol{M}^{-1}\boldsymbol{Y}\eta + \frac{1}{2}\eta^t(\boldsymbol{V} + \frac{1}{4}\boldsymbol{Y}^t\boldsymbol{M}^{-1}\boldsymbol{Y})\eta - \boldsymbol{f}^t\eta \qquad (16.2.20)$$

Hamilton's equations are then found to be

$$\frac{d}{dt}\begin{bmatrix} \boldsymbol{\eta} \\ \boldsymbol{p} \end{bmatrix} = \boldsymbol{C}\begin{bmatrix} \boldsymbol{u} \\ \boldsymbol{p} \end{bmatrix}, \qquad \text{where} \qquad \boldsymbol{C} = \begin{bmatrix} \frac{1}{2}\boldsymbol{M}^{-1}\boldsymbol{Y} & \boldsymbol{M}^{-1} \\ -\boldsymbol{V} + \frac{1}{4}\boldsymbol{Y}\boldsymbol{M}^{-1}\boldsymbol{Y} & \frac{1}{2}\boldsymbol{Y}\boldsymbol{M}^{-1} \end{bmatrix} \qquad (16.2.21)$$

Exercise 16.5 Consider the Lagrangian $\mathcal{L}(\theta, r, \dot{\theta}, \dot{r})$ for a particle that has mass m and is constrained to a circular orbit that has radius r and angular displacement θ and whose origin rotates with angular speed ω:

$$\mathcal{L} = \frac{1}{2}mr^2(\dot{\theta} - \omega)^2 - mr(1 - \cos(\theta - \omega t)) \qquad (16.2.22)$$

Generate a Hamiltonian and decide if it represents the total energy of the system.

First we find the conjugate momentum to the θ coordinate p_θ:

$$p_\theta = mr^2\dot\theta - mr^2\omega \tag{16.2.23}$$

The Hamiltonian $H(\theta, r, p_\theta, p_r)$ is given by

$$H = \frac{1}{2}\frac{p_\theta}{mr^2} + p_\theta\omega + mr(1 - \cos(\theta - \omega t)) \tag{16.2.24}$$

This is not the total energy of the rotating particle. If we fix the origin in place, then $\omega = 0$ and we generate the correct expression for the total energy. This is because the coordinates have an explicit time dependence through the rotation of the origin with an angular frequency.

Exercise 16.6 We know that the equations of motion for Lagrangians with dependence on higher-order time derivatives, such as $\mathscr{L}(q, \dot q, \ddot q, \ldots, q^{(n)})$, are given by the Ostrogradsky equation (see chapter 7 section 7.2). In addition, we know from exercise 11.10, that the corresponding canonical momenta also has a modified expression:

$$\sum_{i=0}^{n}(-1)^i\frac{d^i}{dt^i}\left(\frac{\partial\mathscr{L}}{\partial q^{(i)}}\right) = 0, \qquad p_i = \sum_{j=1}^{n}(-1)^{j-i}\left(\frac{d}{dt}\right)^{j-i}\frac{\partial\mathscr{L}}{\partial q^{(j)}} \tag{16.2.25}$$

Investigate the case where $\mathscr{L}(q, \dot q, \ddot q)$, using the stationary action principle. Derive the conjugate momenta and subsequently the Hamiltonian by performing a Legendre transform. Write Hamilton's equations of motion in the Ostrogradsky formalism to second order.

We begin by computing the variation in the action:

$$\delta S = \int_1^2 dt\left(\frac{\partial\mathscr{L}}{\partial q}\delta q + \frac{\partial\mathscr{L}}{\partial\dot q}\delta\dot q + \frac{\partial\mathscr{L}}{\partial\ddot q}\delta\ddot q\right)$$

$$= \int_1^2 dt\left[\frac{\partial\mathscr{L}}{\partial q} - \frac{d}{dt}\left(\frac{\partial\mathscr{L}}{\partial\dot q}\right) + \frac{d^2}{dt^2}\left(\frac{\partial\mathscr{L}}{\partial\ddot q}\right)\right]\delta q \tag{16.2.26}$$

The conjugate momenta for q and $\dot q$ are given by

$$p_1 = \frac{\partial\mathscr{L}}{\partial\dot q} - \frac{d}{dt}\left(\frac{\partial\mathscr{L}}{\partial\ddot q}\right) \qquad\text{and}\qquad p_2 = \frac{\partial\mathscr{L}}{\partial\ddot q} \tag{16.2.27}$$

The Hamiltonian is then constructed $H(q, p_1, \dot q, p_2)$ as follows, assuming the regularity of the Hessian $\partial^2\mathscr{L}/\partial\ddot q^i\partial\ddot q^j$:

$$H(q, p_1, \dot q, p_2) = p_1\dot q + p_2\ddot q - \mathscr{L}(q, \dot q, \ddot q) \tag{16.2.28}$$

From here, the Hamiltonian is derived by computing the differential of both sides and comparing the coefficients of the dq terms:

$$\frac{\partial H}{\partial p_1} = \dot q \qquad \frac{\partial H}{\partial q} = -\frac{\partial\mathscr{L}}{\partial q} \qquad \frac{\partial H}{\partial p_2} = \ddot q \qquad \frac{\partial H}{\partial\dot q} = \left(p_1 - \frac{\partial\mathscr{L}}{\partial\dot q}\right) \tag{16.2.29}$$

Exercise 16.7 As a brief example of the Routhian method, consider a particle in the xy-plane under a one-dimensional potential $U(x)$. The Lagrangian is written

$$\mathscr{L}(x, \dot{x}, \dot{y}) = \frac{1}{2}m\dot{x}^2 + \frac{1}{2}m\dot{y}^2 - U(x) \tag{16.2.30}$$

We expect the conjugate momentum of y to be conserved ($\dot{p}_y = 0$) to give the following Routhian function:

$$R(x, \dot{x}, p_y) = \frac{1}{2}m\dot{x}^2 + \frac{p_y^2}{2m} - U(x) \tag{16.2.31}$$

Routh's equations then become

$$\frac{d}{dt}\left(\frac{\partial R}{\partial \dot{x}}\right) - \frac{\partial R}{\partial x} = m\ddot{x} + U(x) = 0 \tag{16.2.32}$$

and

$$\dot{p}_y = -\frac{\partial R}{\partial y} = 0, \quad \text{and} \quad \dot{y} = \frac{\partial R}{\partial p_y} = p_y/m \tag{16.2.33}$$

The result agrees entirely with the Newtonian formalism which the reader is invited to convince themselves of.

As another example, consider a particle free to move in the xy-plane and undergoing simple harmonic oscillation in one dimension:

$$\mathscr{L}(x, \dot{x}, \dot{y}) = \frac{1}{2}m\dot{x}^2 + \frac{1}{2}m\dot{y} - \frac{1}{2}kx^2 \tag{16.2.34}$$

Show that the Routhian equations of motion are given by $m\ddot{x} + kx = 0$ and $\dot{y} = p_y/m$.

As a more challenging example, consider the same oscillator in polar coordinates with the Lagrangian $\mathscr{L}(r, \dot{r}, \dot{\theta}) = \frac{1}{2}m(\dot{r}^2 + r^2\dot{\theta}^2) - \frac{1}{2}kr^2$. Show that the Routhian becomes

$$R(r, \dot{r}, \dot{\theta}) = \frac{1}{2}m\dot{r}^2 + \frac{p_\theta^2}{2mr^2} - \frac{1}{2}kr^2 \tag{16.2.35}$$

and that the Routhian equations are given by

$$\dot{p}_\theta = -\frac{\partial R}{\partial \theta} = 0, \qquad \dot{\theta} = \frac{\partial R}{\partial p_\theta} = \frac{p_\theta}{mr^2}, \qquad \frac{d}{dt}\left(\frac{\partial R}{\partial \dot{r}}\right) - \frac{\partial R}{\partial r} = m\ddot{r} - \frac{p_\theta^2}{mr^3} + kr = 0 \tag{16.2.36}$$

Exercise 16.8 Building off exercise 7.10, here we will explore a non-local Hamiltonian formulation, whose action depends on more than one point in time. Generalising our earlier understanding, in the Lagrangian theory, a non-local action functional can be written as

$$A[q, \dot{q}] = \int_1^2 \left(\mathscr{L}(q, \dot{q}, t) + \int_0^{t'} dt' f(q, \dot{q}, t')\right) dt \tag{16.2.37}$$

where f is some function that links points in time. In the case that the integral over f vanishes the action becomes local. A variation in the action can be expressed using functional derivatives:

$$\delta A[q, \dot{q}] = \int_1^2 \left(\frac{\delta A}{\delta q(t)} - \frac{d}{dt} \frac{\delta A}{\delta \dot{q}(t)} \right) \delta q(t) dt \qquad (16.2.38)$$

from which we obtain the Euler-Lagrange equation

$$\frac{\delta A}{\delta q(t)} - \frac{d}{dt} \frac{\delta A}{\delta \dot{q}(t)} = 0 \qquad (16.2.39)$$

defined over the interval $\forall\, t \in [1, 2]$. The conjugate momentum is then defined as

$$p(t) = \frac{\delta A}{\delta \dot{q}(t)} \qquad (16.2.40)$$

and a non-local Hamiltonian functional can then be written using a Legendre transform:

$$\mathcal{H}[q, p] = \sup_{\dot{q}(t)} \left(\int_1^2 dt\, p(t) \dot{q}(t) - A[q, \dot{q}] \right) \qquad (16.2.41)$$

Notice that this is defined using the action, rather than the Lagrangian! Variation of this functional gives us

$$\delta \mathcal{H}[q, p] = \int_1^2 \left(-\frac{\delta A}{\delta q(t)} \delta q(t) - \frac{\delta A}{\delta \dot{q}(t)} \delta \dot{q}(t) + \delta p(t) \dot{q}(t) + p(t) \delta \dot{q}(t) \right) dt$$
$$= \int_1^2 \left(-\frac{\delta A}{\delta q(t)} \delta q(t) + \delta p(t) \dot{q}(t) \right) dt \qquad (16.2.42)$$

Considering the variation in $\mathcal{H}[q, p]$ on the left-hand side:

$$\delta \mathcal{H}[q, p] = \int_1^2 \left(\frac{\delta \mathcal{H}}{\delta q(t)} \delta q(t) + \frac{\delta \mathcal{H}}{\delta p(t)} \delta p(t) \right) dt \qquad (16.2.43)$$

we obtain Hamilton's non-local equations:

$$\dot{p} = -\frac{\delta \mathcal{H}}{\delta q(t)}, \qquad \frac{\delta \mathcal{H}}{\delta p(t)} = \dot{q}(t) \qquad (16.2.44)$$

where we have used the definition of the conjugate momentum and the Euler-Lagrange relation. For local theories, this expression becomes:

$$\mathcal{H}[q, p] = \int_1^2 dt\, H(q, p, t) \qquad (16.2.45)$$

17
Poisson Brackets & Angular Momentum

The **Poisson bracket** was developed by the French mathematician Poisson in the late 19th century and it is again a
reformulation or, at least, a tidying up of Hamilton's equations into one neat package. (see figure 17.1 for a portrait of
Siméon Denis Poisson (1781-1840).) Suppose that we have
a quantity[1] that can be written as a function of phase space
such that $A = A(p_1, \ldots, p_n, q^1, \ldots, q^n, t)$; the whole point of
dynamics is to characterise the time evolution of the system
and we can assume that, along a phase space trajectory, this
quantity can change as time evolves, since it's a function of
the ps and qs. As with any other function, we can take its
time derivative to see its rate of change with time:

Fig. 17.1: Siméon
Denis Poisson
(1781–1840).

$$\dot{A}(\boldsymbol{p}, \boldsymbol{q}, t) = \sum_{j=1}^{n} \frac{\partial A}{\partial q^j} \dot{q}^j + \sum_{j=1}^{n} \frac{\partial A}{\partial p_j} \dot{p}_j + \frac{\partial A}{\partial t} \qquad (17.0.1)$$

This is true of any trajectory in phase space, but, if we now
make an assumption that the curve $\{p, q\}$ is a solution to the equations of motion, we
can substitute Hamilton's equations into this for $\dot{q}$ and $\dot{p}$ and collect the summation
terms to write

$$\dot{A}(\boldsymbol{p}, \boldsymbol{q}, t) = \sum_{j=1}^{n} \frac{\partial A}{\partial q^j} \frac{\partial H}{\partial p_j} - \sum_{j=1}^{n} \frac{\partial A}{\partial p_j} \frac{\partial H}{\partial q^j} + \frac{\partial A}{\partial t}$$

$$= \sum_{j=1}^{n} \left(\frac{\partial A}{\partial q^j} \frac{\partial H}{\partial p_j} - \frac{\partial A}{\partial p_j} \frac{\partial H}{\partial q^j} \right) + \frac{\partial A}{\partial t} \qquad (17.0.2)$$

[1] We don't want to limit ourselves to any particular quantity but you could think of it as some
quantity of the system if you like - maybe the kinetic energy, the angular momentum or even the
position or angular coordinates themselves.

Lagrangian & Hamiltonian Dynamics. Peter Mann, Oxford University Press (2018).
© Peter Mann. DOI: 10.1093/oso/9780198822370.001.0001

It is a general mathematical result that we define the Poisson bracket between any two dynamical functions of phase space F and G as follows in equation 17.0.3:

$$\{F, G\}_{\text{PB}} := \sum_i \left(\frac{\partial F}{\partial q^i} \frac{\partial G}{\partial p_i} - \frac{\partial F}{\partial p_i} \frac{\partial G}{\partial q^i} \right) \qquad \textbf{Poisson bracket} \qquad (17.0.3)$$

Therefore, we can write 17.0.2 as[2]

$$\dot{A}(\boldsymbol{p}, \boldsymbol{q}, t) = \{A, H\} + \frac{\partial A}{\partial t} \qquad (17.0.4)$$

The Poisson bracket of a quantity with the Hamiltonian describes the time evolution of that quantity as we move along a curve in phase space. If the right-hand side of equation 17.0.4 vanishes, then A is conserved for the system and the Poisson bracket is zero if the function is not an explicit function of time. We would say that A is in **involution** with the Hamiltonian, although this holds only if it is a first integral (see chapter 16, section 16.1):

$$0 = \{A, H\} + \frac{\partial A}{\partial t}$$

$$\{A, H\} = -\frac{\partial A}{\partial t}$$

This Poisson bracket contains all the dynamical information of Hamilton's equations thus such is one of the most mathematically elegant and compact forms of classical physics. We can recover both of Hamilton's equations by taking the Poisson bracket of q^j and p_j with the Hamiltonian, as we show below for q^j; for simplicity, let's assume that q^j is not an explicit function of time, so that $\partial q^j / \partial t = 0$:

$$\dot{q}^j = \{q^j, H\}$$

$$= \sum_i \left(\frac{\partial q^j}{\partial q^i} \frac{\partial H}{\partial p_i} - \frac{\partial q^j}{\partial p_i} \frac{\partial H}{\partial q^i} \right)$$

The entire second term disappears since we have independent ps and qs, meaning $\partial q / \partial p = 0$. The first term reduces to zero for all i except for the one case where $i = j$, in which case differentiating q^j by itself gives 1; we then use the delta function δ^{ij}. Therefore, we are left with

$$\dot{q}^j = \frac{\partial H}{\partial p_j} \qquad (17.0.5)$$

which, as we can see, is just one of Hamilton's equations. If we were to repeat this for $\{p_j, H\}$, we would indeed, via a similar argument, obtain the other of Hamilton's

---

[2]For the most part, we aren't going to bother with the little subscript of PB outside the bracket as we would only need to do this if we were going to use Lagrange or Dirac brackets in the same setting. In addition, if we have more than one coordinate system on phase space, we may write them outside, for bookkeeping purposes.

equations. This leads us to rewrite both of Hamilton's equations in terms of the Poisson bracket:

$$\dot{q}^j = \left\{q^j, H\right\} \quad \text{and} \quad \dot{p}_j = \left\{p_j, H\right\}$$

where $j = 1, \ldots, n$. If you are familiar at all with quantum mechanics, then you may recognise these instantly as looking almost exactly the same as **Heisenberg's equations of motion**. Another question we could ask is whether the Hamiltonian itself is conserved, which we do by finding $\{H, H\}$. As with previous discussions, we find it to be

$$\{H, H\} = \frac{\partial^2 H}{\partial q^j \partial p_j} - \frac{\partial^2 H}{\partial p_j \partial q^j} + \frac{\partial H}{\partial t}$$

Using a feature of partial derivatives, we can arbitrarily swap the order of differentiation and so obtain the familiar criterion that the Hamiltonian is always conserved unless is explicitly a function of time:

$$\dot{H} = \frac{\partial H}{\partial t}$$

At this point, there is not really any new physics, and the Poisson bracket is almost just a tidying up of Hamilton's equations, with the benefit of describing the time derivative of any function of phase space, not just q^j or p_j, by taking the Poisson bracket with the Hamiltonian. This is not the only advantage of the Poisson bracket formulation however. The real benefit is how it allows us to connect to geometry and a higher level of thinking in classical mechanics and mathematical machinery.

We don't always have to take the Poisson bracket of a quantity with the Hamiltonian, as we have above. The bracket will no longer give us the time derivative of the quantity, since this is a specific result of taking the bracket of the quantity with H. However, the result will give us an insight into the detail of the system under some transformation or evolution in phase space *generated* by one of the quantities. An important result of the Poisson bracket is that it is invariant under a particular type of coordinate transformation, one that is **canonical**, provided that we describe the system with the correct Hamiltonian. For now don't worry about this (see the end of this section); instead, we will generalise and develop the *axiomatic formulation* of the Poisson bracket, without particular reference to H.

You should note that these relations hold for whichever way around we require them. Taking the Poisson bracket of a quantity with a constant must give zero since the constant doesn't change and hence has no evolution. The first axiomatic rule of Poisson brackets for the functions F, G and I of the ps and qs is the **antisymmetric rule**, or *skew symmetry rule*, also called the non-commutativity rule. It means that we can switch the order of the functions in the bracket by reversing the sign:

$$\boxed{\{F, G\} = -\{G, F\} \qquad \textbf{skew symmetry}} \tag{17.0.6}$$

A corollary of this is that, if we take the Poisson bracket of something with itself, it must equal the negative of itself. For this to hold, it must be zero:

$$\{F, F\} = 0$$

The second rule is that of **linearity**. It tells us that, if we multiply F by a constant λ, then the entire Poisson bracket is multiplied by the same constant:

$$\boxed{\{\lambda F, G\} = \lambda\{F, G\} \qquad \textbf{linearity}} \tag{17.0.7}$$

By the same rule, we can take the Poisson bracket of the sum of two functions $F + I$ with another G:

$$\{(F + I), G\} = \{F, G\} + \{I, G\}$$

$$\{(\lambda F + \gamma I), G\} = \lambda\{F, G\} + \gamma\{I, G\}$$

We have the **Leibniz rule**, which is a consequence of the product rule:

$$\boxed{\{FI, G\} = F\{I, G\} + \{F, G\}I \qquad \textbf{Leibniz}} \tag{17.0.8}$$

We can take the time derivative of a Poisson bracket,

$$\frac{\partial}{\partial t}\{F, G\} = \left\{\frac{\partial F}{\partial t}, G\right\} + \left\{F, \frac{\partial G}{\partial t}\right\} \tag{17.0.9}$$

Finally, we have the **Jacobi identity**:

$$\boxed{\{F, \{G, I\}\} + \{G, \{I, F\}\} + \{I, \{F, G\}\} = 0 \qquad \textbf{Jacobi}} \tag{17.0.10}$$

Interestingly, if two quantities are integrals of motion F and G, we can show that their Poisson bracket is a conserved quantity, too: $\{F, G\}$. This is called **Jacobi-Poisson theorem** and we will prove this here since it isn't as straightforward as the rest of the results above. Start off by rearranging equation 17.0.4 for the partial derivative of the conserved function and then substitute that into equation 17.0.9:

$$\frac{\partial}{\partial t}\{F, G\} = \{-\{F, H\}, G\} + \{F, -\{G, H\}\}$$

$$= \{\{H, F\}, G\} - \{-\{G, H\}, F\}$$

$$= \{\{H, F\}, G\} + \{\{G, H\}, F\}$$

Using the Jacobi identity above,

$$\{\{H, F\}, G\} + \{\{F, G\}, H\} + \{\{G, H\}, F\} = 0$$

Take the middle bracket over to the right-hand side and we arrive at our result:

$$\frac{\partial}{\partial t}\{F,G\} + \{\{F,G\},H\} = 0 \qquad (17.0.11)$$

Therefore, the total time derivative of $\{F,G\} = 0$. We will return to the use of the Jacobi-Poisson theorem below. It is these axiomatic rules that define a **Lie algebra** structure for the Poisson bracket. This is shared by the commutator in quantum mechanics; hence, we have a correspondence between classical and quantum mechanics, due to their similar, underlying mathematical framework.

A set of phase space observables equipped with the Poisson bracket structure is endowed with the above axiomatic rules. These rules happen to give a **Lie algebra structure** to the set of observables. Associated with every Lie algebra is a **Lie group**. The Lie groups of (canonical) infinitesimal transformations are such that they will allow the system to be described by Hamilton's equations; in fact, this is a criterion for a given coordinate transformation to be described as canonical. For the reader not familiar with Lie algebras, it is recommended that they wait until chapter 22 and instead focus on the Poisson bracket for now. It is enough to associate the fact that Poisson brackets demonstrate a hidden formal mathematical structure for a set of observables on the on-shell phase space.

We can use the axioms to greatly simplify the amount of work we do in calculations too. We can use them to develop three so-called **fundamental brackets** that are results of the axioms and pop up all over the place. Three of the fundamental brackets are shown below:

$$\{q^i, q^j\} = 0, \qquad \{p_i, p_j\} = 0, \qquad \{q^i, p_j\} = \delta^i_j$$

The first two are trivial since, when we take the partial derivative of a coordinate with respect to the momentum it is zero and the bracket cancels. So too does the Poisson bracket of the momentum coordinates. This is a general result and we will always get zero unless we have something that is a function of a p on one side and something that is a function of q on the other side.

The third bracket is non-zero in only one situation, which is when $i = j$.

$$\{q^i, p_j\} = \sum_k \left(\frac{\partial q^i}{\partial q^k}\frac{\partial p_j}{\partial p_k} - \frac{\partial q^i}{\partial p_k}\frac{\partial p_j}{\partial q^k} \right) = \sum_k (\delta^{ik}\delta_{jk} - 0) = \delta^i_j$$

The latter term is always zero, since p is independent of q and vice versa. The first delta term is only non-zero when $i = k$ in which case it gives $\partial q^k / \partial q^k = 1$. The same is true of the second delta term, which only gives 1 when $j = k$. Therefore the necessary condition is that $i = k = j$ and we write the **Kronecker delta** as

$$\{q^i, p_j\} = \delta^i_j \begin{cases} 1 & i = j \\ 0 & i \neq j \end{cases}$$

Our final two fundamental brackets are merely to highlight a particular expression:

$$\{F(\boldsymbol{q},\boldsymbol{p}),p_i\} = \frac{\partial F(\boldsymbol{q},\boldsymbol{p})}{\partial q^i} \tag{17.0.12}$$

This tells us that the Poisson bracket of a function of phase space with a momentum coordinate equals the derivative of the function with respect to the conjugate coordinate. The idea is that momentum is the **generator** of translations in position. There is also its analogue for when we take the Poisson bracket of a function with respect to a coordinate; it is the same as differentiating with respect to the canonical momentum, with a negative sign:

$$\{F(\boldsymbol{q},\boldsymbol{p}),q^i\} = -\frac{\partial F(\boldsymbol{q},\boldsymbol{p})}{\partial p_i} \tag{17.0.13}$$

This idea is explored in chapters 18 and 22.

17.1 Poisson Brackets & Angular Momenta

We now move on to an example to see some of the context of our axioms and the power of the Poisson formalism. We know from our elementary theory (see chapter 3) that the angular momentum of a rotating body is defined as $\vec{L} = \vec{r} \times \vec{p}$, but now let's break this up into components using unit vectors $\hat{\boldsymbol{x}}, \hat{\boldsymbol{y}}$ and $\hat{\boldsymbol{z}}$ (see chapter 34) along with the corresponding rules used below (as developed by Hamilton):

$$
\begin{aligned}
\vec{L} &= \vec{r} \times \vec{p} \\
&= (\hat{\boldsymbol{x}}r_x + \hat{\boldsymbol{y}}r_y + \hat{\boldsymbol{z}}r_z) \times (\hat{\boldsymbol{x}}p_x + \hat{\boldsymbol{y}}p_y + \hat{\boldsymbol{z}}p_z) \\
&= \hat{\boldsymbol{x}}r_x\hat{\boldsymbol{y}}p_y + \hat{\boldsymbol{x}}r_x\hat{\boldsymbol{z}}p_z + \hat{\boldsymbol{y}}r_y\hat{\boldsymbol{x}}p_x + \hat{\boldsymbol{z}}r_z\hat{\boldsymbol{x}}p_x + \hat{\boldsymbol{z}}r_z\hat{\boldsymbol{y}}p_y + \hat{\boldsymbol{y}}r_y\hat{\boldsymbol{z}}p_z \\
&= \hat{\boldsymbol{z}}r_xp_y - \hat{\boldsymbol{y}}r_xp_z - \hat{\boldsymbol{z}}r_yp_x + \hat{\boldsymbol{y}}r_zp_x - \hat{\boldsymbol{x}}r_zp_y + \hat{\boldsymbol{x}}r_yp_z \\
&= \hat{\boldsymbol{x}}(r_yp_z - r_zp_y) + \hat{\boldsymbol{y}}(r_zp_x - r_xp_z) + \hat{\boldsymbol{z}}(r_xp_y - r_yp_x)
\end{aligned}
$$

which (for those familiar with vector calculus) is recognised as the determinant of the following, more neatly expressed matrix:

$$\vec{L} = \vec{r} \times \vec{p} = \det \begin{pmatrix} \hat{\boldsymbol{x}} & \hat{\boldsymbol{y}} & \hat{\boldsymbol{z}} \\ r_x & r_y & r_z \\ p_x & p_y & p_z \end{pmatrix} \tag{17.1.1}$$

This concept allows us to write the components of the angular momentum as follows:

$$l_x = (r_yp_z - r_zp_y) \tag{17.1.2}$$
$$l_y = (r_zp_x - r_xp_z) \tag{17.1.3}$$
$$l_z = (r_xp_y - r_yp_x) \tag{17.1.4}$$

Using the **Levi-Civita** notation (see Mathematical Background 2), we can write this as $l_i = \epsilon_{ijk}r_jp_k$. If the system has spherical symmetry about an axis, then the component of angular momentum about that axis is conserved, due to the rotational invariance. The components of angular momentum will prove useful when considering atomic structure, spectra and spin.

> **Mathematical Background 2**
> The **Levi-Civita** notation ϵ_{ijk} is quite similar to the familiar Kronecker delta δ_{ij}. The Kronecker delta can take two values either 1 when $i = j$, or 0 if $i \neq j$. The Levi-Civita can take on three values -1, 0 and 1 but its rules are a little different than those for the Kronecker delta. The ijk corresponds to the three dimensions of space, so each of the indices can only take one of three values: 1, 2 or 3, corresponding to x, y and z.
>
> The Levi-Civita permutation symbol ϵ_{ijk} is zero if any of the two indices are the same and is only non-zero when all three indices are different. There are only six possible configurations where the indices are all different: $\epsilon_{123}, \epsilon_{231}, \epsilon_{312}$, which give a value of 1, and $\epsilon_{213}, \epsilon_{132}, \epsilon_{321}$, which give -1. For a more general discussion, please see Mathematical Background 7 in chapter 31, where we discuss properly how to work with ϵ notation.

Suppose we wish to know how a coordinate changes when we rotate about a particular axis (say, l_z) that has spherical symmetry; we would take the Poisson bracket of a coordinate with l_z:

$$
\begin{aligned}
\{r_x, l_z\} &= \{r_x, (r_x p_y - r_y p_x)\} \\
&= \{r_x, r_x p_y\} + \{r_x, (-r_y p_x)\} \\
&= \left(\frac{\partial r_x}{\partial r_x}\frac{\partial(r_x p_y)}{\partial p_x} - \frac{\partial r_x}{\partial p_x}\frac{\partial(r_x p_y)}{\partial r_x}\right) + \left(\frac{\partial r_x}{\partial r_x}\frac{\partial(-r_y p_x)}{\partial p_x} - \frac{\partial r_x}{\partial p_x}\frac{\partial(-r_y p_x)}{\partial r_x}\right) \\
&= (1 \cdot 0 - 0 \cdot p_y) + (1 \cdot (-r_y) - 0 \cdot 0) \\
&= -r_y
\end{aligned}
$$

where we have used the definition of the Poisson bracket and basic differentiation rules; with a similar analysis we can find $\{r_y, l_z\} = r_x$ and $\{r_z, l_z\} = 0$. Note the axiomatic form tells us we can pull out p_y from the first part and p_x from the second part. The meaning of this will be interpreted later but, for now, we can see we will have nine of these equations for each component of $\vec{L}$ and so we will develop a way to generalise this with some Levi-Civita nifty notation!

We can use this notation to generalise the above Poisson brackets of a coordinate with a component of $\vec{L}$ by writing,

$$\{r_i, l_j\} = \epsilon_{ijk} r_k \tag{17.1.5}$$

In order to derive this, we would need to know a bit about tensors and taking the cross product, but that would detract from the discussion for now. To illustrate this, we will work out the above example, which we can see is much easier using the Levi-Civita notation:

$$\{r_x, l_z\} = \epsilon_{xzy} r_y = \epsilon_{132} r_y = -r_y$$

where we identified i with x, and j with z, meaning that, by default, k equaled y. In the next step, we replaced the letters with their corresponding numbers, which are assigned the same way every time (e.g. $x = 1$ and so on) and then realised it had one pair of numbers swapped, hence, we write a negative sign. (An even number of swaps would be positive 1.) You can now easily work out any combination of coordinates and angular momentum components; relations like this are known as even or odd cyclic permutations. We can do the same with the momentum components too, and again we can write

$$\{p_i, l_j\} = \epsilon_{ijk} p_k \tag{17.1.6}$$

Upon computing some of these, we can see that they have the same form as the set of Poisson brackets equation 17.1.5 have, except its p_i, not r_i. This has a deep-rooted meaning in the symmetry properties of the Poisson bracket. If we calculate the Poisson bracket of a quantity with the angular momentum, we are calculating the change in the quantity when we rotate the coordinates. It allows us to say that the angular momentum is the *generator* of rotations. We can turn this around by saying that the change in a quantity due to a rotation of the system is the Poisson bracket of that quantity with the angular momentum; significantly, it is the canonical angular momentum that is the conserved quantity if there is an invariance to rotational motion in the system.

17.2   Poisson Brackets & Symmetries

It is a general property of the Poisson bracket that, if there is an invariance when we shift some parameter (i.e. translation, rotation, time, etc.), then there is a conserved property. Taking the Poisson bracket of a function of the shifted parameter with the conserved quantity gives us the the rate of change of that function with respect to a change in the parameter. For instance, if we have translational (which varies x) invariance, then the conjugate momentum is conserved; the Poisson bracket of a function of x with the conjugate momentum gives us the rate of change of the function with respect to x. Or, if a system is time-translationally invariant, then the Hamiltonian is the conserved quantity, as we know. Taking the Poisson bracket of a function that varies with time $F(t)$ with the Hamiltonian gives us the time derivative of $F(t)$. Taking the Poisson bracket of a function with a quantity (be it the Hamiltonian, the angular momentum or the linear momentum), we say that the quantity is the *generator* of the resulting transformation of the function (be it in time, rotation or translation, respectively). Is it a coincidence that these relationships are the subject of uncertainty relations in quantum mechanics [3]? We call the quantity the 'generator' of the transformation of phase space because, by definition, it generates small shifts in the parameters of the space. If we have a symmetry of the system, we can read the Poisson bracket in two ways: either how the generator function changes with respect to the parameter transformation, and how the conjugate momentum changes with the transformation generated by the function.

For example, if F is a conserved quantity, then we know that its time derivative is zero, so the Poisson bracket with the Hamiltonian is zero:

$$\{F, H\} = 0$$

If we turn this around, we can say that the Hamiltonian is invariant under a transformation generated by a function and we also have a symmetry in the Hamiltonian. Or F evolves in such a way that the energy is kept constant during the change:

$$\{H, F\} = 0$$

This is a manifestation of **Noether's theorem** in the Poisson formalism and it means the system remains on a **constant energy hypersurface**. In this way, conserved quantities can be viewed as constraints that force the system onto a subsurface of one less dimension (one dimension for each conserved coordinate).

Returning to the Jacobi-Poisson theorem, we can generate new conservation laws for quantities of the system in terms of other conserved quantities. If we know that l_x and l_y are conserved quantities, then their Poisson bracket is also conserved: $\{l_x, l_y\} = l_z$; in this way, we can state that, if any two components of the angular momentum are conserved, then the third component will also be conserved by Jacobi-Poisson theorem:

$$\begin{aligned} \{l_x, l_y\} &= \{r_y p_z - r_z p_y, r_z p_x - r_x p_z\} \\ &= \{r_y p_z, r_z p_x\} + \{r_z p_y, r_x p_z\} \\ &= -r_y p_x + p_y r_x \\ &= l_z \end{aligned}$$

The theorem only ensures that the result of the Poisson bracket is conserved; if the result of the Poisson bracket is trivial, then it might just be a different expression of something we already have accounted for.

---

[3]It's not, the reason being that the uncertainty between two observables is written in terms of their commutator:

$$\langle \Delta \hat{A} \rangle \langle \Delta \hat{B} \rangle \geq \frac{1}{4} |\langle [\hat{A}, \hat{B}] \rangle|$$

which is a residue of the Poisson bracket (or maybe the Poisson bracket is a residue of the commutator, depending on whether you follow historical development or fundamentalness).

From earlier in this chapter, we know that the time derivative of a quantity $F(t)$ of the system is given by its Poisson bracket with the Hamiltonian. If we now expand this quantity about its value at some initial time t_0 by using a Taylor series, we get

$$F(t) = F(t_0) + t\frac{dF}{dt}\bigg|_{t_0} + \frac{t^2}{2!}\frac{d^2F}{dt^2}\bigg|_{t_0} + \cdots \tag{17.2.1}$$

This approximates the value of the quantity at a small time later in terms of its value at the starting time; obviously, keeping more terms would be more accurate but we have enough to make the point. We know that the first term is given by the Poisson bracket evaluated at t_0:

$$\frac{dF}{dt}\bigg|_{t_0} = \{F, H\}_{t_0}$$

We can extend this to the next term, too, and, indeed, subsequent terms:

$$\frac{d^2F}{dt^2}\bigg|_{t_0} = \frac{d}{dt}\left(\frac{dF}{dt}\right)\bigg|_{t_0} = \{\{F, H\}, H\}_{t_0} \tag{17.2.2}$$

This gives us

$$F(t) = F(t_0) + t\{F, H\}_{t_0} + \frac{t^2}{2!}\{\{F, H\}, H\}_{t_0} + \cdots \tag{17.2.3}$$

which means that the time evolution of the quantity can be expressed as a power series of the Poisson bracket at t_0; we call the Hamiltonian the generator of the time evolution of the system. Let's do something rather cool now; let's define the Poisson bracket of F at t_0, with the Hamiltonian as an operator acting on some function $\hat{L}F(t_0) = \{F, H\}_{t_0}$; importantly, you can see that equation 17.2.3 is the exponential function expressed as a power series:

$$e^x = 1 + x + \frac{x^2}{2!} + \frac{x^3}{3!} + \cdots$$

This leads us to define the entire right-hand side of the expansion as

$$F(t) = e^{\hat{L}t}F(t_0) \tag{17.2.4}$$

a result we will return to later (see **classical propagator** in chapter 20).

Before leaving the subject of Poisson brackets, we may also note that, in a more mathematical sense, they take the structure of a determinant of the following two-dimensional matrix of all the partial derivatives:

$$\frac{\partial(F, G)}{\partial(q, p)} = \det\begin{pmatrix} \frac{\partial F}{\partial q} & \frac{\partial F}{\partial p} \\ \frac{\partial G}{\partial q} & \frac{\partial G}{\partial p} \end{pmatrix} = \frac{\partial F}{\partial q}\frac{\partial G}{\partial p} - \frac{\partial F}{\partial p}\frac{\partial G}{\partial q} \tag{17.2.5}$$

This matrix is know as the **Jacobian**, after Jacobi. The determinant of the Jacobian is also confusingly referred to as the Jacobian, and gives important information about the behaviour of a function above a point. Questions regarding the invertibility and differentiability of a function at a point can be understood through its Jacobian determinant. We will return to this idea when we discuss Liouville's theorem and when generalising the Poisson bracket structure. As we start to get a little more advanced with our discussions, we will start talking about Hessians and Jacobians in more detail with regard to the Legendre transform (see chapter 21). The Poisson bracket is explored in appendix D where we offer proof of the axiomatic formulation of the Lie algebra structure, and discuss the Poisson bracket using symplectic notation. We also generalise the Poisson formulation to the **Nambu bracket** in section D.1, although this requires knowledge of chapter 22.

Chapter summary

- The **Poisson bracket** is defined for two phase space functions F, G to be

$$\{F, G\}_{\text{PB}} := \sum_i \left(\frac{\partial F}{\partial q^i} \frac{\partial G}{\partial p_i} - \frac{\partial F}{\partial p_i} \frac{\partial G}{\partial q^i} \right)$$

- In general, Poisson brackets show the manifestation of a transformation induced by a generator of the transformation. The Poisson bracket of a function with the Hamiltonian is exactly the time derivative of the function, implying that the Hamiltonian is the **generator** of time translations. This is the importance of Poisson brackets.

- The Poisson bracket satisfies axiomatic rules that are inherited from its **Lie algebra structure**; these include **antisymmetry** (skew symmetry), **bilinearity**, **Leibniz rule** and the **Jacobi identity**.

- Noether's theorem in the canonical formalism was discussed and it was shown that the Poisson bracket of conserved quantities with the Hamiltonian vanish. The notion of **surfaces of constant energy** was introduced but will not be explored until much later in the book (see chapter 19, section 19.2).

- The Poisson bracket was shown to be the determinant of the **Jacobian matrix**.

- The classical propagator was derived (see chapter 20).

- The properties of angular momentum were explored, to highlight the usefulness of the **Poisson formalism** for solving problems in mechanics and generating handy relations. We also introduced the Levi-Civita tensor which we will use later on (see chapter 31).

Exercise 17.1 In this exercise we extend the Poisson bracket formulation to consider non-local classical theories, building off exercise 16.8. The non-local Poisson bracket for two quantities of the system can be written as

$$\{F, G\} = \int_1^2 dt \left(\frac{\delta F}{\delta q(t)} \frac{\delta G}{\delta p(t)} - \frac{\delta F}{\delta p(t)} \frac{\delta G}{\delta q(t)} \right) \tag{17.2.6}$$

Show that Hamilton's equations of motion can be written as:

$$\dot{q}(t) = \{q(t), \mathcal{H}\}, \qquad \dot{p}(t) = \{p, \mathcal{H}\} \tag{17.2.7}$$

18
Canonical & Gauge Transformations

The chapter is divided into three section. The first provides the reader with enough information on **canonical coordinate transformations** to continue on to future topics without detriment. The harmonic oscillator is then used as an example of a canonical transformation. The second section explicitly explores canonical transformation theory while introducing other types of transformations, including **gauge** transformations. This useful section is closed by a discussion of **action-angle** variables and the notion of **integrability**. The third section looks at **infinitesimal transformations** as a result of functions on phase space. Importantly, time evolution is discussed and the Hamiltonian is shown to be the generator of infinitesimal advancements in time, agreeing with Poisson mechanics (see chapter 17).

18.1   Canonical Transformations I

Changing coordinates is a fundamental operation that is important for any physical theory. We have already seen that the Euler-Lagrange equation of motion is invariant under a **point transformation** which was a change in the generalised coordinates $q \rightarrow Q$ (see chapter 9). We now move to extend this from the Lagrangian to the Hamiltonian phase space description, that is to say, not just $q \rightarrow Q$ but also $p \rightarrow P$:

$$Q = Q(\boldsymbol{q}(t), \boldsymbol{p}(t), t) \qquad P = P(\boldsymbol{q}(t), \boldsymbol{p}(t), t) \qquad (18.1.1)$$

where Q and P are the new coordinates and we assume one-to-one mapping between the two phase space descriptions. This is a more complex transformation since phase space has twice the dimensionality of configuration space; you will note that we only needed to change $q \rightarrow Q$ in the Lagrange formalism. As the old coordinates change under time evolution, the new coordinates change too, but, in general, for any arbitrary transformation these new coordinates won't obey Hamilton's equations. Transformations in phase space that *do* obey and *preserve* Hamilton's equations and the Poisson bracket are called **canonical transformations** for the appropriately modified Hamiltonian K.

There is some confusion as to exactly what conditions must be satisfied in order for a given transformation to be considered canonical. For our purpose, we define the condition according to the rules stated above: the preservation of Hamilton's equations

Lagrangian & Hamiltonian Dynamics. Peter Mann, Oxford University Press (2018).
© Peter Mann. DOI: 10.1093/oso/9780198822370.001.0001

and the fundamental Poisson brackets. However, others may state the definition of a canonical transformation is any that preserves the form of Hamilton's equations but not necessarily the Poisson bracket structure. In this context, transformations which preserve *both* are referred to as **symplectomorphisms**.

In order to sort through all the possible transformations to the coordinates we could perform, we need a method of finding those that *are* canonical and we do this by using four special functions called **generating functions** F_X, where $X = 1, 2, 3, 4$. Every canonical transformation can be identified by a generating function, of which the four types are listed below. The generating function is a combination of the old and new canonical variables, and possibly the independent variable time, too; it is therefore a bridge between the two coordinate descriptions. We require that each function F_X be a function of two of the variables so that it is a function of $2n$ independent variables of phase space. For each of the four mixed variable generator functions, we have a set of two relations and a modified Hamiltonian. There is no point having generating functions of the same variables $F(q, p)$ or $F(Q, P)$, since they don't allow us to generate the mixed expressions that we require in practice to implement the transformation.

The first kind is,

$$F_1 = F_1(q, Q, t), \qquad p_i = \frac{\partial F_1}{\partial q^i}, \qquad P_i = -\frac{\partial F_1}{\partial Q^i}, \qquad K = H + \frac{\partial F_1}{\partial t} \qquad (18.1.2)$$

The second kind is,

$$F_2 = F_2(q, P, t), \qquad p_i = \frac{\partial F_2}{\partial q^i}, \qquad Q^i = \frac{\partial F_2}{\partial P_i}, \qquad K = H + \frac{\partial F_2}{\partial t} \qquad (18.1.3)$$

The third kind is,

$$F_3 = F_3(P, Q, t), \qquad q^i = -\frac{\partial F_3}{\partial p_i}, \qquad P_i = -\frac{\partial F_3}{\partial Q^i}, \qquad K = H + \frac{\partial F_3}{\partial t} \qquad (18.1.4)$$

The fourth kind is,

$$F_4 = F_4(p, P, t), \qquad q^i = -\frac{\partial F_4}{\partial p_i}, \qquad Q^i = \frac{\partial F_4}{\partial P_i}, \qquad K = H + \frac{\partial F_4}{\partial t} \qquad (18.1.5)$$

Each of the four generator functions F_X are related by a Legendre transform with respect to one another; and represent the different combinations of the mixed transformation variables from $q, p \leftrightarrow Q, P$. The transformation is **invertible** (meaning we can go in the opposite direction) and smooth in the calculus sense; recalling our discussion of Jacobian determinants in chapter 32, we see that a necessary condition for a smooth invertible mapping is a non-zero Jacobian determinant.

When solving problems, the objective is to construct the new Hamiltonian $K(Q, P)$ that satisfies the modified Hamilton's equations in the new frame.

$$\dot{Q}^i = \frac{\partial K}{\partial P_i}, \qquad \dot{P}_i = -\frac{\partial K}{\partial Q^i} \tag{18.1.6}$$

From the details of the problem, you will generally be able to identify which of the four generator functions is most appropriate; then, we can use the two middle relations to generate the new Hamiltonian K; proceeding as normal from there.

We are going to have a quick look at canonical transformations in action, using the harmonic oscillator, and then set about understanding where they come from and deriving the above expressions in the next section. Later on, we will prove that the four functions are related via a Legendre transform.

Exercise 18.1 With the above equations in mind, let's have a look at a canonical transformation $(q, p) \rightarrow (Q, P)$, using the harmonic oscillator. We know the Hamiltonian can be expressed as

$$H = \frac{p^2}{2m} + \frac{kq^2}{2} \tag{18.1.7}$$

It would be advantageous to try and simplify this Hamiltonian. If, we could find a way to describe the ps and qs such that

$$p = f(P) \cos Q, \qquad q = \frac{f(P)}{m\omega} \sin Q \tag{18.1.8}$$

with $\omega^2 = k/m$, then the resulting Hamiltonian K can be written as follows:

$$K = \frac{f(P)^2}{2m} \tag{18.1.9}$$

since $\cos^2 x + \sin^2 x = 1$. This is just a function of P rather than two variables (q, p), and hence, is far easier to work with. All that remains now is to find $f(P)$[1].

Our first step is to use one of the generator functions to find a function that mixes up the old and new variables such that one of the old variables is expressed in terms of one of the old and one of the new. To do this, we rearrange both parts of equation 18.1.8 for $f(P)$ and equate them to obtain

$$p = \frac{m\omega q \cos Q}{\sin Q} \tag{18.1.10}$$

---

[1]You may ask what motivated us to particularly define the relations in equation 18.1.8 as we did. How would we know to do that? Well, having spent some time studying the harmonic oscillator, we can account for the physics behind this change, using a bit of intuition and a bit of lucky guesswork. A more truthful answer is presented when we study **action-angle variables** in appendix E, where we derive formulas that generate these sort of relations for us!

This equation is the old variable p in terms of the old variable q and the new variable Q. This matches the first kind of canonical transformation, $F_1(q, \mathbf{Q}, t)$; you therefore know that we need to use the equations in 18.1.2. Importantly, we can see that $p = \partial F_1 / \partial q$ allowing us to write

$$\frac{\partial F_1}{\partial q} = \frac{m\omega q \cos Q}{\sin Q} \tag{18.1.11}$$

and hence determine the identity of F_1 by integration:

$$F_1 = \frac{1}{2} m\omega q^2 \cot Q \tag{18.1.12}$$

We can now use the second equation in 18.1.2:

$$P = -\frac{\partial F_1}{\partial Q} = \frac{m\omega q^2}{2 \sin^2 Q} \tag{18.1.13}$$

where we have just substituted the expression for F_1 in and $\frac{d}{dx} \cot x = -(\sin^2 x)^{-1}$. Now rearrange this for q:

$$q = \left(\frac{2P}{m\omega} \right)^{\frac{1}{2}} \sin Q \tag{18.1.14}$$

Comparing this to equation 18.1.8, we can identify $f(P)$ to be

$$f(P) = (2m\omega P)^{1/2} \tag{18.1.15}$$

The Hamiltonian in equation 18.1.9 is finally written as

$$K = \frac{f(P)^2}{2m} = \omega P \tag{18.1.16}$$

The first thing to note is how much simpler this result is than the original Hamiltonian. It appears that (Q, P) and the new Hamiltonian K, are an advantageous description of the harmonic oscillator compared to (q, p) and H.

The second point to notice is that Q is a cyclic coordinate and hence the conjugate momentum P is conserved. In this particular case, F_1 has no explicit time dependence so we can say that $K = H$ for this system and consequentially, that $P = H/\omega$. A canonical transformation is one that preserves Hamilton's equations so, by definition, we can write $\dot{Q} = \partial H / \partial P = \omega$. Integrating this last expression with respect to time, we get $Q(t) = \omega t + \phi$, where ϕ is a constant determined by the initial conditions. Therefore, at the end of this analysis, we can see that P is time independent while Q is linear in time (pretty straightforward to deal with analytically). We can now solve equation 18.1.14:

$$q = \sqrt{\frac{2H}{m\omega^2}} \sin(\omega t + \phi) \tag{18.1.17}$$

Hence, the maximum amplitude of oscillation is given by $(2H/m\omega^2)^{1/2}$. Using the relations above, you can see that the momentum is given by

$$p = \sqrt{2mH}\cos(\omega t + \phi) \tag{18.1.18}$$

In order to see that this transformation is indeed canonical, we can see if one of the fundamental Poisson brackets holds in the new coordinate system; let's try $\{q,p\}_{QP}$ to see if it equals 1:

$$
\begin{aligned}
\{q,p\}_{Q,P} &= \frac{\partial q}{\partial Q}\frac{\partial p}{\partial P} - \frac{\partial q}{\partial P}\frac{\partial p}{\partial Q} \\
&= \sqrt{\frac{2P}{m\omega}}\cos Q \sqrt{\frac{m\omega}{2P}}\cos Q - \sqrt{\frac{1}{2m\omega P}}\sin Q\left[-\sqrt{2mP\omega}\sin Q\right] \\
&= \cos Q \cos Q + \sin Q \sin Q \\
&= 1
\end{aligned}
$$

Therefore, we have shown that the transformation $(q,p) \to (Q,P)$ is, in fact, canonical. The new description of the system is easier that the original picture for analytical investigation of the dynamics.

18.2 Canonical Transformations II

Having seen how a canonical transformation can be used and how it can emphasise the physics of a system, let us now set about proving the above theory and understand where it comes from.

We will start pretty much from scratch, looking first over the familiar **gauge transformations** (but extending the discussion to Hamiltonian mechanics, see chapter 11) and then on to canonical transformations. If we recall that the Hamiltonian is the Legendre transform of the Lagrangian with respect to the $\dot{q}$s, we can see that

$$H(\boldsymbol{p},\boldsymbol{q},t) = p_i\dot{q}^i - \mathscr{L}(\boldsymbol{q},\dot{\boldsymbol{q}},t)$$

If we add the total time derivative of a function $\varphi = \varphi(\boldsymbol{q},t)$ to the Lagrangian, then we know that the extremals the actions of both the original and this new Lagrangian coincide. So what is the difference between a point transformation and a gauge transformation? The point transformation leaves the Euler-Lagrange equation invariant but, *in addition*, leaves Hamilton's equations invariant, too. Point transformations are a subset of the more general canonical transformation. Gauge transformations are transformations that leave the Lagrange formalism intact but ruin the Hamiltonian description. To qualify this, consider the generalised momentum to the new Lagrangian:

$$L = \mathscr{L} + \frac{d}{dt}\varphi = \mathscr{L} + \sum_i \frac{\partial\varphi}{\partial q^i}\dot{q}^i + \frac{\partial\varphi}{\partial t}$$

The canonical momentum is defined by $p_i' = \partial L/\partial \dot{q}^i$:

$$p_i' = p_i + \frac{\partial\varphi}{\partial q^i}$$

The corresponding Hamiltonian to L is given by its Legendre transform,

$$H' = p_i' \dot{q}^i - L = H - \frac{\partial \varphi}{\partial t}$$

Hamilton's new equations are

$$\dot{q}^i = \frac{\partial H'}{\partial p_i'} = \frac{\partial H}{\partial p_i}, \qquad \dot{p}_i = -\frac{\partial H'}{\partial q^i} = -\frac{\partial H}{\partial q^i} + \frac{\partial}{\partial q^i} \frac{\partial \varphi}{\partial t}$$

We can clearly see an asymmetry in Hamilton's equations here; indeed, this asymmetry runs right throughout our later discussions. Hamilton's equations are not invariant under a gauge transformation in the Lagrangian, meaning gauge transformations are not a subset of canonical transformations. This is due to the definition of the conjugate momentum. Gauge theories belong to their own theoretic structure that extends well beyond classical mechanics, into field theories and quantum mechanics, too. In this way, a Lagrangian system can have multiple portraits in phase space, all corresponding to the same motion.

If we can rewrite the Lagrangian with a total derivative, then we can also rewrite the Hamiltonian, too, since they are related via a Legendre transform. We know that Hamilton's phase space action is given by $S = S(\boldsymbol{q}, \boldsymbol{p})$ but, for a canonical transformation (one that *by definition* corresponds to the same Hamiltonian equations of motion) $qp \to QP$, there must also exist an action integral of the form

$$S'(\boldsymbol{Q}, \boldsymbol{P}) = \int_{t_1}^{t_2} \mathscr{L}_{ph}'(\boldsymbol{Q}, \dot{\boldsymbol{Q}}, \boldsymbol{P}, \dot{\boldsymbol{P}}, t) dt \qquad (18.2.1)$$

$S(\boldsymbol{q}, \boldsymbol{p})$ and $S'(\boldsymbol{Q}, \boldsymbol{P})$, which correspond to the same physical path, are both stationary with variations that vanish at the end points. Just as we may add a total time derivative to the velocity space Lagrangian without consequence, we can do the same to the phase space Lagrangian, too. Further, if we were to transform the coordinates, then both of the actions would need to be the the same up to end point corrections if they describe the same path. To write this mathematically, we can demand

$$\delta S(\boldsymbol{q}, \boldsymbol{p}) = \delta S(\boldsymbol{Q}, \boldsymbol{P}) \qquad (18.2.2)$$

then

$$\mathscr{L}'(\boldsymbol{q}, \dot{\boldsymbol{q}}, \boldsymbol{p}, \dot{\boldsymbol{p}}, t) = \mathscr{L}'(\boldsymbol{Q}, \dot{\boldsymbol{Q}}, \boldsymbol{P}, \dot{\boldsymbol{P}}, t) + \frac{dF_X}{dt} \qquad (18.2.3)$$

In other words, if both coordinate systems describe the same physical path, we can equate their actions. However, there is an uncertainty between the two Lagrangians that disappears when we compute the end point variations but yet is still present in the expression of the action, namely the total time derivative of some function F. We call this a *surface term* in the action (see chapter 11). We can write equation 18.2.3 as $[\dot{q}^j p_j - H(\boldsymbol{q}, \boldsymbol{p}, t)] - [\dot{Q}^j P_j - K(\boldsymbol{Q}, \boldsymbol{P}, t)] = \dot{F}$, which is important result for appendix E and is, in fact, why F is called the generating function, as its time derivative gives a relation between old and new phase space Lagrangians.

A question now arises. Is F a function of the ps and qs, a function of the Ps and Qs or a mixture of both? Well, there are four possible solutions in addition to (QP) and (qp), if we recall from above that each F must be a function of two sets of variables for dimensionality purposes:

$$F_1(\boldsymbol{q}, \boldsymbol{Q}, t), \qquad F_2(\boldsymbol{q}, \boldsymbol{P}, t), \qquad F_3(\boldsymbol{p}, \boldsymbol{Q}, t), \qquad F_4(\boldsymbol{p}, \boldsymbol{P}, t), \qquad (18.2.4)$$

We call these four functions the generating functions. We can now take the time derivative of these by familiar chain ruling:

$$\frac{dF_1}{dt} = \sum_j \left\{ \frac{\partial F_1}{\partial q^j} \frac{\partial q^j}{\partial t} + \frac{\partial F_1}{\partial Q^j} \frac{\partial Q^j}{\partial t} \right\} + \frac{\partial F_1}{\partial t} \qquad (18.2.5)$$

$$\frac{dF_2}{dt} = \sum_j \left\{ \frac{\partial F_2}{\partial q^j} \frac{\partial q^j}{\partial t} + \frac{\partial F_2}{\partial P_j} \frac{\partial P_j}{\partial t} \right\} + \frac{\partial F_2}{\partial t} \qquad (18.2.6)$$

$$\frac{dF_3}{dt} = \sum_j \left\{ \frac{\partial F_3}{\partial p_j} \frac{\partial p_j}{\partial t} + \frac{\partial F_3}{\partial Q^j} \frac{\partial Q^j}{\partial t} \right\} + \frac{\partial F_3}{\partial t} \qquad (18.2.7)$$

$$\frac{dF_4}{dt} = \sum_j \left\{ \frac{\partial F_4}{\partial p_j} \frac{\partial p_j}{\partial t} + \frac{\partial F_4}{\partial P_j} \frac{\partial P_j}{\partial t} \right\} + \frac{\partial F_4}{\partial t} \qquad (18.2.8)$$

Now you can substitute any of these total time derivatives into equation 18.2.3:

$$\mathcal{L}'(\boldsymbol{q}, \dot{\boldsymbol{q}}, \boldsymbol{p}) = \mathcal{L}'(\boldsymbol{Q}, \dot{\boldsymbol{Q}}, \boldsymbol{P}) + \frac{d}{dt} F_1(\boldsymbol{q}, \boldsymbol{Q}) \qquad (18.2.9)$$

$$\mathcal{L}'(\boldsymbol{q}, \dot{\boldsymbol{q}}, \boldsymbol{p}) = \mathcal{L}'(\boldsymbol{Q}, \dot{\boldsymbol{P}}, \boldsymbol{P}) + \frac{d}{dt} F_2(\boldsymbol{q}, \boldsymbol{P}) \qquad (18.2.10)$$

$$\mathcal{L}'(\boldsymbol{q}, \dot{\boldsymbol{p}}, \boldsymbol{p}) = \mathcal{L}'(\boldsymbol{Q}, \dot{\boldsymbol{Q}}, \boldsymbol{P}) + \frac{d}{dt} F_3(\boldsymbol{p}, \boldsymbol{Q}) \qquad (18.2.11)$$

$$\mathcal{L}'(\boldsymbol{q}, \dot{\boldsymbol{p}}, \boldsymbol{p}) = \mathcal{L}'(\boldsymbol{Q}, \dot{\boldsymbol{P}}, \boldsymbol{P}) + \frac{d}{dt} F_4(\boldsymbol{p}, \boldsymbol{P}) \qquad (18.2.12)$$

We know the phase space Lagrangian is given by $\mathcal{L}'_{ph}(\boldsymbol{q}, \dot{\boldsymbol{q}}, \boldsymbol{p}, \dot{\boldsymbol{p}}, t) = \boldsymbol{p} \cdot \dot{\boldsymbol{q}} - H(\boldsymbol{q}, \boldsymbol{p}, t)$, and that $\mathcal{L}'_{ph}(\boldsymbol{Q}, \dot{\boldsymbol{Q}}, \boldsymbol{P}, \dot{\boldsymbol{P}}, t) = \boldsymbol{P} \cdot \dot{\boldsymbol{Q}} - K(\boldsymbol{Q}, \boldsymbol{P}, t)$; therefore, according to equation 18.2.3,

$$\boldsymbol{p} \cdot \dot{\boldsymbol{q}} - H(\boldsymbol{q}, \boldsymbol{p}, t) = \boldsymbol{P} \cdot \dot{\boldsymbol{Q}} - K(\boldsymbol{Q}, \boldsymbol{P}, t) + \frac{dF_1}{dt} \qquad (18.2.13)$$

Now substitute equation 18.2.5 in for the time derivative of F_1 and collect terms to obtain

$$\left(p - \frac{\partial F_1}{\partial q}\right) \cdot \dot{q} - \left[H(\boldsymbol{q}, \boldsymbol{p}, t) + \frac{\partial F_1}{\partial t}\right] = \left(P + \frac{\partial F_1}{\partial Q}\right) \cdot \dot{Q} - K(\boldsymbol{Q}, \boldsymbol{P}, t) \qquad (18.2.14)$$

Since $\dot{q}$ and $\dot{Q}$ are independent variables that run over $2n$-dimensional space, the only way this can hold is if both coefficients vanish independently:

$$p = \frac{\partial F_1}{\partial q}, \qquad P = -\frac{\partial F_1}{\partial Q}, \qquad K = H + \frac{\partial F_1}{\partial t} \qquad (18.2.15)$$

Hence, we arrive at the first expression for a canonical transformation relating the canonical variables and the generator functions. The other generating functions are generated by performing a Legendre transform with respect to the variable we wish to replace. Let's look at $Q \to P$. Using equation 18.2.13, rearrange for dF_1, noticing that the dts cancel on the dotted coordinates:

$$dF_1(\boldsymbol{q}, \boldsymbol{Q}, t) = K\,dt - H\,dt + \sum_{j=1}^{n} p_j dq^j - \sum_{j=1}^{n} P_j dQ^j \qquad (18.2.16)$$

To replace P with Q, add the following derivative to 18.2.16:

$$d\left(\sum_{j=1}^{n} P_j Q^j\right) = \sum_{j=1}^{n} P_j dQ^j + \sum_{j=1}^{n} Q^j dP_j \qquad (18.2.17)$$

$$dF_2(\boldsymbol{q}, \boldsymbol{P}, t) = K\,dt - H\,dt + \sum_{j=1}^{n} p_j dq^j + \sum_{j=1}^{n} Q^j dP_j \qquad (18.2.18)$$

$$F_2(\boldsymbol{q}, \boldsymbol{P}, t) = F_1(q, Q, t) + \sum_{j=1}^{n} Q^j P_j \qquad (18.2.19)$$

As we did with the Hamiltonian when we performed the Legendre transform (see chapter 14), we can expand the left-hand side and compare the terms with the same differential (dt, dq^j, dP_j):

$$dF_2(\boldsymbol{q}, \boldsymbol{p}, t) = \sum_{j=1}^{n} \left(\frac{\partial F_2}{\partial q^j} dq^j + \frac{\partial F_2}{\partial P_j} dP_j\right) + \frac{\partial F_2}{\partial t} dt \qquad (18.2.20)$$

$$p_j = \frac{\partial F_2}{\partial q^j}, \qquad Q^j = \frac{\partial F_2}{\partial P_j}, \qquad K = H + \frac{\partial F_2}{\partial t} \qquad (18.2.21)$$

generating the second of the generating function relations. The rest of the Legendre transforms proceed in the same manner with respect to other variables and it is easy to prove the other relations between the generator functions (see appendix E). Each time we expect to get three relations and an equation relating the two generator functions.

Canonical transformations are useful to highlight alternative aspects of the physics of systems but, in general, there is nothing to say the new coordinates will be any easier to solve. However in certain cases like the harmonic oscillator exercise, there exists a special set of coordinates which *do* simplify the systems description in phase space; these coordinates are called **action-angle variables**.

Action-angle variables trivialise the phase space description of the system and are described as the system's natural units such that they **linearise** the phase space flow lines which we explore later. Not all systems have these action-angle variables and those who do are termed **integrable systems**. In a $2n$-dimensional phase space, an integrable system (by the Liouville definition) is one that has n, globally defined, functionally independent, first integrals (see footnote in section 16.2). A system is **superintegrable** if there exists an additional $n-1$ globally defined first integrals, thus totalling $(2n-1)$ functionally independent, global quantities. N quantities are **functionally independent** if the wedge product of their exterior derivatives is nowhere vanishing.

$$dI_1 \wedge \cdots \wedge dI_N \neq 0$$

In the case of the oscillator, you may remember that Q was a cyclic coordinate, meaning P was conserved where $P = H/\omega$. For such integrable systems, the trajectories in phase space remain close to each other as the system evolves in time. Systems that do not admit n conserved quantities (**first integrals** or **involutive Poisson brackets**) are termed **non-integrable** by the Liouville definition of integrability. The phase space trajectories of non-integrable systems spread exponentially from one another as time evolves. This type of motion is called **chaotic motion** and it introduces an inherent difficulty in predicting the long-term dynamics of systems. In general, two trajectories of a chaotic system are initially separated by $\delta q(0)$, they will have a separation at time t of $|\delta q(t)| \approx e^{\lambda t}|\delta x(0)|$, where λ is the **Lyapunov exponent**.

Integrable systems have a periodic motion in phase space. For instance, the harmonic oscillator forms circles or ellipses of constant energy. In fact, this is a very intuitive result: when thinking about the periodic trade-off between q and p for the oscillator, we can see that, as q increases the momentum conjugate to q decreases. This resembles the trade off between T and U during the oscillation; as the potential increases the kinetic energy decreases and vice-versa all this is very reminiscent of tracing out a circle.

There is indeed a procedure to follow for finding these particular coordinates, and we tackle in appendix F. If you follow the derivation through, you will find that the canonical transformation that gives the action-angle variables Q and P is given by

$$P_i = \frac{1}{2\pi} \oint p_i \, dq^i, \qquad \text{and} \qquad Q^i = \int \frac{\partial p_i}{\partial P_i} \, dq^i \qquad \textbf{action-angle variables}$$

where P is the **action variable** and Q is the **angle variable**. We will also investigate why Q is periodic, which we find is a criterion for the system to be integrable. This is actually a whole new way of solving problems in Hamiltonian dynamics; perform a canonical transformation $(qp) \to (QP)$ such that the Hamiltonian is independent of the new Q coordinates, and their conjugate momenta P are conserved. The new generalised coordinates now evolve linearly in time and are much nicer to solve, both mathematically and graphically. This is the basis of the Hamilton-Jacobi theorem. In the 19th century, it was believed that there were more integrable systems than non-integrable systems; however, this was soon realised as false. We now understand that there are only a limited number of integrable systems, and a great deal of time is spent trying to break otherwise complex models up into integrable subsystems.

18.3 Infinitesimal Canonical Transformations

In this final section regarding the subject of canonical transformations, we consider the infinitesimal time evolution of the system. At each moment, Hamilton's equations are satisfied as the system maps out a trajectory in phase space and, therefore, time evolution is considered to be an **infinitesimal canonical transformation**. There are two ways to think about time evolution: in the first, the system maps out a curve in a fixed axis phase space; in the second, at each moment, from one instant to the next, the coordinate system is changing and we consider the Qs and Ps as solutions to Hamilton's equations at a later time, given the old coordinates as initial conditions:

$$Q^i = q^i + \delta q^i, \qquad P_i = p_i + \delta p_i \qquad (18.3.1)$$

In this description, we picture the Hamiltonian as the generator of *infinitesimal* canonical transformations in the coordinates that all together give us the time evolution of those coordinates for any function of phase space. In this way, we can intimately connect the time evolution of dynamical systems to canonical transformations at every instant; therefore, finding the canonical transformation is akin to solving for the dynamics of the system.

The theory of *infinitesimal* canonical transformations is slightly different from the theory described in our previous discussions in this chapter. To say that the Hamiltonian is an infinitesimal generating function does not mean that it is one of the F_X functions above, but they are very closely linked[2]. We should think of infinitesimal canonical transformations as mapping phase space functions to different coordinates; time evolution fits this description, and so too do rotations.

For any of our transformations F_X, we may take the **identity transformation**, which means $(q, P) \to (q, P)$, using an F_2 for example. In this case, you can work

---

[2]It is perhaps an unfortunate nomenclature that we refer to both G and F as generating functions and there is sometimes confusion.

out a set of trivial relations using equation 18.1.3 (and, of course, the others too). An infinitesimal canonical transformation is close to the identity transformation; as for perturbation theory (see chapter 24), we can write

$$F_2(\boldsymbol{q}, \boldsymbol{P}, t) = q^i P_i + \epsilon G(\boldsymbol{q}, \boldsymbol{P}, t)$$

where G is some function of phase space, called the *generator of the infinitesimal transformation*, and ϵ is a very small smooth parameter. When the small parameter is zero, we have the identity transformation. When we look back at equation 18.1.3 the equations of motion become

$$Q^i = q^i + \epsilon \frac{\partial G}{\partial P_i}, \qquad\qquad p_i = P_i + \epsilon \frac{\partial G}{\partial q^i} \qquad\qquad (18.3.2)$$

With slight rearrangement of the second equation, we have two equations in the new variables (Q, P) in terms of the old two and derivatives of G. By considering equation 18.3.1 we find that

$$\delta q^i = Q^i - q^i = \epsilon \frac{\partial G}{\partial P_i}, \qquad\qquad \delta p_i = P_i - p_i = -\epsilon \frac{\partial G}{\partial q^i}$$

However, since the transformation is infinitesimal, we can express G as a function of any two of (q, p, Q, P) variables in addition to the small, smooth parameter, for instance $G = G(\boldsymbol{q}, \boldsymbol{p}, \epsilon)$. This means

$$\epsilon \frac{\partial G}{\partial P_i} \approx \epsilon \frac{\partial G}{\partial p_i} \qquad\qquad \epsilon \frac{\partial G}{\partial q^i} \approx \epsilon \frac{\partial G}{\partial Q^i}$$

Now, considering the case when $G = H(\boldsymbol{q}, \boldsymbol{p}, t)$, we generate

$$\delta q^i = \epsilon \frac{\partial H}{\partial p_i} = \epsilon \dot{q}^i, \qquad\qquad \delta p_i = -\epsilon \frac{\partial H}{\partial q^i} = \epsilon \dot{p}_i \qquad\qquad (18.3.3)$$

The smooth parameter when G is the Hamiltonian H is just time evolution δt. Therefore, $Q^i = q^i(t + \delta t) = q^i(t) + \dot{q}^i \delta t$ and $P_i = p_i(t + \delta t) = p_i(t) + \dot{p}_i \delta t$, with $\delta q^i = \epsilon\{H, q^i\}$ and momentum equivalent. We conclude that the Hamiltonian is the infinitesimal generator of time translations.

Rotations can be treated with infinitesimal transformations, too, in this case, the generator is called the angular momentum! We soon realise that the generator functions G are the infinitesimal generators of a **Lie group** of continuous translations (in time, space, and so on). These types of continuous transformations are important to Noether's theorem. The relationship between classical observables and infinitesimal canonical transformations is not paralleled by discrete symmetries, as they do not have an infinitesimal form. Hence, discrete symmetries do not have the same mathematical structure as continuous symmetries and are not a part of Noether's theorem.

Some canonical transformations can't be generated by all four types of generator functions but those that can will give equivalent results. One way to check if a given transformation is canonical is by use of the **direct conditions**. Let us consider a transformation that is not an explicit function of time, such that

$$Q^i = Q^i(q, p) \qquad \text{and} \qquad P_i = P_i(q, p) \tag{18.3.4}$$

In this case, any corresponding generator function will not have time dependance, meaning $\partial F_X / \partial t = 0$, so that $K = H$ and, importantly, the Hamiltonian does not change with time. We call transformations like these **restricted transformations**. If we take the time derivative of the new coordinate Q, chain rule as normal, then use Hamilton's equations for $\dot{q}$ and $\dot{p}$, we obtain

$$\dot{Q}^i = \frac{\partial Q^i}{\partial q^k} \frac{\partial H}{\partial p_k} - \frac{\partial Q^i}{\partial p_k} \frac{\partial H}{\partial q^k} \tag{18.3.5}$$

We now consider the inverse transform $q^i = q^i(\boldsymbol{Q}, \boldsymbol{P})$ and $p_i = p_i(\boldsymbol{Q}, \boldsymbol{P})$ and use these to convert the original Hamiltonian into a function of the Qs and Ps such that $H = H\big(q(\boldsymbol{Q}, \boldsymbol{P}), p(\boldsymbol{Q}, \boldsymbol{P})\big)$. At the beginning of this chapter, we commented that canonical transformations are *defined* by preserving Hamilton's equations; we now use this point directly:

$$\dot{Q}^i = \frac{\partial K}{\partial P_i} = \frac{\partial H}{\partial P_i} \tag{18.3.6}$$

Now chain rule the Hamiltonian:

$$\dot{Q}^i = \frac{\partial H}{\partial q^k} \frac{\partial q^k}{\partial P_i} + \frac{\partial H}{\partial p_k} \frac{\partial p_k}{\partial P_i} \tag{18.3.7}$$

Compare equations 18.3.5 and 18.3.7 and match the terms with the same Hamiltonian derivative, which is coloured for ease. With this, we can write two conditions:

$$\left[\frac{\partial Q^i}{\partial q^k} \right]_{q,p} = \left[\frac{\partial p_k}{\partial P_i} \right]_{Q,P} \qquad \text{and} \qquad \left[\frac{\partial Q^i}{\partial p_k} \right]_{q,p} = - \left[\frac{\partial q^k}{\partial P_i} \right]_{Q,P} \tag{18.3.8}$$

You can follow a similar derivation for $\dot{P}$ to get

$$\left[\frac{\partial P_i}{\partial p_k} \right]_{q,p} = \left[\frac{\partial q^k}{\partial Q^i} \right]_{Q,P} \qquad \text{and} \qquad \left[\frac{\partial P_i}{\partial q^k} \right]_{q,p} = - \left[\frac{\partial p_k}{\partial Q^i} \right]_{Q,P} \tag{18.3.9}$$

Any time-independent canonical transformation will obey these **direct conditions** and they prove useful relations to check if transformations are canonical. The little subscripts mean the top variable of the fraction as a function of the subscripted parts, so the left-hand side of equation 18.3.8 would read 'Q_i as a function of $(\boldsymbol{q}, \boldsymbol{p})$'. Importantly if we use the direct conditions to prove a generating function is canonical, then any generating function related by Legendre transform of that function is also canonical!

We can use these direct conditions to investigate whether the Poisson bracket is invariant under a canonical transformation; of course, we already know it is by definition, but let's see some proof for one of the fundamental brackets $\{Q, P\}$:

$$\{Q^i, P_j\}_{q,p} = \frac{\partial Q_i}{\partial q^k}\frac{\partial P_j}{\partial p_k} - \frac{\partial Q^i}{\partial p_k}\frac{\partial P_j}{\partial q^k} \tag{18.3.10}$$

$$= \frac{\partial Q^i}{\partial q^k}\frac{\partial q^k}{\partial Q^j} + \frac{\partial Q^i}{\partial p_k}\frac{\partial p_k}{\partial Q^j} \tag{18.3.11}$$

$$= \frac{\partial Q^i(\boldsymbol{q}, \boldsymbol{p})}{\partial Q^j} \tag{18.3.12}$$

$$= \delta^{ij} \tag{18.3.13}$$

In the last step, we just canceled out the variables in the fractions. In this way, we have shown that the fundamental brackets are invariant under a canonical transformation. Thus, we can write $\{Q^i, P_j\}_{q,p} = \{q^i, p_j\}_{Q,P} = \delta^i_j$.

Chapter summary

- Canonical transformations are those that preserve Hamilton's equations.
- **Gauge transformations** were found to not preserve the Hamiltonian formalism and are therefore not canonical transformations. They are associated with redundant degrees of freedom of the system.
- **Infinitesimal canonical transformations** were explored and applied to time evolution.

Exercise 18.2 Considering exercise 16.6, show for the second-order Lagrangian $\mathscr{L}(q, \dot{q}, \ddot{q})$, how the Ostrogradsky momentum is modified under the addition of a total time derivative of a function $F(q, v, t)$, where v is a variable that, upon satisfaction of the equations of motion, becomes $v^i = \dot{q}^i$.

For this Lagrangian, the Ostrogradsky momentum is given by p_1:

$$p_1 = \frac{\partial \mathscr{L}}{\partial v} - \frac{d}{dt}\left(\frac{\partial \mathscr{L}}{\partial \dot{v}}\right) \tag{18.3.14}$$

Upon the addition of a total time derivative, the new Lagrangian is denoted by L:

$$L = \mathscr{L} + \frac{\partial F}{\partial q}v + \frac{\partial F}{\partial v}\dot{v} + \frac{\partial F}{\partial t} \tag{18.3.15}$$

The new momentum is denoted by P_1:

$$P_1 = p_1 + \frac{\partial F}{\partial q} + \frac{\partial^2 F}{\partial v \partial q}(v - \dot{q}) \tag{18.3.16}$$

Upon satisfaction of the equations of motion, we see that $v = \dot{q}$ and, hence, the above two equations are written

$$L = \mathscr{L} + \frac{dF}{dt} \quad \text{and} \quad P_1 = p_1 + \frac{\partial F}{\partial q} \tag{18.3.17}$$

Exercise 18.3 Consider our old friend, the one-dimensional harmonic oscillator. We know that there exists a canonical transformation to action-angle coordinates $(q, p) \to (Q, P)$ and we have investigated how we can reduce the complexity of the phase space description. In this example we will consider a variable transformation that is *not* canonical; however from any quantum mechanics class should remind us of the use of ladder operators in the quantum harmonic oscillator problem. For ease, we will write the Hamiltonian with $m = \omega = 1$ but factoring these in is straightforward:

$$H(q, p) = \frac{1}{2}(p^2 + q^2) \tag{18.3.18}$$

Let us define the new variables $(q, p) \to (a, a^*)$, called **holomorphic variables**[3]:

$$a = \frac{1}{\sqrt{2}}(p + iq) \quad \text{and} \quad a^* = \frac{1}{\sqrt{2}}(p - iq) \tag{18.3.19}$$

Our first step will be to compute the inverse transformation $a(q, p), a^*(q, p)$, which is done by some manipulation. First, rearrange both a and a^* for p, set them equal to each other, then re-arrange for q:

$$a\sqrt{2} - iq = a^*\sqrt{2} + iq$$
$$q = -\frac{i}{\sqrt{2}}(a - a^*) \tag{18.3.20}$$

Similarly, rearrange both expressions for q and set equal to each other, solving for p:

$$-p + a\sqrt{2} = p - a^*\sqrt{2}$$
$$p = \frac{1}{\sqrt{2}}(a^* + a) \tag{18.3.21}$$

Because of the non-canonical nature of the coordinates we cannot use Hamilton's equations, instead we must write the phase space Lagrangian for the system and take the functional derivative to first order of the action integral.

---

[3] For further reading, research holomorphic symplectic geometry. We could also swap the $q \to p$ in the definition.

$$S = \int (p\dot{q} - H)\,dt$$

$$= \int dt \left\{ \frac{1}{\sqrt{2}}(a + a^*) \cdot \frac{i}{\sqrt{2}}(\dot{a}^* - \dot{a}) - \frac{1}{2}\left[\left(\frac{1}{\sqrt{2}}(a + a^*) \right)^2 + \left(\frac{i}{\sqrt{2}}(a^* - a) \right)^2 \right] \right\}$$

$$= \int dt \left\{ \frac{1}{2}(a + a^*)[i\dot{a}^* - i\dot{a}] - aa^* \right\}$$

$$= \int dt \left\{ \frac{i}{2}(a + a^*)\dot{a}^* - \frac{i}{2}(a + a^*)\dot{a} - aa^* \right\} \qquad (18.3.22)$$

Because the first two terms are of a different sign, we cannot write this integral using the Poincaré-Cartan formula. We will see a little later (see chapter 23) that this is the defining property of canonical mechanics!

Exercise 18.4 An infinitesimal canonical transformation is generally expressed by equation 18.3.2. Prove that the transformations $\delta p_i = P_i - p_i$ and $\delta q^i = Q^i - q^i$ are canonical by considering the fundamental bracket $\{Q^i, P_j\} = \delta^i_j$.

We begin as follows:

$$\{Q^i, P_i\}_{p,q} = \{q^i, p_j\}_{p,q} + \epsilon \left(\left\{ \frac{\partial G}{\partial p_i}, p_j \right\} - \left\{ q^i, \frac{\partial G}{\partial q^j} \right\} \right) + \mathcal{O}(\epsilon^2)$$

$$= \delta^i_j + \epsilon \left[\sum_k \left(\frac{\partial^2 G}{\partial p_i \partial q^k} \delta_{jk} - 0 \right) - \sum_k \left(\delta_{ik} \frac{\partial^2 G}{\partial q^j \partial p_k} - 0 \right) \right] + \mathcal{O}(\epsilon^2)$$

$$= \delta^i_j + \mathcal{O}(\epsilon^2) \qquad (18.3.23)$$

We therefore recover the fundamental bracket, proving that the infinitesimal transformation is canonical.

19
Hamilton-Jacobi Theory

19.1 Hamilton-Jacobi Theory I

Hamilton-Jacobi theory is another formalism of mechanics; it builds on canonical transformations and is widely regarded as the most powerful tool in analytical mechanics - a so-called *apotheosis* of theoretical physics. We have already seen a brief glimpse of its utility in the chapter 11 and, in addition to the equivalent Lagrangian method (which you might like to reread before proceeding) there are two other approaches to understanding the theory: the canonical transform/generating function method (which builds on chapter 18) and a variational procedure (which builds on section 7.5 in chapter 7). The latter may not be as intuitive without an initial appreciation of the former; hence, we divide this chapter into two sections. The reason why Hamilton-Jacobi theory is so well regarded is because it unites elements from four seemingly different areas of the

Fig. 19.1: Carl Jacobi (1804–1851).

theory of differential equations: a second-order ODE (the Euler-Lagrange equation), a system of coupled first-order ODEs (Hamilton's equations), the calculus of variations and a first-order non-linear PDE (the Hamilton-Jacobi equation).

An overview of the first approach is as follows: express the canonical variables q, p in terms of their value at the initial time t_0, as $q = q(q_0, p_0, t)$ and $p = p(q_0, p_0, t)$ and then develop an equation that uses these initial coordinates and momenta, which are constant in time *by definition*, to give their value at time t. In a way, we have gone full circle to Newton's second law which is an equation that uses initial conditions to describe the motion at a later time. We will then generalise this to any arbitrary constants of the motion and study *integrability*.

The aim is to perform a canonical transformation such that the new set of Ps and Qs are these initial coordinates that are *constant in time* such that the new equations of motion are $\dot{P} = 0$ and $\dot{Q} = 0$. Now the mapping of states can be thought of as $Q, P \to q(t), p(t)$, where $Q = q(t_0)$ and $P = p(t_0)$, allowing us to write

Lagrangian & Hamiltonian Dynamics. Peter Mann, Oxford University Press (2018).
© Peter Mann. DOI: 10.1093/oso/9780198822370.001.0001

$$q^i = q^i(\boldsymbol{Q}, \boldsymbol{P}, t), \qquad \text{and} \qquad p_i = p_i(\boldsymbol{Q}, \boldsymbol{P}, t) \tag{19.1.1}$$

Or, if you prefer, $q^i = q^i(q(t_0), p(t_0), t)$ and $p_i = p_i(q(t_0), p(t_0), t)$, in which case the dynamics are completely solved because, by definition nothing moves! One way to enforce that the new coordinates are stationary is to let the new Hamiltonian K equal zero exactly. We now need to choose the correct generating function, which, for our study, will be an F_2-type function (although an F_1 works fine too). If we explicitly write out the problem, we obtain

$$K = H(\boldsymbol{q}, \boldsymbol{p}, t) + \frac{\partial F_2(\boldsymbol{q}, \boldsymbol{P}, t)}{\partial t} = 0 \tag{19.1.2}$$

We then use the F_2-associated canonical equations to express p in terms of F_2 by $p_i = \partial F_2/\partial q^i$:

$$H\left(\boldsymbol{q}, \frac{\partial F_2(\boldsymbol{q}, \boldsymbol{P}, t)}{\partial \boldsymbol{q}}, t\right) + \frac{\partial F_2(\boldsymbol{q}, \boldsymbol{P}, t)}{\partial t} = 0 \tag{19.1.3}$$

This is the **Hamilton-Jacobi equation of motion**. The Hamilton-Jacobi equation is a first-order PDE in $(n+1)$ variables corresponding to $q^1, q^2, \ldots, q^n, t$, where the $\boldsymbol{P}$s are the constants: $p_i(t_0)$. Thus, it is customary to relabel $F_2(\boldsymbol{q}, \boldsymbol{P}, t)$ as $\mathcal{S}(\boldsymbol{q}, \boldsymbol{P}, t) = \mathcal{S}(q^1, q^2, \ldots, q^n, P_1, P_2, \ldots, P_n, t)$, fixing the initial momenta such that they are just constant parameters of the system. Therefore, writing the Hamilton-Jacobi equation out specifically, we see that it is a PDE for $\mathcal{S}$:

$$H\left(q^1, \ldots, q^n, \frac{\partial \mathcal{S}}{\partial q^1}, \ldots, \frac{\partial \mathcal{S}}{\partial q^n}, t\right) + \frac{\partial \mathcal{S}}{\partial t} = 0$$

We call $\mathcal{S}$ **Hamilton's principal function**: it is the generating function of the new Hamiltonian, and it should not be confused with the action **functional**. The Hamilton-Jacobi equation is a remarkable equation, since it shows that the time dependance of the principal function is such that the new Hamiltonian is exactly zero! Knowing $\mathcal{S}$ for a system is equivalent to knowing the full dynamical details of the motion; the solution to the Hamilton-Jacobi equation encodes all possible dynamics. In other words, the F_2 generating function is the solution to the motion and will give us the time evolution of the coordinates and velocities - in fact, not just one solution, but all possible solutions since, we could Legendre transform between generators if we liked. Hopefully, this is intuitive, since a solution must in some way connect the initial canonical variables to those at some later time. Generating functions connect sets of variables, so, naturally, the solution of the Hamilton-Jacobi equation is the function that maps canonical variables; this is why Hamilton-Jacobi theory is an amazing piece of theory!

Since the Ps and Qs are constants, we usually refer to them as α and β, respectively to emphasise their constancy, meaning that Hamilton's principal function is now

$$S = S(q^1, q^2, \ldots, q^n, \alpha_1, \alpha_2, \ldots, \alpha_n, t) = S(\boldsymbol{q}, \boldsymbol{\alpha}, t)$$

The canonical transformation equations now take the form

$$p_i = \frac{\partial S(\boldsymbol{q}, \boldsymbol{\alpha}, t)}{\partial q^i} \quad \text{and} \quad \beta^i = \frac{\partial S(\boldsymbol{q}, \boldsymbol{\alpha}, t)}{\partial \alpha_i} \tag{19.1.4}$$

We can invert the second of equations 19.1.4 to solve for the dynamics of the system in time in terms of the initial condition such that $p_i = p_i(\boldsymbol{\alpha}, \boldsymbol{\beta}, t)$ and $q_i = q_i(\boldsymbol{\alpha}, \boldsymbol{\beta}, t)$, which represents a full solution to both of Hamilton's equations. The exact process of inversion follows a simple recipe and involves evaluating equation 19.1.4 at t_0 (i.e. finding the initial conditions):

$$p_i(t_0) = \frac{\partial S(\boldsymbol{q}, \boldsymbol{\alpha}, t)}{\partial q^i}\bigg|_{t_0, \boldsymbol{q}(t_0), \boldsymbol{\alpha}} \quad \text{and} \quad \beta^i = \frac{\partial S(\boldsymbol{q}, \boldsymbol{\alpha}, t)}{\partial \alpha_i}\bigg|_{t_0, \boldsymbol{q}(t_0), \boldsymbol{\alpha}} \tag{19.1.5}$$

This allows us to solve for $\alpha_i = p_i(t_0)$ and $\beta^i = q^i(t_0)$. We now use these to solve for $q^i(t)$ and $q_i(t)$ at later times, as indicated above. The right-hand side of equation 19.1.5 is quite a remarkable statement: if we forget for a moment that we are assuming that α and β correspond to the initial conditions, we are saying that the change of Hamilton's principal function when we change some parameter α is constant. Is there some proof that this holds without referring to α and β as the initial conditions? Yes, and, as we do for every conserved quantity, we start by taking the time derivative of the quantity we wish to prove is constant. Recalling that $S = S(\boldsymbol{q}, \boldsymbol{\alpha}, t)$ we obtain

$$\begin{aligned}
\frac{d}{dt}\left(\frac{\partial S}{\partial \alpha_i}\right) &= \frac{\partial}{\partial t}\left(\frac{\partial S}{\partial q^j}\frac{\partial q^j}{\partial \alpha_i}\right) + \frac{\partial}{\partial t}\left(\frac{\partial S}{\partial \alpha_i}\right) + \frac{\partial t}{\partial t} \\
&= \frac{\partial^2 S}{\partial q^j \partial \alpha_i}\left(\frac{\partial q^j}{\partial t}\right) + \frac{\partial}{\partial \alpha_i}\left(\frac{\partial S}{\partial t}\right) \\
&= \frac{\partial^2 S}{\partial q^j \partial \alpha_i}\left(\frac{\partial q^j}{\partial t}\right) - \frac{\partial}{\partial \alpha_i}\left(H(q, \frac{\partial S}{\partial q})\right) \\
&= \frac{\partial^2 S}{\partial q^j \partial \alpha_i}\left(\frac{\partial q^j}{\partial t}\right) - \frac{\partial H}{\partial q^j}\frac{\partial q^j}{\partial \alpha_i} - \frac{\partial H}{\partial(\partial_q S)}\frac{\partial(\partial_q S)}{\partial \alpha_i}
\end{aligned}$$

At this point, we note that q^j is independent of α_i, since they correspond to different coordinates, meaning that $\partial q^j/\partial \alpha_i = 0$. We must also note that $\partial S/\partial q^j = p_j$, which we substitute into the first part only of the last term above:

$$\begin{aligned}
\frac{d}{dt}\left(\frac{\partial S}{\partial \alpha_i}\right) &= \frac{\partial^2 S}{\partial q^j \partial \alpha_i}\left(\frac{\partial q^j}{\partial t}\right) - \frac{\partial H}{\partial p_j}\frac{\partial^2 S}{\partial q^j \partial \alpha_i} \\
&= \frac{\partial^2 S}{\partial q^j \partial \alpha_i}\left[\dot{q}^j - \frac{\partial H}{\partial p_j}\right] \\
&= 0
\end{aligned}$$

In the last step, we see that the term in the bracket is one of Hamilton's equations, which we know equals zero on-shell. This proves, therefore, that the quantity $\partial S/\partial \alpha_i$ is constant in time as long as q^j obeys Hamilton's equation. The term $\partial^2 S/\partial q^j \partial \alpha_i$ defines a **Hessian matrix** $\boldsymbol{S}$ and, to be meaningful, must have a non-zero determinant[1], which, if true, means we can solve for the $\boldsymbol{\alpha}$s in terms of $\partial S/\partial q$; this gives the first of Hamilton's equations as its only solution, $\dot{q}^j = \partial H/\partial p_j$. This means we can now solve each q^j in terms of the initial conditions at time t_0:

$$\boldsymbol{S}\left(\dot{\boldsymbol{q}} - \frac{\partial H}{\partial \boldsymbol{p}}\right) = 0 \tag{19.1.6}$$

So, in this way, Hamilton-Jacobi theory leads to the first of Hamilton's equations of motion, but what about the other? To start off, we recall that $p_j = \partial S/\partial q^j$.

$$\dot{p}_j = \frac{d}{dt}\left(\frac{\partial S}{\partial q^j}\right) = \frac{\partial^2 S}{\partial q^j \partial q^i}\frac{\partial q^i}{\partial t} + \frac{\partial^2 S}{\partial q^j \partial t} = \frac{\partial}{\partial q^j}\left(\frac{\partial S}{\partial q^i}\dot{q}^i + \frac{\partial S}{\partial t}\right) = \frac{\partial}{\partial q^j}(p_i \dot{q}^i - H)$$

which is the phase space Lagrangian. In this way, Hamilton-Jacobi theory leads to the Lagrange equation of motion:

$$\dot{p}_j = \frac{\partial \mathscr{L}'}{\partial q^j} = -\frac{\partial H}{\partial q^j} \tag{19.1.7}$$

Note that $\mathscr{L}'(\boldsymbol{q}, \boldsymbol{p}, t) = \mathscr{L}(\boldsymbol{q}, \dot{\boldsymbol{q}}, t)$ only for physical extremal paths that the system follows and we cannot use this derivation for a non-physical path; this will make a lot more sense from the Jacobi variational principle aspect in section 19.2. We are beginning to see that there is a connection between the Hamilton-Jacobi PDE and Hamilton's ODEs on-shell; this is a duality which we explore in a moment.

We can gain physical intuition as to how we can think of the principal function $S(\boldsymbol{q}, \boldsymbol{\alpha}, t)$ if we do a little computation; starting with S, we take its total time derivative:

$$\frac{dS}{dt} = \sum_{j=1}^{n} \frac{\partial S}{\partial q^j}\dot{q}^j + \frac{\partial S}{\partial t}$$

Recalling our definitions in equations 19.1.3 and 19.1.4, we find that we can express this as

$$\frac{dS}{dt} = \sum_j p_j \dot{q}^j - H = \mathscr{L} \tag{19.1.8}$$

So, integrating this, we can see that S differs from the classical Lagrangian action by only a constant, which, of course, disappears when taking variations δS. Feel free to

---

[1] A Hessian with a determinant of zero would mean we have a singular Lagrangian, which could imply there is some form of constraint on the system and the variables are no longer independent of each other. The Hessian provides a *criterion* for invertibility and it must be satisfied in order for S to be a useful solution to the Hamilton-Jacobi equation. This factor is emphasised in our discussion of constrained Hamiltonian dynamics in chapter 21.

return to the discussion of canonical transformations and look at the relationship between the phase space Lagrangians of different coordinates and the generating functions $\mathscr{L}'(q, \dot{q}, p, \dot{p}, t) = \mathscr{L}'(Q, \dot{Q}, P, \dot{P}, t) + \dot{F}$ or the Hilbert integral invariant of the action principle. We then find that:

$$\int_{t_1}^{t_2} \mathscr{L} dt = \Delta \mathcal{S}(\boldsymbol{\alpha}, t)$$

There are various quantities we call action in mechanics (see chapter 7 section 7.4) and the most important are as follows. The Lagrangian and phase space actions $A[q]$ and $S[q]$ are **functionals** of the path taken by the system in their various spaces. Extrema to these are paths between two fixed boundary points and define the physical path a system takes, which we refer to as the **on-shell action**. Hamilton's principal **function** is a function of the final position, integration constants and time. It is the value of the action on the extremal path from an initial position q_0 at time t_0 to a final coordinate q_1 at time t_1, which are a set of Dirichlet boundary conditions. For extremal paths, the two actions $\mathcal{S}[q]_{cl}$ and $\mathcal{S}(q, \alpha, t)$ are equivalent, since they are two mathematically distinct objects that describe the same physical quantity, namely the on-shell trajectory. However, the principal function doesn't exist for variational trajectories; it is only defined for the on-shell path by construction. We will look at this in a little more detail as an exercise at the end of this chapter.

As a quick summary, we can perform a canonical transformation on the old Hamiltonian to construct a new Hamiltonian that is constant (most easily achieved by setting equal to zero), giving us the Hamilton-Jacobi equation of motion. We then find Hamilton's principal function and compute $\partial \mathcal{S}(q, \alpha, t)/\partial \alpha_i = \beta^i$ and then invert it for $q = q(\alpha, \beta, t)$. We then compute $p = \partial \mathcal{S}(q, \alpha, t)/\partial q$, giving us both q and p as functions of time and hence the problem is solved completely: knowing $\mathcal{S}$ gives us complete knowledge of the dynamics of the system. Hamilton's principal function is the generating function that transforms the state from an initial point in phase space at time t_0 to some other point at a later time t.

A very interesting case is where the Hamiltonian is not an explicit function of time, then, time is a **separable variable** such that the Hamilton-Jacobi equation becomes a separable differential equation:

$$H\left(q, \frac{\partial \mathcal{S}}{\partial q}\right) + \frac{\partial \mathcal{S}}{\partial t} = 0 \tag{19.1.9}$$

The equation is composed of two parts: the left is independent of time and only focuses on how $\mathcal{S}$ varies with the qs, and the right is focused on the time dependency. This suggests that $\mathcal{S}$ can be split up into time-dependant and time-independent parts; indeed, separability is a major theme across physics especially in quantum mechanics. If we rearrange the above we obtain

$$H\left(q, \frac{\partial \mathcal{S}}{\partial q}\right) = -\frac{\partial \mathcal{S}}{\partial t} \tag{19.1.10}$$

Now, its clear that the left side doesn't depend on the time while the right side does; therefore let's call $\mathcal{S}$ something else, say, W (after the German *wirkung*). The right side's separation constant (see chapter 5) we will denote as the product of a time-independent quantity and time, ht:

$$H\left(q, \frac{\partial W}{\partial q}\right) = -\frac{\partial(-ht)}{\partial t} \tag{19.1.11}$$

Therefore, if we rearrange this back to equation 19.1.9, we can see that $\mathcal{S}(q, \alpha, t) = W(q, \alpha) - h(\alpha)t$, where h is the time-independent value of H; it is a function of α and, in the most elementary case, can be chosen to be α itself. In other words, W is the Legendre transform of the principle function with respect to an integration constant and time (we show investigate this later on in this section). This separation is only possible when the Hamiltonian is independent of time, and we can clearly see that $\partial \mathcal{S}/\partial q = \partial W/\partial q$ and $\partial \mathcal{S}/\partial t = \partial ht/\partial t$. Therefore, we can define the **time-independent Hamilton-Jacobi equation** as

$$H\left(q, \frac{\partial W(q, \alpha)}{\partial q}\right) - h = 0 \tag{19.1.12}$$

The function $W(q, \alpha)$ has a special name: **Hamilton's characteristic function**. We know that the Hamiltonian is often the total energy of the system and, since it is independent of time, we know it is a constant of the motion, whatever it represents physically. We can rewrite equation 19.1.12 by remembering that $\alpha - \partial \mathcal{S}/\partial q$, which now turns into $\alpha = \partial W/\partial q$, since their Hessians are equal, so that equation 19.1.12 is now

$$H(q, \alpha) = h(\alpha)$$

The new variables β^i are cyclic coordinates, meaning α_i are all integrals of the motion. We can integrate Hamilton's ODEs for β^i directly to obtain linear functions in time:

$$\beta^j(t) = \frac{\partial h}{\partial \alpha_j}t + \beta^j(t_0) \tag{19.1.13}$$

Using the inverse function theorem, we can invert these for q^j and solve for p_j in terms of α_j and β^j:

$$p_j(t) = p_j(\alpha, \frac{\partial h}{\partial \alpha}t + \beta(t_0)), \quad \text{and} \quad q^j(t) = q^j(\alpha, \frac{\partial h}{\partial \alpha}t + \beta(t_0)) \tag{19.1.14}$$

In small neighbourhoods around a well-behaved point in phase space, we can always find a complete solution to the Hamilton-Jacobi problem. That is to say, we can always find a canonical transformation whose generating function's Hessian determinant is non-zero. This idea, although not a global property of phase space, is called **rectification theorem** and locally trivialises the **flow** of phase space. We will only really understand this properly when we discuss phase flow in terms of differential geometry, but, for now, you could think of flow as a sort of flux through phase fluid that a system follows between regions and which is determined by the Hamiltonian. This is subject to *nice* areas of phase space; we discuss these in terms of KAM theory later (see chapter 24 section 24.2), in terms of which we say that rectification theorem can only apply to non-resonant tori or those which obey degeneracy conditions or regions of non-singular Lagrangians. In these so-termed bad regions, the Hamilton-Jacobi equation can not be solved analytically by simple integration.

We have already seen that S, W and ht are related, but we should be aware that this is actually due to a Legendre transformation between W and S, meaning that one action is the Legendre transform of the other, or, in more familiar terms one generating function is the Legendre transform of the other:

$$W(\boldsymbol{q}, \boldsymbol{\alpha}) = S(\boldsymbol{q}, \boldsymbol{\alpha}, t) + h(\boldsymbol{\alpha})t = S(\boldsymbol{q}, \boldsymbol{\alpha}, t) - \frac{\partial S(\boldsymbol{q}, \boldsymbol{\alpha}, t)}{\partial t}t \qquad (19.1.15)$$

As such, we expect W to be associated with an action, too; taking its time derivative, realising that $\partial W / \partial q^j = p_j$ and integrating with respect to time, we define the **abbreviated action** as:

$$W = \int_1^2 p_j \dot{q}^j \, dt = \int_1^2 p_j \, dq^j$$

where contracted indices are summed over. This is the **Maupertius action principle**, a primitive relative of that of Euler and Lagrange. Similarly, we can define an off-shell functional and an on-shell function. The Maupertius action principle states that W is stationary on the classical path between two fixed end points and a fixed time average of the Hamiltonian over the time interval between the points. In the abbreviated action principle, we we fix our energy surface so that dynamics is confined to that surface. We can determine the classical path, provided we know the configuration boundary points and the energy. In other words, the Maupertuis' action principle is a variational problem that fixes the total energy of all paths; it also fixes the initial and final positions. Maupertuis variations are composed of vertical and horizontal variations of the form $\delta_0 q^i = \delta q^i + \dot{q}^i \delta t$. Note that this can be related to $\mathscr{L}$ and H by using the Legendre transform $p\dot{q} = \mathscr{L} + H$.

19.2 Hamilton-Jacobi Theory II

We now follow a variational approach to Hamilton-Jacobi theory from a more general standing where the concept will hopefully be more intuitive, having previously discussed the generating function approach. This section complements section 7.5 in chapter 7, which discusses boundary conditions in variational principles. We start by constructing a function object $S(q_2, t_1)$ which is a function of the final position and time for a fixed initial condition for one degree of freedom.

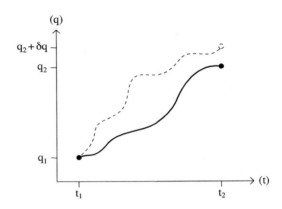

Fig. 19.2: The variational problem associated with the Hamilton-Jacobi equation.

We then ask ourselves, how do the extrema of S change as a function of the final conditions q_2^i, t_2 (for a fixed starting point)? We first investigate its dependance of q_2^i by evaluating the variation in the principal function, keeping t_2 fixed; this is the **Hamilton-Jacobi variational problem** (see figure 19.2). Changing the final boundary point from q^i, t to $q^i + \delta q^i, t$ will change the entire trajectory and so we evaluate the **variation in the action**. Proceeding as usual at first we write

$$\delta S = \int_{t_1}^{t_2} \left[\frac{\partial \mathscr{L}}{\partial q^i} - \frac{d}{dt}\left(\frac{\partial \mathscr{L}}{\partial \dot{q}^i} \right) \right] \delta q^i \, dt + \left[\frac{\partial \mathscr{L}}{\partial \dot{q}^i} \delta q^i \right]_{t_1}^{t_2} \tag{19.2.1}$$

Upon satisfaction of the equations of motion, the Lagrange equation vanishes as usual. However, since we are varying the upper limit, we cannot just set the second term to zero[2], so we write:

$$\delta S = \frac{\partial \mathscr{L}}{\partial \dot{q}^i} \bigg|_{t_2} \delta q_2^i := p_2^i \delta q_2^i \tag{19.2.2}$$

That is, p_2^i is the momentum at the upper limit for the ith coordinate. Therefore, we have found that the difference in S computed along two paths with the same initial conditions is given by the momentum at the end point! Computing the variation with respect to t_2 and keeping q_2^i fixed, we find the following, where we have used $\partial_{\dot{q}} S = \partial_q S$, as per normal:

---

[2] The lower limit is fixed and set to zero, $\delta q_1^i(t_1) = 0$.

$$\frac{dS}{dt_2} = \frac{\partial S}{\partial t_2}\bigg|_{q_2^i} + \frac{\partial S}{\partial q_2^i}\bigg|_{t_2} \dot{q}_2^i = \frac{\partial S}{\partial t_2}\bigg|_{q_2^i} + p_2\dot{q}_2^i = \mathscr{L} \tag{19.2.3}$$

Using equations 19.2.2 and 19.1.8, we can therefore write

$$\frac{\partial S}{\partial t_2}\bigg|_{q_2^i} = \mathscr{L}(\boldsymbol{q}_2, \dot{\boldsymbol{q}}_2, t_2) - p_{2i}\dot{q}_2^i = -H(\boldsymbol{q}_2, \boldsymbol{p}_2, t_2) \tag{19.2.4}$$

We can then write,

$$\boxed{dS = p_i dq^i - H dt \qquad \textbf{Poincaré-Cartan invariant}} \tag{19.2.5}$$

This result is a very important equation whose significance we draw out later (see chapter 23); it is called the **Poincaré-Cartan 1-form** $dS = \theta_H$. For now, we will put it aside and continue the discussion of Hamilton-Jacobi theory. We now have:

$$\boxed{H\left(\boldsymbol{q}_2, \frac{\partial S}{\partial \boldsymbol{q}_2}, t_2\right) + \frac{\partial S}{\partial t_2} = 0} \tag{19.2.6}$$

This is the Hamilton-Jacobi PDE for S as a function of time and the final coordinate. It takes the form of the equations of the generating function of a **canonical transformation**! For multiple degrees of freedom, we can obtain our familiar expression, dropping the '2' notation for brevity:

$$H\left(q^1, \ldots, q^n, \frac{\partial S}{\partial q^1}, \ldots, \frac{\partial S}{\partial q^n}, t\right) + \frac{\partial S}{\partial t} = 0$$

A **complete integral**[3] to this is a solution with $(n + 1)$ arbitrary constants - as many constants as there are **independent** variables - and, since the principal function appears only as a derivative, it provides a complete integral up to an additive constant. These constants are globally defined over the phase space and are Poisson-commuting; in this case, the Hamiltonian depends only on these constants!

Solving this for n generalised coordinates and time introduces $(n + 1)$ integration constants that encode the initial conditions given by α_j in $S(\boldsymbol{q}, \boldsymbol{\alpha}, t)$. In this way, each independent constant of the motion reduces the dimensionality of the phase space by 1; we call these **hypersurfaces** (surfaces of one dimension less than the global space). The point where the hypersurfaces intersect is the point where all the $\boldsymbol{\alpha}$s are constant. We can build up a surface called an **integral surface**, which is the solution to the Hamilton-Jacobi problem. This surface has etched into it curves called its **characteristics** (compare this with a "tube of characteristics" later in chapter 23). When

---

[3]Not all Hamilton-Jacobi equations that we write admit a complete integral.

one talks about choosing the initial conditions, what we really mean is choosing what particular surface we start of from. After chapter 22, we will understand that these surfaces **foliate** the phase space in a tight onion-skin like structure in which each **leaf** is an integral surface. So, how does an integral surface of the Hamilton-Jacobi equation provide solutions to Hamilton's equations?

We start by realising that the complete integral $\mathcal{S}(\boldsymbol{q}, \boldsymbol{\alpha}, t)$ is the generating function from $(\boldsymbol{q}, \boldsymbol{p})$ to $(\boldsymbol{\alpha}, \boldsymbol{\beta})$; with this in place, the new Hamiltonian vanishes everywhere, since the new canonical equations are $\dot{\alpha}_i = \dot{\beta}^i = 0$, where β^i is the position coordinate. Provided that the following Hessian condition holds:

$$\det\left(\frac{\partial^2 \mathcal{S}}{\partial q^i \partial \alpha_j}\right) \neq 0 \qquad (19.2.7)$$

we can find n relations of the form:

$$\frac{\partial \mathcal{S}}{\partial \alpha_j} = \beta^j, \qquad \text{and} \qquad p_j = \frac{\partial \mathcal{S}}{\partial q^j} \qquad (19.2.8)$$

which can be inverted (since the determinant is non-zero and, hence, we can use the implicit function theorem, chapter 32) to find $q^j = q^j(\boldsymbol{\alpha}, \boldsymbol{\beta}, t)$ and $p_i = p_i(\boldsymbol{\alpha}, \boldsymbol{\beta}, t)$. This is known as **Jacobi's theorem**. In this way, the characteristic curves in the integral surface are the solution trajectories to Hamilton's equations of motion. Therefore, the Hamilton-Jacobi equation's solution comprises a family of surfaces that are orthogonal to the flow of an associated flow known as the characteristic flow, and these are precisely the dynamical solutions to Hamilton's equations of motion. We have seen already seen in this chapter that the principal function is closely tied to both of Hamilton's equations of motion. This subject is a little difficult to appreciate just now but after chapter 22 we will return to the Hamilton-Jacobi equation and we see it is constructed in an intimate manner from the Poincaré-Cartan 1-form in chapter 23.

For now we can take away that *a homogeneous non-linear PDE admits a complete integral, which geometrically is an integral surface etched with curves that are the flow curves of a system of coupled first-order ODEs.*

Returning to the notion of separable variables, we say that a particular coordinate q^i is **separable** if we can split $\mathcal{S}$ into two parts, one which is only a function of that variable and a second part that is independent from it:

$$\begin{aligned} \mathcal{S}(q^1, q^2, \ldots, q^i, \ldots, q^n, \boldsymbol{P}, t) &= \mathcal{S}^{(i)}(q^i, \boldsymbol{P}, t) \\ &\quad + \mathcal{S}'(q^1, q^2, \ldots, q^{i-1}, q^{i+1}, \ldots, q^n, \boldsymbol{P}, t) \end{aligned} \qquad (19.2.9)$$

where $\mathcal{S}'$ is not a function of the i-th general coordinate. There are now two Hamilton-Jacobi equations: one for $\mathcal{S}^{(i)}$ and one for the remaining variables $\mathcal{S}'$. If the other $(n-1)$ coordinates are separable in this sense, then the Hamilton-Jacobi equation is **completely separable**. If the principal function is completely separable then the solution is of the form:

$$S = \sum_{j=1}^{n} S^j(q^j, \boldsymbol{P}, t) \tag{19.2.10}$$

The n Hamilton-Jacobi equations are given by

$$H^j\left(q^j, \frac{dS^j}{dq^j}, \boldsymbol{P}, t\right) + \frac{\partial S^j}{\partial t} = 0 \tag{19.2.11}$$

These form a set of first-order ODEs, that are functions of one variable and its derivative. If this is the case, there will exist a set of **separation functions** $S^1, \ldots, S^n$: $\mathbb{R}^{2n} \to \mathbb{R}$ on the region M such that we can write:

$$S^1 = S^1(q^1, p_1), \qquad S^2 = S^2(q^2, p_2, S^1), \qquad S^n = S^n(q^n, p_n, S^1, \ldots, S^{n-1}) \tag{19.2.12}$$

and hence, the Hamiltonian $H(S)$ is only dependent on $S = S^1, \ldots, S^n$. The separation functions are Poisson-commuting constants of the motion; and if they are additionally functionally independent, the system is integrable in the sense we discussed earlier. In that discussion we could identify the separating functions with the constants of the motion. We remark that if the system has a set of action-angle variables then it is both separable and integrable in this sense.

The first studies on separability in classical mechanics were carried out by Liouville and are closely tied to the notion of *integrability*. A differential equation is said to be integrable if it can be solved analytically, even if only in principle. We have discussed integrability in the context of action-angle variables previously in chapter 18, section 18.2 as well as in chapter 16, section 16.2. Integrability implies that the system is separable, but the converse is not necessarily true, as the first integrals need not be functionally independent of one another.

The study of separability and the Hamilton-Jacobi equation was first carried out by T. Levi-Civita, who derived a criterion to be met if a coordinate system is to be separable. The criterion is laid out below and can be thought of as a *test* for the property of separability:

$$\frac{\partial H}{\partial q^i}\frac{\partial H}{\partial q^j}\frac{\partial^2 H}{\partial p_i \partial p_j} + \frac{\partial H}{\partial p_i}\frac{\partial H}{\partial p_j}\frac{\partial^2 H}{\partial q^i \partial q^j} - \frac{\partial H}{\partial q^i}\frac{\partial H}{\partial p_j}\frac{\partial^2 H}{\partial p_i \partial q^j} - \frac{\partial H}{\partial p_i}\frac{\partial H}{\partial q^j}\frac{\partial^2 H}{\partial q^i \partial p_j} = 0 \tag{19.2.13}$$

with $1 \le i < j \le n$, $i \ne j$. Normally, we need to know more than just n first integrals F to solve a system of $2n$ ODEs (counting both $\boldsymbol{p}$ and $\boldsymbol{q}$ to n degrees of freedom); in fact, we generally require $2n$ first integrals. However, if the ODEs are **canonical equations**, then each of the first integrals are in involution (vanishing Poisson bracket $\{F_i, F_j\} = 0$) and each reduces the dimension by two, rather than just one! When this is true, there exists locally a function S of only n first integrals that is a complete integral of the corresponding Hamilton-Jacobi equation. This is the **Arnold-Liouville integrability theorem** and it implies that a set of action-angle variables exist for the system. We will investigate this in detail in chapter 23 where we will conclude that if the surfaces or 'level sets' are *compact*[4] in the phase space, then the phase space consists of (or is *foliated by*) n-tori, and the system has the action-angle property. Such systems are rare.

It is important to realise that *any* complete integral of the Hamilton-Jacobi problem will be sufficient to solve the dynamics. There may be more than one complete integral corresponding to different functional dependence of S but all will correspond to equivalent dynamics; therefore the principal function is not unique. This is a residue of the fact that only the derivative of S appears in the Hamilton-Jacobi equation of motion. Therefore, any S whose gradients coincide are fit to describe the same dynamics.

Some Hamilton-Jacobi equations also admit **general integrals**, which are solutions that aren't dependent on a particular set of integration constants but rather are found from a particular starting integral surface. General integrals are therefore solutions to the **Cauchy problem** associated with the Hamilton-Jacobi equation. General integrals can be constructed from a complete integral by choice of initial starting surface. A corollary of this is that any practical application to systems of particles will require us to implement a statistical element where the different initial conditions are sampled to build up a collection of classical trajectories.

---

[4]A compact manifold is a manifold without a boundary. As an example, consider a circle, a sphere or a torus.

Chapter summary

- The **Hamilton-Jacobi equation of motion** was introduced first, via a type 2 generating function, then from a variational problem. It is a first-order, non-linear PDE:

$$H\left(q^1, \ldots, q^n, \frac{\partial S}{\partial q^1}, \ldots, \frac{\partial S}{\partial q^n}, t\right) + \frac{\partial S}{\partial t} = 0$$

Hamilton's principal function S was introduced as a solution to the Hamilton-Jacobi problem.

- The notion of **separability** was developed for time-independent Hamiltonians, and a **complete integral** solution was discussed as well as the **Levi-Citiva** criterion of separability.

- The Hamilton-Jacobi equation was derived using a **variational problem**, therefore proving the on-shell equivalence between a first-order PDE, a system of first-order coupled ODEs, the calculus of variations and a second-order ODE.

- Solutions to the Hamilton-Jacobi equation of motion provide us with an **integral surface**, upon which curves are generated that are the solutions curves to Hamilton's equations in phase space. In this way, obtaining the integral surface of the Hamilton-Jacobi problem also solves Hamilton's ODEs. The opposite is also true; we can build an integral surface from solution curves to Hamilton's coupled ODEs.

Exercise 19.1 We can consider the Hamilton-Jacobi equation for a particle of mass m under the influence of a conservative potential $U(x)$ with a Hamiltonian given by $H = p^2/2m + U(x)$, and, by using $p = \partial S/\partial q$, we can write the following Hamilton-Jacobi equation:

$$\frac{1}{2m}(\nabla_x S)^2 + U(x) + \frac{\partial S}{\partial t} = 0 \qquad (19.2.14)$$

The above system can be solved by using the characteristic function giving h in terms of x and $\partial_x W$:

$$h = \frac{1}{2m}\left(\frac{\partial W}{\partial x}\right)^2 + U(x)$$

where $W = \int p\, dx$ so

$$W(x) = W(x_0) + \int_{x_0}^{x_1} \sqrt{2m(h - U(x'))}\, dx'$$

Now using the above relations,

$$p = \frac{\partial W}{\partial x} = \sqrt{2m(h - U(x))}, \qquad \text{and} \qquad \beta = \frac{\partial W}{\partial h} = -t + \int_{x_0}^{x_1} \frac{1}{2}\sqrt{\frac{2m}{h - U(x')}}\, dx'$$

Inverting the second equation would give us $x(t)$, which we could substitute into the first one to get $p(t)$. Therefore, we have the dynamics of the system by using W to get the two integration constants h and β. You must note that, since we are dealing with a time-independent system, we can choose one of the n integration constants to be the energy of the system, since it is a constant of the motion; hence, h is an integration constant.

Exercise 19.2 In this example, we show that the classical limit of the Schrödinger equation is the classical Hamilton-Jacobi equation for a particle under a conservative potential.

Taking a **Wentzel-Kramers-Brillouin** wavefunction to be a function of position and time, split it into amplitude A and phase S components $\psi(x,t) = A(x,t)e^{iS(x,t)/\hbar}$ and substitute it into the **time-dependant Schrödinger equation**, we obtain

$$i\hbar\frac{\partial\psi}{\partial t} = -\frac{\hbar^2}{2m}\vec{\nabla}^2\psi + U\psi \tag{19.2.15}$$

$$i\hbar\frac{\partial\psi}{\partial t} = e^{iS/\hbar}\left(i\hbar\frac{\partial A}{\partial t} - A\frac{\partial S}{\partial t}\right)$$

We now turn to the right-hand side of equation 19.2.15 to calculate the second derivative of the wavefunction with respect to x; to do this, we use the product rule twice:

$$\frac{\partial\psi}{\partial x} = e^{\frac{iS}{\hbar}}\left(\frac{\partial A}{\partial x} + \frac{i}{\hbar}A\frac{\partial S}{\partial x}\right)$$

$$\frac{\partial^2\psi}{\partial x^2} = \left(\frac{\partial A}{\partial x} + \frac{i}{\hbar}A\frac{\partial S}{\partial x}\right)\frac{\partial}{\partial x}e^{\frac{iS}{\hbar}} + e^{\frac{iS}{\hbar}}\frac{\partial}{\partial x}\left(\frac{\partial A}{\partial x} + \frac{i}{\hbar}A\frac{\partial S}{\partial x}\right)$$

$$= \left(\frac{\partial A}{\partial x} + \frac{i}{\hbar}A\frac{\partial S}{\partial x}\right)\frac{i}{\hbar}\frac{\partial S}{\partial x}e^{\frac{iS}{\hbar}} + e^{\frac{iS}{\hbar}}\left(\frac{\partial^2 A}{\partial x^2} + \frac{i}{\hbar}A\frac{\partial^2 S}{\partial x^2} + \frac{i}{\hbar}\frac{\partial S}{\partial x}\frac{\partial A}{\partial x}\right)$$

$$= e^{\frac{iS}{\hbar}}\left[\vec{\nabla}^2 A - \frac{1}{\hbar^2}A(\vec{\nabla}S)^2 + \frac{i}{\hbar}(2\vec{\nabla}S\vec{\nabla}A + A\vec{\nabla}^2 S)\right]$$

We can now substitute these results into equation 19.2.15, collecting imaginary terms and real terms, cancelling the exponential and multiplying the entire imaginary portion by $2A$:

$$2\frac{\partial A}{\partial t}A = -\frac{1}{m}(2A\vec{\nabla}A\vec{\nabla}S + AA\vec{\nabla}^2 S)$$

This gives us equation 19.2.16 for the imaginary component, and 19.2.17 for the real part:

$$\text{Im}\left\{\frac{\partial A^2}{\partial t} = -\vec{\nabla}\cdot\left[A^2\frac{\vec{\nabla}S}{m}\right]\right\} \tag{19.2.16}$$

$$\text{Re}\left\{-A\frac{\partial S}{\partial t} = -\frac{\hbar^2}{2m}\left[\vec{\nabla}^2 A - \frac{A}{\hbar^2}(\vec{\nabla}S)^2\right] + VA\right\} \tag{19.2.17}$$

Combining these results, we are left with equation 19.2.18, which, when taking the **classical limit** $\hbar \to 0$, gives us the Hamilton-Jacobi equation for a free particle, as in equation 19.2.15:

$$\frac{1}{2m}(\vec{\nabla}\mathcal{S})^2 + U + \frac{\partial \mathcal{S}}{\partial t} = \frac{\hbar^2}{2m}\frac{\vec{\nabla}^2 A}{A} \tag{19.2.18}$$

This result indicates that the phase of the wavefunction can be identified as the real Hamilton's principal function, which we know we can associate with the action of the physical path. The function $(A(x,t))^2 = \rho = \psi^*\psi$ is associated with a complex quantum amplitude and, when squared gives a probability density, according to a **quantum equilibrium hypothesis**. The significance of the right-hand side of the above equation is that it depends on the probability distribution ρ. This means that the quantum "trajectory" dictated by $\mathcal{S}$ is undergoing a non-local interaction depending on the distribution. This non-local information exchange has a major impact on time-dependent quantum dynamics.

This is the starting point for **Bohm's model** of quantum theory, where the links to classical mechanics are self-evident through the Hamilton-Jacobi formalism. It is of fundamental importance to note how the phase is inherently coupled to the amplitude in quantum mechanics. Later, when we discuss the classical *Koopman von-Neumann theory* in chapter 20, we will investigate this relationship and its importance to the differences between classical and quantum theory; we will also formulate a field theoretic model much much later when we discuss elementary field theory in part IV. Note, the phase-amplitude interaction is responsible for the inherently quantum effects seen in, for example, the double slit experiment. The imaginary equation is the continuity equation for ρ with velocity field $v = (\partial \mathcal{S}/\partial q)/m$ and hence $p = \partial \mathcal{S}/\partial q$:

$$\frac{\partial \rho(x,t)}{\partial t} + \vec{\nabla} \cdot (\rho v) = 0 \tag{19.2.19}$$

Therefore, the probability density satisfies a continuity equation; a probability density times a velocity is called a **current**.

Exercise 19.3 Consider a particle that is free to move in one dimension and whose Hamiltonian is given by $H = p^2/2m$. Using $p = \partial \mathcal{S}/\partial q$, we construct the Hamilton-Jacobi problem as follows:

$$\frac{\partial \mathcal{S}}{\partial t} + \frac{1}{2m}\left(\frac{\partial \mathcal{S}}{\partial x}\right)^2 = 0 \tag{19.2.20}$$

We assume that the solution is separable and takes the form of $\mathcal{S} = Q(q) + T(t)$ such that the Hamilton-Jacobi equation becomes,

$$\frac{\partial T}{\partial t} + \frac{1}{2m}\left(\frac{\partial Q}{\partial q}\right)^2 = 0 \tag{19.2.21}$$

As is standard in the separation of variables procedure we notice that both terms are dependent on one of the two variables and are therefore equal to a separation constant; in addition, that constant happens to equal the total energy, as can be seen from the relation on the right:

$$\frac{\partial T}{\partial t} = -h, \qquad \frac{1}{2m}\left(\frac{\partial Q}{\partial q}\right)^2 = h$$

Solving these for $Q(q)$ and $T(t)$ we obtain

$$Q(q) = \sqrt{2mhq} + b, \qquad T(t) = -ht + c \tag{19.2.22}$$

where b and c are integration constants. From this, we construct the solution to the principal function:

$$S = \sqrt{2mhq} - ht + (b + c) \tag{19.2.23}$$

Then,

$$\beta = \frac{\partial S}{\partial h} = \sqrt{\frac{m}{2h}}q - t \tag{19.2.24}$$

Inverting this expression and setting $p = \partial S/\partial q$, we obtain

$$q(t) = \sqrt{\frac{2h}{m}}(\beta + t), \qquad p(t) = \sqrt{2mh} \tag{19.2.25}$$

Setting $t = 0$, we see that the initial position is given by $q(0) = \dot{q}\beta$.

Exercise 19.4 Consider the harmonic oscillator using the Hamilton-Jacobi method:

$$H = \frac{1}{2m}(p^2 + m^2\omega^2 q^2) \tag{19.2.26}$$

The Hamilton-Jacobi equation is constructed using $p_i = \partial S/\partial q^i$:

$$\frac{1}{2m}\left[\left(\frac{\partial S}{\partial q}\right)^2 + m^2\omega^2 q^2\right] + \frac{\partial S}{\partial t} = 0 \tag{19.2.27}$$

This equation is separable and we consider the time-independent portion setting $h(\boldsymbol{\alpha}) = \alpha$. With $\omega^2 = k/m$, this becomes

$$\frac{1}{2m}\left[\left(\frac{\partial W}{\partial q^i}\right)^2 + m^2\omega^2 q^2\right] = \alpha \tag{19.2.28}$$

This equation can be rearranged for $W(\boldsymbol{q}, \boldsymbol{\alpha})$, noticing $\int(\partial W/\partial q)dq = W$:

$$W(\boldsymbol{q}, \boldsymbol{\alpha}) = \int dq\sqrt{2m\alpha - kq^2} + C_0 \tag{19.2.29}$$

We are free to set the integration constant to be zero: $C_0 = 0$. Then, by the chain rule, so that we let $u = 2m\alpha - kq^2$ and cancel $(\sqrt{m}\sqrt{m})/\sqrt{m} = \sqrt{m}$, we obtain

$$\beta = \frac{\partial S}{\partial \alpha} = \frac{\partial W}{\partial \alpha} - t = \int\sqrt{\frac{m}{2\alpha - kq^2}}dq - t = \text{constant} \tag{19.2.30}$$

$$\beta = \sqrt{\frac{m}{k}}\int\frac{1}{\sqrt{(2\alpha/k) - q^2}}dq - t \tag{19.2.31}$$

Using the identity $\int\frac{dx}{\sqrt{b^2 - x^2}} = \sin^{-1}\frac{x}{b}$,

$$\beta = \sqrt{\frac{m}{k}} \sin^{-1}\left(q\sqrt{\frac{k}{2\alpha}}\right) - t \qquad (19.2.32)$$

By some simple rearranging, we invert this expression for the generalised coordinate as a function of time:

$$q(t) = \sqrt{\frac{2\alpha}{m\omega^2}} \sin(\omega t + \beta\omega) \qquad (19.2.33)$$

which is the expected expression for the generalised coordinate. The momentum is more simple to obtain:

$$p = \frac{\partial W}{\partial q} = \sqrt{2m\alpha - m^2\omega^2 q^2} = \sqrt{2m\alpha} \cos(\omega t + \omega\beta) \qquad (19.2.34)$$

where we have substituted for q and let $\cos^2\theta = 1 - \sin^2\theta$.

Exercise 19.5 By this point, there may be some confusion over the different quantities that we are calling the action of the system. We remark in chapter 7 section 7.4, that there is an off-shell action functional $S[q]$ and an on-shell action function $S(q_i, q_f, t_i, t_f)$; now, we have another on-shell function $S(q, \alpha, t)$. So, what's going on here?

The action functional is used in the variation problem and includes virtual variational paths in addition to the extremal path. The unique extremal/classical path is only dependent on its Dirichlet boundary conditions of initial and final time and position q_i, q_f, t_i, t_f. This makes sense because when we change the boundary conditions (i.e. starting somewhere/sometime else), we expect the trajectory to change. We will now show that, when evaluated for the same classical path, with the same Dirichlet boundary conditions, the on-shell action function is equal to the difference in Hamilton's principal function evaluated on the boundary points:

$$S(q_i, q_f, t_i, t_f) = \Delta S(q, \alpha, t)\big|_{q_i, t_i}^{q_f, t_f} \qquad (19.2.35)$$

Consider a free particle with a kinetic energy Hamiltonian $H = p^2/2m$. The Hamilton Jacobi equation is

$$\frac{\partial S}{\partial t} + \frac{1}{2m}\left(\frac{\partial S}{\partial x}\right)^2 = 0 \qquad (19.2.36)$$

which we have solved before in exercise 19.3, the principal function being $S(q, h, t) = \sqrt{2mh}q - ht$ with $\alpha = h$ and $q_f \geq q_i$. The on-shell action $S(q_i, q_f, t_i, t_f)$ is given by the total energy h:

$$S(q_i, q_f, t_i, t_f) = \int_{t_i}^{t_f} \frac{1}{2}m\left(\frac{q_f - q_i}{t_f - t_i}\right)^2 dt = \frac{1}{2}m\frac{(q_f - q_i)^2}{t_f - t_i} = h \qquad (19.2.37)$$

The difference in Hamilton's principal function $S(q, h, t)$ evaluated between the two boundary points is then computed using this value of integration constant and found to be the on-shell action function:

$$\Delta S(q, h, t) = S(q_f, h, t_f) - S(q_i, h, t_i) = \frac{1}{2}m\frac{(q_f - q_i)^2}{t_f - t_i} = S(q_i, q_f, t_i, t_f) \qquad (19.2.38)$$

The only reason we particularly mention the on-shell action here is because of its use in path integrals and propagators in semiclassical mechanics. For an analogy to this system, consider a rotating rigid body $\mathscr{L} = \frac{1}{2}I\dot{\theta}$. Show that the principal function $S(\theta, L, t)$ is given by

$$\mathcal{S}(\theta, L, t) = L\theta - \frac{L^2}{2I}t \qquad (19.2.39)$$

Further, show that the on-shell action is given by the difference in the principal function:

$$\mathcal{S}(\theta_i, \theta_f, t_i, t_f) = \mathcal{S}(\theta_f, L, t_f) - \mathcal{S}(\theta_i, L, t_i) \qquad (19.2.40)$$

and that the angular momentum L plays a similar role to h above.

Exercise 19.6 Show that the Hamilton-Jacobi theory for the second-order Lagrangian $\mathcal{L}(q, \dot{q}, \ddot{q}, t)$ is given by the following conditions, where we have used the Ostrogradsky definition of the momentum (see exercise 16.6).

$$\frac{\partial S}{\partial t} + H\left(t, q, \frac{\partial S}{\partial q}\right) = 0, \qquad H = p_1 \dot{q} + p_2 \ddot{q} - \mathcal{L} \qquad (19.2.41)$$

with,

$$p_1 = \frac{\partial \mathcal{L}}{\partial \dot{q}^i} - \frac{d}{dt}\left(\frac{\partial \mathcal{L}}{\partial \ddot{q}^i}\right), \qquad p_2 = \frac{\partial \mathcal{L}}{\partial \ddot{q}^i}, \qquad S = S(q, \dot{q}, t) \qquad (19.2.42)$$

Exercise 19.7 Here we will consider a practical use of the Hamilton-Jacobi equation in computer visualisation and image processing. The **eikonal equation** was first introduced by Hamilton to describe wave propagation as a kinematic approximation to the scalar wave equation:

$$\left(\frac{\partial T}{\partial x}\right)^2 + \left(\frac{\partial T}{\partial y}\right)^2 + \left(\frac{\partial T}{\partial z}\right)^2 = n^2 \qquad (19.2.43)$$

It is the usual tool in computer visualisation, imaging and shading techniques and is derived by inserting a plane wave $\psi(x, y, z)$ in the wave equation and cancelling imaginary terms:

$$\psi(x, y, z) = \psi_0 e^{-i\omega(t-T)} \qquad (19.2.44)$$

The solution to the eikonal equation represents wavefronts of constant phase. The wavefronts are propagated in time, normal to their surfaces. These surfaces can be used to construct images in real time and track movements in three dimensions. Often in this field, the motion will develop singularities as solutions propagate to branching points; consequently, this a nontrivial problem to solve.

Through Fermat's principle of geometric optics, it is well known that light travelling between two points minimises the total travel time elapsed, or, τ; and it is through this that we can connect the eikonal equation to a mechanical picture:

$$\tau = \int_1^2 dt = \int_1^2 \frac{n}{c} dl \qquad (19.2.45)$$

Where $n(x, y, z)$ is the refractive index of the medium, c is the speed of light and dl is the displacement element along the path the ray travels. If we let $\mathcal{S} = cT$ and $dl = (\dot{x}^2 + \dot{y}^2 + \dot{z}^2)^{1/2} dt$, the line integral becomes a time integral:

$$S = \int_1^2 n\sqrt{\dot{x}^2 + \dot{y}^2 + \dot{z}^2}\, dt \qquad (19.2.46)$$

From this, it is clear that we can create a Lagrangian $\mathscr{L} = n\sqrt{\dot{x}^2 + \dot{y}^2 + \dot{z}^2}$. Using the momentum for the system, $p_i = \partial \mathscr{L}/\partial q^i$ show that the following relation holds:

$$p_x^2 + p_y^2 + p_z^2 = n^2 \qquad (19.2.47)$$

Further, using Hamilton-Jacobi theory, we know that $p_i = \partial S/\partial q^i$ and, hence, it is clear that the eikonal equation is written

$$\left(\frac{\partial S}{\partial x}\right)^2 + \left(\frac{\partial S}{\partial y}\right)^2 + \left(\frac{\partial S}{\partial z}\right)^2 = n(x, y, z)^2 \qquad (19.2.48)$$

This is nothing other than the time-independent Hamilton-Jacobi equation:

$$\frac{1}{2m}|\vec{\nabla}S|^2 + V - E = 0 \implies |\vec{\nabla}S| = p \qquad (19.2.49)$$

Hence, we can think of S as describing the phase of a field at a point and therefore contour lines of constant S or phase as wavefronts. The dynamics is then perpendicular to these fronts, since the gradient (and hence momentum) is perpendicular to the surfaces.

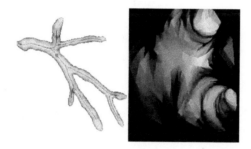

Fig. 19.3: A virtual recreation of the inner lumen of an artery kindly provided by K. Siddiqi.

The reader is referred to the work of K. Siddiqi *et al.*, (2002) for their work on Hamiltonian skeletons and medical imaging. There they created sharp biomedical images by converting the (difficult-to-solve) eikonal equation into (easier-to-solve) Hamilton's equations of motion, using the correspondence between the former and the Hamilton-Jacobi equation. Pictured are two virtual endoscopic images created using this method, one modelling the intestinal tract (see figure 19.4) and the other the interior of arteries (see figure 19.3)!

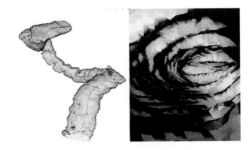

Fig. 19.4: A virtual recreation of the inner lumen of a colon section, kindly provided by K. Siddiqi.

Exercise 19.8 Consider a two-dimensional isotropic harmonic oscillator with the following Hamiltonian:

$$H = \frac{1}{2}(p_x^2 + p_y^2 + x^2 + y^2) \tag{19.2.50}$$

Given that the system belongs to the $SU(2)$ symmetry group, show that the Hamiltonian can be written as the following matrix:

$$H = \begin{pmatrix} 0 & 0 & 1 & 0 \\ 0 & 0 & 0 & 1 \\ -1 & 0 & 0 & 0 \\ 0 & -1 & 0 & 0 \end{pmatrix} \tag{19.2.51}$$

Show that the eigenvalues of this Hamiltonian are $\lambda \pm i$ and that the normalised eigenvectors are:

$$a^+ = \frac{1}{\sqrt{2}}(p_x - ix), \qquad b^+ = \frac{1}{\sqrt{2}}(p_y - iy), \qquad a^- = \frac{1}{\sqrt{2}}(p_x + ix), \qquad b^- = \frac{1}{\sqrt{2}}(p_y + iy)$$

Is the system integrable? If yes, can you find a set of first integrals for this system?

20

Liouville's Theorem & Classical Statistical Mechanics

20.1 Liouville's Theorem & the Classical Propagator

Given an initial configuration at time t_0, we know that our system will occupy a given point of phase space and we can write $\boldsymbol{\xi} = (q^1, \ldots, q^n, p_1, \ldots, p_n)$ as the **phase space vector**. As time evolves, this then traces out a trajectory as a function of time $\boldsymbol{\xi}(t)$. We know that Hamilton's equations describe the evolution of this curve $\dot{\boldsymbol{\xi}} = (\dot{q}^1, \ldots, \dot{q}^n, \dot{p}_1, \ldots, \dot{p}_n)$. If we consider all the points around this system, which only differ slightly from the original ones, we then have a volume of phase space evolving according to Hamilton's equations. As time evolves, the shape of this volume can change; however, the volume it occupies must stay the same. This is **Liouville's theorem**: *the time evolution of a Hamiltonian system conserves the phase space volume occupied by the system*. Liouville's theorem is a defining feature of canonical mechanics and leads to many of the properties of phase space we will come to explore in the next few section. (For a picture of Joseph Liouville, see figure 20.1.)

Fig. 20.1: Joseph Liouville (1809–1882).

It is important to note that, if canonical transformations can portray temporal evolution, then they too must preserve the volume a system occupies; hence, the Jacobian is exactly 1. For a single degree of freedom, we would talk of area preservation; this would be as follows:

$$\det \begin{pmatrix} \frac{\partial Q}{\partial q} & \frac{\partial Q}{\partial p} \\ \frac{\partial P}{\partial q} & \frac{\partial P}{\partial p} \end{pmatrix} = \frac{\partial(Q, P)}{\partial(q, p)} = \frac{\partial Q}{\partial q}\frac{\partial P}{\partial p} - \frac{\partial Q}{\partial p}\frac{\partial P}{\partial q} = 1 \qquad (20.1.1)$$

This is because, for any general change of coordinates, we can write the new volume element in terms of the old volume element multiplied by the Jacobian of the transformation between two regions $\mathcal{R} \rightarrow \mathcal{R}'$ (see Mathematical Background 6):

Lagrangian & Hamiltonian Dynamics. Peter Mann, Oxford University Press (2018).
© Peter Mann. DOI: 10.1093/oso/9780198822370.001.0001

$$\iint_{\mathcal{R}} dQ \, dP = \iint_{\mathcal{R}'} \frac{\partial(Q,P)}{\partial(q,p)} \, dp \, dq \tag{20.1.2}$$

The only way that we can preserve volume is by setting the Jacobian determinant to unity.

We can develop the Poisson bracket into a differential $\hat{L}$ operator called the **Liouville operator** using $i = \sqrt{1}$ as the imaginary number as follows:

$$\hat{L} := i\{H, \ \} = i \sum_j \left(\frac{\partial H}{\partial q^j} \frac{\partial}{\partial p_j} - \frac{\partial H}{\partial p_j} \frac{\partial}{\partial q^j} \right) \tag{20.1.3}$$

such that if $\hat{L}$, acts on a function A, we write $\hat{L}A = i\{H, A\} = -i\{A, H\}$; the i just makes $\hat{L}$ Hermitian $\hat{L}^\dagger = \hat{L}$ and doesn't matter in classical mechanics, since we know that we only want real eigenvalues. In fact, apart from an i, this is the same as the *divergence* of the phase space velocity vector $\dot{\boldsymbol{\xi}} = (\dot{q}^1, \dot{q}^2, \ldots, \dot{q}^n, \dot{p}_1, \dot{p}_2, \ldots, \dot{p}_n)$:

$$\vec{\nabla} \cdot \dot{\boldsymbol{\xi}} = \left(\frac{\partial}{\partial q^1}, \frac{\partial}{\partial q^2}, \ldots, \frac{\partial}{\partial q^n}, \frac{\partial}{\partial p_1}, \frac{\partial}{\partial p_2}, \ldots, \frac{\partial}{\partial p_n} \right) \cdot (\dot{q}^1, \dot{q}^2, \ldots, \dot{q}^n, \dot{p}_1, \dot{p}_2, \ldots, \dot{p}_n)$$

$$= \frac{\partial \dot{q}^1}{\partial q^1} + \cdots + \frac{\partial \dot{q}^n}{\partial q^n} + \frac{\partial \dot{p}_1}{\partial p_1} + \cdots + \frac{\partial \dot{p}_n}{\partial p_n}$$

$$= \sum_{j=1}^{n} \left(\frac{\partial^2 H}{\partial q^j \partial p_j} - \frac{\partial^2 H}{\partial q^j \partial p_j} \right)$$

$$= 0 \quad \forall \, i \tag{20.1.4}$$

where we have collected terms with the same indices and used Hamilton's equations; note that, in order to prove this, you have to assume Hamilton's equations are valid for the system (on-shell). We can now use the phase space vector to rewrite Hamilton's equations into one neat package, namely $\dot{\boldsymbol{\xi}} = \hat{L}\boldsymbol{\xi}$. If we rearrange this, and integrate between two times (initial and final), we obtain the following:

$$\int_{t_0}^{t_1} \frac{1}{\boldsymbol{\xi}} \frac{d\boldsymbol{\xi}}{dt} \, dt = \int_{t_0}^{t_1} \hat{L} \, dt$$

$$\int_{\boldsymbol{\xi}(t_0)}^{\boldsymbol{\xi}(t_1)} \frac{1}{\boldsymbol{\xi}} d\boldsymbol{\xi} = \int_{t_0}^{t_1} \hat{L} \, dt \tag{20.1.5}$$

$$\boldsymbol{\xi}(t_1) = e^{\hat{L}t} \boldsymbol{\xi}(t_0)$$

Those familiar with quantum mechanics will recognise this as the **classical propagator**, evolving a state from t_0 to t_1 which we denote by $\hat{U}(t) = e^{\hat{L}t}$ (see section

17.2). Since $\hat{L}$ is Hermitian, the propagator is a **unitary operator**, meaning that $\hat{U}^{\dagger}(t)\hat{U}(t) = \mathbb{I}$, where $\mathbb{I}$ is the identity and $\hat{U}^{\dagger}(t) = e^{-\hat{L}^{\dagger}t}$ is the conjugate.

Moving on, we can define a **phase space probability density function**[1] $\rho = \rho(q^1, \ldots, q^n, p_1, \ldots, p_n, t)$ such that $\rho(q^1, \ldots, q^n, p_1, \ldots, p_n, t)dq^1 \ldots dq^n dp_1 \ldots dp_n$ defines a volume of phase space that corresponds to finding the particle in a state between $(q^1, \ldots, q^n, p_1, \ldots, p_n)$ and $(q^1 + dq^1, \ldots, q^n + dq^n, p_1 + dp_1, \ldots, p_n + dp_n)$ for a small time interval $t \to t + dt$. Therefore, the probability density is the description of a **mixed state**. By Liouville's theorem, we know that, if the divergence of this probability density fluid is zero, then the probability density must be conserved. Therefore, if we take the time derivative of ρ, it should be zero, as for any conservation law. We know that, to take the time derivative of a quantity, we take the Poisson bracket of that quantity with the Hamiltonian and the partial derivative with respect to time:

$$\frac{d}{dt}\rho = \{\rho, H\} + \frac{\partial \rho}{\partial t} \quad \longrightarrow \quad \frac{\partial \rho}{\partial t} = \{H, \rho\} \quad (20.1.6)$$

where we have used the antisymmetric rule (see chapter 17). We may also consider the left-hand side to generate a continuity equation for the probability density. The velocity field when associated with the Hamilton-Jacobi problem is $v = (\partial \mathcal{S}(\xi)/\partial q)/m$:

$$\frac{d\rho}{dt} = \frac{\partial \rho}{\partial t} + \vec{\nabla} \cdot (\rho v) = 0 \quad (20.1.7)$$

We can now express this in terms of our Liouville operator $\hat{L}$ to give the **Liouville equation**[2]:

$$i\frac{\partial}{\partial t}\rho(\boldsymbol{q}, \boldsymbol{p}, t) = \hat{L}\rho(\boldsymbol{q}, \boldsymbol{p}, t) \qquad \textbf{Liouville equation} \quad (20.1.8)$$

This equation forms the basis of classical statistical mechanics and will be the focus of much of this chapter.

The Liouville probability density is an N-particle function on the phase space of a system of identical particles in a box. Sometimes, we are interested in a subset of n

---

[1] Some authors define a configuration space distribution function and introduce momentum integrals later to form the Liouville equation:

$$f(\xi, t) = \rho(\boldsymbol{q}, t)\delta[\boldsymbol{p} - \partial_q \mathcal{S}(\boldsymbol{q}, t)]$$

For further reading, please see P. Holland's (1993) *The Quantum Theory of Motion.*

[2] The quantum analogue is the von Neumann equation. This result is of the utmost importance to both classical molecular dynamics and quantum density operators. A fascinating and useful resource for this is D. Tanner, (2007) *Introduction to Quantum Mechanics A Time-Dependent Perspective.* In addition, some authors define the probability density to include the division by the partition function already, and some include h^n and $N!$ factors, too, so there is a lot of terminology to learn. We will introduce this complexity later however; whenever we compute any particular quantity, all components come together in the end.

particles, where $n < N$. This may be because we only require a certain level of detail for our problem or we just want to make the Liouville equation a little easier to handle. In this case, we define a reduced phase space probability distribution ρ_n by integrating out the remaining $n + 1 \rightarrow N$ redundant degrees of freedom $(6(N - n))$:

$$\rho_n(q^1, \ldots, q^n, p_1, \ldots, p_n, t) = \frac{N!}{(N-n)!} \int \rho_N(\boldsymbol{q}, \boldsymbol{p}, t) \prod_{i=1}^{N-n} dq^i dp_i \qquad (20.1.9)$$

Like its full-dimensioned counterpart, the reduced density gives the probability of finding n particles in a reduced region of the phase space. There are $N!/(N-n)!$ ways to choose the n particles out of the N total, assuming they are identical; in fact, this is a standard result known as the **permutation formula**. In order to use such a result, we have to integrate the Liouville equation and we start by using equations 20.4.2 and 20.1.8 (see section 20.4). We decompose the total rate of change of momentum into external forces $\vec{F}_i^{(e)}$ and a pairwise interaction between particles in the system $\vec{F}_{ij}$:

$$\frac{\partial \rho_N}{\partial t} + \sum_{i=1}^{N} \frac{p_i}{m} \frac{\partial \rho_N}{\partial q^i} + \sum_{i=1}^{N} \vec{F}_i^{(e)} \frac{\partial \rho_N}{\partial p_i} + \sum_{i=1}^{N} \sum_{j=1}^{N} \vec{F}_{ij} \frac{\partial \rho_N}{\partial p_i} = 0 \qquad (20.1.10)$$

Integrating the unwanted particles $n + 1 \rightarrow N$ by writing $\int \prod_{i=n}^{N-n} dq^i dp_i$ rather than $\int \prod_{i=1}^{N} dq^i dp_i$ as normal.

$$\int \frac{\partial \rho_N}{\partial t} + \sum_{i=1}^{N} \frac{p_i}{m} \frac{\partial \rho_N}{\partial q^i} + \sum_{i=1}^{N} \vec{F}_i^{(e)} \frac{\partial \rho_N}{\partial p_i} \prod_{i=n}^{N-n} dq^i dp_i$$

$$+ \int \sum_{i=1}^{N} \sum_{j=1}^{N} \vec{F}_{ij} \frac{\partial \rho_N}{\partial p_i} \prod_{i=n}^{N-n} dq^i dp_i = 0 \qquad (20.1.11)$$

We have split the integral up for ease and we now take the second portion to the other side of the equality:

$$\int \frac{\partial \rho_N}{\partial t} + \sum_{i=1}^{N} \frac{p_i}{m} \frac{\partial \rho_N}{\partial q^i} + \sum_{i=n}^{N} \vec{F}_i^{(e)} \frac{\partial \rho_N}{\partial p_i} \prod_{i=n}^{N-n} dq^i dp_i$$

$$= - \int \sum_{i=1}^{N} \sum_{j=1}^{N} \vec{F}_{ij} \frac{\partial \rho_N}{\partial p_i} \prod_{i=n}^{N-n} dq^i dp_i \qquad (20.1.12)$$

Inside this integral, every term where $i > n$ will have a probability density of zero (observe that we are integrating over differentiated objects). To see this, consider the following component part of the above integral and use the fundamental theorem of calculus for integrating differentiated objects:

$$\int \frac{\partial \rho_N}{\partial p_i} dp^{N-n} = \rho_N \big|_{-\infty}^{\infty} = 0 \tag{20.1.13}$$

There is zero probability density outside of the boundary of the box the particles are contained in, so this term is exactly zero. Similarly for the $\partial \rho_N / \partial q^i$ term, every degree of freedom we integrate over vanishes. This is all we need to do for the left-hand side of equation 20.1.12, since the only index is i. However, the right-hand side runs over j, too, since it is the inter-particle term and, therefore, the sum over j still runs over the redundant particles:

$$\left(\frac{\partial}{\partial t} + \sum_{i=1}^{n} \frac{p_i}{m} \frac{\partial}{\partial q^i} + \sum_{i=1}^{n} \vec{F}_i^{(e)} \frac{\partial}{\partial p_i} \right) \rho_n = -\sum_{i=1}^{n} \sum_{j=1}^{N} \int \vec{F}_{ij} \frac{\partial \rho_N}{\partial p_i} \prod_{i=n}^{N-n} dq^i dp_i \tag{20.1.14}$$

To make progress with this, we have to split the sum over j into two parts:

$$\left(\frac{\partial}{\partial t} + \sum_{i=1}^{n} \frac{p_i}{m} \frac{\partial}{\partial q^i} + \sum_{i=1}^{n} \vec{F}_i^{(e)} \frac{\partial}{\partial p_i} \right) \rho_n = -\left[\sum_{i=1}^{n} \sum_{j=1}^{n} \int \vec{F}_{ij} \frac{\partial \rho_N}{\partial p_i} \prod_{i=n}^{N-n} dq^i dp_i \right.$$

$$\left. + \sum_{i=1}^{n} \sum_{j=n+1}^{N} \int \vec{F}_{ij} \frac{\partial \rho_N}{\partial p_i} \prod_{i=n}^{N-n} dq^i dp_i \right] \tag{20.1.15}$$

The first sum over j can be pulled outside of the integral, as they have nothing to do with the integration. For convenience, we ship them off to the right-hand side of the equation replacing $\rho_N \to \rho_n$ for the above vanishing reasons:

$$\left(\frac{\partial}{\partial t} + \sum_{i=1}^{n} \frac{p_i}{m} \frac{\partial}{\partial q^i} + \sum_{i=1}^{n} \left(\vec{F}_i^{(e)} + \sum_{j=1}^{n} \vec{F}_{ij} \right) \frac{\partial}{\partial p_i} \right) \rho_n$$

$$= -\sum_{i=1}^{n} \sum_{j=n+1}^{N} \int \vec{F}_{ij} \frac{\partial \rho_N}{\partial p_i} \prod_{i=n}^{N-n} dq^i dp_i \tag{20.1.16}$$

The sum over j on the right-hand side is actually just $(N-n)$ of the same term, since we assign j arbitrarily to any of the particles, swapping these particles (permuting them) makes no difference. Hence, we replace the sum over j with $(N-n)$ for the number of remaining particles:

$$\left(\frac{\partial}{\partial t} + \sum_{i=1}^{n} \frac{p_i}{m} \frac{\partial}{\partial q^i} + \sum_{i=1}^{n} \left(\vec{F}_i^{(e)} + \sum_{j=1}^{n} \vec{F}_{ij} \right) \frac{\partial}{\partial p_i} \right) \rho_n$$

$$= -(N-n) \sum_{i=1}^{n} \int \vec{F}_{i,n+1} \frac{\partial \rho_{n+1}}{\partial p_i} \prod_{i=n}^{N-n} dq^i dp_i \tag{20.1.17}$$

To finalise our expression, we can substitute the force for the negative of a potential gradient, $\vec{F}_i^{(e)} = -\partial U_i^{(e)}/\partial q^i$ and the pair potential $\vec{F}_{ij} = -\partial U_{ij}/\partial q^i$:

$$\left(\frac{\partial}{\partial t} + \sum_{i=1}^{n} \frac{p_i}{m} \frac{\partial}{\partial q^i} - \sum_{i=1}^{n} \left(\frac{\partial U_i^{(e)}}{\partial q^i} + \sum_{j=1}^{n} \frac{\partial U_{ij}}{\partial q^i} \right) \frac{\partial}{\partial p_i} \right) \rho_n$$

$$= (N - n) \sum_{i=1}^{n} \int \frac{\partial U_{i,n+1}}{\partial q^i} \frac{\partial \rho_{n+1}}{\partial p_i} \prod_{i=n}^{N-n} dq^i dp_i \qquad (20.1.18)$$

This result is known as the **BBGKY hierarchy**, after Bogoliubov, Born, Green, Kirkwood and Yuan. It is an expression independent of any index greater than n and we have a relation between a reduced phase space density n and an $(n+1)$ density. We have shown that the partial integration of the Liouville equation yields an expression for an n-body distribution function on a reduced phase space. However, the utility of this function is limited, all we have really done is swap a single equation in $6N$ variables into a family (or *hierarchy*) of coupled PDEs in two unknowns ρ_n and ρ_{n+1}. This means that there is an implicit dependence on the redundant particles because ρ_{n+1} will depend on ρ_{n+2} and so on. For this reason, we say that the BBGKY hierarchy is not **closed**, and, there is a great deal of focus on finding a closure relation for the system at hand so that ρ_n can be expressed only in terms of functions within the confined patch of phase space. The above derivation is an adaptation from J. P. Hansen and I. R. McDonald's (2013) *Theory of Simple Liquids* and an excellent text on reduced phase space density functions is D. A. McQuarrie's (2000) *Statistical Mechanics*. For further reading, research *mean field closure relations*.

Using the above concepts, we define a one-particle density ρ_1 for a canonical ensemble by integrating over $(N-1)$ generalised coordinates *and* N conjugate momenta, leaving ρ_1 as only a function of the position of the particular particle $\rho_1(q)$. The integration over all momenta is a way of removing excess kinetic information from the configurational integral. In this way, we now have a measure defined on the configurations of the particles such that $\rho_1(q)dq$ is a probability (probability = probability density $\times$ volume element) of finding the particular particle in volume dq in configuration space. In choosing the particular particle, we have exactly N choices because the particles are indistinguishable, or we could use $N = N!/(N-1)!$:

$$\left\langle \sum_{i=1}^{N} \delta(q^1 - q^i) \right\rangle = N \langle \delta(q^1 - q^i) \rangle = N \cdot \frac{1}{N! h^{3N} Z} \int \delta(q^1 - q^i) e^{-\beta H} \prod_{i=1}^{N} dq^i dp_i \quad (20.1.19)$$

We exclude q^1 from the integral and, hence, we have a one-particle density function ρ_1:

$$N \cdot \frac{1}{N! h^{3N} Z} \int \delta(q^1 - q^i) e^{-\beta H} \prod_{i=1}^{N} dq^i dp_i$$

$$= N \cdot \frac{1}{N! h^{3N} Z} \int e^{-\beta H} dq^2 \dots dq^N d\boldsymbol{p} \qquad (20.1.20)$$

One may also work backwards from ρ_1 to show that it is equivalent to the ensemble average of the delta. Notice that integrating ρ_1 over q^1 will normalise to 1; it is clear, therefore, that $N \frac{1}{\mathcal{V}} \int \rho_1 dq^1 = N/\mathcal{V}$ is a number density. In addition, we are wrapping the partition function up into the definition of the probability density for normalisation purposes, just for our ease. We can generalise this concept into higher-body density functions by just adding more Dirac delta functions; for example, a two body density function is,

$$\rho_2 = \left\langle \sum_{i=1}^{N} \sum_{j \neq i}^{N} \delta(q^1 - q^i) \delta(q^2 - q^j) \right\rangle = N(N-1) \langle \delta(q^1 - q^i) \delta(q^2 - q^i) \rangle \qquad (20.1.21)$$

Alternatively, an n-body interaction density over non-repeating indices is

$$\rho_n = \left\langle \sum_{i,j,\dots,n}^{N} \prod_{i=1}^{n} \delta(q' - q^i) \right\rangle \qquad \text{normalised to} \qquad \int \rho_n dq^1 \dots dq^n = \left\langle \frac{N!}{(N-n)!} \right\rangle$$

Of all the possible densities we could consider, it is arguably the choice of $n = 2$ that is the most useful. We have already seen how density functions can be used to compute meaningful properties of the system, and this *pairwise* density function is of fundamental importance to **radial distribution functions** g_N^2, with the understanding that *distribution* is distinct from *density*. The two-body radial distribution or *pair correlation function* is defined by

$$g_N^2 = \frac{N!}{(N-2)!} \frac{\rho_2}{N^n / \mathcal{V}^n} \qquad (20.1.22)$$

while the n-body correlation function is

$$g_N^n = \frac{N!}{(N-n)!} \frac{\rho_n}{\rho^n} = \frac{N!}{(N-n)! \cdot \rho^n} \left\langle \sum_{i,j,\dots,n}^{N} \prod_{i=1}^{n} \delta(q' - q^i) \right\rangle \qquad (20.1.23)$$

A distribution/correlation function sets a particular molecule as the coordinate origin and gives the average number of particles found radially outwards from that origin as a function of distance *relative* to the average number obtained if we were to ignore interactions. Therefore, for an ideal gas with no interactions, we expect no deviation;

for a liquid, we expect a periodic decaying graph peaking as we move through solvent layers. If the system is *homogeneous*, it means that there is no privileged particle to choose as the origin. In this case, the two body distribution function is a function of the inter-particle separation $g_N^2 = g_N^2(|q^1 - q^2|)$ and, hence, $\rho_2 = \rho_2(|q^1 - q^2|)$. The difficulty arises when the system is not homogeneous (maybe an amorphous solid or colloid, etc.) and long-range order breaks down; in this case, writing such a distribution function can be very tricky. Although we do not have the space to develop such concepts here, the obvious extension of this example is to discuss scattering experiments and various closure relations appropriate for X-ray or neutron diffraction techniques. We also mention that the correlation function is directly related via Fourier transform to the structure factor of a crystal.

20.2   Koopman-von Neumann Theory

Fig. 20.2: John von-Neumann (1903–1957).

The Liouville equation of motion is a first-order PDE in the probability density. It does not imply that the the probability density is uniform throughout phase space or that the probability density is the same over a given trajectory, but rather that, over a given trajectory, it remains constant and doesn't change from what it was/is, so it is stable, $\rho(t_0) = \rho(t_1)$. This means that the explicit time dependance of the probability density is exactly compensated by the Poisson bracket. The solution takes the form $\rho(t) = e^{\hat{L}t}\rho(t_0)$, as usual. In quantum mechanics, we are familiar with the Born interpretation, where the probability density is given by the square modulus of a wavefunction.

It is the **Koopman von-Neumann theory**, formulated in 1931 that posits that the Liouville probability density is also generated from an underlying wavefunction, the **Koopman von-Neumann wavefunction**, $\psi(\boldsymbol{\xi})$ such that $\rho(\boldsymbol{\xi}) = |\psi(\boldsymbol{\xi})|^2$ is the probability density of a particle occupying a point in phase space. (For a picture of John von Neumann, see figure 20.2.) Just as in quantum mechanics, these KvN classical wavefunctions are complex, square integrable functions of the canonical variables and reside in a complex **Hilbert space**, a vector in which represents the state of a system and, like any Hilbert space, there has to be an inner product $\langle \psi_i | \psi_j \rangle$ conserved under evolution and whose form integrated over the whole of phase space is as follows (the asterix denotes the complex conjugate):

$$\langle \psi_i | \psi_j \rangle := \int \psi_i^*(\boldsymbol{\xi}) \psi_j(\boldsymbol{\xi}) d\boldsymbol{q} d\boldsymbol{p}$$

We can also assume that the wavefunctions are bounded, meaning that $\psi_i(\boldsymbol{\xi}) \mapsto 0$ as $(\boldsymbol{\xi}) \mapsto \pm\infty$. We can introduce bra-ket notation from quantum mechanics to make this all look a bit more familiar: $\psi^*(\boldsymbol{\xi}) = \langle \psi | \boldsymbol{\xi} \rangle$ and $\psi(\boldsymbol{\xi}) = \langle \boldsymbol{\xi} | \psi \rangle$, with a complete orthonormal basis set $|\boldsymbol{\xi}\rangle = |\boldsymbol{q}, \boldsymbol{p}\rangle$, satisfying

$$\int |\boldsymbol{\xi}\rangle\langle\boldsymbol{\xi}| d\boldsymbol{q} d\boldsymbol{p} = 1, \quad \text{and} \quad \langle\boldsymbol{\xi}|\boldsymbol{\xi}'\rangle = \delta(q - q')\delta(p - p')$$

Position and momentum observables are now represented by commuting Hermitian operators such that $[\hat{q}, \hat{p}] = 0$, where $\hat{q}|\boldsymbol{\xi}\rangle = q|\boldsymbol{\xi}\rangle$ and $\hat{p}|\boldsymbol{\xi}\rangle = p|\boldsymbol{\xi}\rangle$. Since we always assume that the two operators commute, they share the same common eigenvectors $|\boldsymbol{\xi}\rangle$.

From these axioms, it can be shown that the classical wavefunction evolves according to the Liouville equation of motion or the **Koopman von-Neumann equation**:

$$i\frac{\partial}{\partial t}\psi(\boldsymbol{\xi}) = \hat{L}\psi(\boldsymbol{\xi}) \qquad \text{or} \qquad i\frac{d}{dt}|\psi\rangle = \hat{L}|\psi\rangle \tag{20.2.1}$$

At this point, we see the first deviation between the classical and quantum theories. The Schrödinger equation is a second-order PDE while the Liouville equation only contains first-order derivatives. This difference in the equations of motion for ψ and ρ leads to a decoupling of the *phase* from the *amplitude* of the system's wavefunction, as we investigate below. Koopman von-Neumann theory generalises classical mechanics by formulating it in terms of a wavefunction in a Hilbert space; however, if we restrict the allowed solutions to those of an algebra of **commuting observables**[3], then it is entirely equivalent to any other formalism we have discussed. A solution to the Liouville equation of motion takes the form of equation 20.2.2, where $A(\boldsymbol{q}, \boldsymbol{p})$ and $S(\boldsymbol{q}, \boldsymbol{p})$ are amplitude and phase functions, respectively and $A^2 = \rho = \psi^*\psi$, such that $A = \sqrt{\rho}$ is given only by the positive square root:

$$\psi(\boldsymbol{\xi}) = A(\boldsymbol{\xi})e^{iS(\boldsymbol{\xi})} \tag{20.2.2}$$

By computing some partial derivatives and collecting real and imaginary terms, we can understand a deeper evolutionary property of the classical wavefunction:

$$\frac{\partial\psi(\boldsymbol{\xi})}{\partial q^j} = e^{iS}\left[iA\frac{\partial S}{\partial q^j} + \frac{\partial A}{\partial q^j}\right], \qquad \frac{\partial\psi(\boldsymbol{\xi})}{\partial p_j} = e^{iS}\left[iA\frac{\partial S}{\partial p_j} + \frac{\partial A}{\partial p_j}\right]$$

$$\text{Re}\left\{i\frac{\partial}{\partial t}\psi = \hat{L}\psi\right\} \Rightarrow \frac{\partial S}{\partial t} = \frac{\partial H}{\partial q^j}\frac{\partial S}{\partial p_j} - \frac{\partial H}{\partial p_j}\frac{\partial S}{\partial q^j} = \left\{\frac{\partial S}{\partial t} = \hat{L}S\right\} \tag{20.2.3}$$

$$\text{Im}\left\{i\frac{\partial}{\partial t}\psi = \hat{L}\psi\right\} \Rightarrow \frac{\partial A}{\partial t} = \frac{\partial H}{\partial q^j}\frac{\partial A}{\partial p_j} - \frac{\partial H}{\partial p_j}\frac{\partial A}{\partial q^j} = \left\{\frac{\partial A}{\partial t} = \hat{L}A\right\} \tag{20.2.4}$$

This indicates that the phase is **decoupled** from the amplitude. The evolution equations are truly separated in classical mechanics. *In this way, the fundamental difference*

---

[3]We would say that we choose a commutative algebra of observables.

between quantum and classical mechanics can be understood in the phase-amplitude interaction of the quantum wavefunction, which is responsible for historic effects in quantum theory (i.e the double-slit experiment, conjugate variables and, hence, uncertainty relations). This is something we are already familiar with from the Bohmian mechanics (see exercise 19.2). These two ideas can be linked into one underlying field model; however, this discussion must wait until much later, since we require Lagrange field equations. Do not confuse the Koopman-von Neumann formulation of *classical mechanics* with the De Broglie-Bohm formulation of *quantum mechanics*. We will return to the Koopman-von Neumann formulation in chapter 29 where we discuss classical path integrals.

The formulation of classical mechanics in Hilbert space is discussed at length in D. Mauro's *Topics in Koopman von-Neumann Theory* as well as other truly fascinating papers by both E. Gozzi and D. Mauro on associated topics (Gozzi and Pagani, 2010; Carta *et al.*, 2011). It should be remembered that, although the formal mathematical structures of classical and quantum mechanics are very similar, the interpretations of states and vectors in the classical and quantum Hilbert spaces are very different, as there is a disparity between the informal interpretations of the mathematical languages; the main benefit of the formulation is to probe the differences and similarities between quantum and classical mechanics.

20.3 Classical Statistical Mechanics

It is an amazing property of our natural world, that complexity arises out of the interaction of many individual units. In this section, we will investigate how the Liouville formulation can be extended to consider an ensemble of particles. This forms the basis of **classical statistical mechanics**, the branch of physics that describes the behaviour of a large number of particles using classical mechanics. This field was pioneered by Boltzmann (figure 20.3) at the close of the nineteenth century, despite much criticism of his work. We begin by using the Liouville probability density to describe averages of the system's quantities. Depending on the precise form of ρ, this concept is used to investigate a classical description of an ideal gas (see exercise 20.1), a paramagnetic crystal (see exercise 20.2), the dipole moment of a molecule (see exercise 20.3) among many other, relevant examples! Interestingly, it is in this formulation that we must introduce quantum mechanical remnants to complement our classical theory (see the H-theorem and the Gibbs paradox).

Fig. 20.3: Ludwig Boltzmann (1844–1906).

The probability density must satisfy normalisation conditions such that, when integrated over the entire available phase space, the probability is equal to 1 by Liouville's theorem, for all times $\int \rho(\boldsymbol{\xi}) d\boldsymbol{q} d\boldsymbol{p} = 1$ (i.e it is just a unitless number). As with the quantum analogue, we can define expectation values or **ensemble averages** of functions of the phase space variables:

$$\langle A(\boldsymbol{\xi}) \rangle := \frac{\int A(\boldsymbol{\xi}) \rho(\boldsymbol{\xi}, t) d\boldsymbol{q} d\boldsymbol{p}}{\int \rho(\boldsymbol{\xi}) d\boldsymbol{q} d\boldsymbol{p}} \qquad \textbf{ensemble average} \qquad (20.3.1)$$

Note that $\int d\boldsymbol{p} d\boldsymbol{q} = \int_{p_1} \int_{p_2} \cdots \int_{p_n} \int_{q_1} \int_{q_2} \cdots \int_{q_n} dp_1 dp_2 \cdots dp_n dq^1, dq^2 \cdots dq^n$; for those that like mathematical notation, we could also write this as $\int \prod_{i=1}^{n} dp_i dq^i$. In **classical statistical mechanics**, the branch of physics that describes the behaviour of a large number of atoms using classical mechanics, our normalisation will instead be the number of particles N rather than 1, since we would be counting the number of ensemble members, (e.g. $dN = \rho(\boldsymbol{q}, \boldsymbol{p}, t) d\boldsymbol{p} d\boldsymbol{q}$ which integrates to $N = \int \rho(\boldsymbol{q}, \boldsymbol{p}, t) d\boldsymbol{p} d\boldsymbol{q}$). In either case, a point in phase space is represented by $3N$ coordinates and $3N$ momenta, or just the $6N$ dimensional ξ. In this case the Liouville probability density can be used to give the probability of finding N particles in a region of phase space $\rho(\boldsymbol{q}, \boldsymbol{p}) d\boldsymbol{q} d\boldsymbol{p}$.

Earlier, we mentioned that, while the volume of a blob in phase space is preserved, the shape can vary to fill any region it is permitted to roam. That is to say, if the system is confined to a time-independent energy surface that limits the regions it can explore (**bounded**), it will just fill this surface. In general, however, we can assume that the constant volume blob will stretch itself over all available phase space; this type of system is called an **ergodic** system. They are characterised by having their time averages equal their ensemble averages, while non-ergodic systems have inaccessible regions of phase space. In general, an ergodic system may be classed as an integrable system, with a small perturbing portion of the Hamiltonian; in this case, the canonical perturbation theory we discuss in chapter 24 becomes fundamental. Ergodicity gives rise to **Poincaré's recurrence theorem**, which states that, for a bounded region of phase space where Liouville's theorem holds, a trajectory will, given long enough, revisit the neighbourhood of a point it once occupied.

This seemingly violates the Clausius statement of increasing entropy, and the only somewhat satisfactory answer lies in the difference in timescales. The second law is satisfied in a matter of seconds (think of mixing gases in a box), while the recurrence theorem has timescales larger than the age of the universe, meaning that the box of gas will unmix in about 10^{20} years after you first put them together. A quantum version also holds; since the mathematics of ergodicity is powerful, this means that a concrete proof of the second law in statistical thermodynamics (where a set of initial conditions (ξ) can be thought of as a microstate) is on thin ice. This doesn't mean it isn't a useful law, Newton's second law doesn't hold for quantum systems but it is still useful. The second law of thermodynamics becomes more accurate, the larger the ensemble becomes. The recurrence theorem also indicates that we cannot have regions of a *bounded* phase space that attract and lock up the Hamiltonian flow. In mathematics, such structures are called **attractors** and they act a bit like a plug hole of a sink; once a set of trajectories ends up in the attractor, it can't leave.

The second law indicates that the total entropy either increases or stays the same; well, this can be related to the volume of phase space. Entropy, however, is not just the volume of phase space the system occupies; it is the volume of phase space the observer *thinks* the system occupies. In this way, if we lose track of the exact position and velocity of a mole of particles, then the region we *think* the system occupies is uncertain and we end up estimating a larger volume of phase space is occupied at any given time. So, we see that the entropy either is constant (i.e. a *fine-grained entropy*) or it increases (this is known as *coarse graining*). This is known as the **Gibbs H-theorem**; it does not tie in well with the recurrence theorem, and this is still a big open problem for mathematicians.

The mathematical nature of the probability distribution $\rho(\xi, t)$ is entirely dependent on how we let the system interact with the environment. Let us assume that the system is entirely isolated from the environment and cannot interact at all. We know that the energy of the system is conserved, thereby reducing the accessible phase space to a $(6N-1)$ hypersurface given by $H(\xi) = E$. In this case, we may consider the probability distribution to be

$$\rho(\xi, E) = \delta[H(\xi, E) - E]$$

We notice that the probability density is zero when we do not limit the Hamiltonian to the constant energy hypersurface. The **Dirac delta function** is an extension of the Kronecker delta and has the properties that

$$\int_{-\infty}^{\infty} f(x)\delta(x - c)dx = \begin{cases} f(c) & x - c \\ 0 & x \neq c \end{cases}$$

. The ensemble average of a quantity using this density is $A(\xi, E)$:

$$\langle A(\xi, E) \rangle = \frac{\int A(\xi)\delta[H(\xi, E) - E]d\mathbf{q}d\mathbf{p}}{\int \delta[H(\xi, E) - E]d\mathbf{p}d\mathbf{q}}$$

We call such a system a **microcanonical ensemble**.

By letting the system exchange heat with the environment, the probability density is proportional to the **Boltzmann factor**, and we call such a system a **canonical ensemble** $\rho(\xi) \propto e^{-\beta H(\xi)}$. A fundamental postulate of Boltzmann statistics is that all regions of phase space on a constant energy hypersurface are equally likely to contain the system.

We should be very tempted to identify the probability density as being equal to the Boltzmann factor; however, doing this gives $\int \rho d\mathbf{q}d\mathbf{p}$ units of (energy·time)^n for n coordinates. Therefore, we must divide by some factor that has these units and we denote this factor by h^n. In this context, h is some arbitrary constant whose units are energy·time, action. The correct expression for equation 20.3.1 using the Boltzmann factor is therefore written with h^n:

$$\langle A(\boldsymbol{\xi}) \rangle = \frac{\frac{1}{h^n} \int A(\boldsymbol{\xi}) e^{-\beta H(\boldsymbol{\xi})} dq dp}{\frac{1}{h^n} \int e^{-\beta H(\boldsymbol{\xi})} dq dp} \tag{20.3.2}$$

carefully noting that the h^n cancels! As an example, consider the average energy:

$$\langle H \rangle = \frac{1}{Z} \int H(q,p) e^{-\beta H(q,p)} dq dp$$
$$= -\frac{1}{Z} \frac{\partial}{\partial \beta} \int e^{-\beta H(q,p)} dq dp$$
$$= -\frac{1}{Z} \frac{\partial Z}{\partial \beta}$$
$$= -\frac{\partial \ln Z}{\partial \beta}$$

So why the introduce h in the first place? We should re-examine the importance of $\int \rho dp dq$. This expression is the smooth adding-up of individual p, q volumes or **microstates**, with each *weighted* by a probability distribution. The addition of microstates does not take a conceptual meaning unless we define a standard volume of a state (i.e. we normalise it). Therefore, we note two things: firstly, that the result should just be a number with no units and second, a classical microstate is just a region of phase space of a given volume, namely h^n. We call this integral a system's **partition function**, Z. Classical physics relies upon quantum mechanics sometimes, since we identify h as Planck's constant 6.626×10^{-34} Js^{-1}; however, you must note that this is still classical mechanics, even though we have discretised phase space. The interval $\int dp dq$, will always have the same number of states inside it, by Liouville's theorem.

In addition, we may not be finished with our expression for the classical partition function, due to the **Gibbs paradox**. In essence, the paradox is as follows: a box of gas molecules is divided down the middle, reducing the possible number of configurations of any given molecule by two, since half the box is now unavailable; thus, entropy should decrease, violating the second law of thermodynamics. However, we over-counted the actual number of states of the system. The number of ways a set of N objects can be permuted (swapping them around between themselves) is $N!$. For example, if we have two electrons (1, 2) and two nuclei (A, B) then two possible configurations of the electrons are A(1)B(2) and A(2)B(1), which is a permutation of electrons between the two nuclei.

Division by this factor un-counts those states that are related by permutation of their elements. This leads to the notion of **indistinguishable particles**, which is an odd feature in classical mechanics and is actually a residue of quantum theory. The indistinguishability of particles in quantum mechanics, arises due to particles being considered as modes of excitation of a field rather than as little balls with individualistic identity. Saying a box has x particles in it is really just saying how excited their quantum field is and, since excitations are discrete, we interpret that as x number of particles. Hence, for this reason, we may add a factor of $1/N!$ to the classical partition function:

$$Z = \frac{1}{h^n \cdot N!} \int \rho \prod_{i=1}^{n} dp_i dq^i \qquad \textbf{partition function}$$

Each type of statistical ensemble will have its own expression for the partition function, depending on the form of the probability density. In general, partition functions are related via a Laplace transform (see exercise 20.6). In order to connect with macroscopic thermodynamics, we would need to define thermodynamic potentials for each ensemble. They are related via Legendre transform to one another just like Lagrangian and Hamiltonian formulations. This is more than coincidence and hints that there will be a geometric formulation of thermodynamics on a manifold. Although this is a very rich theory we will develop the symplectic geometry of classical mechanics in this book rather than the contact geometry of thermodynamics.

The classical probability density evolves according to the Liouville equation of motion and a special case of this is when ρ is not an explicit function of time, such that $\partial_t \rho = 0$, meaning that the Poisson bracket with the Hamiltonian vanishes too $\{\rho_{eqm}, H\} = 0$ so it a constant of the motion. This probability density would be referred to as being at **statistical equilibrium**. The ensemble average will yield a time-independent value for a given observable. Equilibrium is a state whereby the macroscopic quantities of the system have settled to constant values and, when we perform a measurement on the equilibrated system, the value of the quantity is given by the weighted expectation value.

The whole field of equilibrium and non-equilibrium thermodynamics can be understood in terms of the time evolution of the probability density. The study of **near-equilibrium** systems uses **fluctuations** to model slight perturbations from equilibrium. In addition **stochastic** models, which add a little randomness to the system, are fundamental tools in computational chemistry/biology and physics. We discuss Boltzmann statistics further in appendix G.

Returning to the partition function Z, we can easily compute these integrals for individual modes of motion $Z_{\text{trans}}, Z_{\text{vib}}, Z_{\text{rot}}$, and so on, then multiply the resulting functions together to give the overall partition function $Z = Z_{\text{trans}} \cdot Z_{\text{vib}} \cdot Z_{\text{rot}}$, with the choice of multiplication following from the addition properties of the energy.

Exercise 20.1 Calculating expressions for partition functions is a fun exercise to describe the statistical mechanics of systems of particles. Here, we will derive a molecular partition function that describes the distribution of particles and their degrees of freedom. Using the

Hamiltonian that describes free translation $H = p_x^2/2m$ and the Boltzmann density, compute Z_{trans} by using a Gaussian integral $\int_{-\infty}^{+\infty} e^{ax^2} dx = (\pi/a)^{1/2}$.

$$Z_{\text{trans,1D}} = \frac{1}{h^n} \int_{-\infty}^{+\infty} \int e^{-\frac{\beta p_x^2}{2m}} dq_x dp_x$$

Generalising to three dimensions: $\int \int \int dq_x dq_y dq_z = \mathcal{V}$ (volume), results in equation 20.3.3. Interestingly, the position integral must therefore be a bounded integral (so not $\pm\infty$) to the volume of the container; the momentum integral is free, however:

$$Z_{\text{trans,3D}} = \frac{\mathcal{V}}{h^{3n}} \left(\frac{2\pi m}{\beta} \right)^{3/2} \tag{20.3.3}$$

You can try the same with the harmonic oscillator Hamiltonian $H = \frac{1}{2}p^2/\mu + \frac{1}{2}kq^2$:

$$Z_{\text{vib}} = \frac{1}{h^n} \int_{-\infty}^{+\infty} e^{-\beta \frac{p^2}{2m}} dp \int_{-\infty}^{+\infty} e^{-\beta \frac{kq^2}{2}} dq = \frac{2\pi}{h^n \beta} \sqrt{\frac{\mu}{k}} \tag{20.3.4}$$

The rigid rotor is a little more involved, so we can do that one together. The rotational Lagrangian is given by $\mathcal{L}_{\text{rot}} = \frac{1}{2}I(\dot{\theta}^2 + \sin^2\theta \dot{\phi}^2)$, where $I = \mu r^2$ is the moment of inertia. We can calculate the conjugate momenta with respect to $p_\theta = I\dot{\theta}$ and $p_\phi = I \sin^2\theta \dot{\phi}$ and construct the rotational Hamiltonian from $H_{\text{rot}} = \dot{\theta}p_\theta + \dot{\phi}p_\phi - \mathcal{L}$, to obtain $H_{\text{rot}} = \frac{1}{2I}\left(p_\theta^2 + \frac{p_\phi^2}{\sin^2\theta} \right)$. The rotational partition function is given by the integral over these coordinates and momenta over the appropriate range for each coordinate (and $\pm\infty$ for the momenta):

$$Z_{\text{rot}} = \frac{1}{h^n} \int_0^\pi \int_0^{2\pi} \int_{-\infty}^\infty \int_{-\infty}^\infty \exp\left\{ -\beta \left(\frac{p_\theta^2}{2I} + \frac{p_\phi^2}{2I \sin^2\theta} \right) \right\} dp_\phi dp_\theta d\phi d\theta$$

Performing the integral over the momenta yields two Gaussian integrals:

$$Z_{\text{rot}} = \frac{1}{h^n} \int_0^\pi \int_0^{2\pi} \sqrt{\frac{2\pi I}{\beta}} \sqrt{\frac{2\pi I \sin^2\theta}{\beta}} d\phi d\theta$$

$$Z_{\text{rot}} = \frac{2\pi^2 I}{h^n \beta} \int_0^\pi \sin\theta d\theta$$

where we have noted that another 2π arises due to the lack of dependancy on ϕ; using $\int_0^\pi \sin\theta d\theta = -\cos\theta \big|_0^\pi$, we obtain equation 20.3.5:

$$Z_{\text{rot}} = \frac{8\pi^2 I}{h^n \beta} \tag{20.3.5}$$

In real systems, there will be interactions between the molecules, and these are what give chemistry its fascinating properties. We can write the **configurational partition function** Q for $H = p^2/2m - U(\boldsymbol{q})$. Using

$$\int \int F(x)G(y)\,dx\,dy = \int F(x)\,dx \cdot \int G(y)\,dy$$

we obtain

$$Z = \frac{V}{N!h^{3n}} \left(\frac{2\pi m}{\beta} \right)^{3/2} \underbrace{\int e^{-\beta U}\,d\boldsymbol{q}}_{Q}$$

For N particles with interacting potentials, there is a summation in the Hamiltonian $H = \sum p_i^2/2m + U(\boldsymbol{q})$:

$$Z = \frac{1}{N!} \left(\frac{\mathcal{V}}{\Lambda^3} \right)^N Q, \qquad \text{with} \qquad \Lambda = \left(\frac{h^2\beta}{2\pi m} \right)^{1/2} \qquad Q = \mathcal{V}^{-N} \int \prod_{i=1}^{N} dq^i e^{-\beta \sum_{ij} U(\boldsymbol{q}_{ij})}$$

Therefore, the total partition function is viewed as the product of a multi-particle non-interacting term and the configuration integral over positions whose potential encodes the interaction. Evaluation of the potential interaction term is the focus of much modern theoretical chemistry. The prefactor Λ is the **thermal de Broglie wavelength**. This is a very similar topic to the moment closure of the BBGKY hierarchy we discussed at the start of this chapter.

There are many avenues we could follow here to find an analytical evaluation of Q; the most travelled is the **Mayer-Montroll cluster expansion**, but others include low-density approximations for fluids with choice of potential functions from elementary chemical theory. The cluster method involves taking the sum out of the argument of the exponent and instead multiplying the exponentials together (e.g. $e^{(A+B)} = e^A \cdot e^B$), then defining a **Mayer function** $f_{ij} = \exp\{-\beta U(r_{ij})\} - 1$, with the -1 preventing wild divergence, to obtain

$$\exp\left\{ -\beta \sum_{i>j}^{N} U(r_{ij}) \right\} = \prod_{i>j}^{N} \exp\left\{ -\beta U(r_{ij}) \right\} = \prod_{i>j}^{N} (1 + f_{ij})$$

with $U(r_{ij})$ being a two-particle interaction energy or **pair potential** that is a function of the particle's separation: $r_{ij} = r_i - r_j$. The Mayer function is small at large inter-particle separations but becomes more significant the closer the particles get. When you multiply this out, you can collect terms with the same number of particle interactions; the simplest example is taking $N = 3$ and multiplying the brackets, we obtain

$$\prod_{i>j}^{N=3} (1 + f_{ij}) = (1 + f_{21})(1 + f_{31})(1 + f_{32}) = 1 + \sum_{i>j} f_{ij} + \sum_{i>j>k} f_{ij} f_{jk} + \sum_{i>j>k>l} f_{ij} f_{kl} + \dots$$

Each successive summation is essentially just adding another particle to the interaction term; hence, it is called the **cluster expansion** and the indices prevent over counting. The Mayer function f_{ij} is almost zero unless the particles are close together. The low-density approximation will truncate this series early on; if we truncate all but the first term, we have an ideal gas with no interaction terms. The configuration partition function to second order will account for all interactions that have a pairwise topology and will look something like

$$Q = \mathcal{V}^{-N} \int \prod_{i=1}^{N} dq^i \left(1 + \sum_{i>j} f_{ij}\right) = 1 + \mathcal{V}^{-N} \int \sum_{i>j} f_{ij} \prod_{i=1}^{N} dq^i$$

There are, in total, $\frac{1}{2}N(N-1)$ of these terms in $\sum f_{ij}$, due to the different ways we can pick pairs of gas particles, but they are all equal and differ only in labelling:

$$\mathcal{V}^{-N} \int \sum_{i>j} f_{ij} \prod_{i=1}^{N} dq^i = \mathcal{V}^{-N} \frac{N(N-1)}{2} \int f(r_{12}) \prod_{i=1}^{N} dr_i \qquad (20.3.6)$$

Integrating over $r_3, \ldots, r_N$ gives a factor of $\mathcal{V}^{N-2}$. If we then integrate over r_1 and r_2 by resetting the origin to coincide with particle 1 and taking the second particle's relative coordinate, we obtain another factor of $\mathcal{V}$:

$$\int \sum_{i>j} f_{ij} \prod_{i=1}^{N} dq^i = \frac{\mathcal{V}^{N-2} \cdot \mathcal{V}}{\mathcal{V}^{-N}} \frac{1}{2} N(N-1) \int f(r) dr \qquad (20.3.7)$$

We tidy this up by assuming that N is so large that $N - 1 \approx N$:

$$\int \sum_{i>j} f_{ij} \prod_{i=1}^{N} dq^i = \frac{N^2}{2\mathcal{V}} \int f(r) dr \qquad (20.3.8)$$

Let us now symbolise the integral by $\mathcal{I} = \int f(r) dr = \int [\exp\{-\beta U(r)\} - 1] dr$. The second-order configurational partition function is written

$$Q = 1 + \frac{N^2}{2\mathcal{V}} \mathcal{I} \qquad (20.3.9)$$

The partition function for a classical gas of particles interacting to second-order collisions is written

$$Z = \frac{1}{N!} \left(\frac{\mathcal{V}}{\Lambda^3}\right)^N \left\{1 + \frac{N^2}{2\mathcal{V}} \mathcal{I}\right\} \qquad (20.3.10)$$

The integral $\mathcal{I}$ (well, more precisely, $-\frac{1}{2}\mathcal{I}$) is the second **virial coefficient** of the gas. The form of the potential $U(r)$ is open to interpretation, and a popular choice might be a particular of the Mie potential, namely the **Lennard-Jones** (12-6) potential:

$$U(r) = 4\epsilon \left[\left(\frac{\sigma}{r}\right)^{12} - \left(\frac{\sigma}{r}\right)^{6}\right] + \frac{e_1 e_2}{r}$$

where ϵ is the potential well depth, σ is the distance where the inter-particle potential vanishes and r is the distance between particles. For further information and derivation of equations of state (like the van der Waals equation) using the cluster expansion method please see Mandl's (1988) *Statistical Physics* (second edition). At any point in the above derivation, we are free to make different approximations, whether that be retaining higher-order terms in the cluster expansion or using a different potential energy function.

We now have an explicit expression of the partition function:

$$Z = \frac{1}{N! h^N} \int e^{-\beta H} \prod_{i=1}^{N} dp_i dq^i \qquad (20.3.11)$$

We have already seen why the prefactors are required for a consistent model. In statistical mechanics, we can not hope to solve for trajectories of large systems as we have normally done in dynamics. Instead, we must search for descriptions of the ensemble system. The classical **virial theorem** is a powerful tool, all about ensemble averaging. The quantities of interest are the components of the phase space vector ξ, which is denoted ξ_i, and can be either q^i or p_i; we do not for now limit ourselves to a specific choice. We consider the following average:

$$\left\langle \xi_i \frac{\partial H}{\partial \xi_j} \right\rangle = \frac{1}{Z} \int \xi_i \frac{\partial H}{\partial \xi_j} e^{-\beta H} \prod_k dq^k dp_k$$

$$= -\frac{1}{Z\beta} \int \xi_i \frac{\partial}{\partial \xi_j} (e^{-\beta H}) \prod_k dq^k dp_k \qquad (20.3.12)$$

We now integrate by parts with respect to ξ_j:

$$\left\langle \xi_i \frac{\partial H}{\partial \xi_j} \right\rangle = -\frac{1}{Z\beta} \int' \xi_i e^{-\beta H} \prod_k dq^k dp_k \bigg|_{\xi_j(1)}^{\xi_j(2)}$$

$$+ \frac{1}{Z\beta} \int \left(\frac{\partial \xi_i}{\partial \xi_j} \right) e^{-\beta H} \prod_k dq^k dp_k \qquad (20.3.13)$$

If the component is chosen to be the momentum p_i, then evaluation at its boundaries is an evaluation at $\pm\infty$, since it is an unbounded integral. In this case,

$$\lim_{p_i \to \pm\infty} H = +\infty, \qquad \text{then} \qquad \lim_{H \to +\infty} e^{-\beta H} = 0 \qquad (20.3.14)$$

where we have assumed that $H \propto p_i^2$. In this case, the average reduces to

$$\left\langle p_i \frac{\partial H}{\partial p_j} \right\rangle = \frac{1}{Z\beta} \delta_{ij} \int e^{-\beta H} \prod_k dq^k dp_k = \delta_{ij}/\beta \qquad (20.3.15)$$

This result is known as the **virial of Clausius**. It is a more general statement of the virial theorem of kinetic energies, the results of which we can recover if we let $H = \frac{p^2}{2m} + U(q)$:

$$\left\langle \frac{1}{2} p_i \frac{\partial H}{\partial p_i} \right\rangle = \left\langle \frac{p_i^2}{2m} \right\rangle = \frac{\delta_{ij}}{2\beta} \qquad (20.3.16)$$

Both sides have been halved just for convenience of dealing in T rather than $2T$. If we sum over each degree of freedom $p_i(\partial H/\partial p_i)$ we obtain

$$\sum_{i=1}^{3N} \left\langle \frac{1}{2}p_i \frac{\partial H}{\partial p_i} \right\rangle = \sum_{i=1}^{3N} \left\langle \frac{p_i^2}{2m} \right\rangle = \frac{3}{2} \frac{N\delta_{ij}}{\beta} \tag{20.3.17}$$

If we had taken ξ_i to be the position coordinate instead, then the vanishing of the term evaluated at the boundaries would be in question, since the integral over q^i is bounded. If we have a sharp confining potential, then the Hamiltonian will rise to infinity and the probability distribution will approach zero, regenerating the virial.

There is a slight difference between the **equipartition theorem** and the classical virial theorem: the former is the sum of the averages, while the latter is the average of the sums of $p_i(\partial H/\partial p_i)$. However, the equipartition theorem is just a specific example of the more general virial theorem, since the former assumes that each mode is described by the same term. Equipartition theorem is saying that each mode contributes $1/2\beta$ to the average energy, so $3N$ modes gives a total average of $3N/2\beta$. The virial theorem is saying that the average of the sum of all modes is $3N/2\beta$. The virial theorem applies to systems where thermal equilibrium has not been reached and hence different modes may not have equal contributions. There are definitely cases where each mode is not equal: if we had an electric field along a specific axis, equipartition theorem would not be valid since the z axis would be inequivalent to the x and y axes. However, the virial theorem is valid in this case. If we are confident that we have a state of equilibrium, then the virial theorem reduces to the equipartition theorem.

20.4   Symplectic Integrators

We use the Liouville operator and Hamilton's equations in the subject of **molecular dynamics**, where the aim is to study an ensemble of particles classically. If we consider the Hamiltonian $H(\boldsymbol{q}, \boldsymbol{p})$ for N interacting particles to be the kinetic energy plus the potential energy, we can summarise the equations of motion as follows:

$$H(\boldsymbol{p}, \boldsymbol{q}) = \sum_{j=1}^{n} \frac{p_j^2}{2m_j} + U(\boldsymbol{q}) \tag{20.4.1}$$

We can easily see that the Hamiltonian is conserved, since it is not an explicit function of time, so the system evolves on a constant energy surface, as is fully explored in the aforementioned ergodic hypothesis (see section 20.3). These types of Hamiltonians are really favourable to work with since they contain no mixed p and q terms and any time we want to canonically quantise them, we have no commutation issues at play. The Liouville operator for this Hamiltonian is given by equation 20.4.1, if we take $\hat{L} = -i\{\cdot, H\}$:

$$\hat{L} = i \sum_{j=1}^{N} \left(\frac{p_j}{m_j} \frac{\partial}{\partial q^j} + \vec{F}_j(\boldsymbol{q}) \frac{\partial}{\partial p_j} \right) \tag{20.4.2}$$

The Liouville operator can be expanded in terms of components and a basis, like any vector field:

$$\hat{L} = \sum_{j=1}^{N} \dot{\xi}^j \frac{\partial}{\partial \xi_j} = iL_1 + iL_2 \tag{20.4.3}$$

where we identify the components by comparison:

$$iL_1 = \frac{p_j}{m_j} \frac{\partial}{\partial q_j}, \qquad \text{and} \qquad iL_2 = \vec{F}_j(\boldsymbol{q}) \frac{\partial}{\partial p_j}$$

For any function of phase space $f(\boldsymbol{\xi})$, we can see that iL_1 and iL_2 do not commute, so $(iL_1 \cdot iL_2 - iL_2 \cdot iL_1)f(\boldsymbol{\xi}) \neq 0$:

$$[L_{12}, L_{21}]f(\boldsymbol{\xi}) = L_1 L_2 - L_2 L_1 = \frac{p}{m} \frac{\partial}{\partial q} \left(\vec{F}_j(\boldsymbol{q}) \frac{\partial f}{\partial p} \right) - \vec{F}_j(\boldsymbol{q}) \frac{\partial}{\partial p} \left(\frac{p}{m} \frac{\partial f}{\partial q} \right) \neq 0$$

This, in practice, makes solving classical systems rather difficult, since, in order to solve this equation, the classical propagator defined above must be factorised and we can't just use $e^{x+y} = c^x c^y$, since due to the non-commutation

$$\exp\{iL_1 t\} \exp\{iL_2 t\} \neq \exp\{(iL_1 + iL_2)t\} \tag{20.4.4}$$

we don't obtain unique solutions. In order to maintain time-reversal symmetry, we use the **Trotter's theorem**. This theorem states that the exponential can be factorised according to equation 20.4.5. That is to say, we factorise the classical propagator as the product of exponentials:

$$e^{(iL_1+iL_2)t} \approx \lim_{x \mapsto \infty} \left[e^{iL_2 t/2x} \cdot e^{iL_1 t/x} \cdot e^{iL_2 t/2x} \right]^x \tag{20.4.5}$$

where x is a large and finite parameter. If we define a small time interval $\delta t = t/x$ (since x is large), then we can approximate a small stepwise time evolution δt of the classical propagator by direct application of Trotter's theorem (re-arranging the above):

$$e^{\hat{L}\delta t} \approx e^{iL_2 \delta t/2} \cdot e^{iL_1 \delta t} \cdot e^{iL_2 \delta t/2} \tag{20.4.6}$$

This means that we should apply iL_2 for half a time interval, then iL_1 for a full interval and then iL_2 again for another half a time interval. Returning to the classical propagator, we can see that $\exp\{iL_1\delta t\}$ is a translation operator acting on the particles positions, $q^j(t_1 + \delta t) \approx (1 + \hat{L}t)q^j(t_1)$,

$$e^{iL_1\delta t}q^j(t_1) \mapsto q^j(t_1) + \delta t \frac{p_j(t_1)}{m_i}$$

while $\exp\{iL_2\delta t\}$ acts on the momentum, translating it from p_j to

$$e^{iL_2\delta t}p_j(t_1) \mapsto p_j(t_1) + \delta t \vec{F}_j(q(t_1))$$

If we apply equation 20.4.6 to q, then, starting from the right, we can see that the first term $\exp\{iL_2\delta t/2\}$ has no effect on q^j:

$$e^{\hat{L}\delta t}q = e^{iL_2\delta t/2} \cdot e^{iL_1\delta t} \cdot e^{iL_2\delta t/2}q$$

The second term acts on q through its partial derivative, as above, to give the following:

$$e^{\hat{L}\delta t}q = e^{iL_2\delta t/2}\left(q^j + \delta t \frac{p_j}{m_j}\right)$$

The final operator can act on the p in the brackets to finalise the expression for the approximation to the small evolution of the propagator:

$$e^{\hat{L}\delta t}q = q^j + \frac{\delta t}{m_j}\left(p_j + \frac{\delta t}{2}\vec{F}(q)\right)$$

If we take a closer peek at this, we actually note that this is a second-order Taylor series in terms of the initial conditions $(q(t_1), p(t_1))$:

$$q^j(\delta t) = q^j(t_1) + \frac{\delta t}{m_j}p_j(t_1) + \frac{\delta t^2}{2m_j}\vec{F}_j(q(t_1)) \qquad (20.4.7)$$

This is the **Verlet integrator**, the starting point of most molecular dynamics simulations. Application of the evolution operator to p rather than q produces the velocity Verlet integrator by exactly the same method:

$$\dot{q}^j(\delta t) = q^j(t_1) + \frac{\delta t^2}{2m_j}\left[\vec{F}_j(t_1) + \vec{F}_j(\delta t)\right] \qquad (20.4.8)$$

The Liouville equation of motion therefore defines the basics of classical molecular mechanics, where computationally, we calculate the positions and velocities of a system of N interacting particles after a small time step from the initial conditions, by application of the classical propagator and the Liouville operator. These Verlet integrators are actually particulars of a family of **symplectic integrators** that provide a general way to integrate the canonical equations.

It is important to note that the modelling of dynamical chemical systems that involve bond breaking or optical excitations, charge transfer, and so on, in general, require quantum mechanical descriptions. Frustratingly, chemical phenomena are often described by quantities that are governed by quantum laws, not classical ones. This in itself would not pose too much of a problem if the two theories predicted similar results; however, the results of each could not be more different and, in most cases, are almost complete opposites of each other. However, the drive to reduce computational cost by including classical mechanics in combination with quantum approaches is at the forefront of molecular mechanics theory, of which the Liouville operator is the starting point for most avenues of research. Classical mechanics is very useful for describing the molecular processes of extended systems, since we expect a linear increase in computational cost, depending on the size of the system.

Deriving the Verlet integrators in this way is totally different from doing so in a Newtonian framework. This way has the benefit of the canonical framework of mechanics behind it and hence provides a basis for further algorithm development. In addition, we can be assured that the motion, despite being approximate, will evolve on a constant energy hypersurface and hence be true to the Hamiltonian we have used.

The Liouville formalism is also the starting point for **response theory** in classical mechanics. Response theories study the behaviour of an equilibrated system under small perturbations. The perturbation is so small that a linear term in an expansion will be a good approximate. Later we will see that Liouville mechanics has a hidden field theoretic structure together and, together with the Hamilton-Jacobi equation, encapsulates classical mechanics in one Lagrangian function; however, we hold that discussion for the time being!

Chapter summary

- **Liouville's theorem** states that the volume of phase space occupied by the system is maintained under canonical transformations and therefore time evolution. This was derived in two ways: via a Jacobian determinant and by taking the divergence of a **phase space velocity vector** $\dot{\xi}$.

- The **classical propagator** was derived and the **Liouvillian** $\hat{L}$ operator is defined for time evolutions.

- The **classical probability density** was introduced as well as **classical Koopman-von Neumann wavefunctions**, the phase-amplitude properties of which were investigated.

- **Classical statistical mechanics** was discussed and the **partition function** was utilised to generate expressions for translational, vibrational and rotational modes.

- Classical **molecular dynamics** was approached from a Liouville operator standpoint to generate a **Verlet algorithm**. Although this topic was easier to understand using Newtonian physics, the implications that come with this derivation give us a deeper insight into the underlying theory and structure of phase space.

Exercise 20.2 Consider a paramagnetic crystal that, at each lattice site, has an identical independent magnet that may rotate (see figure 20.4).

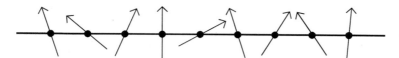

Fig. 20.4: A paramagnetic crystal model.

The magnetic dipole moment $\boldsymbol{\mu}$ at angle θ to an applied field $\boldsymbol{B}$ has the energy $-\boldsymbol{\mu} \cdot \boldsymbol{B}$ and, correspondingly, the lowest energy conformation is achieved with spins aligned with the applied field. The Hamiltonian is written

$$H = \left(\frac{p_\theta^2}{2I} + \frac{p_\phi^2}{\sin^2 \theta} \right) - mB \cos \theta \tag{20.4.9}$$

The partition function is formed by entering this Hamiltonian into the exponent of the Boltzmann factor

$$Z = \left(\frac{2\pi I}{h^n \beta} \right) 2\pi \int_0^\pi d\theta \sin \theta \, e^{\beta mB \cos \theta} \tag{20.4.10}$$

The rotational portion takes the form as previously determined and the portion due to the magnetic moment is made easier by realising that it is the antiderivative of the following:

$$Z = \left(\frac{2\pi I}{h^n \beta}\right) 2\pi \int_0^\pi d\theta \left(\frac{-1}{\beta\mu B}\right) \frac{\partial}{\partial\theta} e^{\beta\mu B \cos\theta}$$

$$= \left(\frac{2\pi I}{h^n \beta}\right) 2\pi \left(\frac{-1}{2\beta\mu B}\right) e^{\beta\mu B \cos\theta}\Big|_0^\pi$$

$$= \left(\frac{8\pi^2 I}{h^n \beta}\right)\left(\frac{\sinh\beta\mu B}{\beta\mu B}\right)$$

The average energy of the system is given by $N\langle H\rangle$, where N is the number of particles in the spin-lattice array (n is the dimension of the phase space):

$$\frac{\partial\ln Z}{\partial\beta} = \frac{\partial}{\partial\beta}\left(\ln 8\pi^2 I - \ln h^n\beta + \ln\sinh\beta\mu B - \ln\beta\mu B\right)$$

$$= -\left(\frac{h^n}{h^n\beta}\right) + \left(\frac{1}{\sinh\beta\mu B}\cosh\beta\mu B \cdot \mu B\right) - \frac{\mu B}{\beta\mu B}$$

Where we have used the chain rule and $\frac{d}{dx}\ln\sinh f(x) = \frac{1}{\sinh f(x)}\cosh f(x)\cdot f'(x)$ and the fact that $\frac{d}{dx}\sinh x = \cosh x$ as well as $\cosh x/\sinh x = \coth x$.

$$N\langle H\rangle = -\frac{\partial\ln Z}{\partial\beta} = NB\mu\left[\frac{2}{\beta\mu B} - \coth\beta\mu B\right] \tag{20.4.11}$$

Note that h^n does not feature in this expression. The **magnetisation** M of the system is a quantification of the materials response to the application of the field by gauging the magnetic moment density:

$$M = N\frac{1}{\beta}\frac{\partial\ln Z}{\partial B} \tag{20.4.12}$$

Using this expression and the above ideas, can you derive the following equation for the magnetisation?

$$N\frac{1}{\beta}\frac{\partial\ln Z}{\partial B} = N\mu\left[\coth\beta\mu B - \frac{1}{\beta\mu B}\right] \tag{20.4.13}$$

In the limit of $\beta \to 0$ or considering only infinitesimally small perturbations in B, we observe that $\beta\mu B \to 0$. Therefore, we can Taylor expand $\coth x = x^{-1} + \frac{x}{3} + \mathcal{O}(x^3)$ about the equilibrium value, to derive the **Curie law**, which is named after P. Curie. It was originally an empirically derived relation between the magnetisation of a paramagnetic solid and a small perturbation in the external magnetic field (although it does not hold at low temperature or strong applied fields):

$$\lim_{\beta,B\to 0}\left(N\frac{1}{\beta}\frac{\partial\ln Z}{\partial B}\right) = \frac{N\mu^2\beta B}{3} \tag{20.4.14}$$

This corresponds to a strong alignment of the lattice spins, also known as a saturation point. The statistical mechanics of paramagnetic materials commonly features $\coth x - x^{-1}$ functions which are called **Langevin functions**. Importantly, Langevin functions are non-singular upon $x \to 0$.

Sparing a thought for the experimentalists, we can divide by B and multiply by the permeability of free space μ_0, then we can define the **magnetic susceptibility** χ. The susceptibility of a material indicates the degree of magnetisation a material exhibits for a certain applied field. Hence, we regenerate an equation, which, together with its hysteresis graph, is familiar to any material scientist:

$$\chi = \frac{M\mu_0}{B} \tag{20.4.15}$$

Now we hit upon a *slight* snag in our classical paradigm: its name is the **Bohr-van Leeuwen theorem**. The theorem *assures* us that classical statistical mechanics can not account for para-, dia- or ferro-magnetic phenomena in equilibrium Drude-Lorentz solids, making magnetism entirely a quantum mechanical theory. Proofs of the theorem all involve the vanishing of the unbounded momentum integral. A straightforward proof can be found by considering the Hamiltonian for an N-charged-particle system in an applied magnetic field $\boldsymbol{B} = \vec{\nabla} \times \boldsymbol{A}$. Let us now re-write this Hamiltonian by letting $\boldsymbol{p}_i \to \bar{\boldsymbol{p}}_i = \boldsymbol{p}_i + e\boldsymbol{A}$:

$$H(\boldsymbol{q}, \boldsymbol{p}) = \frac{1}{2m} \sum_{i=1}^{N} (\boldsymbol{p}_i + e\boldsymbol{A})^2 + eU(\boldsymbol{q}) = \sum_{i=1}^{N} \frac{\bar{\boldsymbol{p}}_i^{\,2}}{2m} + eU(\boldsymbol{r}) \tag{20.4.16}$$

Note that $\boldsymbol{p}$ commutes with $\boldsymbol{A}$ (not true in quantum systems), allowing us to completely remove the vector potential from the Hamiltonian and, therefore, the partition function, too, by a change of variables! Computation of the magnetisation will now show that it is exactly zero, since $\partial \ln Z / \partial B = 0$. This often overlooked theorem is actually one of the major factors in the motivation for developing quantum mechanics, akin to the ultraviolet catastrophe and interference experiments of the early twentieth century. It was developed in general by Hendrika van Leeuwen (1887-1974) in her doctoral thesis at the dawn of the quantum revolution; however, it had first been suggested by Niels Bohr (1885-1962) some years earlier. It clearly motivates a non-classical theory of magnetism. It does not, however, say that classical magnetism is a null concept, since Maxwell's electrodynamics is one of the cornerstones of classical physics, as we see in chapter 27; only that the origins of magnetism in condensed matter theory is quantum mechanical.

The above localised spin lattice system already incorporates quantum mechanical concepts, due to the quantisation of the spins on discrete lattice points, therefore we can see that classical statistical mechanics can still prove extremely useful to material science, provided we ask the right questions!

Exercise 20.3 Under the application of an electric field $\boldsymbol{E}$, a material may exhibit two responses, depending on its properties: (i) the external field distorts the molecular charge distributions and *induces* a dipole moment in each molecule which then aligns with the field; and (ii) the external field aligns permanent dipoles in polar molecules that were initially randomised due to thermal fluctuations. The dipole moment is defined by $\mu = e\boldsymbol{q}$, where e is the charge and $\boldsymbol{q}$ is the displacement from equilibrium.

To model an induced polarisation, we assume that the charge e experiences a harmonic restoring force when perturbed from equilibrium by the external field applied along the z axis:

$$m\omega^2 \boldsymbol{q} = e\boldsymbol{E} \tag{20.4.17}$$

The Hamiltonian is then given by

$$H = \frac{p^2}{2m} + \frac{m\omega^2 q^2}{2} - eEq_z \qquad (20.4.18)$$

Calculate the average molecular polarisation exhibited by the material in the z direction, using the phase space average:

$$\langle \mu_{\text{induced}} \rangle = \frac{1}{Z} \int dp\, dq\, (eq_z \rho) \qquad (20.4.19)$$

We can incorporate the effects of dipole alignment by adding an additional term to the Hamiltonian. Here, $\boldsymbol{D}$ is the permanent dipole in the molecule; it has a non-zero product at angles θ from the applied field $\boldsymbol{E}$:

$$H_{\text{align}} = -\boldsymbol{D} \cdot \boldsymbol{E} \qquad (20.4.20)$$

Averaging over this Hamiltonian and expanding exponentials, show that the total dipole moment can be written as the sum of both the induced and aligned contributions:

$$\langle \mu \rangle = \frac{e^2}{m\omega^2} E + \frac{1}{3}\beta D^2 E \qquad (20.4.21)$$

The **polarisability** α is defined by $\mu = \alpha E$, and the **polarisation density** is $\mathcal{R} = \mu/\nu$, where ν is a volume element; these are two important characteristics of a material.

Exercise 20.4 The microcanonical ensemble is a closed statistical ensemble with fixed energy. The canonical ensemble has a little more freedom by allowing energy to be exchanged. The **grand canonical ensemble** allows the exchange of both heat and *particle number*. We now have two reservoirs: a heat bath and a particle store. In the microcanonical ensemble, we don't have any free parameters other than the Hamiltonian. In the canonical ensemble, there appears a variable of the heat bath, β, that models the heat exchange and, therefore, in the grand canonical ensemble, we expect there to be another parameter to model the particle exchange. We call it the **chemical potential** μ.

The grand canonical system is a little more tricky to model, since in addition to another variable μ, we no longer have a fixed number of phase space coordinates, so we have to explicitly choose to describe the system when it has N particles. The Hamiltonian for N particles will be the sum of single-particle Hamiltonians when the particles are non-interacting ideal gas particles. The grand canonical partition function Z_{G} is written in terms of the canonical partition function Z:

$$Z_{\text{G}} = \sum_{N=0}^{\infty} \frac{1}{h^n \cdot N!} e^{\beta \mu N} \int e^{-\beta H_N(\boldsymbol{\xi})}\, d\boldsymbol{q}\, d\boldsymbol{p} = \sum_{N=0}^{\infty} e^{\beta \mu N} Z \qquad (20.4.22)$$

where $n = 3N$. The probability density ρ_{G} is given by

$$\rho_{\text{G}} = \frac{1}{h^n \cdot N!} \frac{1}{Z_{\text{G}}} e^{\beta \mu N} e^{-\beta H_N(\boldsymbol{\xi})} \qquad (20.4.23)$$

where we include Z_{G} in the definition for later normalisation ease. In an ideal (non-interacting) gas, the canonical partition function is derived from $H = p^2/2m$, giving $Z = \frac{1}{N!}\left(\frac{V}{\Lambda^3}\right)^N$, where Λ is the thermal de Broglie wavelength. The grand canonical partition function is then

$$Z_G = \sum_{N=0}^{\infty} e^{\beta\mu N} \frac{1}{N!} \left(\frac{\mathcal{V}}{\Lambda^3}\right)^N = \sum_{N=0}^{\infty} \frac{1}{N!} \left[e^{\beta\mu} \frac{\mathcal{V}}{\Lambda^3}\right]^N = \exp\left\{e^{\beta\mu} \frac{\mathcal{V}}{\Lambda^3}\right\} \qquad (20.4.24)$$

Using this information, show that the average energy $\langle E \rangle$ and the average number of particles $\langle N \rangle$ are given by

$$\langle E \rangle = e^{\beta\mu} \frac{\mathcal{V}}{\Lambda^3} \frac{3}{2\beta} \qquad (20.4.25)$$

$$\langle N \rangle = \frac{\partial \ln Z}{\partial(\beta\mu)} = e^{\beta\mu} \frac{\mathcal{V}}{\Lambda^3} \qquad (20.4.26)$$

The grand canonical ensemble is an unnatural distribution on the phase space, since the dimensionality of the manifold is varying. This difficulty gives rise to some interesting equations where the h factor no longer cancels out! Most notable of this is the **Saha equation** describing ionisation populations of high-temperature atomic species, the classical physics of which leads to natural discretisation of the phase space.

Building on our previous work on the BBGKY hierarchy, we now develop a classical density functional theory. Consider a system of particles in a box of fixed volume $\mathcal{V}$ in contact with heat and particle baths and under the influence of an external one-particle potential V_{ext} (i.e it depends only on q^i). Consider a one-particle density function defined with the partition function present for later convenience (we also emit the subscripted "1" notation). We recall from section 20.1 that such a density function is normalised to unity:

$$\rho = \frac{1}{Z_G} e^{-\beta(H-\mu N)} \qquad (20.4.27)$$

With a brief rearrangement, we arrive at the following expression, of which we then take the average of denoted by $\Omega[\rho]$:

$$-\frac{1}{\beta} \ln Z_G = H - \mu N + \frac{1}{\beta} \ln \rho \qquad (20.4.28)$$

$$\Omega[\rho] = \sum_{N=0}^{\infty} \frac{1}{h^{3N} N!} \int dq dp (T + U + V_{\text{ext}} - \mu N + \frac{1}{\beta} \ln \rho)\rho \qquad (20.4.29)$$

This is then split up into two parts: the first is an auxiliary functional $\mathcal{F}[\rho]$, called the **Helmholtz free energy**, while the second consists of terms independent of the momentum:

$$\Omega[\rho] = \mathcal{F}[\rho] + \int dq (V_{\text{ext}} - \mu)\rho \qquad \text{where} \qquad \mathcal{F}[\rho] = \sum_{N=0}^{\infty} \frac{1}{h^{3N} N!} \int dq dp (T + U + \frac{1}{\beta} \ln \rho)\rho$$

$$(20.4.30)$$

where $\rho = \rho_0$ corresponds to the equilibrium density, and $\Omega[\rho]$ is minimised to a local minimum, with $\Omega[\rho] > \Omega[\rho_0]$. The Euler-Lagrange equation finds the extrema of functionals and therefore allow us to find a unique equilibrium probability density:

$$\left.\frac{\delta\Omega[\rho]}{\delta\rho}\right|_{\rho=\rho_0} = \left.\frac{\delta\mathcal{F}[\rho]}{\delta\rho}\right|_{\rho=\rho_0} + V_{\text{ext}} - \mu = 0 \qquad (20.4.31)$$

In this way, the two functionals $\Omega[\rho_0]$ and $\mathcal{F}[\rho_0]$ give us a complete knowledge of the system. Unfortunately, finding the exact functional derivative of the Helmholtz free energy is rather tricky and, in practice, and needs to be approximated by **perturbative techniques** or **mapping schemes**. The method outlined here was developed by N. Mermin and R. Evans soon after the original Hohenberg-Kohn theorems for coarse-grained studies of liquids. For a recent survey of classical density functional theory see Evans *et al.*'s (2016) article, 'New developments in classical density functional theory'.

Exercise 20.5 We are aware that the partition function for free translations of a single particle is given by

$$Z = \frac{\mathcal{V}}{h^{3n}}\left(\frac{2\pi m}{\beta}\right)^{3/2} \tag{20.4.32}$$

We can compute the average energy:

$$\langle E \rangle = -\frac{\partial}{\partial \beta}\log Z = \frac{\partial}{\partial \beta}\left(\frac{3}{2}\log \beta\right) = \frac{3}{2\beta} \tag{20.4.33}$$

The result is in agreement with the equipartition theorem. The harmonic oscillator partition function was agreed to be

$$Z = \frac{2\pi}{h^n \omega \beta}, \qquad \text{with} \qquad \omega = \sqrt{\frac{k}{\mu}} \tag{20.4.34}$$

with $H = p^2/2m + m\omega^2 q^2/2$. Application of the equipartition theorem results in $\langle E \rangle = 1/\beta$. Show that the same result is obtained by considering $\langle H \rangle$, firstly with $\langle \frac{1}{2}m\omega^2 q^2 \rangle = 1/2\beta$ and then with $\langle p^2/2m \rangle = 1/2\beta$, using the ensemble average formula.

To begin, recall the indefinite Gaussian integrals $\int dx \exp\{-x^2\} = \sqrt{\pi}$ and $\int dx \exp\{-x^2\}x^2 = \sqrt{\pi}/2$, we obtain

$$\left\langle \frac{p_i^2}{2m} \right\rangle = \frac{\int \prod dp(\frac{p^2}{2m})e^{-\beta\frac{p^2}{2m}}}{\int \prod_i dp_i e^{-\beta\frac{p^2}{2m}}} = \frac{1}{2\beta} \tag{20.4.35}$$

Similarly,

$$\left\langle \frac{1}{2}m\omega^2 q^2 \right\rangle = \frac{1}{2}m\omega^2\frac{\int \prod dq(q^2)e^{-\beta\frac{1}{2}m\omega^2 q^2}}{\int \prod dq e^{-\beta\frac{1}{2}m\omega^2 q^2}} = \frac{1}{2\beta} \tag{20.4.36}$$

Then we see that $\langle H \rangle = 1\beta$, by adding both terms together.

Exercise 20.6 Let us prove that the canonical partition function $Z_C(\beta)$ is given by the Laplace transform of the microcanonical partition function $Z_M(E)$. The result of a **Laplace transform** of a function $f(x)$ is another function $F(\beta)$:

$$F(\beta) = \int dx e^{-\beta x} f(x) \tag{20.4.37}$$

We start from the following:

$$\frac{\partial \ln Z_M}{\partial E} = \beta \tag{20.4.38}$$

Then $\partial_E(\ln Z_M - \beta E) = 0$ and the Legendre transform of $\ln Z_M$ is $\ln Z_C$. Hence we have $\ln Z_M - \beta E = \ln Z_M$, which upon exponentiation and integration over E becomes:

$$Z_C(\beta) = \int Z_M(E)e^{-\beta E}dE. \tag{20.4.39}$$

In fact, it is a general result of mathematics that, if one takes a **saddle point approximation**, then the Laplace and Legendre transforms have the following intimate relationship:

$$F(\beta) = \int dx f(x)e^{-x\beta} = \int dx \exp\left\{ \ln[f(x)] - x\beta \right\} \tag{20.4.40}$$

In the saddle point approximation, the integral is evaluated at the point where it is a maximum. Physically, this is not too bad an approximation, as the system will settle into one particular state at equilibrium, so the approximation becomes more valid the larger the number of particles in the ensemble. The maximum of the integrand is found by setting the first derivative to zero:

$$\frac{d}{dx}\left[\ln[f(x)] - x\beta \right] = 0 \tag{20.4.41}$$

With this approximation, we find $\frac{d}{dx}\ln[f(x)]_{x_0} = \beta$.

Exercise 20.7 Here, we will investigate the nature of the classical state in more depth. We know that the algebra of smooth commuting observables $\mathcal{C}^\infty(M)$ on the phase space M forms a Lie algebra under the Poisson bracket $\{\cdot, \cdot\} : \mathcal{C}^\infty(M) \times \mathcal{C}^\infty(M) \to \mathcal{C}^\infty(M)$. A **classical state** σ assigns to each observable $F \in \mathcal{C}^\infty(M)$ a probability distribution or **measure** on the reals $\mathbb{R}$; we denote this by $0 \le \sigma_F(E) \le 1$ for a set E on $\mathbb{R}$. If the state can be written as a convex combination of probability measures, then it is a **mixed state**; otherwise, it is a **pure state**. The expectation value of an observable F in state σ is given by $\langle F_\sigma \rangle$:

$$\langle F_\sigma \rangle = \int F(q,p)\rho_\sigma(q,p)dqdp \tag{20.4.42}$$

where $\rho_\sigma(q,p)$ is a density function and $\rho dqdp$ is a measure on M and is normalised to unity. The pure state is a single point in phase space (e.g it is a Dirac delta function at a particular point (q',p')):

$$\rho(q,p) = \delta(q - q')\delta(p - p') \tag{20.4.43}$$

Hence, for a pure state, we can write

$$F(q',p') = \langle F_\sigma \rangle = \int F(q,p)\delta(q - q')\delta(p - p')dqdp \tag{20.4.44}$$

A simple mixed state σ could be a linear combination of pure states σ_1 and σ_2 with weighting coefficients c_i that sum to 1:

$$\rho(q,p) = c\delta(q - q_1)\delta(p - p_1) + (1 - c)\delta(q - q_2)\delta(p - p_2) \tag{20.4.45}$$

More complex mixed states involve higher-order linear combinations. The fluctuation of the probability density function $(\Delta F_\sigma)^2$ (as defined in appendix G) is zero exactly for pure states and increases with mixing, which you may wish to prove.

Exercise 20.8 Show that the ensemble average for a phase space function A can be recovered from the Koopman-von Neumann formulation:

$$\langle A \rangle = \frac{\langle \psi | A(\hat{x}, \hat{p}) | \psi \rangle}{\langle \psi | \psi \rangle} = \frac{1}{Z} \int A(\boldsymbol{\xi}) \rho(\boldsymbol{\xi}, t) d\mathbf{q} d\mathbf{p} \qquad (20.4.46)$$

Exercise 20.9 Consider the following action for a system of N particles:

$$A = \frac{1}{T} \int_1^2 dt \left\{ \frac{1}{2} \sum_{i=1}^N m_i (\dot{q}^i)^2 - U(q^1, \dots q^N) \right\} \qquad (20.4.47)$$

where T is a time normalisation factor $(2 - 1)$. Derive the virial theorem

$$\langle m_i (\dot{q}^i)^2 \rangle - \langle q^i \cdot \frac{\partial U}{\partial q^i} \rangle = 0 \qquad (20.4.48)$$

from this action using a linear transformation $q^i \to q^i(1 + \epsilon)$; note the following time average notation:

$$\langle * \rangle := T^{-1} \int_1^2 * dt \qquad (20.4.49)$$

21

Constrained Hamiltonian Dynamics

Fig. 21.1: Paul Dirac (1902–1984).

The standard approach to Hamiltonian mechanics is to treat all the variables as being independent, write a Lagrangian for the system, perform a Legendre transform to the Hamiltonian, and then solve the canonical equations. If all the generalised velocities can be inverted, we say that the Lagrangian is **regular**. In the constrained case, we have a constraint function ϕ linking the variables so they are no longer independent. Dirac's theory of constrained Hamiltonians, also known as **Dirac-Bergmann theory**, applies to those situations where we cannot invert all of the generalised velocities, that is, when we have what a mathematician would call a **singularity**, and the implicit function theorem no longer holds. (For a picture of Paul Dirac, see figure 21.1.) The generalised velocities which we can't invert can be treated as constraints. If we recall our discussion in chapter 8, we concluded that the constraints force the system to lie on a **constraint surface**, restricting the allowed areas of phase space.

To begin we consider the Euler-Lagrange equation and expand the time derivative:

$$\frac{d}{dt}\left(\frac{\partial \mathscr{L}(\dot{q}, q)}{\partial \dot{q}^i}\right) - \frac{\partial \mathscr{L}}{\partial q^i} = \left(\frac{\partial^2 \mathscr{L}}{\partial \dot{q}^j \partial \dot{q}^i}\ddot{q}^i + \frac{\partial^2 \mathscr{L}}{\partial \dot{q}^j \partial q^i}\dot{q}^i\right) - \frac{\partial \mathscr{L}}{\partial q^j} = 0 \qquad (21.0.1)$$

The aim of Lagrangian mechanics is to solve for the accelerations $\ddot{q}$; this is only possible if the term in square brackets is non-zero:

$$\left[\frac{\partial^2 \mathscr{L}}{\partial \dot{q}^j \partial \dot{q}^i}\right]\ddot{q}^i = \frac{\partial \mathscr{L}}{\partial q^j} - \sum_i \frac{\partial^2 \mathscr{L}}{\partial \dot{q}^j \partial q^i}\dot{q}^i \qquad (21.0.2)$$

We call the term in the square brackets the **Hessian**[1]. The component-wise Hessian is actually a square $n \times n$ array of second order derivatives, that is to say, it is a matrix $\boldsymbol{W}$; we denote it in components by:

---

[1]You can also think of it as the Jacobian of the conjugate momenta, but in literature it is known as the Hessian condition.

Lagrangian & Hamiltonian Dynamics. Peter Mann, Oxford University Press (2018).
© Peter Mann. DOI: 10.1093/oso/9780198822370.001.0001

$$W_{ji} = \frac{\partial^2 \mathscr{L}}{\partial \dot{q}^j \partial \dot{q}^i} \qquad \textbf{Hessian condition} \qquad\qquad (21.0.3)$$

The matrix only has an inverse $(\boldsymbol{W})^{-1}$ if its determinant is non-zero $(\det(\boldsymbol{W}) \neq 0)$, that is to say it's rank is n. What this means is that we can only solve for the accelerations if the Hessian is invertible. Therefore, we can solve Lagrange's equation of motion and invert the generalised velocity if and only if the Hessian determinant is non-zero; an awful lot depends on this condition being met!

If the condition is not met, then we describe the Lagrangian as having a singularity, and some of the variables are linearly dependant, meaning we can not find a unique solution. In this case, some of Lagrange's equations correspond to constraints on what values the qs and $\dot{q}$s can take, restricting the accessible region of phase space to the system. When this is the case, there will exist relations between the linearly dependent variables, called **constraint equations** $(\phi(\boldsymbol{p}, \boldsymbol{q}, t) = 0)$, one for each un-invertible velocity.

This indicated to Dirac that there were more general equations to be found to describe canonical mechanics and we need to modify the Hamiltonian to account for these constraint equations. In direct analogy to the Lagrangian instance, we add each of the constraint equations multiplied by an undetermined Lagrange multiplier λ:

$$\mathscr{H}(\boldsymbol{p}, \boldsymbol{q}, \boldsymbol{\lambda}) = H(\boldsymbol{q}, \boldsymbol{p}) + \sum_k \lambda_k \phi_k(\boldsymbol{q}, \boldsymbol{p}) \qquad\qquad (21.0.4)$$

Note that H is the Hamiltonian we obtain if we perform a Legendre transform to the Lagrangian of incomplete rank (some velocities that can't be inverted): it is sometimes called the **canonical Hamiltonian**.

To derive our equations of motion, we can use the stationary action principle with vanishing boundary points. The Lagrangian is given as usual by $\mathscr{L} = p_i \dot{q}^i - \mathscr{H}$:

$$\delta S = \delta \int_{t_0}^{t_1} \left[p_i \dot{q}^i - H(\boldsymbol{q}, \boldsymbol{p}) - \lambda_n \phi_n(\boldsymbol{q}, \boldsymbol{p}) \right] dt$$

$$= \int_{t_0}^{t_1} \left[\delta p_i \dot{q}^i + p_i \delta \dot{q}^i - \delta H(\boldsymbol{q}, \boldsymbol{p}) - \delta \lambda_n \phi_n - \lambda_n \delta \phi_n(\boldsymbol{q}, \boldsymbol{p}) \right] dt$$

The highlighted term is an integration by parts, with vanishing terms evaluated at the boundary points:

$$\int_{t_0}^{t_1} p_i \delta \dot{q}^i dt = p_i \delta q^i \Big|_{t_0}^{t_1} - \int_{t_0}^{t_1} \dot{p}_i \delta q^i dt$$

We then substitute this in and collect like variations:

$$\delta S = \int_{t_0}^{t_1} \left[\delta p_i \dot{q}^i - \dot{p}_i \delta q^i - \frac{\partial H}{\partial q^i} \delta q^i - \frac{\partial H}{\partial p_i} \delta p_i - \delta \lambda_n \phi_n - \lambda_n \left(\frac{\partial \phi_n}{\partial q^i} \delta q^i + \frac{\partial \phi_n}{\partial p_i} \delta p_i \right) \right] dt$$

$$= \int_{t_0}^{t_1} \left[\left(\dot{q}^j - \frac{\partial H}{\partial p_j} - \lambda_n \frac{\partial \phi_n}{\partial p_j} \right) \delta p_j + \left(-\dot{p}_j - \frac{\partial H}{\partial q^j} - \lambda_n \frac{\partial \phi_n}{\partial q^j} \right) \delta q^j - (\phi_n) \delta \lambda_n \right] dt$$

When we impose the stationary action principle, the variations vanish independently and each bracket is equal to zero, giving us three equations describing the time evolution. If we had performed the action principle without the Lagrange multipliers, then the variations wouldn't be independent. We thus obtain

$$\dot{q}^i = \frac{\partial H}{\partial p_i} + \lambda_n \frac{\partial \phi_n}{\partial p_i}, \qquad \dot{p}_i = -\frac{\partial H}{\partial q^i} - \lambda_n \frac{\partial \phi_n}{\partial q^i}, \qquad \phi_n = 0$$

which tidies up nicely if we substitute for $\mathscr{H}$ from above:

$$\boxed{\dot{q}^i = \frac{\partial \mathscr{H}}{\partial p_i}, \qquad \dot{p}_i = -\frac{\partial \mathscr{H}}{\partial q^i}, \qquad \phi_n = 0} \tag{21.0.5}$$

Importantly, the addition of the multiplier means that the solution is not unique, since, if $H + \lambda \phi$ satisfies the equations of motion, then so will $H + \lambda' \phi$, although the underlying physics is the same, our descriptions are not. We can take the time derivative of a function $F(q, p)$:

$$\dot{F} = \frac{\partial F}{\partial q^j} \dot{q}^j + \frac{\partial F}{\partial p_j} \dot{p}_j$$

$$= \frac{\partial F}{\partial q^j} \left(\frac{\partial H}{\partial p_j} + \lambda_n \frac{\partial \phi_n}{\partial p_j} \right) + \frac{\partial F}{\partial p_j} \left(-\frac{\partial H}{\partial q^j} - \lambda_n \frac{\partial \phi_n}{\partial q^j} \right)$$

$$= \left(\frac{\partial F}{\partial q^j} \frac{\partial H}{\partial p_j} - \frac{\partial F}{\partial p_j} \frac{\partial H}{\partial q^j} \right) + \lambda_n \left(\frac{\partial F}{\partial q^j} \frac{\partial \phi_n}{\partial p_j} - \frac{\partial F}{\partial p_j} \frac{\partial \phi_n}{\partial q^j} \right) \tag{21.0.6}$$

We can now write this more neatly by making use of Poisson bracket notation:

$$\dot{F} = \{F, H\} + \sum_n \left(\lambda_n \{F, \phi_n\} \right) \tag{21.0.7}$$

If we derive this from our fundamental axioms of the Poisson bracket, then we obtain a slightly different result, $\dot{F} = \{F, \mathscr{H}\}$:

$$\dot{F} = \{F, H + \lambda_n \phi_n\}$$

$$= \{F, H\} + \{F, \lambda_n \phi_n\}$$

$$= \{F, H\} + \{F, \lambda_n\} \phi_n + \lambda_n \{F, \phi_n\} \tag{21.0.8}$$

However, just above we said that, when the constraint equations are satisfied: $\phi_n = 0$; we assumed this implicitly in the stationary action principle, which gives us the equations of motion directly. However, you should be aware of this extra term.

Dirac now imposes a condition upon the constraints that, on the constraint surface their time derivatives are zero; they are constant on the submanifold. It does not hold in the general phase space; only when the system is restricted to the submanifold are they satisfied, and they do not evolve in time. This is known as a **consistency condition** and is denoted $\dot{\phi} \approx 0$, where the $\approx$ notation does not signify 'approximately' but means equality when on the constraint surface.

Dirac uses the consistency conditions to derive a generalised bracket to describe constrained motion; this is called the *Dirac bracket*. If we let the function be the constraint, then we know on the constraint surface the time derivative vanishes. Let $F = \phi$, then, $\dot{\phi} \approx 0$:

$$\{\phi_n, H\} \approx -\sum_l \{\phi_n, \phi_l\} \lambda_l$$

We can solve for the undetermined multipliers λ_n if and only if the **constraint matrix** $\{\phi_n, \phi_l\}$ has an inverse $(\{\phi_n, \phi_l\})^{-1}$, or, if you prefer its determinant is non-zero. Since we cannot assume this, we define two types of constraints: those which have an inverse, and those that don't. The nomenclature of these suffers from horrendous confusion. We have, in total, four types of constraints (two of which are not related directly to the constraint matrix), which are termed primary, secondary, first-class and second-class constraints.

- A **first-class constraint** is one whose Poisson bracket with all the other constraints vanishes on the constraint surface. First-class constraints have a column of zeros in their constraint matrix, meaning the inverse cannot be taken. Later, we will understand that the **symplectic 2-form** is **degenerate** when its inverse doesn't exist.
- A **second-class** constraint is one that is not a first-class constraint.
- A **primary constraint** is a function of q and p and holds without using the equations of motion.
- A **secondary constraint** is one that is not primary; it need not hold off-shell.

First- and second-class constraints can be either primary or secondary. In this elementary text, we will focus on secondary constraints[2]; ones that have a non-zero constraint matrix, and we can take their inverse and solve for the undetermined multipliers:

$$\lambda_n = -\sum_l \left(\{\phi_n, \phi_l\}\right)^{-1} \{\phi_l, H\}$$

With these solved, we plug them back into the general expression for $\dot{F}$; they aren't going to change just because we choose to study a different function. We thus obtain

---

[2]First-class constraints are canonical gauge theories and are highly important for canonical quantisation procedures. First-class constraints generate gauge transformations.

$$\dot{F} = \{F, H\} + \sum_n \left(\{F, \phi_n\} \lambda_n \right)$$

$$= \{F, H\} - \sum_{n,l} \left(\{F, \phi_n\} (\{\phi_n, \phi_l\})^{-1} \{\phi_l, \mathcal{H}\} \right) \qquad (21.0.9)$$

This is the **Dirac bracket** of F, with the Hamiltonian $\mathcal{H}$ denoted $\{F, \mathcal{H}\}_{\text{DB}}$ and it describes the time evolution of the function on the submanifold; for certain scenarios, we are free to write $\{F, H\}_{\text{DB}}$ (see exercise 21.1). We can now say, just like we did with the Poisson bracket, that the bracket applies equally for functions of phase space F and G, giving us the final expression

$$\{F, G\}_{\text{DB}} = \{F, G\}_{\text{PB}} - \sum_{n,l} \{F, \phi_n\}_{\text{PB}} (\{\phi_n, \phi_l\})^{-1} \{\phi_l, G\}_{\text{PB}} \qquad (21.0.10)$$

We call this the Dirac bracket for F with the function G. For two functions in the larger manifold, the Dirac bracket describes their motion in the submanifold defined by the image of the constraint surface under the Legendre transform. Their motion in the subsurface is written in terms of their Poisson brackets in the larger manifold; in this way, we can either extend or reduce motion to higher or lesser dimensions. As with the Poisson bracket, if $G = \mathcal{H}$, the Dirac bracket describes the time derivative of F.

The Dirac bracket has the Lie algebra axiomatic rules of the Poisson bracket, including antisymmetry, the Leibniz identity and the Jacobi identity, although it does not have the fundamental brackets like the Poisson bracket. To be more formal, for m second-class constraints on a $2n$ dimensional phase space, we would say that the Dirac bracket is a non-invertible Poisson structure of dimension $2(n - m)$ on the larger manifold.

If we take the Dirac bracket of a constraint and any arbitrary function of the phase space, the bracket will vanish (e.g let $F = \phi$, and we get $\{\phi, G\} - \{\phi, G\} = 0$), indicating that the flow is constrained to a submanifold of the phase space described by the Poisson bracket.

Importantly, since the constraints are always satisfied for any physical motion of the system, it is precisely these subsurfaces that correspond to the physical phase space we consider normally in Hamiltonian dynamics. It is the aim of Dirac-Bergmann theory to find this subsurface upon which we can treat the problem with our existing canonical equations and Poisson brackets and this is achieved via the Dirac bracket where possible.

Note, the discussion of constraints in the Dirac formalism is independent of holonomic versus non-holonomic constraints that appear in the Lagrangian formulation. The Dirac bracket can still be constructed for non-holonomic Lagrangian constraints, since the phase space does not depend on the $\dot{q}$ coordinate. Constraints in the Hamiltonian formulation instead arise due to singularities (as we have discussed), or via gauge transformations (as we have not discussed).

The Hessian is actually the determinant of the Jacobian of the Legendre transformation $(q^j, \dot{q}^j) \mapsto (q^i, p_i)$ and this is why we cannot invert velocities for conjugate momenta when it is singular. In this case, some of the generalised accelerations are not uniquely determined by their positions and velocities at a given instant of time and the rank of the Hessian represents the invertible velocities:

$$\frac{\partial(q^i, p_i)}{\partial(q^j, \dot{q}^j)} = \begin{pmatrix} \frac{\partial q^i}{\partial q^j} & \frac{\partial p_i}{\partial q^j} \\ \frac{\partial q^i}{\partial \dot{q}^j} & \frac{\partial p_i}{\partial \dot{q}^j} \end{pmatrix} = \begin{pmatrix} \delta^{ij} & \frac{\partial^2 \mathscr{L}}{\partial q^j \partial \dot{q}^i} \\ 0 & \frac{\partial^2 \mathscr{L}}{\partial \dot{q}^j \partial \dot{q}^i} \end{pmatrix}$$

The Jacobian is non-singular if and only if the Hessian condition holds. In this case, the Lagrangian is said to be regular:

$$\det\left(\frac{\partial^2 \mathscr{L}}{\partial \dot{q}^j \partial \dot{q}^i}\right) \neq 0 \tag{21.0.11}$$

The discussion of constrained Hamiltonian dynamics will be continued when we discuss the *Hamilton-Nambu formulation* in appendix D and we comment on some modern methods in the theory of constrained dynamics in chapter 22.

We mention that a simplification of the Dirac-Bergmann formalism for first order Lagrangians (not quadratic) was introduced by R. Jackiw and L. Faddeev in the late 1980s. The **Jackiw-Fadeev formulation** tackles singular Lagrangians without the need to classify first- or second-class constraints but still gives equivalent equations of motion. The main usage of any such constraint theory is focused towards canonical quantisation, that is to say, turning classical systems into quantum ones; in fact this was the motivation for Dirac's theory, too! In general, any second-order Lagrangian $\mathscr{L}(q, \dot{q})$ can be reduced to first order by Legendre transforming to $\mathscr{L}(q, p, \lambda, \dot{q}, \dot{p}, \dot{\lambda}) = p\dot{q} - H(q, p, \lambda)$. In this way, the phase space Lagrangian will have crossed terms $p\dot{q}$ rather than second-order terms in $\dot{q}^2$; hence, the Jackiw-Fadeev formalism is very useful; for more information see L. Faddeev and R. Jackiw' (1988) *Hamiltonian Reduction of Unconstrained and Constrained Systems*.

Finally, we investigate **Güler's method** for a constrained Hamilton-Jacobi theory; unlike the Jackiw-Faddeev approach, forces us to distinguish between the type of constraints we use. In the case of a non-singular Lagrangian $\mathscr{L}(q^i, \dot{q}^i)$ with $i = 1, \dots, n$, the unconstrained Hamilton-Jacobi equation is written

$$H_0(t, q^i, p_i) + \frac{\partial S}{\partial t} = 0, \qquad p_i = \frac{\partial S}{\partial q^i} \tag{21.0.12}$$

according to chapter 19. The principal function is then found as a solution to this equation. For singular systems, the determinant of the Hessian vanishes and some of the generalised velocities cannot be inverted; they are dependent. Let us suppose that the first $b = 1, \ldots, m$ velocities cannot be inverted but the remaining $a = m + 1, \ldots, n$ can be: $\dot{q}_a = \dot{q}_a(q^i, p_a, \dot{q}^b)$; the index i runs from 1 to n. The momenta corresponding to the dependent velocities $p_b = p_b(q^i, p_a)$ are functions of all the positions and the inverted momenta. We then impose the primary constraints of $\phi_b(q^i, p_i) = 0$, where

$$\phi_b = p_b - \left. \frac{\partial \mathscr{L}}{\partial \dot{q}^b} \right|_{\dot{q}^a = \dot{q}^a(q^i, \dot{q}^b, p_a)} = p_b + H_b \qquad (21.0.13)$$

We then obtain a set of $(m + 1)$ coupled differential equations:

$$\frac{\partial S}{\partial t} + H_0\left(t, q^i, \frac{\partial S}{\partial q^a}\right) = 0, \qquad \frac{\partial S}{\partial q^b} + H_b\left(t, q^i, \frac{S}{\partial q^a}\right) = 0 \qquad b = 1, \ldots m \quad (21.0.14)$$

It is customary to split the configuration space into two parts: $q^I = q^A + q^B$, where q^A corresponds to all the q^i variables that are independent, and q^B corresponds to all the dependent variables. The second of these equations can be rewritten as

$$H'_b = \frac{\partial S}{\partial q^b} + H_b\left(t, q^i, \frac{S}{\partial q^a}\right) = 0 \qquad (21.0.15)$$

The system is then said to be integrable if the commutator of any two of these equations vanishes. If any of the H'_b do not vanish under variations (e.g. $dH'_b = 0$) then this becomes a new constraint and the procedure is rewritten until the variations $[H'_\alpha, H'_\beta] \; \forall \; \alpha, \beta \in B$, vanish and the algebra of Hamiltonians closes. The characteristic differential equations of an integrable system can be solved via quadrature in the usual fashion, so the integrability criterion becomes important. It is unfortunate, however, that this is not valid for secondary, second-class constraints since the Hamiltonians H'_b are not in involution. A possible shortcut is to embed the second-class system into a larger first-class phase space through a **BFT embedding procedure**, after Batalin, Fradkin and Tyutin, thereby introducing gauge freedom to the system.

We have followed Dirac's (1964) *Lectures on Quantum Mechanics*, which has provided the staple of our discussion but, for more information, I would suggest the seminal work on the subject, *Classical and Quantum Dynamics of Constrained Hamiltonian Systems* by J. J. Rothe and K. D Rothe.

Chapter summary

- **Singularities** in the **Hessian** prevent the Legendre transformation of the Lagrangian to the Hamiltonian formalism. The Lagrange multiplier method is extended to the Hamiltonian, which was shown to be non-unique, in order to develop **constrained canonical** equations.

- The **Dirac bracket**, which describes the time evolution of constrained Hamiltonians, and replaces the inadequate Poisson bracket, was derived. The notion of **constraint submanifolds** was introduced as well as the restriction of the system's motion to that surface while constrained.

- **Consistency conditions** were introduced, as well as **first-class** and **second-class** constraints.

- We then investigated **Güler's method** for a constrained Hamilton-Jacobi theory and briefly consider the **Jackiw-Faddeev formalism**.

Exercise 21.1 In this example, we will revisit example 8.1, where we discussed the dynamics of a particle confined to the surface of a sphere. We are aware that a naive construction of Hamilton's equations without reference to the constraint will not be sufficient to describe the system. The unconstrained Lagrangian in spherical coordinates is given by

$$\mathscr{L} = \frac{1}{2}m(\dot{r}^2 + r^2\dot{\theta}^2 + r^2\dot{\varphi}^2 \sin^2\theta) - V(r,\theta,\varphi) \tag{21.0.16}$$

We can perform a Legendre transform to the canonical Hamiltonian H to give the following:

$$H = \frac{1}{2m}\left(p_r^2 + \frac{p_\theta^2}{r^2} + \frac{p_\varphi^2}{r^2\sin^2\theta}\right) + V(r,\theta,\varphi) \tag{21.0.17}$$

We then introduce the constraint equation $\phi_1 = (r - l)$, which fixes the radius of the sphere. From this, we generate a further constraint to ensure consistency conditions are met:

$$\phi_2 = \dot{\phi}_1 = \{\phi_1, H\} \tag{21.0.18}$$

Importantly, we can see that we have second-class constraints by the non-vanishing Poisson bracket:

$$\{\phi_1, \phi_2\} = \{\phi_1, \{\phi_1, H\}\} \neq 0 \tag{21.0.19}$$

The overall Hamiltonian $\mathscr{H}$ is then constructed to account for the constraint equations and the undetermined multipliers λ_i, are introduced:

$$\mathscr{H} = H + \lambda_1(r - l) + \lambda_2(p_r/m) \tag{21.0.20}$$

The motion of the system constrained to the submanifold is given by the following Dirac brackets:

$$\dot{q}^i = \{q, H\}_{\text{DB}}, \quad \text{and} \quad \dot{p}_i = \{p_i, H\}_{\text{DB}} \tag{21.0.21}$$

In this example (second-class, secondary constraints), use of either $\mathcal{H}$ or H will give the same result. Constructing the Dirac brackets requires a fairly straightforward but quite fiddly set of calculations:

$$\{r, H\}_{\text{DB}} = \{r, H\} + \frac{\{r, \phi_1\}\{\{\phi_1, H\}, H\}}{\{\phi_1, \{\phi_1, H\}\}} - \frac{\{r, \{\phi_1, H\}\}\{\phi_1, H\}}{\{\phi_1, \{\phi_1, H\}\}} \tag{21.0.22}$$

For the r coordinate, we realise that the Dirac bracket vanishes, indicating that $\dot{r} = 0$, a result that ties in with exercise 8.1.

Exercise 21.2 Consider the following Lagrangian for a free particle that has two coordinates (x, y):

$$\mathcal{L} = y\dot{x} - \frac{y^2}{2m} \tag{21.0.23}$$

By considering the conjugate momentum to each coordinate, show that $p_x = y$ and $p_y = 0$. Try to construct a Hamiltonian of the form:

$$H = p_x\dot{x} + p_y\dot{y} - \mathcal{L} \tag{21.0.24}$$

Upon realising that we cannot invert p_x and p_y for $\dot{x}$, construct a primary constraint that allows the Hamiltonian to be written as

$$H \approx \frac{p_x^2}{2m} \tag{21.0.25}$$

on the primary constraint submanifold.

Exercise 21.3 Consider the following Lagrangian $\mathcal{L}$ with the primary constraint $\phi_1 = p_y = 0$, and show that the secondary constraint is given by $\phi_2 = y - \frac{1}{2}(p_x + x)$ and that together the constraints are second-class:

$$\mathcal{L} = \frac{1}{2}\dot{x}^2 + \dot{x}y - \frac{1}{2}(x - y)^2 \tag{21.0.26}$$

The solution involves finding the equations of motion, the Hamiltonian, producing the secondary constraint from consistency conditions $\dot{\phi}_1 = 0$ and then taking the Poisson bracket of both constraints to show that it is non-vanishing.

The equations of motion for the x and y coordinates are given by,

$$\ddot{x} + \dot{y} + (x - y) = 0 \quad \text{and} \quad \dot{x} + (x - y) = 0 \tag{21.0.27}$$

It is good practice to write the conjugate momentum at the same time: $\partial\mathcal{L}/\partial\dot{x} = \dot{x} + y$ and $\partial\mathcal{L}/\partial\dot{y} = 0$. By taking the time derivative of the second one, we can show that $\ddot{x} = \dot{y} - \dot{x}$

which is entered into the second equation of motion to give $2\ddot{x} = 0$ and, hence, $\ddot{x} = 0$. The Hamiltonian is constructed in the usual manner:

$$H = \frac{1}{2}(p_x - y)^2 + \frac{1}{2}(x - y)^2 \qquad (21.0.28)$$

The secondary constraint is then found using $\{\phi_1, H\} = 0$:

$$\{\phi_1, H\} = \frac{\partial p_y}{\partial y}\frac{\partial H}{\partial p_y} - \frac{\partial p_y}{\partial p_y}\frac{\partial H}{\partial y}$$

$$= -2(y - \frac{1}{2}(p_x + x)) \qquad (21.0.29)$$

The constraints are shown to be secondary by evaluation of $\{\phi_2, \phi_1\} = $ non-zero:

$$\{\phi_2, \phi_1\} = \frac{\partial \phi_2}{\partial y}\frac{\partial \phi_1}{\partial p_y} - \frac{\partial \phi_2}{\partial p_y}\frac{\partial \phi_1}{\partial y} = 1 \qquad (21.0.30)$$

22

Autonomous Geometrical Mechanics

Geometry is at the heart of mathematical physics and is key to understanding classical mechanics in depth. The underlying framework of differential geometry is explored in chapter 38, it is assumed the reader will be familiar with that content before proceeding. We will develop the stationary action principle in a geometrical setting before defining a 1-form θ_H which encapsulates classical mechanics through various operations. We will look at *symplectic manifolds* and their generalisation as leaves in a *Poisson manifold*. In this chapter, it is hoped that the reader will fully understand the importance of the Hessian condition to solving the dynamics, the phase space action and the Legendre transform, as well as many more unanswered questions. The geometrical formalism was developed by V. Arnold, K. Meyer, J. E. Marsden, S. Smale, R. Abraham and A. Weinstein, among others in the mid-1970s.

The position data of a system is defined by a point on a smooth[1], n-dimensional manifold M whose n localised coordinates are called the **generalised coordinates** $[q^i]$,[2] with one dimension for each degree of freedom; it is the set of all allowed positions.

The *state* of a system is defined by two pieces of information, namely a point on $p \in M$ and a tangent vector to that point, $X_p \in T_pM$. Lagrangian mechanics is formulated therefore on the **tangent bundle** to a **configuration manifold**, which we have been calling velocity space. A **Lagrangian** $\mathscr{L}$ is a smooth function on TM and is written $\mathscr{L} : TM \mapsto \mathbb{R}$, mapping to the reals; the configuration manifold and tangent bundle need not have a metric specified, and it is limiting the power of the Lagrangian formalism to assume so.

Placing an atlas $\{\varphi_i : U_i \mapsto \mathbb{R}^n\}$ for each of the open sets on M induces a chart on the tangent $\pi : TM \to M$ and cotangent $\pi^* : T^*M \to M$ bundles to M (see figure 22.1). If the local coordinates on M are given by $[q^i] = (q^1, \ldots, q^n)$ defined over a smooth time line $t \in \mathbb{R}$ such that $q : \mathbb{R} \to M$, then the induced coordinates on the tangent bundle are $[q^i, v^i] = (q^1, \ldots, q^n, v^1, \ldots, v^n)$ and those of the cotangent bundle are $[q^i, p_i] = (q^1, \ldots, q^n, p_1, \ldots, p_n)$, meaning that the cotangent bundle is a phase space.

---

[1]Manifolds that correspond to dynamical systems are generally always differentiable, so we can usually assume smoothness as a given.

[2]We previously used curly brackets in the maths chapter $\{q^i\}$ but this might get confusing with the Poisson bracket later on!

Lagrangian & Hamiltonian Dynamics. Peter Mann, Oxford University Press (2018).
© Peter Mann. DOI: 10.1093/oso/9780198822370.001.0001

A point in TM is specified using the local coordinates of point p by

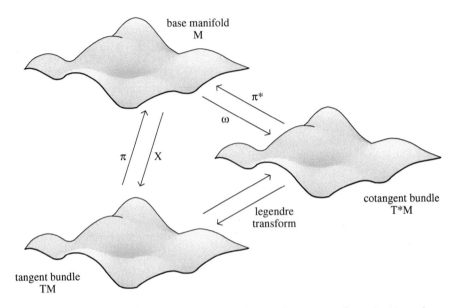

Fig. 22.1: All three spaces used within the geometrical picture of mechanics: the configuration manifold, the tangent bundle and the cotangent bundle, the latter are linked via the Legendre transform.

$(q^1, \ldots, q^n, v^1, \ldots, v^n)$, with $q, v \in T_pM$ representing the generalised position and velocity to point $p \in M$. In this chapter, we will consider autonomous systems despite the reduction in generality. We refer to the time-dependant product spaces $T^*M \times \mathbb{R}$ and $\mathbb{R} \times TM$ as an **extended phase space** and **extended velocity space**, which we discuss in chapter 23. We begin by looking at the calculus of variations in a more abstract setting; this topic is rather distinct from those discussed in the remainder of the chapter.

An **observable** is a real-valued function $F(q, p)$ on the phase space $F : T^*M \to \mathbb{R}$, sending a point in phase space to a numerical value. The set of all smooth functions on any manifold M is given by $\mathcal{C}^\infty(M)$, which forms a **commutative algebra**, that is to say, associative, commutative (e.g a product $(F, G) = FG$), satisfying the Leibniz condition as well as scalar multiplication. If and only if the set of smooth functions is equipped with a Lie bracket that is bilinear, antisymmetric and satisfies the Jacobi relation then we say it is a **Lie algebra**. If it has both a commutative algebra and a Lie bracket, then we call it a **Poisson algebra** structure. The Poisson structure of a canonical system should also be retained under time evolution (e.g. be an automorphism).

Incidentally, in quantum mechanics, we note that observables are not commutative and therefore we require a modification to the study of non-commutative rings.

A physical quantity can be calculated by evaluating the observable function at a point in phase space. For instance, the energy of the system in a **pure state** (q_0, p_0) is computed by evaluating the Hamiltonian at the point (q_0, p_0) in phase space. To evaluate the quantity, we require a functional expression, as discussed in chapter 20. For a 1-dimensional system the energy is given by a localised distribution - the Dirac Delta.

$$E = \int dq\, dp\, H(q, p) \delta(q - q_0) \delta(p - p_0)$$

The evolution equations for the pure state are given by Hamilton's equations of motion. **Mixed states** are interpreted as non-local probability measures on the phase space $\rho(q, p)$, not Dirac deltas, and their evolution equation is given by the Liouville equation of motion.

We now examine the **calculus of variations** in a geometrical setting. The motion of a system traces a curve γ in M that joins the two points $q_0(t_0)$ and $q_1(t_1)$. The set of smooth paths from $q_0(t_0)$ to $q_1(t_1)$ is denoted $\mathcal{P}(q_0, q_1, [t_0, t_1])$. Given a curve $\gamma \in \mathcal{P}(q_0, q_1, [t_0, t_1])$, we denote its **tangent lift** to TM by $\tilde{\gamma}$, where the tangent lift is the curve in the tangent bundle; this describes the position *and* velocity over the original curve and is a vector field over γ and, for any smooth function $f \in \mathcal{C}^\infty(M)$,

$$\tilde{\gamma}(t)(f) = \frac{d}{dt}(f \circ \gamma)(t)$$

The important point to note is that a general curve in TM will just be described by some combination $(q(t), v(t))$ but the lifted curve is precisely $(q(t), \dot{q}(t))$. We can write $\tilde{\gamma}$ in coordinates on TM such that $\tilde{\gamma}(t) = (q^i(t), v^i(t))$ or, in terms of the curves, $\tilde{\gamma}(t) = (\gamma(t), \gamma'(t))$. A **variation** $\gamma_\alpha(\alpha)$ to γ is a curve in $\mathcal{P}(q_0, q_1, [t_0, t_1])$ or, flipped around: $\mathcal{P}(q_0, q_1, [t_0, t_1])$ is a family of curves that represent variations to the curve on M over the time interval $\forall\, t \in [t_0, t_1]$. Note that these variations all vanish at the end points and are parametrised by α, such that if $\alpha = 0$, then $\gamma_0(t, 0) = \gamma(t)$.

The tangent space $T_\gamma \mathcal{P}(q_0, q_1, [t_0, t_1])$ for a curve $\gamma \in \mathcal{P}(q_0, q_1, [t_0, t_1])$ is the set of infinitesimal variations $\delta\gamma$ or vector fields that are induced by γ_α such that $\delta\gamma \in T_\gamma \mathcal{P}(q_0, q_1, [t_0, t_1])$:

$$\delta\gamma(t) = \left. \frac{d\gamma_\alpha(\alpha, t)}{d\alpha} \right|_{\alpha=0}$$

We define a functional on the space of curves $A : P(q_0, q_1, [t_0, t_1]) \mapsto \mathbb{R}$; a critical point of which, is an extremal curve. We could flip this around and say γ is a critical point of a functional if and only if $\delta A(\gamma) \cdot \delta\gamma = 0$. In the literature, this condition is usually expressed as either of the following:

$$\left. \frac{d(A \circ \gamma_\alpha)}{d\alpha} \right|_{\alpha=0} = 0, \qquad \delta A(\gamma) = 0, \qquad \frac{\delta A(\gamma)}{\delta\gamma} = 0 \qquad (22.0.1)$$

The stationary action principle states that, out of the space of curves, the one that describes the motion of the system is the one that makes the action stationary, that is the $\gamma \in \mathcal{P}(q_1, q_0, t_0, t_1)$ such that $\delta A(\gamma) = 0$ for every infinitesimal variation we could make. The action functional that is relevant to mechanics takes the form of an integral over the Lagrangian over the interval $[t_0, t_1]$:

$$A(\gamma) = A[q^i, v^i] = \int_{t_0}^{t_1} \mathcal{L}(\tilde{\gamma})dt = \int_{t_0}^{t_1} \mathcal{L}(q^i(t), v^i(t))dt \qquad (22.0.2)$$

From here, we can derive in the standard way the Euler-Lagrange equation of motion:

$$\delta A(\gamma) \cdot \delta\gamma = \int_{t_0}^{t_1} \left(\frac{\partial \mathcal{L}}{\partial q^i} \cdot \delta\gamma^i(t) + \frac{\mathcal{L}}{\partial v^i} \cdot \delta\dot{\gamma}^i(t) \right) dt$$

$$= \int_{t_0}^{t_1} \left(\frac{\partial \mathcal{L}}{\partial q^i} \circ \tilde{\gamma} - \frac{d}{dt}\left(\frac{\partial \mathcal{L}}{\partial v^i} \circ \tilde{\gamma} \right) \right) dt$$

Thus, we obtain the Euler-Lagrange equation in the familiar form.

We can define a **Euler-Lagrange vector field** $X_{\mathcal{L}} \in TM$ on the tangent bundle whose integral curves are the solutions to the Lagrange equation, as generated by a given Lagrangian, curves that make the action stationary. That is a very important point: *Lagrangians generate vector fields which are the solutions to differential equations!* This vector field has a flow $\Psi_{\mathcal{L}}$ associated with it on the tangent bundle over the interval $[t_0, t_1]$, which is a mapping into the tangent bundle with an advancement in the time parameter. The aim of Lagrangian mechanics is to find those vector fields whose lifts correspond to the curves in the tangent or cotangent bundles and hence, they become *integral curves* of the Lagrangian and Hamiltonian vector fields, respectively. To recap (since this is important), the dynamical system will map some curve through the tangent bundle, say, and correspondingly a curve on the configuration manifold, too. The vector field that induces this "dynamical curve" is the Lagrange vector field on $X_{\mathcal{L}} : TM \rightarrow T(TM)$, and its integral curve the dynamical curve on TM. Finding the vector field on M which has as it's integral curve precisely the dynamical curve when lifted from the configuration manifold to the tangent bundle is the aim of the game.

The Lagrangian vector field can be expanded in terms of components about the basis fields, like any general vector field, to obtain

$$X_{\mathcal{L}} = \sum_{i=1}^{n} \dot{q}^i \frac{\partial}{\partial q^i} + \sum_{i=1}^{n} \dot{v}^i \frac{\partial}{\partial v^i} \qquad \textbf{Lagrangian vector field} \qquad (22.0.3)$$

For any scalar function on the tangent bundle, $F(\boldsymbol{q}, \boldsymbol{v}) \in \mathbb{R}$, applying the Lagrangian vector field to it gives its time derivative $\dot{F} = X_{\mathcal{L}}(F)$, employing summation conventions, we obtain

$$X_{\mathscr{L}}(F) = \dot{q}^i \frac{\partial F}{\partial q^i} + \ddot{q}^i \frac{\partial F}{\partial \dot{q}^i}$$

This expression can be understood better if we use the Lagrange equation to solve for the generalised accelerations $\dot{v}^i$. In the one-dimensional case, we can take the total derivative term and expand it:

$$\frac{d}{dt} \frac{\partial \mathscr{L}}{\partial v^i}(q, v) = \frac{\partial^2 \mathscr{L}}{\partial v^i \partial v^j} \frac{dv^j}{dt} + \frac{\partial^2 \mathscr{L}}{\partial q^j \partial v^i} \frac{dq^j}{dt}$$

Then substitute this back into the Lagrange equation and rearrange:

$$\frac{\partial^2 \mathscr{L}}{\partial v^i \partial v^j} \frac{dv^j}{dt} = \frac{\partial \mathscr{L}}{\partial q^i} - \frac{\partial^2 \mathscr{L}}{\partial q^j \partial v^i} \frac{dq^j}{dt}$$

We now solve for the accelerations if and only if $(\partial^2 \mathscr{L}/\partial v^i \partial v^j)$ has an inverse; this is true if the Lagrangian is non-degenerate (i.e. for each fibre, there is a non-zero element $x \in T_p M$), then there exists another element w such that the following is true:

$$\sum_{i,j=1}^{n} \frac{\partial^2 \mathscr{L}}{\partial v^i \partial v^j} x^i w^j \neq 0$$

If this is true, then we can invert the above to find dv^j/dt and substitute the result into $X_{\mathscr{L}}$. It is the invertibility of the Hessian that gives us the vector field and the implicit function theorem allows us to invert the generalised velocities for the generalised momenta. Note the coloured terms are highlighted for later use; they are the components of the basis vectors ∂_q and ∂_v:

$$X_{\mathscr{L}} = \dot{q}^i \frac{\partial}{\partial q^i} + \left[\frac{\partial^2 \mathscr{L}}{\partial v^j \partial v^k} \right]^{-1} \left(\frac{\partial \mathscr{L}}{\partial q^j} - \frac{\partial^2 \mathscr{L}}{\partial q^k \partial v^j} \dot{q}^k \right) \frac{\partial}{\partial v^i} \qquad (22.0.4)$$

We can use the **fibre derivative** of the Lagrangian function on the tangent space and convert it into a function defined on the cotangent space. We actually realise that the fibre derivative is the **Legendre transform** $\mathbb{L}_{\mathscr{L}} : TM \to T^*M$, carried out from fibre to fibre, defining the **Hamiltonian** H as a scalar function on the cotangent space that maps to the reals $H : T^*M \mapsto \mathbb{R} : [q^i, p_j] \to H([q^i, p_j])$. There is no reason why we can not perform the fibre derivative with respect to the Hamiltonian such that $\mathbb{L}_H : T^*M \to TM$ is a map that takes the coordinates from $[q^i, p_j] \mapsto [q^i, v^j]$. Note that $\mathbb{L}_{\mathscr{L}} \circ \mathbb{L}_H$ would take us back to where we started. This is subject to regularity conditions, as we previously discussed for Dirac-Bergmann theory in chapter 21 and singularities in the Hessian indicate that T^*M is no longer isomorphic to TM. A more mathematical way of expressing this is that the mapping is regular if it is surjective.

In the same way that a Lagrangian vector field characterised the dynamical flow for the Lagrangian system, the Hamiltonian defines a **Hamiltonian vector field** X_H on the cotangent bundle $X_H : T^*M \to T(T^*M)$, complete with a Hamiltonian flow Ψ_H, a one-parameter family of diffeomorphisms. On the cotangent bundle, we recall (see chapter 38) that we didn't use vector fields to get from the base space; instead, we used differential forms. The aim of geometrical mechanics in the Hamiltonian formalism is to find the 1-form on M such that, when it is composed with the configuration space curve, it gives us the integral curve of the Hamiltonian vector field:

$$X_H = \sum_{i=1}^{n} \dot{q}^i \frac{\partial}{\partial q^i} + \sum_{i=1}^{n} \dot{p}_i \frac{\partial}{\partial p_i} \qquad \textbf{Hamiltonian vector field} \qquad (22.0.5)$$

A curve is an integral curve for X_H exactly if Hamilton's equations are satisfied for $\dot{q}$ and $\dot{p}$, which, if so, means that the Hamiltonian vector field is $\{\cdot, H\}$, the Poisson bracket. Indeed, we can relate $X_{\mathscr{L}}$ to X_H by utilising the pushforward $(\mathbb{L}_{\mathscr{L}}\mathscr{L})_* X_{\mathscr{L}} = X_H$. The integral curves of the Lagrangian vector field maps to integral curves of the Hamiltonian vector field and both map to the original base space's integral curves via their lifts. We say that the Lagrangian flow and the Hamiltonian flow are **conjugated**.

For a function other than the Hamiltonian $F \in \mathcal{C}^\infty(T^*M)$, with the unique vector field X_F, the one-parameter family of diffeomorphisms Ψ_F is exactly the unique solution to the **Cauchy problem** associated with the vector field. That is, for a point in phase space $\xi \subset T^*M$ and a smooth parameter $s \in \mathbb{R}$, we have the following Cauchy problem:

$$\frac{d\xi}{ds} = X_F(\xi(s)) \qquad \text{with} \qquad \xi(0) = (q_0, p_0)$$

When F is the Hamiltonian (i.e. $F(\xi(s)) = H(q(t), p(t))$) the Cauchy problem is nothing other than Hamilton's equations describing time evolution! An integral curve is a solution to the system such that $(q(0), p(0)) \mapsto (q(t), p(t))$ and, by definition, they satisfy Hamilton's equations:

$$\dot{H}(\xi) = X_H(H) = \frac{\partial H(q, (t), p(t))}{\partial q} \dot{q}(t) + \frac{\partial H(q(t), p(t))}{\partial p} \dot{p}(t) = 0 \qquad (22.0.6)$$

As expected, the time derivative of a time-independent Hamiltonian vanishes when Hamilton's equations are satisfied. In other words, for a time-independent system, we cannot have a Hamiltonian that is not constant along a physical path.

The phase space itself can be treated like a manifold; that is, the cotangent bundle to the configuration manifold is also a manifold in its own right. Just as we defined a Lagrangian on the tangent bundle to the configuration manifold, we can also define a Lagrangian on the tangent bundle to the phase space $\mathscr{L}' : T(T^*M) \to \mathbb{R}$. We may recall that we call this the **phase space Lagrangian**, and earlier we derived

Hamilton's equations using it in a stationary action principal (see chapter 15). Since the tangent bundle to a manifold has twice the dimensionality to the original space and is spanned by the coordinates and their derivatives, we expect the tangent bundle to a cotangent bundle to be spanned by $p, q, \dot{p}$ and $\dot{q}$. It is precisely when Hamilton's equations are satisfied that the variables assume their relationships as defined by the Legendre transform, and the phase space and velocity space Lagrangians share the same critical points.

As we recall, the Hamiltonian of an autonomous system represents the total energy, and the Jacobi energy function J represents the total energy in the Lagrange formalism, defined, $J : TM \to \mathbb{R}$ below:

$$J(q, v) = \mathbb{L}_{\mathscr{L}}(v^i) \cdot v^i - \mathscr{L}(q, v)$$
$$= \frac{\partial \mathscr{L}}{\partial v^i} v^i - \mathscr{L}(q, v)$$

It is useful to calculate the exterior derivative of J; we can put this to one side for later application:

$$dJ = d\left(\sum_i v^i \frac{\partial \mathscr{L}}{\partial v^i} - \mathscr{L}(q, v) \right)$$

$$= d\left(\sum_i v^i \frac{\partial \mathscr{L}}{\partial v^i} \right) - d(\mathscr{L}(q, v))$$

$$= \sum_i \left[d(v^i) \frac{\partial \mathscr{L}}{\partial v^i} + v^i d\left(\frac{\partial \mathscr{L}}{\partial v^i} \right) \right] - \sum_j \left[\frac{\partial \mathscr{L}}{\partial q^j} dq^j + \frac{\partial \mathscr{L}}{\partial v^j} dv^j \right]$$

$$= \sum_{i,j} \left(\frac{\partial v^i}{\partial v^j} dv^j \right) \frac{\partial \mathscr{L}}{\partial v^i} + \sum_{i,j} v^i \left[\frac{\partial^2 \mathscr{L}}{\partial v^i \partial q^j} dq^j + \frac{\partial^2 \mathscr{L}}{\partial v^i \partial v^j} dv^j \right] - \sum_j \frac{\partial \mathscr{L}}{\partial q^j} dq^j - \sum_j \frac{\partial \mathscr{L}}{\partial v^j} dv^j$$

$$= \sum_{i,j} v^i \frac{\partial^2 \mathscr{L}}{\partial v^i \partial v^j} dv^j + \sum_{i,j} \left[v^i \frac{\partial^2 \mathscr{L}}{\partial v^i \partial q^j} - \frac{\partial \mathscr{L}}{\partial q^j} \right] dq^j \qquad (22.0.7)$$

The coloured term gives the delta function δ^{ij}, meaning the indices are just for bookkeeping. It is important to note that, while it is not necessary, we can define a metric on the tangent bundle. In this case, there exists an isomorphism linking the metric to the cotangent bundle, therefore defining an equivalent structure on the phase space.

At this point, we have a geometrical description that can be satisfied in both TM and T^*M, and (in the regular case), switching between the two descriptions is just expressing the same problem in different ways. In either case, we seek either vector fields or forms whose lift of a configuration curve correspond to precisely the integral curves of the Lagrangian and Hamiltonian vector fields. We now need a method of actually finding said objects and we do this using the Hamilton-Jacobi method. We begin by looking at the geometrical structure on the cotangent bundle and we then pull this structure back to TM, that is to say, the structure inherent to T^*M *induces* a structure on TM (provided the Hessian is regular). Apart from this inherited structure, TM does not have a structure of its own. In addition, it should be noted that the tangent bundle is a place where second-order differential equations are well defined, while this is not possible in the cotangent bundle, where instead, we get a set of first-order equations; we will investigate why this is and so, in doing so, discover why Lagrange's equations are second order while Hamilton's are first order.

22.1 A Coordinate-Free Picture

We start by defining the **Poincaré-Cartan 1-form** θ_H on T^*M; this is a Pfaffian 1-form that treats the cotangent bundle as its base space and maps elements to the cotangent space of the cotangent bundle of the manifold M; in essence $\theta_H : T^*M \rightarrow T^*(T^*M)$. It is so defined that, for a 1-form α defined on the configuration manifold $\alpha : M \rightarrow T^*M$, the Poincaré-Cartan 1-form cancels the pullback $\alpha^*\theta_H = \alpha$. Recalling that a 1-form can be written in terms of components and a basis set in local coordinates, we write the Poincaré-Cartan 1-form in terms of the local coordinates of the cotangent bundle:

$$\theta_H := \sum_{i=1}^{n} p_i dq^i \qquad \textbf{Poincaré-Cartan 1-form} \qquad (22.1.1)$$

This 1-form does not depend on any specific choice of Hamiltonian function; rather it is an implicit result of the cotangent bundle structure - from this 1-form, all of time-independent mechanics can be recovered. We can use θ_H to define the **symplectic 2-form** $\overset{2}{\omega}$ on T^*M by taking the negative of its exterior derivative $\overset{2}{\omega} = -d\theta_H$; in this way the cotangent bundle to a configuration manifold *naturally* has a symplectic structure:

$$\omega := -d\theta = -d(p_i dq^i) = -[dp_i \wedge dq^i - p_i \wedge d(dq^i)] = dq^i \wedge dp_i \qquad (22.1.2)$$

$$\omega = dq^i \wedge dp_i \qquad \textbf{symplectic 2-form}$$

where we have dropped the 2 notation for brevity, used the summation convention (so there should be a sum over i), used $d(dx) = 0$, and used the asymmetry of the wedge product $\alpha \wedge \beta = -\beta \wedge \alpha$.

The fact that we can always choose local coordinates such that the 2-form can be written is a result of **Darboux' theorem**. Darboux' theorem is a foundational theorem that states that, *locally, all cotangent bundles look the same (are isomorphic), so the global properties must be used to understand them.* It also indicates that we can cover the manifold with charts, **Darboux charts**, whose transition functions are canonical transformations between open subsets. A key result of Darboux' theorem is that Hamiltonian vector fields may be **rectified**, that is, that, locally the flow of an autonomous system is trivialised (provided the vector field is regular); therefore, global structure characterises Hamiltonian systems. If the system is integrable, then the global flow may be rectified.

If we recall the definition of the Hamiltonian vector field $X_H = \dot{q}\partial_q + \dot{p}\partial_p$, importantly, we will use the following two results later on:

$$\iota_{X_H} dq = \frac{\partial H}{\partial p}, \qquad \iota_{X_H} dp = -\frac{\partial H}{\partial q}$$

The meaning of the first of these is as follows: we are literally saying *the component of X_H in the direction of the ∂_q basis vector is $\dot{q}$ or, when Hamilton's equations are satisfied, is $\partial H/\partial p$*. So, we therefore generate both of Hamilton's equations from the definition of the vector field and coordinates on the cotangent bundle.

Symplectic geometry is a class of differential geometry that itself comes with many of its own unique characteristics. We briefly consider some key points of symplectic geometry in the mathematical background 3. Before we do, consider the following mathematical *commutative diagram* for autonomous mechanics. Some of the terms may not have yet been discussed but the diagram will provide a reference for later work.

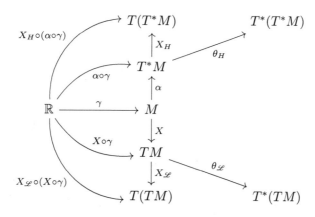

We start from $\mathbb{R}$ with a smooth curve γ on the configuration space M. The vector field X is exactly the vector field whose lifts provide integral curves for $X_{\mathscr{L}}$, and α is the 1-form whose composition with the configuration space curve is an integral curve of X_H. We note that the following equalities hold for the diagram: for the Lagrangian side, $X \circ \gamma = \dot{\gamma}$, $X_{\mathscr{L}} \circ \dot{\gamma} = \overline{X \circ \gamma}$, and, for the Hamiltonian, side $X_H \circ (\alpha \circ \gamma) = \overline{\dot{\alpha \circ \gamma}}$.

Mathematical Background 3

Symplectic geometry concerns smooth, even-dimensional manifolds Q equipped with the **closed, non-degenerate, skew-symmetric bilinear** tensor $\omega_p : T_pQ \times T_pQ \to \mathbb{R}$, where $\omega \in \Omega^2(Q)$ (the set of all 2-forms on Q). Closed means that ω does not change under flow, or that it's Lie/exterior derivative is zero: $\mathcal{L}_X\omega = d\omega = 0$. The non-degenerate condition $\omega(v, \cdot) \neq 0$ for each non-zero $v \in TQ$ on a manifold Q ensures that the vector field associated with the differential of a function is unique and, further, that $\omega^n \neq 0$ at any point.

The wedge product of the two 1-forms α and β is a 2-form, for two tangent vectors $x, y \in T_pQ$ that are the representatives of two vector fields X, Y on Q at $p \in Q$:

$$(\alpha \wedge \beta)_p(x, y) = \alpha_p(x)\beta_p(y) - \alpha_p(y)\beta_p(x)$$

The nth exterior power ω^n is a volume form on Q and there exists a basis in T^*Q such that $\omega = \frac{1}{2}\omega_{ij}e^i \wedge e^j$, where

$$\omega_{ij} = \begin{pmatrix} 0 & \mathbb{I} \\ -\mathbb{I} & 0 \end{pmatrix} = -\omega_{ji}$$

Where 0 is the null matrix, and $\mathbb{I}$ is the identity matrix, the inverse matrix is written ω^{-1} with components ω^{ij} such that $\omega_{ij}\omega^{jk} = \delta_i^k$ and $\omega_{ij} = -\omega^{ij}$; these are the **symplectic matrices**. Notice that the diagonal consists of zeros which give it the skew-symmetry property and that the determinant is $+1$:

$$\omega^{ij} = \begin{pmatrix} 0 & -\mathbb{I} \\ \mathbb{I} & 0 \end{pmatrix}$$

Non-degeneracy would mean that there were no zero-valued eigenvalues in the matrix, so it has full rank; degeneracies would imply the opposite. We could also write this (significantly) as $\omega = \sum_{i=1}^n e^i \wedge e_{n+i}$. Together, the pair (Q, ω) comprise a **symplectic manifold**. A Hamiltonian vector field is any vector field that satisfies $\mathcal{L}_X\omega = 0$ and, locally, we can use this and Cartan's formula to show that $d(\iota_X\omega) = 0$. This means that $\iota_X\omega$ is a closed 1-form and, by the Poincaré lemma, there exists *locally* a function $f : T^*Q \mapsto \mathbb{R}$ such that $\iota_X\omega = \omega(X_f, \cdot) = df$. All global Hamiltonian vector fields are locally Hamiltonian; however, the converse is not always true. The set of Hamiltonian vector fields is a Lie algebra equipped with a Lie bracket. If ω is non-degenerate, then the vector field X has a unique dual $\iota_X\omega$.

The 2-form has the property of antisymmetry, meaning that, for any two vector fields, $\omega(X, Y) = -\omega(Y, X)$; since a skew-symmetric matrix can only be inverted for even-dimensional spaces, T^*Q cannot be odd dimensional. In classical mechanics, we say that the cotangent bundle associated to a configuration manifold is a symplectic manifold (e.g $Q = T^*M$). For two symplectic manifolds (Q_1, ω_1) and (Q_2, ω_2), a smooth mapping $\phi : Q_1 \mapsto Q_2$ is a **symplectomorphism** if and only if $\phi^*\omega_2 = \omega_1$; the Jacobian of the transformation is unity, so it is an invertible transformation, meaning that, by the inverse function theorem, it is a diffeomorphism. A **Lagrangian subspace** of a symplectic manifold Q is a submanifold $\mathcal{Q}$ of half the dimension of Q such that the symplectic form, when restricted to $\mathcal{Q}$, vanishes (e.g $\omega|_{\mathcal{Q}} = 0$ for $\mathcal{Q} \subset Q$ or $\omega(v, w) = 0$ for $v, w \in \mathcal{Q}$).

In order for a manifold to be classified as symplectic, it must satisfy certain conditions. As stated in chapter 38, it must be Hausdorff and second countable, even dimensional, compact and orientable. These are criteria that ensure that the space may is "well behaved" enough to be considered a bounded phase space and be equipped with the forms we require. The meaning and implications of these criteria however, are beyond the scope of our introduction.

Finally, **Kähler manifolds** K have **almost complex, Riemannian** and symplectic structures that are all mutually compatible to give a triple (J, g, ω). That is to say, there is a metric g on K (hence Riemannian), an almost complex structure on each tangent space (hence almost complex) and a symplectic form ω. In more detail almost complex, means that there is a tensor field $J : TK \to TK$ of degree $(1, 1)$ such that $J^2 = -1$; almost complex structures become complex structures if J is an integrable tensor.

Symplectic structure is the result of a more general structure called a **contact structure** that arises naturally in the language of jets (see chapter 23).

The cotangent bundle to a differentiable manifold is a natural setting for a symplectic manifold, where we now consider the cotangent bundle to be a manifold upon which we can define things just like we did before, except one level up. A symplectic manifold is the *pair* (T^*M, ω), a fibre bundle/manifold *and* a symplectic form (note that it is not the most general case to consider every symplectic manifold a cotangent bundle to a configuration manifold, but every cotangent bundle is symplectic; in any case, what we have here will suffice).

The symplectic 2-form ω is a closed, non-degenerate differential 2-form on T^*M and is a bilinear map acting on pairs of vector fields such that $\omega : T(T^*M) \times T(T^*M) \to \mathbb{R}$. The non-degeneracy condition indicates that the Hamiltonian vector field is uniquely determined by the Hamiltonian. If there were degeneracies, then there would be a non-unique evolution and a non-fixed gauge freedom to the motion; remember back to constrained dynamics and Dirac theory in chapter 21; in this case, we would consider some constraint surface within the total phase space. We mention here that there exist **musical isomorphisms** that reverse each other when applied in sequence above the same point. They are vector bundle isomorphisms, defined as

$$\flat : TM \to T^*M, \qquad \sharp : T^*M \to TM \tag{22.1.3}$$

and called the flat and sharp maps, respectively. These two are natural isomorphisms between a symplectic space and its dual, provided the 2-form is non-degenerate, which we discuss later in section 22.4.

We can also define an inverse of the 2-form ω^{-1} such that $(\omega^{-1} \circ \omega) = \mathbb{I}$; note that, if ω acts on vector fields then its inverse or **dual** will act on the dual of a vector field; where the latter dual is a 1-form. Therefore, $\omega(X, Y) = \omega^{-1}(dX, dY)$, the dual as applied to classical mechanics is called the **Poisson bi-vector field** η:

$$\eta = \omega^{-1} = \eta^{ij} \frac{\partial}{\partial q^i} \wedge \frac{\partial}{\partial p_i} \tag{22.1.4}$$

You can see where this is going: the Hamiltonian has a differential dH which is a 1-form on T^*M. If and only if we could take physical trajectories to be the integral curves of the vector field associated with dH, we would be looking pretty good in terms of our geometric theory! It so happens that the 2-form is exactly the tool that we need to implement this.

Taking the interior product of the Hamiltonian vector field and the 2-form is exact and will lead us to an equation that has come to define Hamiltonian mechanics:

$$\iota_{X_H} \omega := \omega(X_H, \cdot) \tag{22.1.5}$$

Taking the left-hand side and using the asymmetry of the wedge product in the final step:

$$\iota_{X_H} \omega = \sum_{i=1}^{n} \iota_{X_H} (dq^i \wedge dp_i)$$

$$= \sum_{i=1}^{n} (\iota_{X_H} dq^i) \wedge dp_i - \sum_{i=1}^{n} dq^i \wedge (\iota_{X_H} dp_i)$$

$$= \sum_{i=1}^{n} \frac{\partial H}{\partial p_i} dp_i + \sum_{i=1}^{n} \frac{\partial H}{\partial q^i} dq^i$$

which is nothing other than the total derivative of the Hamiltonian[3], dH, which is a closed 1-form; it makes sense that the interior derivative of a 2-form is a 1-form. The so-called **fundamental identity of dynamics** is written as follows and we can see that it can be immediately be integrated to give the Hamiltonian:

$$\iota_{X_H} \omega = dH \qquad \textbf{Geometric Hamilton-Jacobi equation} \tag{22.1.6}$$

---

[3]It is a general result of symplectic geometry that $\iota_{X_f} \omega = df$, where $f : T^*M \to \mathbb{R}$ is a scalar function defined on the symplectic manifold, X_f is a Hamiltonian vector field and ω is a bilinear, closed, non-degenerate form on the manifold.

In some literature it is also called the Hamilton-Jacobi equation, but this might cause confusion in some of our earlier chapters, so we will refer to this equation as the *fundamental equation*, or the geometrical Hamilton-Jacobi equation, when there can be no confusion. This equation contains all of the structure of the Hamiltonian formalism, and we will soon see that it can be used to describe the Lagrangian formalism, too, at least in the regular case, that is! Reading this equation, we say (from the left to the right) that the interior product of the symplectic 2-form with a Hamiltonian vector field is equal to the exterior derivative of the corresponding Hamiltonian function of the vector field. The vector field satisfying this relation is *unique*, as you will be invited to prove at the end of this chapter. The fundamental equation of dynamics allows us to view the symplectic 2-form as a tool to relate vector fields to their vector dual field: for vector field X on a vector space, $\iota_X \omega$ is the dual field spanning the vector duel space.

We can now define a **Hamiltonian system** to be the triplet (M, ω, H), that is, a symplectic manifold and a function on it that is Hamiltonian. The dynamics of H is such that all phase space trajectories are integral curves of the Hamiltonian vector field X_H. This means that the Hamiltonian is a conserved quantity along the flow of the system. While there is nothing new about this in terms of dynamics it gives us an amazing way to think about dynamics on manifolds in a coordinate-free way.

Arnold's theorem states that the symplectic form is preserved during the flow Ψ_H of the system, meaning that Hamilton flow is symplectic. This is a deeper understanding of Liouville's theorem, which states that the volume element of an n-dimensional phase space is given by $\Lambda = \frac{1}{n!} \omega^n$ (some people take a different definition of the $(1/n!)$ factor depending on orientation, or omit it entirely), where $\omega^n = \omega \wedge \cdots \wedge \omega = \rho \, dq^1 \wedge \cdots \wedge dq^n \wedge dp_1 \wedge \cdots \wedge dp_n$ the n-fold wedge product. Note that ρ is the **volume density** and is given by the Pfaffian $\mathrm{Pf}(\omega_{IJ})$, so we may regard ω as providing a measure on the phase space. We can prove Arnold's theorem using Cartan's formula for the Lie derivative and the non-degeneracy condition for the symplectic form:

$$\mathcal{L}_{X_H} \omega = \iota_{X_H} d\omega + d(\iota_{X_H} \omega) = \iota_{X_H} 0 + d(dH) = 0$$

where we use the antisymmetric property of the wedge product $\alpha \wedge \beta = -\beta \wedge \alpha$:

$$d(dH) = d\left(\frac{\partial H}{\partial q^i} dq^i + \frac{\partial H}{\partial p_i} dp_i \right)$$

$$= \frac{\partial^2 H}{\partial q^i \partial p_i} dp_i \wedge dq^i + \frac{\partial^2 H}{\partial p_i \partial q^i} dq^i \wedge dp_i$$

$$= \left(\frac{\partial^2 H}{\partial q^i \partial p_i} - \frac{\partial^2 H}{\partial p_i \partial q^i} \right) dq^i \wedge dp_i$$

$$= 0$$

Therefore, for a family of diffeomorphisms generated by the Hamiltonian vector field $\Psi_H : (M, \omega) \to (M, \omega)$, the symplectic form is preserved for the motion. The family of diffeomorphisms generated by the Hamiltonian, however, is precisely the *time evolution*, so, in essence, we are saying that ω is preserved under time evolution. Then, by the Leibniz rule on Lie derivatives, we can say that the nth fold wedge product will also vanish too!

A **symplectic transformation** ϕ is a transformation that takes place between manifolds and preserves the symplectic form: if (M, ω) and (N, β) were two symplectic manifolds, then the map $\phi : M \mapsto N$ is symplectic if the pullback of β onto M is equal to ω, namely $\phi^* \beta = \omega$; in this case, we say ω is **integral invariant**. If the mapping is onto the same symplectic manifold, then we would call it a 'symplectomorphism' or a **canonical transformation**, so there are symplectic transformations that are not canonical. A vector field is symplectic if its flow preserves ω. We could write $dQ \wedge dP = dq \wedge dp$, up to some constant. For proof, consider the intersection on M of two charts, $q(Q)$ and $Q(q)$, which give rise to the natural coordinates $(q, \dot{q})$ and $(Q, \dot{Q})$ on their tangent bundles. By the chain rule, $\dot{q}^i = (\partial q^i / \partial Q^j) \dot{Q}^j$ and $\dot{Q}^i = (\partial Q^i / \partial q^j) \dot{q}^j$, with $\mathscr{L} : TM \to \mathbb{R}$ written $\mathscr{L}(Q, \dot{Q}) = \mathscr{L}(q(Q), (\partial q / Q)\dot{Q})$. We compute the conjugate momenta as follows:

$$P_i = \frac{\partial \mathscr{L}}{\partial \dot{Q}^i} = \frac{\partial \mathscr{L}}{\partial q^j}\frac{\partial q^j}{\partial \dot{Q}^i} + \frac{\partial \mathscr{L}}{\partial \dot{q}^j}\frac{\partial \dot{q}^j}{\partial \dot{Q}^i} = \frac{\partial \mathscr{L}}{\partial \dot{q}^j}\frac{\partial q^j}{\partial Q^i} = \frac{\partial q^i}{\partial Q^i}p_j$$

The first term is zero, since $q \neq q(\dot{Q})$, and the blue term cancels the time derivative at the top and at the bottom. Then,

$$dQ^i \wedge dP_i = \left(\frac{\partial Q^i}{\partial q^j}dq^j\right) \wedge \left(\frac{\partial q^k}{\partial Q^i}dp_k\right) = \left(\frac{\partial Q^i}{\partial q^j}\frac{\partial q^k}{\partial Q^i}\right)dq^j \wedge dp_k = \delta^{jk}dq^j \wedge dp_k$$

Note that the Hamiltonian can (and probably will) change form upon a symplectic mapping $\phi : H \to K$, but it maps the trajectories of X_H onto the trajectories of X_K, and the phase flow diffeomorphisms Ψ_H or Ψ_K are canonical transformations. This is important: *the flow of a Hamiltonian vector field is a symplectomorphism!*

For two functions F and G in the space of smooth functions (which describe the observables) on the cotangent bundle $\mathcal{C}^\infty(T^*M)$, their **Poisson bracket** is given by

$$\{F, G\} := \omega(X_F, X_G) = \eta(dF, dG)$$

It inherently endows a Lie algebra structure, in addition to the commutative algebra. For proof of the above relation, we simply have to enter the expressions for the Hamiltonian vector fields of F and G as follows,

$$\omega(X_F, X_G) = dq^i \wedge dp_i\left(\frac{\partial F}{\partial q^i}\frac{\partial}{\partial p_i} - \frac{\partial F}{\partial p_i}\frac{\partial}{\partial q^i}, \frac{\partial G}{\partial q^i}\frac{\partial}{\partial p_i} - \frac{\partial G}{\partial p_i}\frac{\partial}{\partial q^i}\right)$$

$$= \left(-\frac{\partial F}{\partial q^i}\frac{\partial G}{\partial p_i}\right) - \left(\frac{\partial G}{\partial q^i}\right)\left(-\frac{\partial F}{\partial p_i}\right)$$

$$= \{F, G\}$$

Here, we have used $(\alpha \wedge \beta)(X_F, X_G) = \alpha(x)\beta(y) - \alpha(y)\beta(x)$, where the vector fields can be expanded in terms of components $X = x^i \partial_i$, and α and β are 1-forms. We then write in the component of the vector fields for the basis specified by the 1-form (e.g. $dq^i(X_F)$ is the component of the $\partial/\partial q^i$ field in the X_F vector field, and so on). Similarly, the Poisson bi-vector field follows:

$$\eta(dF, dG) = \frac{\partial}{\partial q^i} \wedge \frac{\partial}{\partial p_i}\left(\frac{\partial F}{\partial q^i}dq^i + \frac{\partial F}{\partial p_i}dp_i, \frac{\partial G}{\partial q^i}dq^i + \frac{\partial G}{\partial p_i}dp_i\right)$$
$$= \left(\frac{\partial F}{\partial q^i}\frac{\partial G}{\partial p_i} - \frac{\partial F}{\partial p_i}\frac{\partial G}{\partial q^i}\right)$$

If we take the Lie bracket of two vector fields X_F and X_G and the vector field of their Poisson bracket $X_{\{F,G\}}$, then we can use the Jacobi identity to show that $\{X_F, X_G\} = -X_{\{F,G\}}$ such that we have the **Poisson structure** $\{\cdot, \cdot\} : \mathcal{C}^\infty(T^*M) \times \mathcal{C}^\infty(T^*M) \mapsto \mathcal{C}^\infty(T^*M)$; the two spaces are used because it is a bilinear operation on two functions. There is further information on the Poisson bracket as well as Poisson tensors $\omega^{-1} \equiv \eta \in \bigwedge^2 T\mathcal{P}$ and their generalisation to **Nambu structures** in appendix D.

Before we move on, consider the coordinates $(q^1, \ldots, q^n, p_1, \ldots, p_n)$ and two functions $F(\boldsymbol{q}, \boldsymbol{p})$ and $G(\boldsymbol{q}, \boldsymbol{p})$. We can formulate the **Lagrange bracket** as follows:

$$\{F, G\}_{\text{LB}} = \omega\left(\frac{\partial}{\partial F}, \frac{\partial}{\partial G}\right) \qquad \textbf{Lagrange bracket} \qquad (22.1.7)$$

We can evaluate this expression in a way that is similar to the method we use to evaluate the Poisson bracket, using the chain rule to obtain the bracket in local coordinates:

$$\{F, G\}_{\text{LB}} = (dq^i \wedge dp_i)\left(\frac{\partial q^i}{\partial F}\frac{\partial}{\partial q^i} + \frac{\partial p_i}{\partial F}\frac{\partial}{\partial p_i}, \frac{\partial q^i}{\partial G}\frac{\partial}{\partial q^i} + \frac{\partial p_i}{\partial G}\frac{\partial}{\partial p_i}\right)$$
$$= \left(\frac{\partial q^i}{\partial F}\frac{\partial p_i}{\partial G} - \frac{\partial p_i}{\partial F}\frac{\partial q_i}{\partial G}\right) \qquad (22.1.8)$$

22.2 Poisson Manifolds & Symplectic Reduction

A **Poisson manifold** $(\mathcal{P}, \{\cdot, \cdot\})$ is a differentiable manifold whose set of functions $\mathcal{C}^\infty(\mathcal{P})$ is equipped with a Poisson algebra structure $\mathcal{C}^\infty, \{\cdot\cdot\}$. Every symplectic manifold is a Poisson manifold, since the Poisson bracket is defined by the symplectic 2-form, as in section 22.1. Every Poisson manifold is the disjoint union of even-dimensional Poisson submanifolds, which are also known as **symplectic leaves** or symplectic manifolds, which can now be thought of as single-leaf Poisson manifolds. We say that the Poisson manifold is **foliated** into symplectic submanifolds, each of which is the leaf of the foliation. If all the leaves have the same dimension, then they are called **regular leaves**. As we have seen, the symplectic manifolds are where the solutions to Hamilton's equations live, so we say that the symplectic leaves are solution spaces to differential equations. So, why introduce the Poisson manifold? Well, while symplectic manifolds are required to be of even dimension, Poisson manifolds are not required to be; thus, their symplectic forms can be degenerate. Therefore, systems with **constraints** are described using Poisson manifolds and we will look at some key points on this topic a bit later. This is also an important point for generalising canonical Hamiltonian mechanics to **non-canonical** Hamiltonian mechanics! Non-canonical Hamiltonian systems have a Poisson structure rather than a symplectic structure. Their coordinates need not be canonical; for example, for an n-dimensional system, the phase space is $2n$ dimensional, but there need not be n general coordinates q and n momenta coordinates p; instead, they can be arbitrary, and the phase space may well have a mixed coordinate frame. This is perhaps a more natural way to understand systems and, frequently, if one wishes to study a symplectic manifold, it is often best practice to embed it within a Poisson manifold.

A continuous **symmetry** of a Hamiltonian system (M, ω, H) is a vector field X whose flow Ψ_X on M comprises a family of diffeomorphisms that satisfy $\mathcal{L}_X \omega = 0$ and $\mathcal{L}_X H = 0$, preserving the symplectic form, the Hamiltonian and, thereby, the structure on M. A set of Noether symmetries in the Lagrangian framework is a set of vector fields that are present on the configuration space and which, upon a canonical lift to the cotangent bundle, generate Hamiltonian symmetries; they form a subgroup of the permissible symplectomorphisms on the phase space. By the Poincaré lemma, if $\iota_X \omega$ is closed, then there exists locally a scalar function $F : M \to \mathbb{R}$, meaning $\iota_X \omega = dF$, that is, if X is symplectic, then it is locally (in the neighbourhood) Hamiltonian, and [4] $X_F(H) = \{H, F\} = 0$, meaning F is in involution and is constant along integral curves of X_H, so it is a conserved quantity: Noether's theorem. Functions that satisfy this are called **Casimir functions** of the Poisson structure and they commute with all other members of the algebra; they are constant while on a particular symplectic leaf. Casimirs can be seen as constraints on a system, and Poisson submanifolds are generated by the family of Casimirs, as we shall see in a moment:

$$\{F, H\} = \omega(X_F, X_H) = (\iota_{X_F} \omega)(X_H) = dF(X_H) = X_H(F) = \mathcal{L}_{X_H} F$$

The Poisson bracket is therefore the Lie derivative of a function along a vector field, $X(F)$. It must be understood that this is the importance of the Poisson bracket, that the generator of a transformation induces a flow that changes an observable/function as seen by the Lie dragging going on.

From now until Mathematical Background 4, the text can be skipped upon first reading without too much detriment. It may be useful to read chapter 23 in its entirety before attempting this portion. What follows is loosely based on appendix 5 of Arnold's (1989) *Mathematical Methods of Classical Mechanics*, although more modern terminology has been used where appropriate.

---

[4]Or $X_H(F) = 0$.

For a smooth phase space (M, ω) with a Lie group of symmetry transformations $\mathcal{G}$ with Lie algebra $\mathfrak{g}$ and dual $\mathfrak{g}^*$, the Lie group $\mathcal{G}$ acts on M in such a way as to preserve ω as well as creating a pairing $\langle \cdot, \cdot \rangle : \mathfrak{g}^* \times \mathfrak{g} \to \mathbb{R}$; in other words, the transformations are assumed to be symplectomorphisms. An element of the Lie algebra $\xi \in \mathfrak{g}$ induces a vector field $X(\xi)$ on M and is associated with the infinitesimal action of $\mathfrak{g}$; further, since the transformations are symplectomorphisms, the interior product $\iota_{X_\xi} \omega$ is closed $\forall\ \xi \in \mathfrak{g}$. As stated in section 22.1, in classical mechanics, we also assume ω is exact, meaning there exists some function F on M such that $\iota_{X_\xi} \omega = dF$ with $F : M \to \mathbb{R}$. Associated with the action of $\mathcal{G}$ is a unique function on M into the dual, $\mu : M \to \mathfrak{g}^*$, which we call the **momentum map**. The momentum map is a very important object in classical mechanics, as we shall see in a moment, and its values are vectors κ in $\mathfrak{g}^*$. We remark that $\iota_X \omega$ is closed and exact and, in such a situation (we say that the action of $\mathcal{G}$ is *Hamiltonian*), then the exterior derivative of the pairing is equal to the interior product of the 2-form and the vector field corresponding to $\xi \in \mathfrak{g}$:

$$d(\langle \mu, \xi \rangle) = \iota_{X_\xi} \omega$$

Due to this (as we show in a second), there exists a linear mapping $\xi \to F_\xi$, or $\mathfrak{g} \to (\mathcal{C}^\infty(M), \{\cdot, \cdot\})$, known as a **Lie algebra homomorphism**, taking elements of the Lie algebra into the space of smooth functions on M. This allows us to identify the pairing of μ and ξ with the function F_ξ such that $\langle \mu, \xi \rangle = F_\xi$. Vector fields that are symplectic are a subalgebra of the Lie algebra of all vector fields on M; those that are Hamiltonian are a further subalgebra or **ideal** of the symplectic ones. Equivalently, we may say that, for two Hamiltonian vector fields X_F and X_G, then their Lie bracket $[X_F, X_G]$ is also a Hamiltonian vector field. So, it is the Hamiltonian subalgebra of a Lie algebra $\mathfrak{g}$ of a Lie group of transformations $\mathcal{G}$ that neatly embeds itself into the Poisson algebra of smooth observables on M and defines the fundamental equation of motion $\iota_X \omega = dF$, all through the Lie algebra homomorphism. We can equivalently write this as

$$\Lambda^2 \mathfrak{g} \ni \xi_i \wedge \xi_j \xrightarrow{\tilde{\mu}} \{F_{\xi_i}, F_{\xi_j}\} \in \mathcal{C}^\infty(M) \tag{22.2.1}$$

The momentum map is the dual map to $\tilde{\mu}$, which is called the **Hamiltonian action** (in the group action sense, not the action functional sense). We can also set up a Lie algebra homomorphism $\tilde{\mu} : (\mathfrak{g}, [\cdot, \cdot]) \to (\mathcal{C}^\infty(M), \{\cdot, \cdot\})$. We can use the Hamiltonian action to understand Noether's theorem in more detail on the symplectic manifold M. The Lie algebra homomorphism can be written $\rho : \mathfrak{g} \to \Gamma(TM)$ where the map ρ is the **anchor** and $\Gamma(TM)$ is the space of smooth vector fields on M and is itself an infinite Lie algebra. We have the Lie algebra homomorphism $\mathcal{C}^\infty \ni F \to X_F \in \Gamma(TM)$, and the Hamiltonian action $\tilde{\mu} : \mathfrak{g} \to \mathcal{C}^\infty(M)$ that form a commutative diagram:

$$\mathfrak{g} \xrightarrow{\tilde{\mu}} \mathcal{C}^\infty(M)$$
$$\rho \searrow \qquad \downarrow X$$
$$\Gamma(TM)$$

This allows us to concretely write that $\forall\ \xi \in \mathfrak{g}$, we have $\rho(\xi) = X_F$ and, hence, $\xi \to F_\xi$, meaning symmetries embed themselves into the smooth algebra of functions on M, as we claimed the homomorphism does.

If the function F is invariant under the Hamiltonian action of a Lie group $\mathcal{G}$ on M, then the momentum $\kappa \in \mathfrak{g}^*$ is a first integral of the system with F. Then the inverse image of the momentum (i.e a level set in M) is denoted $M_\kappa = \mu^{-1}(\kappa)$ and is itself a manifold of dimension $(2n - 1)$ for $\dim(M) = 2n$. The action of the Lie group on M is generally to take elements of M_p into one another; however, there is a subgroup of $\mathcal{G}$, denoted $\mathcal{G}_\kappa$, that leaves M_κ stationary upon its group action, mapping the M_κ into themselves. The symmetry-reduced phase space is then the quotient manifold of M_κ, with the subgroup $\mathcal{G}_\kappa$ denoted $\mu^{-1}(\kappa)/\mathcal{G}_\kappa$, and is called the **Marsden-Weinstein symplectic quotient**. It is endowed with a symplectic structure inherited from M such that the pullback to $\mu^{-1}(\kappa)$ is the restriction of the 2-form ω on M to $\mu^{-1}(\kappa)$ and is of dimension $(2n - 2)$. As we have divided out the group action, the reduced phase space is now invariant under the flow of a Hamiltonian vector field Ψ_X, giving us the surfaces within the total space where the Noether currents are conserved. Note that the submanifold is characterised by the smooth functions in its algebra $\mathcal{C}^\infty(M_\kappa)$. In the larger space, the ideal of the functions which are Casimirs on M_p are surjective submersions of smooth functions on M and are restricted to the submanifold. We can divide $\mathcal{C}^\infty(M)$ by the ideal to get $\mathcal{C}^\infty(M_\kappa)$; we call the ideal the **centre** of the Poisson algebra. Routhian reduction is a special case of symplectic reduction such that, when the Euler-Lagrange equation is evaluated on the reduced space, it is the Routhian function that leads to the correct dynamics.

Symplectic reduction can also be seen in theories of constrained dynamics. Suppose there are $2m$ constraints ϕ^I, where $I = 1, \ldots, 2m$, with $m \leq n$. The submanifold $\mathcal{M}$ is given by[5]

$$\mathcal{M} := \{(q, p) \in M | \phi^I(q, p) = 0\} \subseteq M \qquad (22.2.2)$$

and, on this submanifold, $\{\phi^i, \phi^j\}$ is an invertible matrix $\forall\ (q, p) \in \mathcal{M}$. In this case, there exists a non-invertible Poisson structure of dimension $2(n - m)$ on M, called the Dirac bracket. The constraints are not derived via the Noether procedure but rather using the Dirac-Bergmann procedure, as detailed earlier in chapter 21.

A Lie group $\mathcal{G}$ may be considered as a manifold in its own right. Let us now consider the case where the base configuration manifold is a smooth Lie group with a Lagrangian

---

[5]For readers unfamiliar with this notation, the expression reads "the submanifold $\mathcal{M}$ is defined to be all the points in M for which the constraints are satisfied".

on the tangent bundle, $\mathscr{L} : T\mathcal{G} \to \mathbb{R}$. Recall that the Lie algebra of $\mathcal{G}$ is regarded as a tangent space of $\mathcal{G}$ when restricted to the identity element $\mathfrak{g} \equiv T_{\mathbb{I}}\mathcal{G}$ for the identity element $\mathbb{I} \in \mathcal{G}$. For an element of the dual of the Lie algebra $\kappa \in \mathfrak{g}^*$, the **Lie-Poisson bracket** for $F : \mathfrak{g}^* \to \mathbb{R}$ and $H : \mathfrak{g}^* \to \mathbb{R}$ is given by the pairing $\langle \cdot, \cdot \rangle : \mathfrak{g}^* \times \mathfrak{g} \to \mathbb{R}$:

$$\{F, H\}_{\pm}(\kappa) = \pm \left\langle \kappa, \left[\frac{\delta F}{\delta \kappa}, \frac{dH}{d\kappa} \right] \right\rangle \tag{22.2.3}$$

where $[\cdot, \cdot] : \mathfrak{g} : \mathfrak{g} \to \mathfrak{g}$ is a Lie bracket on the Lie algebra. We may think of this bracket as the canonical Poisson bracket (the normal one we use) on $T^*\mathcal{G}$ restricted to the cotangent space at the identity $T_{\mathbb{I}}^* \equiv \mathfrak{g}^*$. The theme of restricting the canonical/symplectic structure carries through the theory of Lie-Poisson mechanics and is called **Lie-Poisson reduction** $\mathfrak{g}^* = T^*\mathcal{G}/\mathcal{G}$. In a similar manner, we have a **Euler-Poincaré** system on the Lie algebra; this is regarded as a reduced Lagrangian system on the tangent bundle $T\mathcal{G}/\mathcal{G}$. When restricted to the dual Lie algebra, Hamilton's equations are called the **Lie-Poisson equations**, while the restricted Euler-Lagrange equation is known as the **Euler-Poincaré** equation. The evaluation of Hamilton's equations (Euler-Lagrange equations) on the Lie group are exactly the Lie-Poisson (Euler-Poincaré) equations and, via a Legendre transform, can be converted to the Euler-Poincaré (Lie-Poisson) equations:

$$\frac{d\kappa}{dt} = \mp \mathrm{ad}^*_{\delta H/\delta \kappa} \kappa \qquad \textbf{Lie-Poisson equation} \tag{22.2.4}$$

For $\xi = \xi^i e_i \in \mathfrak{g}$ for basis e_k with $l = \mathscr{L}|_{\mathfrak{g}}$ representing the reduced Lagrangian (restricted) such that $l : \mathfrak{g} \to \mathbb{R}$,

$$\frac{d}{dt}\left(\frac{dl}{d\xi} \right) = \pm \mathrm{ad}^*_\xi \frac{dl}{d\xi} \qquad \textbf{Euler-Poincaré equation} \tag{22.2.5}$$

Note that $\delta l/\delta \xi \in \mathfrak{g}^*$ is the derivative of the restricted Lagrangian with respect to an element of the Lie algebra (defined via the fibre derivative) and is the conjugate momentum to ξ; it is the dual to $\xi \equiv \delta H/\delta \kappa \in \mathfrak{g}$. Recall in both cases that the **adjoint action** of ξ on $\mathfrak{g}$ for $\xi \in \mathfrak{g}$ is a homomorphic map $\mathrm{ad}_\xi : \mathfrak{g} \to \mathfrak{g}$ with $\mathrm{ad}_\xi(\eta) = [\xi, \eta]$, $\forall \eta \in \mathfrak{g}$, and ad_ξ^* is the dual. These equations are used in rigid body dynamics, spinning tops and fluid dynamics.

Mathematical Background 4

The concept of infinitesimal generators stretches further than classical mechanics and is also seen in the position representation of the **Schrödinger picture** of quantum mechanics. The momentum eigenstates of the wavefunction are $\psi_p(x) = e^{ipx}$. Then consider an infinitesimal translation $\psi(x) \to \psi(x + c)$,

$$\psi(x + c) = e^{ip(x+c)} = e^{ipx}e^{ipc} = e^{ipx}(1 + icp + \cdots)$$

which is a series expansion around the momentum. However, this should also correspond to a Taylor expansion of the initial state by $\psi(x + c)$:

$$\psi(x + c) = \psi(x) + c\frac{d}{dx}\psi(x) + \cdots$$

When comparing the two, we can see that:

$$\psi(x + c) = \psi(x) + ic(-i\frac{d}{dx})\psi(x) + \cdots$$

Therefore, we identify the infinitesimal generator of momentum values as the operator $-i\frac{d}{dx}$, which is brought about by an infinitesimal translation, with the minus sign being necessary since $i^2 = -1$. Compare this result to the classical propagator in chapter 20 section 20.1.

22.3 Geometrical Lagrangian Mechanics

So that is pretty much all of autonomous Hamiltonian dynamics in a geometrical framework. Here we will demonstrate how the Poincaré-Cartan 1-form can also define most of Lagrangian mechanics, too, by way of a pullback to the tangent bundle. This will be the focus of the latter portion of this chapter before we qualitatively considering constrained systems in section 22.4.

For a smooth configuration manifold M, we can pull the symplectic 2-form back to the tangent bundle, using the Legendre transform $\mathbb{L}$, if the Lagrangian is regular (non-zero Hessian), to give a symplectic 2-form called the **Poincaré-Cartan 2-form** $(TM, \Omega, \mathscr{L})$ on TM, such that $\Omega = \mathbb{L}\omega$. Like ω on the cotangent bundle, Ω is closed $(d\Omega = 0)$ and exact. The fundamental equation of motion in the Lagrangian framework (the **Lagrangian-Hamilton-Jacobi equation**) is then

$$\iota_{X_{\mathscr{L}}}\Omega = \Omega(X_{\mathscr{L}}, \cdot) = dJ \qquad \textbf{Lagrangian-Hamilton-Jacobi equation} \qquad (22.3.1)$$

where $J \in \mathcal{C}^{\infty}(TM)$ is the energy function $J : TM \to \mathbb{R}$, as before. As for the Hamiltonian case, we aim to find the unique vector field whose integral curve lifted to TM is an integral curve of the Lagrangian vector field $X_{\mathscr{L}}$. We can assume the non-degeneracy of Ω if $\mathscr{L}$ is regular, so that unique solutions exist and the Hamilton-Jacobi problem can be solved. We can express Ω using the Darbouxian coordinates that define ω:

$$\Omega = \mathbb{L}(\sum_i dq^i \wedge dp_i)$$

$$= \sum_i d(\mathbb{L}^* q^i) \wedge d(\mathbb{L}^* p_i)$$

$$= \sum_{i,j} d(q^i) \wedge d\left(\frac{\partial \mathscr{L}(q,v)}{\partial v^j}\right)$$

$$= \sum_{i,j} dq^i \wedge \left(\frac{\partial^2 \mathscr{L}}{\partial q^j \partial v^i} dq^j + \frac{\partial^2 \mathscr{L}}{\partial v^i \partial v^j} dv^j\right)$$

$$= \sum_{i,j} \left(\frac{\partial^2 \mathscr{L}}{\partial q^j \partial v^i} dq^i \wedge dq^j + \frac{\partial^2 \mathscr{L}}{\partial v^i \partial v^j} dq^i \wedge dv^j\right) \qquad (22.3.2)$$

where we have used $d\phi = (\partial\phi/\partial x^i)dx^i$. In essence, we have pulled back θ_H by $\theta_{\mathscr{L}} = \mathbb{L}^* \theta_H$ and then taken $\Omega = -d\theta_{\mathscr{L}}$, where $\theta_{\mathscr{L}}$ is the **Poincaré-Cartan 1-form** $\theta_{\mathscr{L}}$. In local coordinates, it is given by

$$\theta_{\mathscr{L}} = \frac{\partial \mathscr{L}}{\partial v^i} dq^i$$

The time derivative of $\theta_{\mathscr{L}}$, is given by $d\mathscr{L}$ with components of the generalised force and the generalised momentum:

$$\frac{d}{dt}\left(\frac{\partial \mathscr{L}}{\partial \dot{q}^i} dq^i\right) = \left(\frac{d}{dt}\frac{\partial \mathscr{L}}{\partial \dot{q}^i}\right) dq^i + \frac{\partial \mathscr{L}}{\partial \dot{q}^i} d\dot{q}^i = d\mathscr{L} \qquad (22.3.3)$$

where, in the last equality, we have used the Euler-Lagrange equation. The exterior derivative of this must vanish, since $d^2 = 0$, and, commuting it with the time derivative, we obtain

$$\frac{d}{dt}\left(d\frac{\partial \mathscr{L}}{\partial \dot{q}^i} \wedge dq^i\right) = d^2 \mathscr{L} = 0 \qquad (22.3.4)$$

This shows us that the flow induced by $\theta_{\mathscr{L}}$ is symplectic, just as in the Hamiltonian case, where $\mathcal{L}_{X_{\mathscr{L}}}\theta_{\mathscr{L}} = d\mathscr{L}$, and hence provides us with a Lagrangian version of Arnold's theorem.

A key question should be how the above equation can lead to a family of second-order equations while the Hamiltonian analogue leads to a family of first-order differential equations? We will see that this is one of the main differences between Hamiltonian and Lagrangian formulations which is a residue of their respective spaces and the definition of their conjugate momenta.

It now becomes a matter of evaluating the components! Returning to equation 22.0.4, we see the component of the ∂v basis and the component of the ∂q basis (note that the last equality only holds for on-shell, integral curves):

$$\iota_{X_{\mathscr{L}}} dq = \dot{q}, \qquad \iota_{X_{\mathscr{L}}} dv = \dot{v} = \left[\frac{\partial^2 \mathscr{L}}{\partial v^i \partial v^j}\right]^{-1} \left(\frac{\partial \mathscr{L}}{\partial q^i} - \frac{\partial^2 \mathscr{L}}{\partial q^j \partial v^i} \dot{q}^j\right) \qquad (22.3.5)$$

As we have discussed several times previously, this is the differential form of the Lagrange equation of motion. So, all we have to do is take the interior product of Ω and substitute the on-shell expression in, and we should prove the above theorem, just as we did for the Hamiltonian analogue! Thus,

$$
\begin{aligned}
\iota_{X_{\mathscr{L}}}\Omega = \iota_{X_{\mathscr{L}}} & \left(\sum_{i,j} \frac{\partial^2 \mathscr{L}}{\partial q^j \partial v^i} dq^i \wedge dq^j + \sum_{i,j} \frac{\partial^2 \mathscr{L}}{\partial v^i \partial v^j} dq^i \wedge dv^j\right) \\
= \sum_{i,j} & \left(\frac{\partial^2 \mathscr{L}}{\partial q^j \partial v^i}(\iota_{X_{\mathscr{L}}} dq^i) \wedge dq^j - \frac{\partial^2 \mathscr{L}}{\partial q^j \partial v^i} dq^i \wedge (\iota_{X_{\mathscr{L}}} dq^j)\right. \\
& \left. + \frac{\partial^2 \mathscr{L}}{\partial v^i \partial v^j}(\iota_{X_{\mathscr{L}}} dq^i) \wedge dv^j - \frac{\partial^2 \mathscr{L}}{\partial v^i \partial v^j} dq^i \wedge (\iota_{X_{\mathscr{L}}} dv^j)\right) \\
= \sum_{i,j} & \left(\frac{\partial^2 \mathscr{L}}{\partial q^j \partial v^i} \dot{q}^i dq^j - \frac{\partial^2 \mathscr{L}}{\partial q^j \partial v^i} \dot{q}^j dq^i + \frac{\partial^2 \mathscr{L}}{\partial v^i \partial v^j} \dot{q}^i dv^j - \frac{\partial^2 \mathscr{L}}{\partial v^i \partial v^j} \dot{v}^j dq^i\right) \\
= \sum_{i,j} & \frac{\partial^2 \mathscr{L}}{\partial v^i \partial v^j} \dot{q}^i dv^j - \sum_{i,j} \frac{\partial^2 \mathscr{L}}{\partial v^i \partial v^j}\left(\frac{\partial^2 \mathscr{L}}{\partial v^i \partial v^j}\right)^{-1}\left[\frac{\partial \mathscr{L}}{\partial q^j} - \dot{q}^i \frac{\partial^2 \mathscr{L}}{\partial v^i \partial q^j}\right] dq^j \qquad (22.3.6)
\end{aligned}
$$

where in the first , we have used the product rule on the wedge product $\iota_X(\alpha \wedge \beta) = (\iota_X\alpha) \wedge \beta + \alpha \wedge (\iota_X\beta)$ twice, since we have two wedges. In the second step, we just evaluated each interior product using the above relations from the Lagrange vector field. In the final step, we assumed that the equations of motion have indeed been satisfied, so we can substitute for the accelerations; in addition, we are assuming the Hessian can be inverted. All that remains is to cancel the two Hessians $W_{ij}(W_{ij})^{-1} = \mathbb{I}$ and we arrive at the previously calculated dJ in equation 22.0.7:

$$\iota_{X_{\mathscr{L}}}\Omega = \sum_{i,j} v^i \frac{\partial^2 \mathscr{L}}{\partial v^i \partial v^j} dv^j + \sum_{i,j}\left[v^i \frac{\partial^2 \mathscr{L}}{\partial v^i \partial q^j} - \frac{\partial \mathscr{L}}{\partial q^j}\right] dq^j = dJ \qquad (22.3.7)$$

Therefore, the equation $\iota_{X_{\mathscr{L}}}\Omega = dJ$ holds. Vector fields that satisfy this are called **Lagrangian vector fields**; they are the solutions to the equation of motion for the base curve γ, meaning $X \circ \gamma = \dot{\gamma}$. Two Lagrangians are **equivalent** if their Lagrangian vector fields are equal:

$$\frac{d}{dt}\left(\frac{\partial \mathscr{L}}{\partial v^i}\right) - \frac{\partial \mathscr{L}}{\partial q^i} = \frac{\partial^2 \mathscr{L}}{\partial v^i \partial q^j}\dot{q}^j + \frac{\partial^2 \mathscr{L}}{\partial v^i \partial v^j}\dot{v}^j - \frac{\partial \mathscr{L}}{\partial q^i} = 0$$

You can recover the Lagrange equation of motion directly by equating to dJ prior to substituting for the acceleration and comparing like coefficients; you will note that terms in dv^j show that $\dot{q}^j = v^j$, while terms in dq^j show that, when the Lagrangian is regular, we generate the Euler-Lagrange equation. Later, when we discuss jet theory in chapter 23, we will understand how a second-order differential equation is actually an equivalence class between the base curve on the configuration manifold, the tangent vectors (the velocities) and the second derivatives (the accelerations). It may not seem like a big deal but the fact that $\dot{q}^j = v^j$ is actually the reason why the tangent bundle structure of mechanics gives a set of second-order equations while the cotangent bundle gives a set of two first-order equations. It is the Lagrange vector field really that gives us this property.

We can write a Poisson bracket on $\mathcal{C}^\infty TM$ by evaluating the Poisson bi-vector field on $F, G \in \mathcal{C}^\infty(TM)$ in the following way:

$$\eta(dF(q,v), dG(q,v)) = \left(\frac{\partial F}{\partial q^i}\left(\frac{\partial^2 \mathscr{L}}{\partial v^i \partial v^j}\right)^{-1}\frac{\partial G}{\partial v^i} - \frac{\partial G}{\partial q^i}\left(\frac{\partial^2 \mathscr{L}}{\partial v^i \partial v^j}\right)^{-1}\frac{\partial F}{\partial v^i}\right) \qquad (22.3.8)$$

where $\eta \in \chi(\bigwedge^2(TM))$ on the space of sections:

$$\eta = \left(\frac{\partial^2 \mathscr{L}}{\partial q^i \partial v^j}\right)^{-1}\frac{\partial}{\partial q^i} \wedge \frac{\partial}{\partial q^j} + \left(\frac{\partial^2 \mathscr{L}}{\partial v^i \partial v^j}\right)^{-1}\frac{\partial}{\partial q^i} \wedge \frac{\partial}{\partial v^j} \qquad (22.3.9)$$

This only exists if the Lagrangian is regular and hence the Hessian is invertible. It is not a natural structure but an induced one, since the Poisson algebra is natural to T^*M.

We must note the importance of the Hessian condition; whenever its inverse does not exist, we say that the 2-form is **presymplectic**, meaning that it is not closed (you will note that the symplectic 2-form was required to be closed and non-degenerate). An **almost symplectic form** contains degeneracies and it is not of full rank. If there is a non-zero X for which $\iota_X \omega = 0$, then ω is no longer invertible and there are no bundle isomorphisms between the tangent and cotangent fibre bundles. We can try to reduce the dimensionality of the symplectic manifold to space where the 2-form is non-degenerate through the process of symplectic reduction, as detailed earlier in section 22.2 and which may sound familiar from our discussion of constraints. The reader must note that, if the form is not closed, then the algebra of observables no longer satisfies the Jacobi identity of the Lie algebra on the Poisson manifold. Further, Darboux' theorem no longer applies, so we struggle to express the form in local coordinates and we certainly can't induce structures on TM globally.

Therefore, we would use a presymplectic form to describe constrained dynamics whose Lagrangians have singularities. We can't solve the fundamental equations of motion $\iota_{X_H}\omega = dH$ and $\iota_{X_{\mathcal{L}}}\Omega = dJ$ globally in this case, but we suspect that their solution exists in some regions. Theories of constrained dynamics aim to find the sub-manifolds which have a non-degenerate 2-form, that is to say, vector fields that satisfy the respective fundamental equations.

22.4 Elements of Constrained Geometry

Let us now frame a mechanical system subject to a constraint in the geometrical setting. Constraints define a subbundle of the tangent bundle to which the motion of the system must remain. The Euler-Lagrange equations are defined by a vector field on the tangent bundle that, in the general case, is not compatible with the constraint. Instead, we define a projection from the space of accelerations to the tangent bundle such that the vector field is compatible with the constraints. There is not a unique way of performing this reduction.

A common approach is to formulate the configuration space as a Riemannian man-ifold (see Appendix B). A Riemannian manifold is a smooth manifold equipped with a metric that induces a smooth inner product in the tangent spaces above points; in this case, the Euler-Lagrange equation describes geodesics on the base space. We recall that the constraint force is normal to the surface. The most general type of constraint we may encounter is a time-dependent, non-holonomic constraint (velocity dependent) $\phi(q, v, t) = 0$. We may write the time derivative:

$$\dot{\phi} = \frac{\partial\phi}{\partial q^i}v^i + \frac{\partial\phi}{\partial v^i}\dot{v}^i + \frac{\partial\phi}{\partial t} = 0 \qquad (22.4.1)$$

The acceleration for the unconstrained system is written using Cristoffel symbols:

$$a^j = \dot{v}^j + \Gamma^j_{kl}v^k v^l \qquad (22.4.2)$$

If we substitute this into the equation for $\dot{\phi}$, we obtain an affine relation for the accel-erations:

$$\frac{\phi}{\partial v^k}a^k + (\phi_k v^k - \frac{\partial\phi}{\partial v^i}\Gamma^i_{kl}v^k v^l + \frac{\partial\phi}{\partial t}) = 0 \qquad (22.4.3)$$

The acceleration of the constrained problem a_ϕ must be made to lie in the affine plane. One way to achieve this is to subtract the orthogonal projection $a_\perp$ component from the unconstrained acceleration a, the result being the acceleration of the constraint; this is **d'Alembert's principle**, in the language of affine connections:

$$a_\phi = a - a_\perp \qquad (22.4.4)$$

Restriction of the integrable portion, a, would yield Lagrange's equation.

There are however, several approaches that investigate constrained dynamics: the **Dirac-Bergmann** method, most notably, but also the **Gotay-Nester** theory, **Hamilton-Nambu dynamics** (appendix D), **vakonomic mechanics** and the **Skinner-Rusk formalism**. Most try to find phase space regions to which the 2-form can be restricted to and in which they can be inverted locally, and will include some method of including what we have been calling Lagrange multipliers from the space of smooth functions. What follows is a *qualitative* discussion of a very tricky topic.

The process of finding the reduced phase space is called **symplectic reduction** and we have seen that the **momentum map** is a function on the phase space that tells us how the Lie algebra of a symmetry embeds into the Poisson algebra of the phase space observables. It can be used to find a surface where the symmetry is invariant under the flow of a Hamiltonian.

The **Skinner-Rusk formalism** combines all the aspects of geometrical mechanics into one neat package by sort of bunching up (not a technical term) of the tangent and cotangent bundles. To be more precise about it, we say that the *evolution space* is the **Whitney sum** of the tangent and cotangent bundles. A Whitney sum $\oplus$ is an operation that takes two vector bundles over a manifold and makes a new one over the base space. A fibre in the Whitney sum space is the direct sum of the fibres of both constituent spaces over a point in the base space. For instance, we write $\mathbb{T}M = TM \oplus T^*M$; it is referred to as the **Pontryagin bundle** in some literature. There are projections that map from the sum space into the respective constituent bundles, allowing us to regenerate either description of mechanics. The use behind this approach is to unite singular and regular dynamics together. The important object of this formalism is the **Courant bracket** $\{\cdot,\cdot\}_{CB}$ that acts on the set of sections of a vector bundle, together with the extension of the symplectic structure to a **Dirac structure**.

If we recall, a vector field assigns to a point on M a vector in the tangent bundle. The generalisation of this to any vector bundle E is called a **section**, such that it assigns a vector from the attached vector bundle. Therefore, a section of the tangent bundle is a vector field on the base space that is a smooth function of the coordinates of a manifold. We denote the set of all sections at $p \in M$ as $\Gamma_p E$ for our constructed bundle; in our case, $E = \mathbb{T}M$. The Courant bracket maps two sections in the direct sum space to a third: $\{\cdot,\cdot\}_{CB} : \Gamma(\mathbb{T}M) \otimes \Gamma(\mathbb{T}M) \to \Gamma(\mathbb{T}M)$. It satisfies all the usual properties that come with brackets, apart from the Jacobi identity, meaning it does not constitute a Lie algebra structure but rather a **Courant algebroid** structure; this is the first main difference. We define a **Dirac subbundle** L as a subbundle of the Whitney sum bundle with the Courant bracket on its sections $L \subset TC \oplus T^*C$ (recalling that we defined C as a leaf submanifold of the Poisson manifold; we now take that leaf to be a constraint manifold).

We close by recalling that non-holonomic mechanics is best described by the Gibbs-Appell equations of motion. The geometry of these equations can be found in J. F.

Cariñena and J. Fernández-Núñez's (2010) article, A geometric approach to the Gibbs-Appell equations in Lagrangian mechanics. For a detailed exposition of constrained dynamics via the Dirac algebroid approach to dynamics please see K. Grabowska and J. Grabowski' Dirac algebroids in Lagrangian and Hamiltonian mechanics. For a discussion of Dirac manifolds and the introduction of the Courant bracket, see T. J. Courant's (1990) article, Dirac manifolds. The reader may also be interested in the geometry of symplectic reduction; in that case, I recommend J. E. Marsden, T. S. Ratiu and J. Scheurle, *Reduction theory and the Lagrange-Routh equations* which gives a fantastic in-depth account, as well as other papers in the reference section.

This concludes our discussion of geometrical mechanics for time-independent systems. It is hoped that the reader will now appreciate why we treat $\dot{q}^i = v^i$ as being independent from the q^is, and why the phase space Lagrangian strictly depends on four variables. It is hoped that a firm understanding of the respective spaces employed in classical mechanics have been investigated and how they lead to second-order differential equations in the Lagrange case, and first-order in the Hamiltonian case. Last, we can also appreciate that constrained dynamics is a lot more complicated than perhaps first imagined; as well as the importance of the Hessian condition for solving the fundamental equations. (We have not discussed gauge theories in geometry due to their involved nature and principal-bundle structure.)

Perhaps most importantly, we have developed ideas in differential geometry and shown how they can be used to understand physical theory in a different light. Classical mechanics is not the only branch of physics that has a geometric formulation. Quantum mechanics, thermodynamics and general relativity all share in a slice of pure mathematics; personally I think that this can be the most rewarding and exciting area of modern science. We will very briefly consider how electrodynamics can be understood using geometry a little later on in chapter 27.

For further reading on the general formalism of Lagrangian and Hamiltonian mechanics in geometry, please see J. E. Marsden and T. S. Ratiu's (1994) book, *Introduction to Mechanics and Symmetry*, Abraham and Marsden's (1994) book, *Foundations of Mechanics* and Arnold's (1989) book,*Mathematical Methods of Classical Mechanics*.

Chapter summary

- The setting of classical mechanics using the language of differential geometry was developed. The **tangent bundle** to a configuration manifold was found to be a natural setting for the Lagrangian formalism, which is summarised by the **fundamental equation of motion,**

$$\iota_{X_{\mathscr{L}}}\Omega = dJ$$

  The Hamiltonian formalism was developed on the **cotangent bundle,** which is a **symplectic phase space**. It was shown that a single scalar function $H : T^*M \mapsto \mathbb{R}$ determines the dynamics of the system. The fundamental equation of canonical dynamics is then expressed as

$$\iota_{X_H}\omega = dH$$

Exercise 22.1 Given a trajectory $\gamma(t) : (1,2) \to M$ with $\gamma(0) \in M$, that is parameterised by generalised coordinates q^i, such that, $q^i(\gamma(t)) = x^i(t)$, and the tangent vector, $X = X^i \partial_{x^i} \in TM_{\gamma(0)}$ where:

$$X^i = \frac{dx^i(t)}{dt} \tag{22.4.5}$$

How can we write the velocity of the curve?

Exercise 22.2 Consider a Riemannian configuration manifold (M, g) with local coordinates $(q^1, \ldots, q^n)$ on M and induced coordinates $(q^1, \ldots, q^n, v^1, \ldots, v^n)$ on TM and $(q^1, \ldots, q^n, p_1, \ldots p_n)$ on T^*M. Using the Lagrangian for a free particle, $\mathscr{L} \in \mathcal{C}^\infty(TM)$ which, in local coordinates, is given by $\mathscr{L} = \frac{1}{2}\sum \mathrm{g}_{ij}v^iv^j$. Show that the equations of motion are given by:

$$\dot{v}^i + \Gamma^i_{jk}v^jv^k = 0 \tag{22.4.6}$$

where

$$\Gamma^a_{bc} = \frac{1}{2}g^{ad}\left(\frac{\partial g_{bd}}{\partial q_c} + \frac{\partial g_{cd}}{\partial q_b} - \frac{\partial g_{bc}}{\partial q_d}\right) \tag{22.4.7}$$

Solve this equation for a free particle in a flat space, where $g_{ij}(q^k) = \delta_{ij}$.

Exercise 22.3 Continuing from exercise 22.2, we can perform a Legendre transform to obtain the Hamiltonian $H \in \mathcal{C}^\infty(T^*M)$ which in local coordinates is given by $H = \frac{1}{2}\sum \mathrm{g}^{ij}p_ip_j$. Prove that in this context Hamilton's ODEs are equivalent to the geodesic equation.

$$\dot{q}^k = \frac{\partial H}{\partial p_k}, \qquad \dot{p}_k = -\frac{\partial H}{\partial q^k} \qquad \longrightarrow \qquad \ddot{q}^s + \sum_{ji}\Gamma^s_{ji}\dot{q}^j\dot{q}^i = 0 \tag{22.4.8}$$

The first step is to construct some partial derivatives that appear in Hamilton's equations:

$$\frac{\partial H}{\partial p_k} = g^{kj}p_j, \qquad -\frac{\partial H}{\partial q^k} = -\frac{1}{2}\frac{\partial g^{ij}}{\partial q^k}p_i p_j \qquad (22.4.9)$$

Using $p_k = g_{lk}\dot{q}^l$, we can re-write Hamilton's equations[6].

$$\dot{p}_k = -\frac{1}{2}\frac{\partial g^{ij}}{\partial q^k}g_{il}g_{jm}\dot{q}^l\dot{q}^m = \frac{1}{2}\frac{\partial g_{ij}}{\partial q^k}\dot{q}^i\dot{q}^j = \sum_i \frac{\partial g_{ki}}{\partial q^j}\dot{q}^i\dot{q}^j + \sum_{ki} g_{ki}\ddot{q}^i \qquad (22.4.10)$$

Using the delta function and inverse relationship $-\frac{\partial g^{ij}}{\partial q^k}g_{il}g_{jm} = \frac{\partial g_{ij}}{\partial q^k}$ and separating the time and spatial components out,

$$\ddot{q}^i + \sum_{ikj} g^{ik}\left(\frac{\partial g_{ik}}{\partial q^j} - \frac{1}{2}\frac{\partial g_{ij}}{\partial q^k}\right)\dot{q}^i\dot{q}^j = 0 \qquad (22.4.11)$$

Therefore,

$$\ddot{q}^s + \frac{1}{2}\sum_{kij} g^{sk}\left(\frac{\partial g_{ki}}{\partial q^j} + \frac{\partial g_{kj}}{\partial q^i} - \frac{\partial g_{ij}}{\partial q^k}\right)\dot{q}^i\dot{q}^j = 0 \qquad (22.4.12)$$

Exercise 22.4 If a curve $\gamma : \mathbb{R} \to (U, q^1, \ldots, q^n)$ obeys the Euler-Lagrange equations on a chart $U \subset M$, then the fibre derivative or Legendre transform of the tangent-bundle $(TU, q^1, \ldots q^n, v^1, \ldots, v^n)$ lifted curve $\tilde{\gamma}(t) = (\gamma(t), \gamma'(t))$ is an integral curve of the Hamiltonian vector field X_H on the cotangent bundle $(\mathbb{L}(TU), q^1, \ldots, q^n, p_1, \ldots, p_n)$. In order to demonstrate this, we assume that $\tilde{\gamma}(t)$ is a solution to the Euler-Lagrange equations such that we can write:

$$\frac{d}{dt}\left(\frac{\partial \mathscr{L}}{\partial v^i}(\tilde{\gamma}(t))\right) = \frac{\partial \mathscr{L}}{\partial q^i}(\tilde{\gamma}(t)) \qquad (22.4.13)$$

The fibre derivative of the lifted curve is $\mathbb{L}_{\mathscr{L}}(\tilde{\gamma}(t)) = \mathbb{L}_{\mathscr{L}}(\gamma(t), \gamma'(t)) = (\gamma(t), \mathbb{L}_{\mathscr{L}}(\gamma'(t)))$, since the q coordinates are shared by both TU and T^*U. Hamilton's equations for the curve are given by

$$\frac{d}{dt}\gamma^i(t) = \frac{\partial H}{\partial p_i}(\gamma(t)\mathbb{L}_{\mathscr{L}}(\gamma'(t))), \qquad \frac{d}{dt}(\mathbb{L}_{\mathscr{L}}(\gamma'(t)))_i(t) = -\frac{\partial H}{\partial q^i}(\gamma(t), \mathbb{L}_{\mathscr{L}}(\gamma'(t))) \quad (22.4.14)$$

We now need to show that the two descriptions are consistent with each other, in order to complete the proof. We do this by realising that the fibre derivative in the other direction, $\mathbb{L}_H$

---

[6] For any vector $v = v^i e_i$ the contravariant (v^i) and covariant (v_i) components can be switched using a metric tensor $g^{ik}g_{kj} = \delta^i_j$; to switch the components, we invert the relationship $g^{ij}v_j = g^{ij}g_{jk}v^k = \delta^i_k v^k = v^i$. You can then look at this as applying g^{ij} to a vector v_j; doing so will raise its index, while applying g_{ij} to v^i will lower the index.

taking $(T^*U \to TU)$, can be written $\mathbb{L}_H(\gamma(t), \mathbb{L}_{\mathscr{L}}(\gamma'(t))) = \tilde{\gamma}(t)$; in other words, applying the fibre derivative $\mathbb{L}_{\mathscr{L}}$ to $\tilde{\gamma}(t)$ followed by $\mathbb{L}_H$ gets us right back to $\tilde{\gamma}(t)$. Hence,

$$\gamma_i'(t) = \frac{\partial H}{\partial p_i}((\gamma(t), \mathbb{L}_{\mathscr{L}}(\gamma'(t)))) \tag{22.4.15}$$

which is the first of Hamilton's equations. The second follows from,

$$\frac{d}{dt}(\mathbb{L}_{\mathscr{L}}(\gamma'(t)))_i \overset{(1)}{=} \frac{d}{dt}\frac{\partial \mathscr{L}}{\partial v^i}(\tilde{\gamma}(t)) \overset{(2)}{=} \frac{\partial \mathscr{L}}{\partial q^i}(\tilde{\gamma}(t)) \overset{(3)}{=} -\frac{\partial H}{\partial q^i}(\gamma(t), \mathbb{L}_{\mathscr{L}}(\gamma'(t))) \tag{22.4.16}$$

where (1) is by definition, (2) uses the Lagrange equation and (3) follows from what we set out to prove. The reverse logic is now straightforward, although now we denote the cotangent lift of a curve as $\gamma_\sharp$.

Exercise 22.5 Why is the Hamiltonian vector field X_H unique for a non-degenerate symplectic 2-form ω and a Hamiltonian H?

Since the 2-form is non-degenerate there is no non-zero vector field X such that $\iota_X\omega = 0$. The dual vector field for X is $\iota_X\omega$, and both the vector space and its duel space are $2n$ dimensional. The mapping between spaces is therefore bijective with an inverse relationship. This means that, for every vector field X, there is a unique duel field $\iota_X\omega$. It all follows from the non-degeneracy of the 2-form. Conversely, the Hamiltonian $H \in \mathcal{C}^\infty(T^*M)$ for a Hamiltonian vector field is not unique but is defined up to an additive constant.

Exercise 22.6 Consider the cotangent space $T^*(\mathbb{R}^n)$ with coordinates q and p. Let $\omega = dq \wedge dp$ be the symplectic 2-form, and the Hamiltonian $H(q,p) = p^2/2 + V(q)$. The vector field X_H is:

$$X_H = p\partial_q - \partial_q V \partial_p \tag{22.4.17}$$

Show that the flow equations are given by Newton's equation of motion:

$$\dot{q} = p, \qquad \dot{p} = -\partial_q V \tag{22.4.18}$$

Show that the Lie derivative of ω vanishes under this evolution.

Exercise 22.7 Liouville's theorem does not hold on the tangent bundle TM to a configuration manifold, M. This is because there is no natural volume form on TM unlike on the cotangent bundle, T^*M. Taking $\mathscr{L} = q\dot{q}^2/2$ show that

$$dp \wedge dq = q d\dot{q} \wedge dq \tag{22.4.19}$$

Reason, that if the symplectic form is preserved under Hamiltonian flow, then, neither of the quantities on the right, q or $d\dot{q} \wedge dq$ can be constant independently. Therefore, the volume form on the tangent bundle is not conserved during the motion of the system and Liouville's theorem is invalid.

Exercise 22.8 The *non-squeezing theorem* is an important result in the geometry of symplectic transformations. It states that one cannot embed a ball of points in phase space into a cylinder via a symplectic mapping unless the cylinder is narrower than the ball. This is in contrast to the weaker condition of Liouville's theorem, which simply requires a volume preserving map; where a ball of any radius can be squeezed into any cylinder by squeezing it small enough. Hence the theorem shows that symplectic transformations are a special case of volume-preserving maps.

Suppose that we have symplectic spaces $\mathbb{R}^{2n}$ with coordinates $(q^1,\ldots,q^n,p_1,\ldots,p_n)$, a ball of radius R

$$B(R) = \{z \in \mathbb{R}^{2n} \,|\, \|z\| < R\}, \tag{22.4.20}$$

and a cylinder of radius r

$$C(r) = \{z \in \mathbb{R}^{2n} \,|\, q_1^2 + p_1^2 < r^2\}, \tag{22.4.21}$$

each of which is endowed with a symplectic form

$$\omega = dq^i \wedge dp_i \tag{22.4.22}$$

The non-squeezing theorem states that there is no symplectic embedding of $B(R)$ into $C(r)$ unless $r > R$. Therefore, under Hamiltonian flow, there is *no* evolution that squeezes the ball in a direction such that a particular (q^i,p_i)-plane can pass through a hole with an area less than the cross section of the ball. This is stronger than Liouville's theorem and the conservation of the symplectic volume form. Investigate the implications of the non-squeezing theorem on Hamiltonian flow and, to the Heisenberg uncertainty relation.

Exercise 22.9 Let $M \equiv \mathcal{G}$ be a Lie group with Lie algebra $\xi = \xi^i e_i \in \mathfrak{g}$ equipped with a Lie bracket $[\cdot,\cdot]$, a Lagrangian $l : \mathfrak{g} \to \mathbb{R}$ and a dual $\kappa = \kappa_i e^i \in \mathfrak{g}^*$ where the components are given by $\kappa_i = \partial l(\xi)/\partial \xi^i$. The system has a natural pairing $\langle\cdot,\cdot\rangle : \mathfrak{g}^* \times \mathfrak{g} \to \mathbb{R}$, given by $\langle\kappa,d\xi\rangle = dl$. Consider a variation in ξ to be given by $\delta\xi = \dot\eta \pm [\xi,\eta]$ where $\xi,\eta \in \mathfrak{g}$ and the basis vectors satisfy $\langle e^i, e_j\rangle = \delta^i_j$, with $[\xi^i e_i, \eta^j e_j] = \xi^i \eta^j c^k_{ij} e_k = [\xi,\eta]^k e_k$ pulling the components out in the first step and using the structure constants. The variations are subject to η vanishing at the end points and the $\pm$ sign can be written as $+$ for left invariance and $-$ for right invariance; to begin with, follow the derivation through with either a plus or a minus.

Using Hamilton's principle, show that the vanishing of the action integral of the Lagrangian l leads to the Euler-Poincaré equations of motion:

$$\delta S = \int_1^2 dt \left\langle \frac{\partial l}{\partial \xi}, \delta\xi \right\rangle \tag{22.4.23}$$

$$= \int_1^2 dt \left\langle \frac{\partial l}{\partial \xi}, \dot\eta \pm [\xi,\eta] \right\rangle \tag{22.4.24}$$

$$= \int_1^2 dt \left\langle \frac{\partial l}{\partial \xi^m} e^m, \dot\eta^i e_i \pm \xi^j \eta^k c^i_{jk} e_i \right\rangle \tag{22.4.25}$$

$$= \int_1^2 dt \left(\frac{\partial l}{\partial \xi^m} \dot\eta^i e^m e_i \pm \frac{\partial l}{\partial \xi^m} \xi^j \eta^k c^i_{jk} e^m e_i \right) \tag{22.4.26}$$

We next use $\langle e^m, e_i\rangle = \delta^m_i$ to collect the indices and then integrate the first term by parts so we can collect η^i terms with relabelling:

$$\int_1^2 dt \left(-\frac{d}{dt}\frac{\partial l}{\partial \xi^i} \pm \frac{\partial l}{\partial \xi^k}\xi^j c^k_{ji} \right)\eta^i + \left(\frac{\partial l}{\partial \xi^i}\eta^i \right)^2_1 \tag{22.4.27}$$

Setting $\delta \mathcal{S} = 0$ gives us the Euler-Poincaré equation,

$$-\frac{d}{dt}\frac{\partial l}{\partial \xi^i} \pm \frac{\partial l}{\partial \xi^k}\xi^j c^k_{ji} = 0 \tag{22.4.28}$$

We can collect the negative signs by using the asymmetry of the structure constants $c^k_{ji} = -c^k_{ij}$ and the adjoint notation $-\mathrm{ad}^*_\xi = \frac{\partial l}{\partial \xi^k}\xi^j c^k_{ij}$ to give us the usual expression:

$$\frac{d}{dt}\frac{\partial l}{\partial \xi} \mp \mathrm{ad}^*_\xi \frac{\partial l}{\partial \xi} = 0 \tag{22.4.29}$$

Exercise 22.10 We discuss here how a Poisson algebra $(\mathcal{C}^\infty(\mathcal{P}), \{\cdot, \cdot\})$ can be deformed into a quantum mechanical system through a procedure called **deformation quantisation**. Our discussion is quite brief and overviews the main ideas rather than the explicit details. Recall that the observables on the phase space are real-valued functions that are associative and commutative and possess a Lie algebra structure. A state is a point in phase space and is characterised by the evaluation of the observable functions on that point.

Imagine now a picture where the observables are no longer commutative and the states are non-localised distributions on the phase space (e.g. not Dirac delta functions). Therefore, evaluation of the state's observable is the resultant of a non-local contribution about the neighbourhood of the point and, hence, there is an inherent uncertainty associated with evaluating the observables. The non-commutative property is endowed to the algebra of observables by equipping it with a **star product**, $(F, G) = F * G$, in place of its (classical) commutative product, $(F, G) = FG$. The star product is a formal power series about the formal deformation parameter $(\hbar)$ for real-valued functions and can be written

$$F * G = \sum_{k=0}^{\infty}(\hbar)^k C_k(F, G) \tag{22.4.30}$$

The coefficients C_k are bi-differential operators and are hence made up of the derivatives of F and G with respect to the coordinate basis $\xi^i = (q^1, \ldots, q^n, p_1, \ldots, p_n)$, where $i = 1, \ldots 2n$. The product is associative and the $k = 0$ term is simply FG indicating it truly is a deformation of the $\mathcal{C}^\infty(\mathcal{P})$ algebra. We can define a star commutator $[F, G]_* = F * G - G * F$ and set the classical limit as the Poisson bracket:

$$\lim_{\hbar \to 0}\frac{1}{\hbar}[F, G]_* = \{F, G\} \tag{22.4.31}$$

If the phase space is Euclidean $\mathbb{R}^{2n}$ then we can choose the first coefficient of the series C_1 in terms of the coefficients of the Poisson bi-vector field η^{ij}:

$$C_1(F, G) = \eta^{ij}(\partial_i F)(\partial_j G) \tag{22.4.32}$$

In this case, $C_1(F, G) - C_1(G, F) = \{F, G\}$, and the first few terms of the star product are written

$$F * G = FG + \hbar \sum_{i,j} \eta^{ij} \frac{\partial F}{\partial \xi^i} \frac{\partial G}{\partial \xi^j} + \frac{\hbar^2}{2} \sum_{i,j,k,l} \eta^{kl} \eta^{ij} \frac{\partial^2 F}{\partial \xi^i \partial \xi^k} \frac{\partial^2 G}{\partial \xi^j \partial \xi^l} + \cdots \qquad (22.4.33)$$

which can be tidied up into a closed expression known as the **Moyal product**:

$$F * G = F \exp\left\{ \frac{\hbar}{2} \eta^{ij} \overleftarrow{\partial_i} \overrightarrow{\partial_j} \right\} G \qquad (22.4.34)$$

There are, however, other products we may choose for different quantisation protocols. Some are equivalent to the Moyal product on $\mathbb{R}^{2n}$, others generalise to general manifolds; hence, we may see slight variations on the above equation. The general theory of constructing star-product operator algebras as Poisson deformations in $\hbar$ is given by the **Kontsevich quantisation formulas**.

With the star product now defined, the algebra is no longer commutative and we can use it to define a bracket. If the Moyal product has been used, then the **Moyal bracket** is defined as $\{F, G\}_{\text{MB}}$ for two functions $F(q, p)$ and $G(q, p)$ on the phase space and it satisfies the Jacobi identity:

$$\{F, G\}_{\text{MB}} = \frac{1}{\hbar}(F * G - G * F) \qquad (22.4.35)$$

In the same way that a Poisson algebra defines implicitly classical mechanics, a C^***-algebra** can be used to recover a quantum theory. So, we have deformed the Poisson algebra into a non-commutative algebra in an $\hbar$-dependent manner such that, for the classical limit, we regenerate the Poisson algebra. You can also write familiar quantum expressions in this language, too, exchanging the Hermitian linear operators $\hat{F}$ and $\hat{G}$ for smooth functions. For instance,

$$e^{\frac{it}{\hbar}[\hat{G}, \cdot]} \hat{F} = e^{i\hat{G}t/\hbar} \hat{F} e^{-i\hat{G}t/\hbar} \qquad (22.4.36)$$

could be written using the **star exponential**:

$$e^{\frac{it}{\hbar}[G, \cdot]_*} F = e_*^{iGt/\hbar} * F * e_*^{-iGt/\hbar} \qquad \text{where} \qquad e_*^X = 1 + X \sum_{n=1}^{\infty} (*X)^{n-1}/n! \qquad (22.4.37)$$

Hopefully, this brief background is enough to start the reader in their own journey into algebraic quantum mechanics and classical deformations. We have avoided the use of a complex field for simplicity only. In practice, we normally convert the variables to complex-valued functions, an example of which was shown at the end of chapter 18 using the harmonic oscillator (see example 18.3). From there, computation of the Moyal bracket and evaluation of the appropriate star-genvalue equation should lead to the correct quantum description. As an exercise, try showing that the quantum analogue of the Liouville density behaves like a *compressible* fluid!

23

The Structure of Phase Space

In this chapter, we look at the structure of phase space for integrable systems in closer detail. We will use basic ideas from chapter 22, but will not rely too heavily on them. Because of this step, the rest of the canonical mechanics in part III is open to readers who may have glossed over the finer points of the previous chapter. We will explore integrable systems and the implications of action-angle variables, to set the foundations for chapter 24, which covers canonical perturbation and KAM theory and which, in some sense, this chapter is a pre-requisite for.

We agreed in chapter 22 that the phase space of a system can be treated as a manifold. We also agreed that the manifold could be embedded in a higher-dimensional manifold, and we can think of hypersurfaces and submanifolds defined by constraints or constants of the motion. What we explore in this chapter is how the gradient of the Hamiltonian with the action variable P_i parametrises these surfaces, otherwise known as the *frequencies* $\boldsymbol{\omega}_i$. This ω is nothing to do with the symplectic 2-form ω we defined earlier in chapter 22, the values just have similar symbols in the literature.

In general, to integrate a system of $2n$ ODEs we must know $2n$ first integrals, however, for canonical equations it is sufficient to know n first integrals via the **Arnold-Liouville integrability theorem**. We tend to say that the trajectories of $p_i(t)$ and $q^i(t)$ with $(i = 1, \ldots n)$ are uniquely determined if we know the initial conditions $p_i(t_0)$ and $q^i(t_0)$. A corollary to the theorem is that there will exist a function $\mathcal{S}(\boldsymbol{q}, \boldsymbol{c}, t)$ that is a complete integral of the Hamilton-Jacobi equation depending on n integration constants c (previously called α). A function F is a first integral if it is constant along the solution curves to an ODE and its Poisson brackets are in involution, that is to say, if its derivative in the direction of the Hamiltonian vector field is zero:

$$ X_H(F) = \dot{F} = \dot{q}^i \frac{\partial F}{\partial q^i} + \dot{p}_i \frac{\partial F}{\partial p_i} = 0 \qquad (23.0.1) $$

So far, we have been calling these functions 'constants of the motion', and Noether's theorem says that there is a cyclic coordinate associated with them. Each constant of the motion defines a smooth hypersurface of $\dim(2n - 1)$ and their intersection defines a **level set** (submanifold) in the phase space. The trajectory will be confined to the region where the surfaces intersect, creating a bounded region, the level set of constant

Lagrangian & Hamiltonian Dynamics. Peter Mann, Oxford University Press (2018).
© Peter Mann. DOI: 10.1093/oso/9780198822370.001.0001

Fs. If there are m first integrals, then the problem is reduced to $(2n - m)$-dimensions, since the level set would have a codimension of m where they intersect.

If the gradient of the surface corresponding to a first integral is not equal to zero, then it is a **regular** surface. For an autonomous system, the Hamiltonian is conserved and equal to some fixed constant c. We can define the *regular energy surface* as $M_H = \{(q, p) \in T^*M : H(\boldsymbol{q}, \boldsymbol{p}) = c\}$, where $dH \neq 0$ (the 1-forms dH are linearly independent); this means that we can group together all the points $(q, p) \in T^*M$ in the phase space where the Hamiltonian is a constant $H = c$ to make a surface M_H; in this way, the choice of c defines the surface. The corresponding configuration manifold is the space of qs that satisfy any constraint condition $M = \{\boldsymbol{q} \in \mathbb{R}^n : \phi_i(\boldsymbol{q}) = 0, i = 1, \ldots, m\}$. The gradient dH is simply the derivative with, at that point,

$$dH = \left(\frac{\partial H}{\partial q}, \frac{\partial H}{\partial p} \right) \neq (0, 0)$$

When the gradient is zero, we run in to some problems called **singularities** which we discuss in chapter 24. The gradient is everywhere orthogonal to the Hamiltonian vector field for these surfaces, since the Poisson bracket of two integrals vanishes and so too do their vector fields $\{F, G\} = \{X_F, X_G\} = 0$ and their flows. Since you can swap a Poisson bracket around the other way, F being constant along the flows of G means that G is constant along the flows of F for the two first integrals F, G, this is **involution**. This has the effect of making the level sets **invariant** under the flow of Hamiltonian vector fields, and solutions will stay on the surface for all times. Therefore, the invariance of the structure of phase space created by first integrals under the flow of Hamiltonian vector fields leads to the Arnold-Liouville theorem.

If, for example, the energy is conserved, then we define a constant energy surface, since it is a first integral of the motion. If we have another function F that is conserved, it will define a surface where it is constant, too. The place where the level sets of F cross the energy surface will be the region where both constraints are satisfied.

From now on, we will consider a $2n$-dimensional phase space with n first integrals such that it is completely integrable in the Arnold-Liouville sense, and motion is therefore restricted to an n-dimensional surface in the phase space.

If the surface M_F defined by the intersecting level sets is *bounded* and *compact* (mathematical terms we won't worry about)[1] in addition to having linearly independent differentials, then it is diffeomorphic to the n-dimensional **torus** $\mathbb{T}^n = \mathbb{R}^n/\mathbb{Z}^n$ for n first integrals. The n-torus can be viewed as the cartesian product of n independent 1-dimensional topological circles S^1, one for each integrable parameter. We say that the submanifold tori *foliate* the phase space such that each torus is a symplectic leaf

---

[1] A bounded set is a set whose elements have limits on it's maximum and minimum values. A space is compact if all of its open covers have finite subcovers; basically, there is no boundary to the manifold (e.g. a sphere or a torus).

in the Poisson manifold of phase space. We have already said that the flow is invariant on these surfaces and so we call them **invariant tori**. Therefore, we get tori, due to having first integrals, and they are nested within phase space like layers of an onion. Each layer corresponds to the possible values we could choose for the constants of the motion; however, once they have been chosen, the flow remains on a given torus. Note that the layers do not cross over each over, since a point in phase space is unique.

Wherever there is a torus, there will be action-angle coordinates in the region of phase space (P_i, Q^i) with $i = 1, \ldots n$ and $0 \leq Q^i \leq 2\pi$, that is to say, we get *periodic flow*: each n-torus has n periodic angular variables. We know from appendix F and earlier discussions in chapter 18 that

$$\frac{dP_i}{dt} = 0, \qquad \frac{dQ^i}{dt} = \frac{\partial H(\boldsymbol{P})}{\partial P_i} = \nabla H(\boldsymbol{P}) = \boldsymbol{\omega}_i$$

The solutions to this are easily found to be $P_i(t) = P_i(t_0)$ and $Q^i(t) = \boldsymbol{\omega}_i t + Q^i(t_0)$, note that the flow is linear in time and will therefore be a straight line (i.e. rectified in a plot of Q^is). To be straight in charts means that, on the torus, the flow will wrap and wind around. Since the P_is do not vary over the torus, they can be used as a label such that each of the tori are parameterised by their action variables. The action-angle variables now form Darbouxian coordinates.

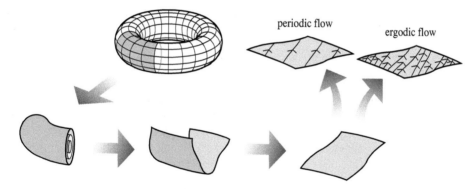

Fig. 23.1: Following left to right, we visualise the flow over the surface of the torus.

The **frequencies** defined by $\boldsymbol{\omega}_i$ are constant, since the angular coordinates are cyclic, and we can identify each torus with a unique frequency $\mathbb{R}^n \ni \boldsymbol{\omega} = (\boldsymbol{\omega}_1, \boldsymbol{\omega}_2, \ldots, \boldsymbol{\omega}_n)$. The angle coordinates provide a coordinate basis on each torus and we can see how different initial conditions will impact the torus we consider; this is a big point in molecular mechanics (see chapter 20 section 20.4.

As an example, we consider an energy hypersurface of a two-dimensional integrable system in a four-dimensional phase space spanned by (Q^1, Q^2, P_1, P_2) and which is diffeomorphic to the 2-torus (see figure 23.1). Each of the angle variables is labelled with a frequency ω_1 for Q^1, and ω_2 for Q^2. We take a little cut out of the torus and flatten it out to a square. We then see that the flow lines exit one side of the square and will appear on the other side, since, in reality, they are right next to each other[2]. If the flow goes off the edge of the square and rejoins where it started, creating an infinite cycle, then the ratio of their frequencies will be a **rational number** $\omega_1/\omega_2 \in \mathbb{Q}$ and we will have **periodic flow**, since the motion repeats itself after a while (its period) and therefore not all points on the tori are explored. If the flow never reconnects with itself, then every cycle will fill the torus more and more densely with flow lines, and the ratio of their frequencies will be **irrational** $(\omega_1/\omega_2 \in \mathbb{Q}\backslash\mathbb{R})$; we call this **ergodic flow**.

The frequency $\omega = (\omega_1, \omega_2, \ldots, \omega_n)$ of the torus is a vector where each frequency vector ω_i is associated with each angle variable. For another vector $\mathbb{Z}^n \ni k$, where $k = (k_1, k_2, \ldots, k_n)$, whose components are integers $k_i \in \mathbb{Z}$, which we assume are non-zero we have:

$$k \cdot \omega = \sum_i k_i \omega_i \qquad (23.0.2)$$

If $k \cdot \omega = 0$ with each $k_i \neq 0$, then ω_i is termed **resonant**; if not, then it is called **non-resonant**. This is the generalisation of the discussion just above about the ratio of frequencies. If we have a resonant relation between frequencies, then

$$\omega_1 k_1 + \omega_2 k_2 = 0 \quad \text{and} \quad \frac{\omega_1}{\omega_2} = -\frac{k_2}{k_1}$$

So, resonant tori have rationally related frequencies; doing the same for non-resonant frequencies shows they have irrational ratios. We would label the frequency, flow and entire torus by this property. Therefore, we can talk about *(non)-resonant tori*. The property of resonance will be important later when we discuss canonical perturbation theory. In particular, we will ask a question later regarding the distribution of resonant and non-resonant tori amongst phase space. The answer is that the resonant tori are **dense** amongst the non-resonant tori. To understand this, we return to the analogy of rational and irrational numbers. Any rational number can be written as a repeating decimal like $0.585000\bar{0}$. The probability of writing this number randomly is very low since, the probability of getting that many successive zeros is slim when we could write any number. Therefore, rational numbers are dense within the set of irrational ones.

We will understand the significance of this in chapter 24, which is on canonical perturbation theory and the KAM theorem. There, we will introduce the frequencies as $\omega = \nabla H_0$ but, from their definition above, they are equivalent. We discuss the

---

[2]Kind of like how on Google maps, you can go so far right past Russia that you come to America again, and so on.

importance of resonant tori in phase space and how they can be troublesome to treat analytically.

23.1 Time-Dependent Geometrical Mechanics

At this point, it would be fruitful to extend the Poincaré-Cartan 1-form θ_H defined in equation 22.1.1 for autonomous systems to those where time is an explicit variable in the **extended phase space** $T^*M \times \mathbb{R}$. Time now changes from a smooth parameter of the flow to its own dimension within a larger space. As well as this, we are going to stop saying that the $\dim(2n)$ phase space is the cotangent bundle to a $\dim(n)$ configuration manifold M and restrict its expression to simply $\mathbb{R}^n \times \mathbb{R}^n = \mathbb{R}^{2n}$. Therefore, the expression for the extended phase space is $T^*M \times \mathbb{R} = \mathbb{R}^{2n} \times \mathbb{R} = \mathbb{R}^{2n+1}$ with coordinates $(\boldsymbol{q}, \boldsymbol{p}, t)$. You might note that this manifold is odd dimensioned, while previously we said symplectic manifolds are *even dimensioned*! The observant reader may also have queried how one can foliate phase space with hypersurfaces, which are, by definition, of one dimension less? We start by extending the Poincaré-Cartan 1-form:

$$d\mathcal{S} = \theta_H = p_i dq^i - H dt \tag{23.1.1}$$

with summation over indices assumed and $H(\boldsymbol{q}, \boldsymbol{p}, t) : \mathbb{R}^{2n+1} \mapsto \mathbb{R}$. We can in similar fashion to previous work construct an extended 2-form which is *closed* ($d\boldsymbol{\Omega}_H = 0$) by taking the exterior derivative of the Poincaré-Cartan 1-form $\boldsymbol{\Omega}_H = d\theta_H$, remembering the Leibniz rule, and that, in any case, $d^2 f = 0$.

$$\boldsymbol{\Omega}_H = d\theta_H = dp_i \wedge dq^i - dH \wedge dt \tag{23.1.2}$$

If we now substitute for dH recalling $df \wedge df = 0$ and then collect like terms together, we obtain

$$
\begin{aligned}
\boldsymbol{\Omega}_H &= dp_i \wedge dq^i - \left(\frac{\partial H}{\partial p_i} dp_i + \frac{\partial H}{\partial q^i} dq^i + \frac{\partial H}{\partial t} dt \right) \wedge dt \\
&= dp_i \wedge dq^i - \frac{\partial H}{\partial p_i} dp_i \wedge dt - \frac{\partial H}{\partial q^i} dq^i \wedge dt - \frac{\partial H}{\partial t} dt \wedge dt \\
&= \left(dp_i + \frac{\partial H}{\partial q^i} dt \right) \wedge \left(dq^i - \frac{\partial H}{\partial p_i} dt \right)
\end{aligned}
\tag{23.1.3}
$$

This is just the product of two 1-forms $\boldsymbol{\Omega}_H = \omega_1 \wedge \omega_2$. You will notice that these 1-forms are Hamilton's equations, which are equal to zero for integral curves of the **extended Hamiltonian vector field** $\bar{X} := X_H + \partial t$. This means that $\boldsymbol{\Omega}_H = d\theta$ vanishes, too! We have just shown that, when $d\theta_H$ is evaluated on the surface of trajectories where Hamilton's equations are satisfied, it vanishes, thus proving θ_H gives rise to Hamilton's equations. This is also telling us that, whenever Hamilton's equations are satisfied, there will be a function $\mathcal{S}$ that encodes the dynamics. If we integrate this function (see Stokes' theorem in mathematical background 5) we obtain

$$\mathcal{S}(\boldsymbol{q}, \boldsymbol{p}, t) = \int \theta_H = \int p_i dq^i - H dt \tag{23.1.4}$$

Expanding $d\mathcal{S}$,

$$p_i dq^i - H dt = d\mathcal{S} = \frac{\partial \mathcal{S}}{\partial q^i} dq^i + \frac{\partial \mathcal{S}}{\partial p_i} dp_i + \frac{\partial \mathcal{S}}{\partial t} dt \tag{23.1.5}$$

Comparing the derivatives on each side, we arrive at our old friend, the Hamilton-Jacobi equation of motion:

$$\frac{\partial \mathcal{S}}{\partial t} + H(q, \frac{\partial \mathcal{S}}{\partial q}, t) = 0 \tag{23.1.6}$$

We can also see that, if the time derivative of a function F is an integral curve of $\bar{X}$, then:

$$\frac{dF}{dt} = \bar{X}(F) = (X_H + \partial_t) F = \{F, H\} + \partial_t F \tag{23.1.7}$$

which is the Poisson bracket of the Hamiltonian and a function F with explicit time dependance. Compare this to the discussion of complete integrals/integral surfaces and etched characteristics from chapter 19. The above section is the "proof" of how the Hamilton-Jacobi PDE contains within it the solutions to Hamilton's coupled ODEs, the geometrical picture being more equipped to highlight this. These geometrical structures define a **contact structure** on $\mathbb{R}^{2n+1}$ and, in some literature, we refer to Ω as the **contact form**. The manifold is "integrable" if its Pfaffian vanishes and it is these submanifolds, where $\Omega = 0$, that we wish to find. Indeed, most of geometrical mechanics can be converted to a Pfaff problem and it is for this reason that it has gained much attention over the last two centuries!

Contact geometry is a branch of differential geometry concerned with smooth, odd-dimensioned manifolds, just as symplectic geometry is concerned with smooth, even-dimensioned manifolds. The time dependent version of the Hamiltonian vector field is known as a **Reeb vector field**, the analogue of a Lagrangian submanifold is called an n-dimensional **Legendrian submanifold** (within $2n+1$) and transformations that preserve the contact structure are called, you guessed it, **contact transformations**. In this way, classical mechanics in extended phase space is a contact structured theory, while time-independent classical mechanics is a symplectic structured theory.

We now ask, what this would look like in a Lagrangian setting? You may recall in chapter 22 we pulled all the symplectic structure back to the tangent bundle to give us the geometrical Hamilton-Jacobi equation in the Lagrangian setting. Here, we will assume that the Lagrangian is regular and that the kernel of $d\theta_{\mathscr{L}}$ is 1. In the time-dependent case, we now factorise θ_H and invert the expression of the momentum to give us the Pfaffian 1-form in the Lagrange setting $\theta_{\mathscr{L}}$:

$$\theta_{\mathscr{L}} = \frac{\partial \mathscr{L}}{\partial \dot{q}^i} dq^i - \left(\frac{\partial \mathscr{L}}{\partial \dot{q}^i} \dot{q}^i - \mathscr{L} \right) dt = \frac{\partial \mathscr{L}}{\partial \dot{q}^i} (dq^i - \dot{q}^i dt) + \mathscr{L} dt \qquad (23.1.8)$$

We then need to take the exterior derivative of this $\Omega_{\mathscr{L}} = d\theta_{\mathscr{L}}$ and observe the criteria under which it vanishes[3]. We change notation for brevity by letting $\partial \mathscr{L} / \partial \dot{q}^i = p_i$:

$$\begin{aligned}
d\theta_{\mathscr{L}} &= d(p_i dq^i) - d(p_i \dot{q}^i dt) + d(\mathscr{L} dt) \\
&= d(p_i \wedge dq^i) - d(p_i \dot{q}^i \wedge dt) + d(\mathscr{L} \wedge dt) \\
&= dp_i \wedge dq^i - d(p_i \dot{q}^i) \wedge dt + d\mathscr{L} \wedge dt \\
&= dp_i \wedge dq^i - (\dot{q}^i dp_i + p_i d\dot{q}^i) \wedge dt + d\mathscr{L} \wedge dt \\
&= dp_i \wedge dq^i - (\dot{q}^i dp_i + p_i d\dot{q}^i - d\mathscr{L}) \wedge dt
\end{aligned}$$

We then note that the differential can be written $d\mathscr{L}(q, \dot{q}, t) = \frac{\partial \mathscr{L}}{\partial q^i} dq^i + \frac{\partial \mathscr{L}}{\partial \dot{q}^i} d\dot{q}^i + \frac{\partial \mathscr{L}}{\partial t} dt$; but, of course, terms in dt cancel with the wedge, anyway. In the third step, we use the Lagrange equation:

$$\begin{aligned}
&= dp_i \wedge dq^i + \left(-\dot{q}^i dp_i - \frac{\partial \mathscr{L}}{\partial \dot{q}^i} d\dot{q}^i + d\mathscr{L} \right) \wedge dt \\
&= dp_i \wedge dq^i + \left(-\dot{q}^i dp_i + \frac{\partial \mathscr{L}}{\partial q^i} dq^i \right) \wedge dt \\
&= dp_i \wedge dq^i + (-\dot{q}^i dp_i + \dot{p}_i dq^i) \wedge dt \\
&= dp_i \wedge dq^i - \dot{q}^i dp_i \wedge dt + \dot{p}_i dq^i \wedge dt \\
&= (dq^i - \dot{q}^i dt) \wedge (dp_i - \dot{p}_i dt)
\end{aligned}$$

Yet again, we generate the product of two 1-forms, thus, confirming that $d\theta_{\mathscr{L}}$ vanishes on-shell. The second term is the Euler-Lagrange equation of motion with $\partial \mathscr{L} / \partial \dot{q}^i = p_i$, while the first term is the contact form:

$$d\left(\frac{\partial \mathscr{L}}{\partial \dot{q}^i} \right) - \frac{\partial \mathscr{L}}{\partial q^i} dt = 0$$

For further information please see Delphenich's (2012) article The role of integrability in a large class of physical systems and Dedecker's (1957) article, A property of differential

---

[3]The following proof may not look like much but it took me a while to figure out! Naturally finishing it at about 4 am, it has to be my favourite proof in the book. Before looking, try it as an exercise!

forms in the calculus of variations. As in the Hamiltonian case, we can then construct the Hamilton-Jacobi equation in the Lagrangian setting:

$$\mathcal{A}(q,t) = \int \theta_{\mathscr{L}} = \int \frac{\partial \mathscr{L}}{\partial \dot{q}^i} dq^i - \left(\frac{\partial \mathscr{L}}{\partial \dot{q}^i} \dot{q}^i - \mathscr{L} \right) dt = \int \frac{\partial \mathscr{L}}{\partial \dot{q}^i} (dq^i - \dot{q}^i dt) + \mathscr{L} dt \quad (23.1.9)$$

Hence,

$$\frac{\partial \mathcal{A}}{\partial t} + J\left(t, q, \frac{\partial \mathcal{A}}{\partial q}\right) = 0 \qquad \text{with} \qquad \frac{\partial \mathcal{A}}{\partial q} = \frac{\partial \mathscr{L}}{\partial \dot{q}^i} \quad (23.1.10)$$

We have therefore witnessed how time-dependent mechanics is generated from a contact structure θ_H on an extended phase space and how this structure also gives rise to the Euler-Lagrange equation via $\theta_{\mathscr{L}}$ when we can pull the structure back. The **complete solution** to the Hamilton-Jacobi problem is the set of vector fields $\bar{X}_{\mathscr{L}}$ whose integral curves supply all possible integral curves of the Lagrangian vector fields and, for some function F, we have $\dot{F} = \bar{X}_{\mathscr{L}}(F)$:

$$\bar{X}_{\mathscr{L}} = \dot{q}^i \frac{\partial}{\partial q^i} + \ddot{q}^i \frac{\partial}{\partial \dot{q}^i} + \frac{\partial}{\partial t} \quad (23.1.11)$$

The differential geometry of autonomous mechanics is fairly similar to time-independent mechanics and best portrayed using the language of **jets**, since jet bundles admit **contact structures**. We take the base manifold with coordinates (t, q^i) to be a vector bundle[4], $\pi : \mathbb{R} \times M \equiv E \to \mathbb{R}$ and the extended velocity phase space is the first-order **jet manifold** to E, which we symbolise by $\mathbb{R} \times TM \equiv J^1 E \to E$ with coordinates (t, q^i, v^i). The jet manifold is the disjoint union of the **jet spaces** above a point on the base space with vector fields $X : E \to J^1 E$ and can be embedded into TE. The Lagrangian is a function on the jet manifold, $C^\infty(J^1 E) \ni \mathscr{L} : J^1 E \to \mathbb{R}$, and the Pfaffian 1-form is also defined on $J^1 E$. We can also extend the definition of the Legendre transform, $\widehat{\mathbb{L}} : \mathbb{R} \times TM \to T^* M \times \mathbb{R}$, to an extended phase space with coordinates (t, q, p). In this way, the jet manifold is a general object that contains all manner of vectors and covectors. Its target projection leads to fibres in the cotangent bundle while its source results in fibres of the tangent bundle. Curves $\gamma(t)$ on M may be lifted to the extended velocity space and Legendre transformed into the extended phase space in much the same way as in autonomous mechanics. We have already seen that the Pfaffian form on $\mathbb{R} \times TM$ is the pullback of the Pfaffian form on $T^* M \times \mathbb{R}$ under the Legendre mapping and that, in both cases, we may form the Poincaré-Cartan integral invariant. We may also obtain local expressions for the forms. due to Darboux' theorem holding for contact charts.

Interestingly, the **Gibbs-Appell equations** can also be described using the language of jets. Since we must use accelerations, we now require the second jet bundle

---

[4] In more advanced texts, whenever we describe a bundle, we always write it with the projection acting on it to its base space.

J^2E with natural coordinates (t, q^i, v^i, a^i). The extremal curve is hence parametrised as $j^2\gamma(t) = (t, q, \dot{q}, \ddot{q})$. The second-order jet bundle is an immersed submanifold of TTE, acceleration vectors in the tangent space to the first-order jet. We can define the dynamics of the system in three ways: a mapping from E to J^1E, a variation δ as a vector field on the state's image in $T(J^1E)$, and with a 1-form on J^1E. The local expression for the variation vector field on the first jet manifold is

$$\delta = \frac{\partial}{\partial t}\delta t + \frac{\partial}{\partial q^i}\delta q^i + \frac{\partial}{\partial v^i}\delta v^i. \tag{23.1.12}$$

The 1-form θ_F is known as the *generalised force* and will hence have three components:

$$\theta_F = x_i(t, q, v)dt + F_j(t, q, v)dq^j + p_k(t, q, v)dv^k \tag{23.1.13}$$

The first term vanishes if we do not consider explicit time dependence. The second is the (applied) **force 1-form** $F_j(t, q, v)dq^j : \mathbb{R} \times TM \to T^*M$, which describes the external forces on the system, while the third is the forces arising from generalised momenta. If θ_F is exact, then all three components must be equal to a 1-forms df, where the f are functions on the jet manifold. In this case, the force must be derived from a potential and not have dependence on velocity or time, as is evident from the exterior derivative:

$$dF_i(t, q, v)dq^i = df_i \wedge dq^i = \left(\frac{\partial F_i}{\partial t}dt + \frac{\partial F_i}{\partial q^j}dq^j + \frac{\partial F_i}{\partial v^j}dv^j\right) \wedge dq^i$$

If θ_F is exact, then there must exist a function on J^1E such that $\theta = d\mathscr{L}(t, q, v)$:

$$d\mathscr{L} = \frac{\partial \mathscr{L}}{\partial q^i}dq^i + \frac{\partial \mathscr{L}}{\partial v^i}dv^i = F_i dq^i + p_i dv^i \tag{23.1.14}$$

Integrating $p_i dv^i$ by parts in an action principle leads to $-\dot{p}_i dq^i$ and, hence, we observe that an extremal curve on M with the correct boundary conditions would satisfy d'Alembert's principle:

$$(F_i - \dot{p}_i)dq^i = 0 \tag{23.1.15}$$

For further reading please see (Cariñena and Fernández-Núñez's (2010) article, A geometric approach to the Gibbs-Appell equations in Lagrangian mechanics).

Very little literature on the geometric theory of **Ostrogradsky's formulation** exists at the time of writing; however, it would seem natural to follow a jet bundle approach. The similarity between this method and the Gibbs-Appell equations is still unknown.

Moving on, we can treat the Pfaffian form θ_H using *Stokes' lemma*, so you will need to read about *Stokes' theorem* before proceeding.

Mathematical Background 5

We consider in brief **Stokes' theorem** and how we might apply in to manifolds; this will be in preparation for applying it to the Pffafian 1-form on extended phase space. A sheet of paper is a two-dimensional manifold whose edge is a one-dimensional boundary. Therefore, a **bounded manifold** is an n-manifold with an $(n-1)$-manifold acting as its edge. We would denote the bounded manifold as M and the boundary as ∂M.

Stokes' theorem is as follows. Let M be a smooth, oriented n-manifold with boundary ∂M and $\omega \in \bigwedge^{n-1} M$ being an $(n-1)$-form that is *compactly supported* on M; then, the following integrals are equal:

$$\int_M d\omega = \int_{\partial M} \omega \qquad (23.1.16)$$

where we take ω as restricted to the boundary on the right-hand side. Normally, M is a submanifold of a larger manifold.

A corollary to Stokes' theorem is **Stokes' lemma**. We will summarise this in a nutshell (enough to make a mathematician cringe with the lack of technicality), but here goes. In an extended phase space, we imagine integral flow curves running along the q, p and t axes. If we surround some of these integral curves with a closed curve γ_1 at some point on the time axis and then again at another time γ_2, then the integral curves passing through will form a little tube which we call the **tube of characteristics**. Stokes' lemma says that, for a *non-singular* differential form ω,

$$\oint_{\gamma_1} \omega = \oint_{\gamma_2} \omega \qquad (23.1.17)$$

The proof of this is straightforward and uses Stokes' theorem in the final step. If we take a length of the tube σ, then the boundary is $\partial\sigma = \gamma_1 - \gamma_2$:

$$\oint_{\gamma_1} \omega - \oint_{\gamma_2} \omega = \int_{\partial\sigma} \omega = \int_\sigma d\omega = 0$$

Now, since ω is closed (by assumption), we prove the the Stokes lemma.

If we now apply this to classical mechanics, we will see that θ_H is exactly such a 1-form that will obey Stokes' lemma between two paths that encircle the flow at different times:

$$\oint_{\gamma_0} p_i dq^i - H dt = \oint_{\gamma_t} p_i dq^i - H dt \qquad (23.1.18)$$

It is because of this that θ_H is called the **integral invariant of Poincaré-Cartan** and it is a coordinate-independent form on extended phase space. For each of the constant-time slices, we note that $dt = 0$ and therefore we drop the $H dt$ term, regenerating the autonomous geometric formalism. We may recall that, by taking the exterior derivative of the 1-form we obtain the symplectic 2-form ω, which we said earlier was invariant under canonical transformations; well, that is just a particular of Stokes' lemma, as is Liouville's theorem of preservation of the phase space volume:

$$\oint_{\partial \sigma} p_i dq^i = \iint_{\sigma} dp_i \wedge dq^i$$

Note that we are using Stokes' theorem and we have the symplectic 2-form on the right; that is also an absolute integral invariant, so it can be written as a surface or contour integral.

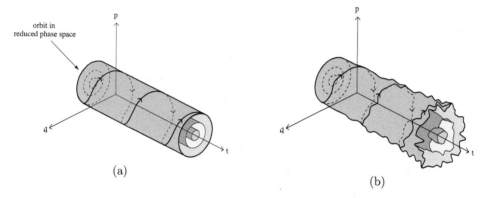

Fig. 23.2: A time-independent Hamiltonian (a), time-dependent Hamiltonian (b).

23.2 Picturing Phase Space

I don't know about you, but I certainly picture phase space in lots of weird ways with all the different things going on! We have Poisson manifolds, symplectic leaves, tori, tangent bundles to configuration manifolds, hyperplanes, level set contours, foliations, integral surfaces, characteristic curves, and so on (granted, some of them are the same thing). So, how then should we picture the extended phase space? Perhaps the most primitive answer is that it is just the space of classical solutions to the equations of motion, independent of the choice of coordinate frame; it is a covariant object associated with a dynamical system. The obvious answer, meanwhile, is that you can't picture very much past the most simple of cases (i.e. a one-dimensional system). We next look at extended phase space and hopefully clarify some of these different views; there are some ways of viewing things that, whilst not perfect, can really help.

When talking about the extended phase space of a one-dimensional system, we can imagine a two-dimensional plane (a slice of constant time) spanned by q and p and with a time axis running orthogonal. For a periodic solution of an autonomous integrable Hamiltonian, the flow will trace a little closed pattern in the qp-plane; this is referred to as its *orbit* for arbitrary choice of t_0. The shape of this orbit will remain constant as we travel up the time axis and will trace out a cylinder, while the flow generated by the Hamiltonian will travel periodically over the surface[5] (see figure 23.2). Since this system has action-angle variables, it will be parametrised by constant a P_i and a periodic angular dependance $0 \leq Q^i \leq 2\pi$.

We now introduce a time-dependant Hamiltonian $H(\boldsymbol{q}, \boldsymbol{p}, t)$. As we evolve along the time axis, the cylinder will not maintain its shape, since time-dependant terms in the Hamiltonian become more significant and the cylinder distorts. As this happens, the action variables are no longer constant but change to new values!

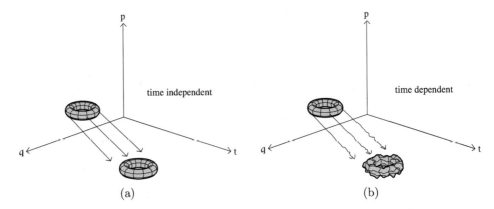

Fig. 23.3: A time-independent Hamiltonian (a), a time-dependent Hamiltonian (b).

If this change is slow with respect to the winding of the flow, or the period of the motion Q^i, then the action won't actually be *that* drastically different than when it started (for small times later, at least), therefore meaning that the shape of the orbit being traced out is pretty similar, too. So, even if the time dependance of the Hamiltonian means that, in the long term, the cylinder looks nothing like how it started, we can say that, *for a short time after it got going*, it wasn't that far off where it started. This is linked to **adiabatic invariance**, the constancy of the action of a torus P_i under a slow change in the Hamiltonian. Adiabatic invariance is not about invariance in general and occurs only when periodic motion explores the torus well (so it is ergodic in some sense) before it changes significantly, and we don't know for sure that a system

---

[5]The trajectories extend over the surface of these cylinders; in fact the cylinder is just a way of visualising the flow of the curves, so, by design, it is on the surface!

will exhibit adiabatic invariance. The other end of the scale is that it distorts crazily fast compared to the period of the flow. Therefore, we develop the idea of **slow** and **fast** Hamiltonian systems, into which there has been vast research.

Adiabatic invariance is a bit of a tricky problem, since any proof must show invariance where, strictly speaking, there should be none! The key idea behind it is that, in special circumstance two independent quantities become functions of one another such that, when time evolves, they remain almost constant to each other. As an example, if we had a ratio of two quantities and varied one smoothly while keeping the other constant, then we would expect a big change in their ratio. However, if they were adiabatic invariants, then changing the smooth parameter would induce a change in the other such that their ratio didn't change over the evolution.

Under the adiabatic cycling of the parameters of the system, there arises a **geometric phase** associated with the angle coordinate. The angle variable is shifted as the adiabatic parameter is advanced through its period. So, while the ratio of the parameter and the action variable may be constant there is a constant, advancement of the angle as the torus moves in phase space. The shift in the angle variable under the adiabatic change is called the **Hannay angle** and it is the classical counterpart of the Berry phase in quantum mechanics. There is a move to generalise this phenomena to non-adiabatic evolution, but it is well beyond the scope of this book, as complications in the smooth parameter quickly arise.

At this point, we are pretty much finished with our discussion of the canonical formalism. In chapter 24, we investigate the dynamics of those tori with resonant frequencies which are *non-integrable* and how KAM theory addresses some technical issues we come across. For further reading on the present topic, there is no better resource than Arnold's (1989) *Mathematical Methods of Classical Mechanics*. For the geometrical take on time-dependent Hamiltonian mechanics, please see chapter 5 of Mangiarotti and G. Sardanashvily's (1999) *Gauge Mechanics*.

Chapter summary

- A **first integral** F of an ODE is constant along integral curves of a vector field X_F such that $X_F(F) = 0$.

- The **Arnold-Liouville integrability theorem** states that, for an n-dimensional system of $2n$ canonical equations, n independent first integrals will be sufficient to obtain a set of action-angle variables. In this case, there will exist a function $\mathcal{S} = \mathcal{S}(q, c, t)$ that is a complete integral for the Hamilton-Jacobi equation.

- If the **level set** defined by $M_F = \{(q, p) \in T^*M : F_i(q, p) = c_i\}$, where c_i is fixed is bounded and compact and its differentials dF are linearly independent and, equivalently, are orthogonal to the Hamiltonian vector fields, then it is diffeomorphic to the n **torus** $\mathbb{T}^n$.

- The tori **foliate** the phase space in symplectic leaves; they have invariant flow and are hence called **invariant tori**. The tori can be parametrised by the action coordinates, which are constant on each torus. In addition, each torus has a unique **frequency** ω determined by its angle coordinates' frequencies ω_i. The following relation can be used to determine whether a torus is **resonant**:

$$k \cdot \omega = \sum_i k_i \omega_i$$

- The **Poincaré-Cartan 1-form** θ_H is used with the **multidimensional Stokes lemma** to encapsulate canonical mechanics in **extended phase space** $\mathbb{R}^{2n+1}$ with local coordinates (q, p, t):

$$\theta_H = \sum_i p_i dq^i - H dt$$

Time-dependent mechanics is thus shown to possess a **contact structure**.

Exercise 23.1 In this example, we will consider an elementary approach to study adiabatic invariants in classical mechanics; in general, the time dependence of the action variable is tricky to study analytically and any "proof" of adiabatic invariant theory will usually involve some averaging and physical intuition on our part. However, this should give a flavour of the ideas involved.

To set the situation, we imagine that the Hamiltonian H and our action-angle variables (P, Q) are functions of a smooth parameter usually symbolised by λ in the literature[6]. We never specify what this variable is for generality, except that it is some parameter of the system we can evolve continuously and that, when we do so, changes the Hamiltonian and the

---

[6]See Wells and Siklos 's (2006) article, The Adiabatic invariance of the action variable in classical dynamics, for more information and a through derivation.

action label of the torus. The time derivative of the action variable, the angle variable and the instantaneous energy E are given in the usual fashion:

$$\frac{dP}{dt} = \frac{\partial P}{\partial E}\dot{E} + \frac{\partial P}{\partial \lambda}\dot{\lambda}, \qquad \frac{dQ}{dt} = \frac{\partial H}{\partial P} + \frac{\partial Q}{\partial \lambda}\dot{\lambda}, \qquad \frac{dE}{dt} = \frac{\partial H}{\partial \lambda}\dot{\lambda} \qquad (23.2.1)$$

From action-angle variable theory, we are aware that $\partial H/\partial P = \omega$ and that $P = \frac{1}{2\pi}\oint p\,dq$. Therefore, we find that

$$\frac{\partial P}{\partial H} = \frac{1}{\omega} \qquad \text{and} \qquad \frac{\partial P}{\partial \lambda} = \frac{1}{2\pi}\frac{\partial}{\partial \lambda}\oint p\,dq \qquad (23.2.2)$$

We substitute these relations into the time derivative of the action variable above to generate the following expression:

$$\frac{dP}{dt} = \frac{1}{\omega}\frac{\partial H}{\partial \lambda}\dot{\lambda} + \frac{1}{2\pi}\frac{\partial}{\partial \lambda}\oint p\,dq\,\dot{\lambda} \qquad (23.2.3)$$

This expression depends on how λ changes in time. If the parameter can be chosen to vary slow enough over one period of evolution such that it is *constant* in time, then $\dot{\lambda} \approx 0$ and, hence, the average of the time derivative of the action variable over the period also vanishes: $\langle \dot{P} \rangle \approx 0$. Therefore, the action label of the torus is constant for short time evolutions, provided the parameter is slowly varying.

For this particular example, the time dependance of the angle coordinate also gives us an expression for the Hannay angle! By rearrangement of the expression for $\dot{Q}$ from before, we can formulate the general expression observed in the literature (multiply both sides by dt and integrate over a period):

$$\int_1^2 d\theta = \int_1^2 \omega\,dt + \int_1^2 \frac{\partial Q}{\partial \lambda}\frac{d\lambda}{dt}\,dt \qquad (23.2.4)$$

leading to

$$Q(2) = Q(1) + \int_1^2 \omega\,dt + \oint \left\langle \frac{\partial Q}{\partial \lambda} \right\rangle d\lambda \qquad (23.2.5)$$

The first term accounts for the starting point of the angle, the second is the dynamical change and the contour integral is the geometric change or the Hannay angle.

Exercise 23.2 Use the Poisson formalism to asses whether the following Hamiltonian is integrable such that there exist two independent functions F, G such that $\{F, G\} = 0$:

$$H(q_1, q_2, p_1, p_2) = \frac{1}{2}(q_1^2 + q_2^2 + p_1^2 + p_2^2) + \frac{q_1^3}{3} - \frac{q_2^3}{3} \qquad (23.2.6)$$

Your first step should be to compute $\{H, H\}$; then, from that separate the variables $H(1,2) = H(1) + H(2)$.

Lagrangian & Hamiltonian Dynamics

Exercise 23.3 Show that the extended Hamiltonian vector field $\bar{X} = X_H + \partial_t$ and the Pfaffian 1-form on $T^*M \times \mathbb{R}$ satisfy $\iota_{\bar{X}}\theta_H = p_i \frac{\partial H}{\partial p_i} - H$ and $\iota_{\bar{X}}d\theta_H = 0$:

$$\iota\left(X_H + \partial_t\right)\left(p_i dq^i - H dt\right) = p_i \frac{\partial H}{\partial p_i} - H \tag{23.2.7}$$

$$\iota\left(X_H + \partial_t\right)\left(dp_i \wedge dq^i - dH \wedge dt\right) = dH - dH \tag{23.2.8}$$

Exercise 23.4 The adiabatic motion of dynamical systems is associated with the careful, infinitely slow motion of a time-dependent parameter $\lambda(t)$. For such a system, we can modify the Hamiltonian $H_0(q, p, \lambda(t))$, with a **counter adiabatic term**, $H_{CD}(q, p, t)$, that suppresses non-adiabatic fluctuations in the dynamics. Therefore, we can actually drive the dynamics faster through $H(x, p, \lambda(t)) = H_0(x, p, \lambda(t)) + H_{CD}$, yet still retain the invariant action of the original motion H_0. When the equations of motion are evaluated at a particular point, the original Hamiltonian is denoted by the energy surface $H_0 = E_0(\lambda(t))$. Taking the time derivative of this, we find that:

$$\frac{dH_0}{dt} = \frac{dE_0}{d\lambda}\dot{\lambda}(t) \tag{23.2.9}$$

Upon solution of the equations of motion for H, we then have

$$\{H_0, H\} + \frac{\partial H_0}{\partial \lambda}\dot{\lambda} = \dot{\lambda}\frac{dE_0}{d\lambda} \tag{23.2.10}$$

Hence, H is the generator of the evolution characterised at each instant by E_0.

24

Near-Integrable Systems

Many problems in physics and chemistry do not have exact analytical solutions; these systems are in direct opposition to the integrable systems and action-angle variables we are familiar with. The focus of this chapter is *near-integrable systems* and we start by considering tiny perturbations to integrable Hamiltonians to see what happens. Poincaré in 1893 claimed this was the fundamental question of classical mechanics and, fittingly, Hamilton-Jacobi theory is the starting point.

24.1 Canonical Perturbation Theory

As some readers are no doubt interested in the chemical applications of **perturbation theory**, so it is worth considering some interdisciplinary examples to get a feel for the subject. Chemical physicists might likely be concerned with reaction dynamics at saddle points. Although traditionally this is a quantum mechanical domain, we can incorporate tunnelling effects and other potential energy surface phenomena of molecular systems into classical trajectory simulations through the use of a mashup of classical and quantum mechanics; this is termed *semiclassical mechanics*.

Quantum tunnelling is a phenomena utilised in **scanning tunnelling microscopy**, or STM. Here, weak coupling between a surface and a atomistic tip that is dragged across it can be probed to gain atomic resolution of metal surfaces and adsorbates (see figure 24.1). Usually performed under ultra-high vacuum conditions at low temperatures, STM experiments provide surface scientists with a powerful image-based tool.

As well as this, material scientists might take inspiration from the *Toda lattice*, which is a particular of the *Fermi, Pasta and Ulam* (FPU) non-linear problem in lattice modelling. These models simulate chains of molecules connected via springs to their nearest neighbours. In this way, they represent crystal structures and can be used to simulate materials. The FPU system corresponds of N oscillators connected in a one-dimensional chain of length L with positions $x_i(t) = (iL)/(N-1) + \eta_i(t)$, where η_i is the instantaneous displacement from a lattice point. The equation of motion is found to be

$$m\ddot{x}_i(t) = k(x_{i+1} + x_{i-1} - 2x_i)[1 + \alpha(x_{i+1} - x_{j-1})] \qquad (24.1.1)$$

Lagrangian & Hamiltonian Dynamics. Peter Mann, Oxford University Press (2018).
© Peter Mann. DOI: 10.1093/oso/9780198822370.001.0001

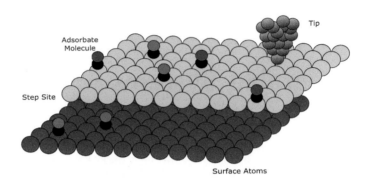

Fig. 24.1: Surface topologies can be mapped during a scanning tunnelling microscopy experiment. Surface adsorbates and novel structures can be studied and catalyst reaction processes investigated. For further information see exercise 12.2.

The Toda lattice is a completely integrable system that exhibits **soliton solutions**, which are self reinforcing wavepackets. Meanwhile, the more general FPU model exhibits chaotic motion. The Toda Hamiltonian is given by a potential that contains exponential displacements:

$$H = \sum_{i=1}^{N} \left(\frac{p^2}{2m_i} + e^{x_i - x_{i+1}} + e^{x_N - x_1} \right) \tag{24.1.2}$$

Perturbation theory provides a unique tool for investigating small changes in energy/frequency/time, and so on, in systems with a familiar solution. We will investigate frequency perturbation, since we can express the Hamiltonian readily using action-angle theory.

The key to perturbation theory is that we can split the system's Hamiltonian H up into two parts:

$$H = H_0 + \epsilon H_1 \tag{24.1.3}$$

where H_0 has an exact solution (integrable) and ϵH_1 is a *small* perturbation term; the control over the size is given by the ϵ such that, if $\epsilon = 0$, we have our integrable system. This means that we view any unsolvable problem as composed of an integrable system and some small variation to it that distorts the action-angle variables. This is **first-order perturbation theory**. In an STM experiment, we might have a surface Hamiltonian and a tip Hamiltonian as a weak coupling perturbation term.

The result of perturbation theory, in general, is a power series in the small parameter whose first term is the exact solution, with all subsequent terms being little shifts to this value:

$$H = H_0 + \epsilon H_1 + \epsilon^2 H_2 + ... + \epsilon^n H_n \tag{24.1.4}$$

Since the parameter ϵ is tiny, the higher-order terms become smaller and smaller and, for the most part, we ignore those of higher-order by truncating the series, keeping terms only of, say, first or second order; hence, the series *converges*. As time evolves, the approximate solution breaks down as time-dependent terms grow and deviate form the physical solution. Therefore, constructing a solution for all times is tricky. There are several methods for determining these perturbed terms; here, we consider time-independent perturbation theory.

As you would expect, we can perform a time-independent canonical transformation on the solvable system H_0 to action-angle coordinates, which means that $H_0 = H_0(\boldsymbol{P})$; we treat the system as being integrable and being a closed torus in phase space with periodic flow. For a single perturbative term ϵH_1, we can write the Hamiltonian of the full system and the resulting canonical equations:

$$H(\boldsymbol{Q}, \boldsymbol{P}, \epsilon) = H_0(\boldsymbol{P}) + \epsilon H_1(\boldsymbol{Q}, \boldsymbol{P}) \tag{24.1.5}$$

$$\dot{P}_i = -\epsilon \frac{\partial H_1}{\partial Q_i}, \qquad \dot{Q}_i = \frac{\partial H_0}{\partial P_i} + \epsilon \frac{\partial H_1}{\partial P_i}$$

At this point, we are going to make a notational change for ease of reading; instead of calling the original coordinates (q, p), we are going to denote them by (Q_0, P_0). The idea behind canonical perturbation theory is finding a canonical transformation *from* one set of action-angle variables to *another* set of action-angle variables; this is denoted $(Q_0, P_0) \to (Q, P)$, and we reserve p and q for the non-action-angle variables. We are also going to be using a type-2 generating function in this introduction and a system that is one-dimensional and autonomous system (no explicit time dependance) as well as being periodic in 2π. We assume that the perturbation is so small that the perturbed system still has these action-angle variables (Q, P), although the validity of this is questionable and will depend entirely on the system itself.

We are looking for a transformation such that the Hamiltonian can be expressed as

$$H(Q_0(Q, P), P_0(Q, P), \epsilon) = K(P) = H_0(P_0) + \epsilon H_1(Q_0, P_0)$$

The first step is to expand the generating function in increasing powers of the parameter ϵ, again by the assumption that the perturbation is tiny. The first term is the generating function of the identity transformation, since we are after solutions about a known system; we can recover these if $\epsilon = 0$:

$$F(Q_0, P) = Q_0 P + \epsilon F^{(1)}(Q_0, P) + \epsilon^2 F^{(2)}(Q_0, P) + \mathcal{O}\epsilon^3 \cdots \tag{24.1.6}$$

We now turn to the general theory of canonical transformations for F_2 functions, to compute the following expressions for P_0 and Q:

$$P_0 = \frac{\partial F}{\partial Q_0} = P + \epsilon \frac{\partial F^{(1)}}{\partial Q_0} + \epsilon^2 \frac{\partial F^{(2)}}{\partial Q_0} + \mathcal{O}\epsilon^3 \cdots \tag{24.1.7}$$

$$Q = \frac{\partial F}{\partial P} = Q_0 + \epsilon \frac{\partial F^{(1)}}{\partial P} + \epsilon^2 \frac{\partial F^{(2)}}{\partial P} + \mathcal{O}\epsilon^3 \cdots \tag{24.1.8}$$

You can now express $K(P)$ using these, noting that the result is the approximated Hamilton-Jacobi equation:

$$K(P) = H_0 \left(P + \epsilon \frac{\partial F^{(1)}}{\partial Q_0} + \epsilon^2 \frac{\partial F^{(2)}}{\partial Q_0} + \mathcal{O}\epsilon^3 + \ldots \right) + \epsilon H_1 \left(Q_0, P + \epsilon \frac{\partial F^{(1)}}{\partial P} + \epsilon^2 \frac{\partial F^{(2)}}{\partial P} + \mathcal{O}\epsilon^3 + \cdots \right)$$

We now expand the Hamiltonians about $P_0 - P$, since we assume that the deviation of the perturbed system is very small from the integrable system:

$$K(P) = H_0(P) + \frac{\partial H_0}{\partial P}(P_0 - P) + \frac{1}{2}\frac{\partial^2 H_0}{\partial P^2}(P_0 - P)^2 + \cdots$$
$$+ \epsilon H_1(P, Q_0) + \epsilon \frac{\partial H_1}{\partial P}(P_0 - P) + \mathcal{O}\epsilon^4 + \cdots \tag{24.1.9}$$

Next, substitute equation 24.1.7 in, multiply it out, tidy it up and keep terms up to second order in ϵ, keep more if you want to be a little more precise in your approximate solution:

$$K(P) = H_0(P) + \epsilon \left[\frac{\partial H_0(P)}{\partial P}\frac{\partial F^{(1)}}{\partial Q_0} + H_1(Q_0, P) \right] + \epsilon^2 \left[\frac{\partial H_0(P)}{\partial P}\frac{\partial F^{(2)}}{\partial Q_0} \right.$$
$$\left. + \frac{1}{2}\frac{\partial^2 H_0}{\partial P^2}\left(\frac{\partial F^{(1)}}{\partial Q_0}\right)^2 + \frac{\partial H_1}{\partial P}\frac{\partial F^{(1)}}{\partial Q_0} \right] + \epsilon^3 \tag{24.1.10}$$

We also expanded the left side of equation 24.1.10; this is written below:

$$K(P) = K_0(P) + \epsilon K_1(P) + \epsilon^2 K_2(P) + \epsilon^3 \mathcal{O} \tag{24.1.11}$$

You then equate terms on the left and right in their respective powers:

$$K_0(P) = H_0(P) \tag{24.1.12}$$

$$K_1(P) = \frac{\partial H_0}{\partial P}\frac{\partial F^{(1)}}{\partial Q_0} + H_1(Q_0, P) \qquad \textbf{Poincaré's fundamental equation}$$

$$\tag{24.1.13}$$

$$K_2(P) = \frac{\partial H_0}{\partial P}\frac{\partial F^{(2)}}{\partial Q_0} + \frac{1}{2}\frac{\partial^2 H_0}{\partial P^2}\left(\frac{\partial F^{(1)}}{\partial Q_0}\right)^2 + \frac{\partial H_1}{\partial P}\frac{\partial F^{(1)}}{\partial Q_0} \tag{24.1.14}$$

This procedure forms the basis of the iterative method of canonical perturbation theory; we always aim to find a series of transformations such as that demonstrated until the resultant Hamiltonian is obtained. There are, unfortunately, more unknowns than equations and, in order to find both the Hamiltonian and the generating function, we need a way of removing some of these unknowns! For instance, in order to determine K_1, we need to know both $F^{(1)}$ and H_1.

At this point, consider the form of what we have written; it appears that, while the left-hand side is only a function of P, the right-hand side looks as if it is a function of both P *and* Q_0. We must note that Q_0 is periodic in 2π and it is this periodicity that we exploit, since all the successive functions can only take values in a period, according to equation 24.1.8. The functions $F^{(i)}$ can be *chosen* in such a way as to remove the dependance on the right-hand side of both 24.1.13 and 24.1.14. In each perturbation order, we have **oscillating functions** $\partial F^{(i)}/\partial Q_0$, whose sign is undetermined, and it is these that we need to tackle in each order we consider in order to maintain equality. They can be removed without losing information about the dynamics of the motion, since they oscillate around a general evolution which is described by the action variables. Remember that it is the action variables that can label a torus.

We start from equation 24.1.12, which we know is the zeroth-order solution (i.e. when $\epsilon = 0$) and, since the generating function is periodic, too, we can average over Q_0 in the expression for the next order term, as in 24.1.13. If we fix P constant (that bit is important so that we get back to the same point) and advance both $F^{(1)} + 2\pi$ and $Q_0 + 2\pi$, we return to $F^{(1)}$ and Q_0; for example, $\frac{1}{2\pi}[F(2\pi) - F(0)] = 0$. We say that, over one period, the average vanishes: $\langle \partial F^{(1)}/\partial Q_0 \rangle = 0$, so that, while the sign of $\partial F^{(1)}/\partial Q_0$ is oscillating, its *average is zero*, denoted by $\oint_{2\pi} \frac{\partial F^{(1)}}{\partial Q_0} dQ_0 = 0$, meaning that the entire coloured term in equation 24.1.15 is equal to zero and disappears, removing the periodic elements of the resulting Hamiltonian. We call this **averaging over the torus**. In general, $\langle H_1(P) \rangle$ is a known function. We thus obtain,

$$
\langle K_1(P) \rangle = \langle H_1(P) \rangle + \left\langle \frac{\partial H_0}{\partial P} \frac{\partial F^{(1)}}{\partial Q_0} \right\rangle
$$

$$
= \frac{1}{2\pi} \int_0^{2\pi} H_1(P, Q_0) dQ_0 + \frac{1}{2\pi} \int_0^{2\pi} \frac{\partial H_0}{\partial P} \frac{\partial F^{(1)}}{\partial Q_0} dQ_0 \qquad (24.1.15)
$$

This now is the first-order correction to the energy (the angled brackets mean averaged over a period) and, having computed the *average* perturbation term $\langle K_1(P) \rangle$, we now return to equation 24.1.13 to compute $\partial F^{(1)}/\partial Q_0$. We realise that, since we are averaging over a variable of which K_1 is not a function of, then $\langle K_1 \rangle = K_1$ and we obtain

$$
\frac{\partial F^{(1)}}{\partial Q_0} = \left(\frac{\partial H_0}{\partial P} \right)^{-1} [K_1(P) - H_1(P, Q_0)] \qquad (24.1.16)
$$

$$
F^{(1)}(P, Q_0) = \left(\frac{\partial H_0}{\partial P} \right)^{-1} \int_0^{Q_0} [K_1(P) - H_1(P, x)] dx \qquad (24.1.17)
$$

where x is just a dummy variable and we are assuming $\partial H_0/\partial P \neq 0$ for any one-dimensional system; in this case, we will always be able to obtain a solution. We therefore have our two unique pieces of information, $F^{(1)}$ and $K_1(P)$, which are required to compute the first iteration to the perturbed system's energy. It is for this reason that equation 24.1.13 is known as *Poincaré's fundamental equation of classical perturbation theory*. This method is termed **Poincaré-Lindstedt theory** and is not an axiomatically defined process; rather, it is a process that repeatedly has success in modern physics.

The second term proceeds in similar fashion, where now $\langle \partial F^{(2)}/\partial Q_0 \rangle = 0$, in this way, the second-order correction to equation 24.1.14 is presented below, using the above solutions for K_1 and $F^{(1)}$:

$$\langle K_2(P) \rangle = \frac{1}{2}\frac{\partial^2 H_0}{\partial P^2}\left\langle \left(\frac{\partial F^{(1)}}{\partial Q_0}\right)^2 \right\rangle + \left\langle \frac{\partial H_1}{\partial P}\frac{\partial F^{(1)}}{\partial Q_0} \right\rangle + \frac{\partial H_0}{\partial P}\left\langle \frac{\partial F^{(2)}}{\partial Q_0} \right\rangle \qquad (24.1.18)$$

Just as in the first-order situation, the energy correction term is now used back in 24.1.14 and rearranged to find the correction to the generating function:

$$\frac{\partial F^{(2)}}{\partial Q_0} = \left(\frac{\partial H_0}{\partial P}\right)^{-1}\left[K_2(P) - \frac{1}{2}\frac{\partial^2 H_0}{\partial P^2}\left(\frac{\partial F^{(1)}}{\partial Q_0}\right)^2 - \frac{\partial H_1}{\partial P}\frac{\partial F^{(1)}}{\partial Q_0}\right] \qquad (24.1.19)$$

Therefore, we can say that, *provided* that $\partial H_0/\partial P$ is not zero and can be inverted and the average of $F^{(i)}$ over Q_0 is zero, we can *always* find a unique solution to the one-dimensional perturbed integrable system for all orders:

$$H(P) = H_0(P) + \sum_{i=1}^{\infty} \epsilon^i H_i(P), \qquad F(P,Q_0,\epsilon) = PQ_0 + \sum_{i=1}^{\infty} \epsilon^i F^{(n)}(P,Q_0)$$

You can see the theme develop for this method: assume that the $(n-1)$th order is known and then average the nth order and solve for K_n. Put this result into the non-averaged equation for that order and solve for the generating function. We then move to the $(n+1)$th level and so on.

At this point, we can attempt to generalise this method for systems of more than one degree of freedom, with the notation $\boldsymbol{P} = (P_1, \ldots, P_n)$, $\boldsymbol{Q_0} = (Q_0^{(1)}, \ldots, Q_0^{(n)})$, $\boldsymbol{\nabla} H_0(\boldsymbol{P}) = (\frac{\partial H_0}{\partial P_1}, \frac{\partial H_0}{\partial P_2}, \ldots, \frac{\partial H_0}{\partial P_n})$ and $\boldsymbol{\nabla_{Q_0}} = (\frac{\partial}{\partial Q_0^{(1)}}, \frac{\partial}{\partial Q_0^{(2)}}, \ldots, \frac{\partial}{\partial Q_0^{(n)}})$; let us now reconsider equation 24.1.10:

$$K(\boldsymbol{P}) = H_0(\boldsymbol{P}) + \epsilon \left[\boldsymbol{\nabla} H_0 \boldsymbol{\nabla}_{\boldsymbol{Q_o}} F^{(1)} + H_1(\boldsymbol{Q_0}, \boldsymbol{P}) \right]$$

$$+ \epsilon^2 \left[\boldsymbol{\nabla} H_0 \boldsymbol{\nabla}_{\boldsymbol{Q_o}} F^{(1)} + \frac{1}{2} \sum_{i,j}^{n} \boldsymbol{\nabla}_{i,j}^2 H_0 \boldsymbol{\nabla}_{i,j\boldsymbol{Q_o}}^2 F^{(1)} \right.$$

$$\left. + \boldsymbol{\nabla} H_1 \boldsymbol{\nabla}_{\boldsymbol{Q_o}} F^{(1)} \right] + \mathcal{O}\epsilon^3 \qquad (24.1.20)$$

As before, we expand $K(\boldsymbol{P})$ into powers of ϵ and match them with the right-hand side terms:

$$K_0(\boldsymbol{P}) = H_0(\boldsymbol{P}) \qquad (24.1.21)$$
$$K_1(\boldsymbol{P}) = \boldsymbol{\nabla} H_0 \boldsymbol{\nabla}_{\boldsymbol{Q_o}} F^{(1)} + H_1(\boldsymbol{Q_0}, \boldsymbol{P}) \qquad (24.1.22)$$
$$K_2(\boldsymbol{P}) = \boldsymbol{\nabla} H_0 \boldsymbol{\nabla}_{\boldsymbol{Q_o}} F^{(1)} + \frac{1}{2} \sum_{i,j}^{n} \boldsymbol{\nabla}_{i,j}^2 H_0 \boldsymbol{\nabla}_{i,j\boldsymbol{Q_o}}^2 F^{(1)} + \boldsymbol{\nabla} H_1 \boldsymbol{\nabla}_{\boldsymbol{Q_o}} F^{(1)} \qquad (24.1.23)$$

We again exploit the periodicity to remove the oscillating function and average $H_1(\boldsymbol{Q_0}, \boldsymbol{P})$ over the angle variables:

$$K_1(\boldsymbol{P}) = \int_0^{2\pi} dQ_0^{(1)} \cdots \int_0^{2\pi} dQ_0^{(n)} \ H_1(\boldsymbol{Q_0}, \boldsymbol{P}) \qquad (24.1.24)$$

Use this expression to compute the first-order correction to the generating function:

$$\boldsymbol{\nabla} H_0 \boldsymbol{\nabla}_{\boldsymbol{Q_o}} F^{(1)} = \sum_{i=1}^{n} \left[\frac{\partial H_0}{\partial P_i} \frac{\partial F^{(1)}}{\partial Q_0^{(i)}} \right] = K_1(\boldsymbol{P}) - H_1(\boldsymbol{Q_0}, \boldsymbol{P}) \qquad (24.1.25)$$

With the condition of averages (see below about this), we obtain

$$\frac{1}{(2\pi)^n} \oint_{(2\pi)^n} \left[\boldsymbol{\nabla}_{\boldsymbol{Q_o}} F^{(1)} \right] dQ_0^{(1)} \dots dQ_0^{(n)} = 0 \qquad (24.1.26)$$

You will notice now that, to proceed, we have to divide the right side of equation 24.1.25 by *the n-dimensional vector* $\boldsymbol{\nabla} H_0$ as in analogue to equation 24.1.16; however, therein lies the problem! Division by vectors is not a well-defined operation, in general. These denominator terms can become arbitrarily close to zero as the order increases and thus can lead to higher-order terms being *more* influential than lower-order terms, leading to no certainty of convergence of the series (thats a problem) and the implication that even the slightest non-integrable perturbation leads to completely chaotic motion. Hence, the difficulty with perturbing systems of more that one dimension is termed the

problem of small divisors, as first encountered by Delaunay in the 1840s; it does not affect one-dimensional systems, since all one-dimensional Hamiltonians are integrable. It's more of a maths problem than a physics problem but it leads on to an important idea called *resonance*, which, in turn, is of the utmost importance to chemists studying reaction dynamics and potential energy surfaces using perturbation theory.

The idea is to have ∇H_0 (which we refer to as a *frequency*, which we referred to in chapter 23 as ω) in the denominator of a convergent Fourier expansion of the generating function. Any periodic function can be written in terms of a Fourier series where the exponent acts like an oscillating phase ($f(x) = \sum f_k e^{ikx}$):

$$F^{(1)} \propto \sum_{k \neq 0}^{n} \frac{1}{k \cdot \nabla H_0} \tag{24.1.27}$$

The denominator $k \cdot \nabla H_0 = \sum k_i \nabla_i H_0$, where we have introduced $k = \{k_1, k_2, \ldots, k_n\}$ as a vector of integers belonging to $\mathbb{Z}^n$ and summed over all distinct combinations, is the problematic term. We *assume* that $k \neq 0$, to define two conditions. Firstly, if $k \cdot \nabla H_0 = 0$, in which case the expression for the generating function diverges, then we say we have **resonance**. If there are resonant terms, then they lead to the problem of small divisors and we cannot solve for $F^{(1)}$; in that case, we say equation 24.1.27 has one or more **singularities**. This is similar to what happens when we perturb an oscillator: if the driving force has resonance with the natural frequency of the system, then the amplitude of oscillation goes crazy; if not, then the resulting amplitude will remain in the approximate region of the original system. Canonical perturbation theory does not work for resonant or **near-resonant** degrees of freedom, where near-resonant is defined as a denominator that is so tiny it is almost zero. Unfortunately, every computer has a minimum precision, a single byte perhaps; this means that in computational simulations the frequencies that we use are rational and hence resonant.

If, on the other hand, we have $k \cdot \nabla H_0 \neq 0$, we say that ∇H_0 is **non-resonant**. If we consider $\nabla H_0 \nabla_{Q_0}$ in 24.1.25 to be a linear operator acting on the function $F^{(1)}$, then the non-resonance condition means [1] that $\nabla H_0 \neq 0$ for all k. When ∇H_0 is non-resonant, $F^{(1)}$ *has* a formal solution for the generating function, meaning it does exist even if the series is still divergent due to near-resonances, a problem that must be resolved in order for these solutions to be in any way useful. This paragraph is of fundamental significance; it is **Poincaré's theorem** which states that *we can formally write the canonical perturbation series for all orders of non-resonant systems.*

In general, the non-resonance condition only holds for a small region of phase space around the integrable system and soon integrability breaks down as the strength of the perturbation increases. In other words, there are regions of phase space which, when subject to a computational simulation, behave in a non-resonant manner, despite

---

[1] The mathematical way of writing it would be $\forall\, k \in \mathbb{Z}^n/\{0\} : k \cdot \nabla H_0 \neq 0$, which means excluding $k = 0$, the trivial solution.

the use or rational numbers. In the above example of the oscillator, the non-resonant perturbation doesn't really alter the amplitude of the oscillation too much; it stays pretty much how it was, with only a tiny change. Non-resonant perturbations mean that the system remains near-integrable, while resonant perturbations wildly change the dynamics of the situation.

Perturbation theory in its current form does not, however, paint a full picture of the true dynamics of all systems. This is because it still assumes that the perturbed Hamiltonian can be described by action-angle coordinates and an invariant torus; it makes no room for the development of chaotic dynamics or the intermediate situation. Even the smallest non-linear perturbation appears to destroy perturbation theory by causing extreme divergence when some actions appear to be hardly changing. Another problem with solving equation 24.1.25 is the averaging process involved in removing the oscillatory functions. Equation 24.1.26 demands that we average over the Q_0 variables of the *preserved tori* or *a completely integrable system* but, if the tori are breaking up, then the equation is no longer sufficient to describe all types of motion. Averaging over a broken torus does not make sense.

To summarise, integrable systems are described by tori in phase space. Perturbation theory concerns itself with the evolution of these tori, whether they are preserved or not. If a system has resonances, then, *according to theory so far*, the tori are drastically distorted or even destroyed, meaning we have chaotic motion, and perturbation theory will give us a drastically wrong answer. There is no happy medium.

Therefore, canonical perturbation theory may be very misleading and its application is hit and miss. We need some way of assessing how many tori are drastically changed and how many are only slightly distorted. This is addressed by Kolmogorov, Arnold and Moser in what has become known as **KAM theory** (for *Kolmogorov-Arnold-Moser*). In its essence, the KAM theory consists of two parts. The first requires that we search for nice, non-resonant regions of phase space, while the second asserts that for sufficiently small ϵ, these regions admit a converging perturbation series for all orders of ϵ.

24.2 KAM Theory & Elements of Chaos

In the KAM theory approach, we start by looking for regions of phase space that consists of *good frequencies* by imposing two conditions on the unperturbed Hamiltonian: the **Hessian** condition and the **Diophantine** condition. The Hessian condition concerns the invertibility of $\partial H_0/\partial P$ and the frequency and the smoothness of the Hamiltonian H_0:

$$\det \left| \frac{\partial^2 H_0}{\partial P_i \partial P_j} \right| \neq 0 \qquad (24.2.1)$$

If a frequency satisfies this condition, then it is **non-degenerate**. If the Hessian condition is not met, then we must keep reducing the area of phase space we examine until we have an agreeable subset product space.

The Diophantine condition further reduces the subset of allowed frequencies by bounding them from below; that is, it gets rid of the near-resonances by saying that $|\boldsymbol{k} \cdot \boldsymbol{\nabla} H_0| \geq$ (a number of our choosing). This affects the rate at which the denominator approaches zero:

$$|\boldsymbol{k} \cdot \boldsymbol{\nabla} H_0| \geq \gamma |\boldsymbol{k}|^{-\chi} \qquad (24.2.2)$$

The numbers γ and χ are positive constants that depend on the strength of perturbation, so that, as ϵ grows, so does γ, thereby reducing the number of acceptable frequencies. As the strength of the perturbation decreases, so too does γ and hence more frequencies are contained in the subset of phase space! So you can view the KAM methods as sieving out the resonant from non-resonant tori, allowing us to properly apply perturbation theory with confidence.

KAM theory states that there is a subset called a **Cantor set** (related to fractals) of the phase space that satisfies the Hessian and Diophantine conditions; when ϵ, γ and χ are suitably chosen, the fraction of frequencies that satisfy these conditions increases as ϵ decreases and, in the limit $\epsilon \mapsto 0$, the phase space is increasingly filled with invariant tori. Kolmogorov's theorem states that there is always some remnant of the integrable system left in the perturbed Hamiltonian over a smooth transition, where the new canonical variables describe the near-integrable system, *provided* the perturbation is not too large. Large perturbations will remove any regular structure that the phase space had. The surviving tori are **quasi-periodic KAM tori**; imagine them as little stable islands dense within a sea of chaotic motion. Importantly, we call them non-ergodic tori; this opposes theories predating the KAM theory, by Fermi stating that all perturbed tori would be destroyed. The resonant loops (the ones that get distorted/destroyed) can mix together and form complex patterns around the invariant tori, in a process known as **Arnold diffusion**, the extent of which depends on the strength of the perturbation (see figure 24.2). In particular, Arnold diffusion addresses questions of near-integrable systems and their long term (in)stability; some solutions remain close to the original action while others are wildly different. KAM theory does not sit well with ergodic theory, and the system can not be both KAM and ergodic. This is because KAM theory suggests the time spent on a particular torus is quasiperiodic. Ergodic theory indicates that the time spent on the torus is proportional to the measure, roughly the Liouville measure $\prod dqdp$, or $\omega^n/n!$, or the volume of the region. Perhaps the most important result of KAM theory is the concept of dividing the phase space into stable and unstable regions.

Does KAM theory guarantee the convergence of the perturbative series? We now look at the second part of KAM theory, known as the **superconvergent** perturbation theory. We have seen that each level provides the basis of the approximation to the next level up. If we were to compute the perturbative series for a Hamiltonian, evaluating it at the *origin*, not the previous result (as is normal, really), then the parameter ϵ increases by one power every time so that, in general, we have

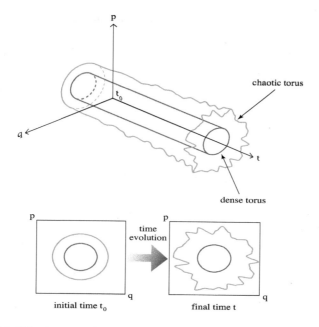

Fig. 24.2: Arnold diffusion visualised in two dimensions for a chaotic and a dense torus.

$$H_0 + \epsilon H_1 + \epsilon^2 H_2 + \cdots + \epsilon^n H_n$$

Kolmogorov's theorem states that we can find canonical transformations such that the order of ϵ increases more rapidly than just by one each time. This is the heart of KAM theory. In the traditional perturbative method, we evaluate the next perturbation from the original unperturbed Hamiltonian. In KAM theory, we are expanding the successive perturbation in terms of the *previous one*, not the starting one. Therefore, the order of ϵ increases rapidly; this is Newton's method:

$$H_0 + \epsilon H_1 + \epsilon^2 H_2 + \epsilon^4 H_3 + \cdots + \epsilon^{2(n-1)} H_n \tag{24.2.3}$$

Therefore, what we have is successive first-order perturbation rather than just increasing orders of a single perturbation from an original Hamiltonian. You perturb H_0 to first-order and then use the result as the starting point to perform first-order perturbation again, thus, KAM theory is first-order perturbation theory. Therefore, KAM theory may be thought of as sifting through solutions that remain close to the original starting action, and those whose action changes vastly.

    KAM theory is much more complicated than this introduction would have you believe and there are many more finer details than would be appropriate to go into here. For more information, any of the papers by Kolmogorov, Arnold and Moser, or their books would provide the best sources and some are listed at the end. In addition to not going into the mathematical details of KAM theory, we have not discussed **Hamiltonian normal forms**, which would be important to anyone applying canonical perturbation theory in practice.

Tori with resonant or near resonant frequencies are normally destroyed and we see the birth of **chaos**. Chaotic dynamics is a vast branch of mathematics and physics that appears in all walks of science, from economics to biology, chemistry, physics, geography, cosmology, and so on. At its heart, it is the study of stable or unstable manifolds under evolution. Most will agree that Henri Poincaré is the father of this branch of mathematics since, he was the first to make real progress in its study.

So far, we have considered both integrable systems and near-integrable systems in phase space; however, there exists a third category: chaotic tori. In particular, we talk about **deterministic chaos**, which is *a system where the present determines the future, but the approximate present does not approximately determine the future.* This means that, if we change, even slightly, the initial conditions we get wildly different results; this does not imply non-determinism, just difficulty in choosing initial conditions. Some initial conditions lead to nice regular paths, while others give chaotic trajectories.

Integrable systems in low dimensions are fairly easy to visualise, since they always take the shape of an n-torus; however, in chaotic dynamics, we are not granted the luxury of knowing the topological shape of the phase space, because the symplectic leaves don't have a regular structure. It is apparent, therefore, that we need a new way of looking at the phase space - one that doesn't make explicit reference to the topological surface. This is achieved by a **Poincaré section**, which we will look at in a moment.

To summarise, if a small non-integrable perturbation is added to an integrable system so that the resulting system is non-integrable, KAM theory can tell us whether the regular motion will be destroyed and dynamical chaos ensue. When the perturbations are small and the frequencies are non-degenerate, then the motion remains in the vicinity of the integrable torus and the dynamical trajectories of the motion will remain regular.

Let us imagine a two-dimensional system in the reduced phase space $\mathbb{R}^{2n}$ with coordinates (q^1, q^2, p_1, p_2) and, for simplicity, imagine an energy surface $H(\boldsymbol{q}, \boldsymbol{p}) = c$ whose flow is represented by a single trajectory. If we place a sheet or plane along the q_2 axis and mark wherever the trajectory intersects the plane from a particular side of our choosing, then we will have a two-dimensional representation of the three-dimensional energy surface. Arbitrarily, we choose to put the plane at $q_2 = 0$ such that the surface of section has the points in phase space for which both $q_2 = 0$ and $H = c$:

$$M_{\text{PS}} = \left\{ (q, p) \in \mathbb{R}^4 : q_2 = 0, H = c \right\} \tag{24.2.4}$$

We now use the surface of section plane to make a **Poincaré map** that characterises the time evolution of the marked points. Interesting motifs appear in Poincaré maps;

these correspond intimately with the system's motion and inherit many of its characteristics. To generalise this, we say that, for an n-dimensional conservative system, the energy hypersurface gives rise to a $(2n-2)$-dimensional surface of section. However, the usefulness of this technique is questionable beyond $n = 2$, since there is no fixed way to define the Poincaré section; our choice of $q_2 = 0$ was arbitrary.

Fig. 24.3: Henri Poincaré (1854–1912).

Poincaré maps are particularly useful when there is periodicity in the flow and they can be selected such as to exploit natural mappings between periods; these are called **stroboscopic maps**. By the application of Stokes' lemma, we know that a tube of integral curves leads to the integral invariant θ_H and, hence, the area of Poincaré maps is also preserved.

We may now wonder if there is some threshold non-integrable perturbation for which the system will transition from remaining close to the regular motion on the torus to exhibiting dynamical chaos? If the strength of the perturbing term is increased, the chaotic threshold is assumed to be reached; we do not know, however, if the motion is strongly chaotic or has only small chaotic components. What we need is a measure of the divergence of trajectories. This can be achieved through the **Lyapunov exponent** of the system; this is a measure of the separation between trajectories $x_i(t)$ and $x_i(t) + \delta x_i$. Another method is to make a **Fourier analysis** of the trajectories for a particular trajectory. Regular non-chaotic motion will be just that - regular, with anharmonic periods. The Fourier spectrum should consist of discrete peaks at the fundamental frequencies and their overtones. If, however, the motion is chaotic, then the Fourier spectrum will contain continuous and discrete components.

If there is **dissipation** in the system, then the system will relax to a local minimum (*point attractor*) or a stationary cycle (*cycle attractor*) in phase space. We will leave our discussion here since this subject is vast but the reader is referred to Ott's (2002) book, *Chaos in Dynamical Systems*, for an entire treatment of non-integrable Hamiltonian systems. This may leave us with strange ideas about determinism within classical mechanics. According to Laplace in the early 19th century if the configuration of a system is known completely at a given time, then it will be determined for all times by simple application of the laws of dynamics. This does not offend deterministic chaos, since, given a perfect description of the current state, the future can indeed, be predicted. The theorem is known colloquially as **Laplace's demon** and is stated below:

We may regard the present state of the universe as the effect of its past and the cause of its future. An intellect which at a certain moment would know all forces that set nature in motion, and all positions of all items of which nature is composed, if this intellect were also vast enough to submit these data to analysis, it would embrace in a single formula the movements of the greatest bodies of the universe and those of the tiniest atom; for such an intellect nothing would be uncertain and the future just like the past would be present before its eyes. Laplace [1814] 1951, p. 4.

However, this cannot be true in the physical world if we take the Copenhagen interpretation of quantum mechanics, since it deals in uncertainties. Even within classical mechanics, there are non-deterministic systems, most notably those that involve breaking of symmetries, such as **Norton's dome**. Even further, **Gödel's first incompleteness theorem** says that it is impossible to construct a complete theory that will predict with certainty the future: either there will be some false statements or there will be incomplete truths or questions that the current theory cannot provide an answer to; an example in language is the *Liar Paradox*. The history of physics somewhat provides an analogy to the incompleteness theorem. First, Newtonian physicists thought that the three laws of motion could provide a description of all material points. Then, Hamilton's principle superseded this, but along with it came the small-divisor problem a lá Poincaré. Replacing this was Schrödinger wave mechanics - which could not describe relativistic theories. Then, Dirac's theory, then quantum field theory, then quantum gravity .... Every new model has an inherent incompleteness or inability to describe the new frontier of science. That isn't to say we shouldn't keep optimising!

Exercise 24.1 Here we consider a perturbed oscillator. These calculations are fun but are quite intricate in parts and there are plenty of places to make a slight error. Consider the following Hamiltonian consisting of an integrable one-dimensional harmonic oscillator and a cubic perturbation term controlled by a small parameter ϵ:

$$H(q, p, \epsilon) = \frac{p^2}{2m} + \frac{1}{2}m\omega^2 q^2 + \epsilon q^3 \tag{24.2.5}$$

We are aware from appendix F that action-angle variables (P, Q) exist for the integrable Hamiltonian and that (p, q) can be expressed as functions of them:

$$p = \sqrt{2m\omega P}\cos Q, \qquad q = \sqrt{\frac{2P}{m\omega}}\sin Q \tag{24.2.6}$$

Hence, the original Hamiltonian can be written $H(Q, P, \epsilon)$:

$$H(Q, P, \epsilon) = \frac{1}{2m}\left[2m\omega P \cos^2 Q\right] + \frac{1}{2}m\omega^2 \frac{2P}{m\omega}\sin^2 Q + \epsilon\left(\frac{2P}{m\omega}\right)^{3/2}\sin^3 Q$$

$$= \omega P + \epsilon\left(\frac{2P}{m\omega}\right)^{3/2}\sin^3 Q$$

$$= H_0 + \epsilon H_1 \tag{24.2.7}$$

We then use the fundamental equation of motion to compute $K_1(P)$:

$$K_1(P) = \frac{\partial H_0}{\partial P}\frac{\partial F^{(1)}}{\partial Q_0} + \left(\frac{2P}{m\omega}\right)^{3/2}\sin^3 Q \tag{24.2.8}$$

We average over the torus to remove the first term and then we use $\int_0^{2\pi}\sin^3 x\,dx = 0$ to generate an expression for $\langle K_1\rangle$, recalling $\langle\partial F^{(1)}/\partial Q_0\rangle = 0$:

$$\langle K_1(P)\rangle = \frac{1}{2\pi}\int_0^{2\pi} dQ\left[\frac{\partial H_0}{\partial P}\frac{\partial F^{(1)}}{\partial Q_0} + \left(\frac{2P}{m\omega}\right)^{3/2}\sin^3 Q\right]$$
$$= 0 \tag{24.2.9}$$

Then, since $\langle K_1\rangle = K_1$,

$$\frac{\partial F^{(1)}}{\partial Q_0} = -H_1\left(\frac{\partial H_0}{\partial P}\right)^{-1} \tag{24.2.10}$$

Therefore, the first-order energy correction vanishes. The fundamental equation is then used to compute the second-order contribution:

$$K_2(P) = \frac{\partial H_0}{\partial P}\frac{\partial F^{(2)}}{\partial Q_0} + \frac{1}{2}\frac{\partial^2 H_0}{\partial P^2}\left(\frac{\partial F^{(1)}}{\partial Q_0}\right)^2 + \frac{\partial H_1}{\partial P}\frac{\partial F^{(1)}}{\partial Q_0} \tag{24.2.11}$$

The average is then computed by using $\langle\partial F^{(2)}/\partial Q_0\rangle = 0$, the identity in equation 24.2.10, and the middle term vanishes, since $\partial^2 H_0/\partial P^2 = 0$:

$$\langle K_2\rangle = \frac{1}{2\pi}\int_0^{2\pi}\frac{\partial H_1}{\partial P}\frac{\partial F^{(1)}}{\partial Q_0}dQ$$
$$= -\frac{1}{2\pi}\int_0^{2\pi} dQ\left[\frac{\partial H_1}{\partial P}H_1\left(\frac{\partial H_0}{\partial P}\right)^{-1}\right] \tag{24.2.12}$$

We then compute $\partial H_1/\partial P$:

$$\frac{\partial H_1}{\partial P} = \frac{3}{2}\left(\frac{2P}{m\omega}\right)^{1/2}\frac{2}{m\omega}\sin^3 Q \tag{24.2.13}$$

Then, piecewise, we construct $\langle K_2\rangle$, labelling $\partial H_0/\partial P = \omega$:

$$\langle K_2\rangle = -\frac{1}{2\pi}\int_0^{2\pi} dQ\left[\frac{3}{2}\left(\frac{2P}{m\omega}\right)^{1/2}\frac{2}{m\omega}\sin^3 Q\right]\cdot\left[\frac{1}{\omega}\left(\frac{2P}{m\omega}\right)^{3/2}\sin^3 Q\right]$$
$$= -\frac{1}{2\pi}\int_0^{2\pi} dQ\left[\frac{3}{2}\frac{1}{\omega}\left(\frac{2}{m\omega}\right)^3 P^2\sin^6 Q\right]$$
$$= -\frac{5}{16}\frac{3}{2}\frac{1}{\omega}\left(\frac{2}{m\omega}\right)^3 P^2$$
$$= -\frac{15}{4}\frac{P^2}{m^3\omega^4} \tag{24.2.14}$$

using $\int_0^{2\pi} dQ \sin^6 Q = 5\pi/8$. The energy can then be written as the following by summing the zeroth-, first- and second-order contributions:

$$H = H_0 + \epsilon K_1 + \epsilon K_2 + \cdots$$

$$= \omega P + \epsilon \cdot 0 - \epsilon \frac{15}{4} \frac{P^2}{m^3 \omega^4} + \cdots \tag{24.2.15}$$

This system actually has an exact solution and can be integrated for q straight away by using elliptic integral. The proof of which is beyond this text but definitely worth researching if you are interested in Jacobi elliptic functions or cubic anharmonics.

Exercise 24.2 Consider now a slightly more involved oscillator with a quartic perturbation and compute the second-order correction to the Hamiltonian, using the perturbation theory outlined previously. The procedure is exactly the same, however, this time $\partial F^{(1)}/\partial Q_0$ does not vanish as conveniently as with the previous example:

$$H = \frac{p^2}{2m} + \frac{1}{2} m\omega^2 q^2 + \epsilon q^4 \tag{24.2.16}$$

The Hamiltonian in action-angle variables (P, Q) is given by,

$$H = \omega P + \epsilon \left(\frac{2P}{m\omega} \right)^2 \sin^4 Q \tag{24.2.17}$$

The first-order energy term is calculated as above:

$$K_1(P) = \frac{\partial H_0}{\partial P} \frac{\partial F^{(1)}}{\partial Q} + H_1$$

$$= \frac{\partial H_0}{\partial P} \frac{\partial F^{(1)}}{\partial Q} + \left(\frac{2P}{m\omega} \right)^2 \sin^4 Q \tag{24.2.18}$$

$$\langle K_1(P) \rangle = \frac{1}{2\pi} \int_0^{2\pi} dQ K_1$$

$$= \frac{3}{8} \left(\frac{2P}{m\omega} \right)^2 \tag{24.2.19}$$

We use equation 24.1.16 to compute $\partial F^{(1)}/\partial Q_0$ and piecewise construct terms in $\langle K_2(P) \rangle$:

$$\frac{\partial F^{(1)}}{\partial Q} = \frac{1}{\omega} \left[K_1 - H_1 \right]$$

$$= \frac{1}{\omega} \left[\frac{3}{8} \left(\frac{2P}{m\omega} \right)^2 - \left(\frac{2P}{m\omega} \right)^2 \sin^4 Q \right]$$

$$= \frac{1}{\omega} \left(\frac{2P}{m\omega} \right)^2 \left[\frac{3}{8} - \sin^4 Q \right] \tag{24.2.20}$$

Next, we construct $\partial H_1 / \partial P$:

$$
\begin{aligned}
\frac{\partial H_1}{\partial P} &= \frac{\partial}{\partial P}\left[\left(\frac{2P}{m\omega}\right)^2 \sin^4 Q\right] \\
&= \frac{4}{m\omega}\left(\frac{2P}{m\omega}\right) \sin^4 Q
\end{aligned}
\tag{24.2.21}
$$

these results are then multiplied:

$$
\begin{aligned}
\frac{\partial H_1}{\partial P}\frac{\partial F^{(1)}}{\partial Q_0} &= \frac{4}{m\omega}\left(\frac{2P}{m\omega}\right)\sin^4 Q \cdot \frac{1}{\omega}\left(\frac{2P}{m\omega}\right)^2\left[\frac{3}{8}-\sin^4 Q\right] \\
&= \frac{12}{8m\omega^2}\left(\frac{2P}{m\omega}\right)^3 \sin^4 Q - \frac{4}{m\omega^2}\left(\frac{2P}{m\omega}\right)^3 \sin^8 Q
\end{aligned}
\tag{24.2.22}
$$

$$
\begin{aligned}
\frac{1}{2\pi}\int_0^{2\pi} dQ\left[\frac{\partial H_1}{\partial P}\frac{\partial F^{(1)}}{\partial Q_0}\right] &= \frac{1}{2\pi}\int_0^{2\pi} dQ\left[\frac{12}{8m\omega^2}\left(\frac{2P}{m\omega}\right)^3 \sin^4 Q\right. \\
&\qquad\left. - \frac{4}{m\omega^2}\left(\frac{2P}{m\omega}\right)^3 \sin^8 Q\right]
\end{aligned}
\tag{24.2.23}
$$

Using $\int_0^{2\pi}\sin^4 x\,dx = 3\pi/8$ and $\int_0^{2\pi}\sin^8 x\,dx = 35\pi/64$, we obtain

$$
\begin{aligned}
&= \frac{1}{2\pi}\frac{12}{8m\omega^2}\left(\frac{2P}{m\omega}\right)^3\frac{3\pi}{8} - \frac{1}{2\pi}\frac{4}{m\omega^2}\left(\frac{2P}{m\omega}\right)^3\frac{35\pi}{64} \\
&= \frac{1}{m\omega^2}\left(\frac{2P}{m\omega}\right)^3\left[\frac{36}{128}-\frac{35}{32}\right] \\
&= -\frac{1}{m\omega^2}\left(\frac{2P}{m\omega}\right)^3\cdot\frac{13}{16}
\end{aligned}
\tag{24.2.24}
$$

Hence, the energy corrections to second-order for the oscillator with a quartic perturbation is given by

$$
H = \omega P + \epsilon\frac{3}{8}\left(\frac{2P}{m\omega}\right)^2 - \epsilon\frac{1}{m\omega^2}\left(\frac{2P}{m\omega}\right)^3\frac{13}{16} + \cdots
\tag{24.2.25}
$$

We assume at all times that, since the system is one-dimensional, the frequencies are non-resonant and, therefore, that the series can be converged by numerical means, if necessary.

Part IV

Classical Field Theory

Up to this point, mechanics has concerned itself with the description of discrete particles and, for the most part, this is very convenient. However, in large arrays of molecules, such as in a crystal, for example, we can treat the system as a single entity, using the idea of continuous properties that all members will share. There are many such systems we could consider, an infinite rod of oscillators being perhaps the most rewarding and pedagogically informative. In physics, the concept of a field is not the same as the mathematical definition; here, we define a field as a quantity defined at every point of space and time. There is a degree of freedom associated with each point in space. Importantly, we see a shift from ODEs to PDEs, and the treatment of space and time on equal footing.

Classical field theory is the natural generalisation of classical mechanics. It is used across many disciplines of physics, with direct applicability to vibrating chemical systems and sound waves passing through gases and, of course, it is the precursor to quantum field theory. Here, we shall derive a Lagrangian density which replaces discrete Lagrangians in a continuum limit and consider how to extend the theory to a Hamiltonian description. During this, we will take a little time out to look at electromagnetism in a Lagrangian setting. We will then culminate our discussions with the field theoretic version of Noether's theorem, symmetries in physics and the classical path integral. The following also nicely happens to review discrete mechanics along the way!

25
Lagrangian Field Theory

The motion of systems with a very large number of degrees of freedom requires us to specify an almost infinite number of discrete coordinates; think of all the vibrating lattice points in a three-dimensional crystal and how much data we would require to specify even one-dimensional oscillations. This is quite problematic; yet, we can simplify the situation by taking the continuum limit, which replaces the individual coordinates with a continuous function that describes a **displacement field**, which assigns a displacement vector[1] to each position the system could occupy relative to an equilibrium configuration. In other words, it assigns to a point on a **spacetime manifold** (more on this later) an amplitude, which could be a scalar, a spinor, a vector, a complex number or, most generally, a tensor. The field takes a point in the spacetime manifold and assigns it a value corresponding to whatever the field represents.

To begin with, we consider a one-dimensional *elastic* rod of length l vibrating longitudinally[2] between fixed end points, please note that this is a *toy model* which is used to introduce the general theory, and the actual expression of the Lagrange density will depend on the system under study, just as a Lagrangian would.

We can model this system as a series of identical masses connected by massless springs with an equilibrium separation of Δx (the lattice spacing) and a force constant of k. The total length of the rod is given by equation 25.0.1, which you can envisage from drawing a few repeats of the system:

$$l = (n+1)\Delta x \qquad (25.0.1)$$

For the ith particle, the displacement from the equilibrium position in the x dimension is given by the quantity $\phi_i(x, t)$, which forms a discrete set of coordinates for all the masses in the system (don't confuse the notation with constraints). This then becomes our coordinate, which we can use to define the kinetic and potential energies:

---

[1] A scalar field would assign a scalar to every point in space; a tensor field assigns a tensor to each point of the space.

[2] The displacement from the equilibrium positions is in the same direction as the propagation of the wave. This is in contrast to a transverse wave, which oscillates in a plane that is orthogonal to the direction of propagation.

Lagrangian & Hamiltonian Dynamics. Peter Mann, Oxford University Press (2018).
© Peter Mann. DOI: 10.1093/oso/9780198822370.001.0001

$$T = \frac{1}{2}\sum_{i=1}^{n} m\dot{\phi}_i^2, \qquad U = \frac{1}{2}\sum_{i=0}^{n} k(\phi_{i+1} - \phi_i)^2 \qquad (25.0.2)$$

Note that the potential is a sum over $(n+1)$ springs, and the kinetic is a sum over n identical masses. We can now write the Lagrangian as $\mathscr{L} = T - U$:

$$\mathscr{L}(\phi, \dot{\phi}) = \frac{1}{2}\sum_{i=1}^{n} m\dot{\phi}_i^2 - \frac{1}{2}\sum_{i=0}^{n} k(\phi_{i+1} - \phi_i)^2 \qquad (25.0.3)$$

The potential term can be a little difficult and you need to expand the ith term: $(\phi_{i+1} - \phi)^2 = (\phi_{i+1}^2 + \phi_i^2 - 2\phi_i\phi_{i+1})$. Then take the partial derivative with respect to ϕ_i to get $2\phi_i - 2\phi_{i+1}$. The same procedure for the $(i-1)$th term gives us $-2\phi_{i-1} + 2\phi_i$. Therefore, the final term is $-\frac{1}{2}k(-2\phi_{i-1} + 4\phi_i - 2\phi_{i+1})$, leaving us with the equation of motion,

$$m_i\ddot{\phi}_i - k(\phi_{i+1} - 2\phi_i + \phi_{i-1}) = 0 \qquad (25.0.4)$$

which has the form of $m\ddot{\phi}_i = F_i$, or Newton's second law. So far, everything is very discrete and finite and thus this discussion resembles our previous discussions on oscillations and crystal vibrations. At this point, we take the *continuum limit* while *keeping the length of the rod fixed at l*. That is to say, we let the number of masses go to infinity $(n \to \infty)$ and hence let the distance between the masses, Δx, tend to zero $(\Delta x \to 0)$. It is from here that we use continuous distributions of matter and so have to deal with continuous properties.

In order to do this, it is fruitful to express the Lagrangian slightly differently by multiplying the top and bottom by Δx:

$$\mathscr{L} = \frac{1}{2}\sum_{i=1}^{n} \Delta x \left(\frac{m}{\Delta x}\right)\dot{\phi}_i^2 - \frac{1}{2}\sum_{i=0}^{n} \Delta x(k\Delta x)\left(\frac{\phi_{i+1} - \phi_i}{\Delta x}\right)^2 \qquad (25.0.5)$$

As we increase the number of particles, we would expect the total kinetic energy to diverge and, similarly, as we decrease the inter-particle spacing, we expect the equilibrium force on each spring,$k\Delta x$, to vanish. We take the unusual step of *fixing* the *linear density* μ, which is defined as the mass per unit length $(m/\Delta x)$ such that it is constant (i.e the ratio is constant). Further, we fix the *elasticity modulus* $\mathcal{Y}$, defined as the equilibrium force per unit extension, $k\Delta x$.

The discrete label for each mass is now replaced by a continuous position coordinate x such that $\phi(x,t)$ (i.e a function of x such that $\phi_i \to \phi(x)$) defines the field of displacements:

$$\lim_{\Delta x \to 0}\frac{\phi_{i+1} - \phi_i}{\Delta x} = \lim_{\Delta x \to 0}\frac{\phi(x + \Delta x) - \phi(x)}{\Delta x} = \frac{d\phi}{dx} = \nabla_x\phi = \partial_x\phi \qquad (25.0.6)$$

The last three relations just clarify the different notations. Further, we replace the sum over discrete particles by an integral over the length of the rod:

$$\sum_i^n \longrightarrow \int_0^l dx$$

The Lagrangian in 25.0.5 now becomes

$$\mathscr{L}(\phi(x,t), \dot\phi(x,t), \partial_x\phi(x,t)) = \frac{1}{2}\int_0^l \left[\mu\dot\phi^2 - \mathcal{Y}(\nabla_x\phi)^2\right]dx = \int_0^l \mathcal{L}dx \qquad (25.0.7)$$

It is important to note that, for this particular chapter $\partial_x\phi$ or $\partial_t\phi$ represent the total derivatives $d\phi/dx$ and $d\phi/dt$, respectively. Equation 25.0.7 is now an integral over x of the one-dimensional **Lagrangian density**, which, for this system, is given by $\mathcal{L} = \frac{1}{2}(\mu\dot\phi - \mathcal{Y}(\nabla\phi)^2)$ and can be thought of as the Lagrangian per unit length. As such, the action functional is now dependent on the field rather than on the discrete coordinates and is written as an integral over space and time of the Lagrangian density:

$$S[\phi(x,t)] := \int_{t_1}^{t_2} \mathscr{L}[\phi]dt = \int_{t_1}^{t_2}\int_0^l \mathcal{L}(\phi, \dot\phi, \partial_x\phi)dxdt \qquad (25.0.8)$$

The Lagrange density may also depend on $\partial_t\phi, \partial_x\phi, \phi_x$ or t in general, noting that $\dot\phi = \partial_t\phi$, which we treat independently of ϕ; note that the square brackets indicate the functional relationship for both S and $\mathscr{L}$, while $\mathcal{L}$ is a function of $\phi : \mathbb{R}^2 \to \mathbb{R}$, which depends on a space dimension and time. It is from this action principle that we will

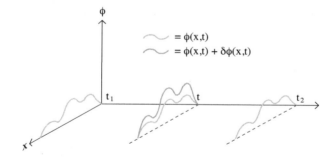

Fig. 25.1: The stationary action principle for fields.

derive the Lagrange equation of motion by examining an infinitesimal change in the field such that $\phi(x,t) \to \phi(x,t) + \delta\phi(x,t)$, and similar variations in both $\partial_t\phi$ and $\partial_x\phi$ (see figure 25.1); note we are treating ϕ as a generalised coordinate. Correspondingly, we see that the extremal of the action is given by $\delta S[\phi(x,t)] = S[\phi(x,t)+\delta\phi(x,t)] - S[\phi(x,t)]$.

We now follow the standard procedure of the calculus of variations for finding the extremals of functionals:

$$\delta S[\phi(x,t)] = \int_{t_1}^{t_2} dt \int_0^l dx \left\{ \frac{\partial \mathcal{L}}{\partial \phi(x,t)} + \frac{\partial \mathcal{L}}{\partial(\partial_t \phi(x,t))} \frac{\partial}{\partial t} \right.$$

$$\left. + \frac{\partial \mathcal{L}}{\partial(\partial_x \phi(x,t))} \frac{\partial}{\partial x} \right\} \delta\phi(x,t) \tag{25.0.9}$$

At this point, it should be pretty obvious what we need to do now: integrate both coloured terms by parts:

$$\int_{t_1}^{t_2} \frac{\partial \mathcal{L}}{\partial(\partial_t \phi)} \frac{\partial}{\partial t} \delta\phi \, dt = \left. \frac{\partial \mathcal{L}}{\partial(\partial_t \phi)} \delta\phi \right|_{t_1}^{t_2} - \int_{t_1}^{t_2} \frac{d}{dt}\left(\frac{\partial \mathcal{L}}{\partial(\partial_t \phi)} \right) \delta\phi \, dt$$

$$\int_0^l \frac{\partial \mathcal{L}}{\partial(\partial_x \phi)} \frac{\partial}{\partial x} \delta\phi \, dx = \left. \frac{\partial \mathcal{L}}{\partial(\partial_x \phi)} \delta\phi \right|_{t_1}^{t_2} - \int_0^l \frac{d}{dx}\left(\frac{\partial \mathcal{L}}{\partial(\partial_x \phi)} \right) \delta\phi \, dx$$

As usual, the terms evaluated at the boundary of the region of integration vanish, since we are considering variation in the field between fixed end points $\delta\phi(x,t_1) = \delta\phi(x,t_2) = \delta\phi(0,t) = \delta\phi(l,t) = 0$, so we are left in both cases with only the second part of each equation. Later, in Noether's variational problem, we don't cancel these and allow variation at the boundary points of space and time! We thus obtian

$$\delta S[\phi(x,t)] = \int_{t_1}^{t_2} \int_0^l \left\{ \frac{\partial \mathcal{L}}{\partial \phi} - \frac{d}{dt}\left(\frac{\partial \mathcal{L}}{\partial(\partial_t \phi)} \right) - \frac{d}{dx}\left(\frac{\partial \mathcal{L}}{\partial(\partial_x \phi)} \right) \right\} \delta\phi \, dx \, dt \tag{25.0.10}$$

Imposing the stationary action principle, we can define the Lagrange equation for a one-dimensional continuous system as in equation 25.0.11. Note that this single continuous Lagrange equation replaces the discrete case where each mass has its own Lagrange equation. This is the **Lagrange field equation**:

$$\frac{d}{dt}\left(\frac{\partial \mathcal{L}}{\partial(\partial_t \phi)} \right) + \frac{d}{dx}\left(\frac{\partial \mathcal{L}}{\partial(\partial_x \phi)} \right) - \frac{\partial \mathcal{L}}{\partial \phi} = 0 \tag{25.0.11}$$

Notice how there is a spatial derivative in addition to the time derivative! It is easy to generalise this to higher dimensions by the following relations:

$$S[\phi(x,y,z,t)] = \int_{t_1}^{t_2} \int_{x_1}^{x_2} \int_{y_1}^{y_2} \int_{z_1}^{z_2} \mathcal{L} \, dx \, dy \, dz \, dt \tag{25.0.12}$$

where

$$\mathcal{L} = \mathcal{L}\big(\phi_x, \phi_y, \phi_z, \partial_x \phi, \partial_y \phi, \partial_z \phi, \partial_t \phi, t\big) \tag{25.0.13}$$

Therefore, the equations of motion becomes

$$\frac{d}{dt}\left(\frac{\partial \mathcal{L}}{\partial(\partial_t \phi)}\right) + \frac{d}{dx}\left(\frac{\partial \mathcal{L}}{\partial(\partial_x \phi)}\right) + \frac{d}{dy}\left(\frac{\partial \mathcal{L}}{\partial(\partial_y \phi)}\right) + \frac{d}{dz}\left(\frac{\partial \mathcal{L}}{\partial(\partial_z \phi)}\right) - \frac{\partial \mathcal{L}}{\partial \phi} = 0 \quad (25.0.14)$$

Writing this out every time we want to talk about spacetime gets rather annoying, so we can compress this big expression into one neat equation by letting $\partial_\mu = (\partial_t, \vec{\nabla})$, and we define the **four vector** x^μ, which has the four components t, x, y and z over a region $\mathcal{R}$ of spacetime; note $\partial^\mu = (\partial_t, -\vec{\nabla})$ and $\phi : \mathbb{R}^4 \to \mathbb{R}$. This reversal of the spatial components but not the time stems from the signature of the metric $g = \text{diag}(1, -1, -1, -1)$ and assuming time is the zeroth component:

$$S = \int_{\mathcal{R}} \mathcal{L}(\phi, \partial_\mu \phi, x^\mu) d^4 x \qquad \textbf{field action principle} \qquad (25.0.15)$$

with the Euler-Lagrange equation

$$\frac{\partial \mathcal{L}}{\partial \phi} - \partial_\mu \left(\frac{\partial \mathcal{L}}{\partial(\partial_\mu \phi)}\right) = 0 \qquad (25.0.16)$$

A classical solution to the Lagrange field equation is called an **instanton**, and solving the field equation should give us as a general result all the modes of oscillation of the field as a superposition:

$$\phi(\boldsymbol{x}, t) = \sum_k a_k e^{-i(\omega t - \boldsymbol{k} \cdot \boldsymbol{x})}$$

Just like the Lagrangian function, a Lagrange density exhibits gauge invariance, since we can add a total divergence $\partial_\mu \mathcal{K}^\mu = (\partial \mathcal{K}/\partial \boldsymbol{x} + \partial \mathcal{K}/\partial t)$ to the end without altering the equations of motion. Adopting one field configuration is known as **gauge fixing**, while transforming to another configuration is known as a gauge transformation. We will explore gauge fields in chapter 28.

Just as in the discrete case, we can consider Lagrange densities with higher-order field dependance $\mathcal{L}(x^\mu, \phi, \partial_\mu \phi, \partial_\nu \partial_\mu \phi, \dots)$; in the second-order case, the Euler-Lagrange equation would be the following Ostrogradsky equation:

$$\frac{\partial \mathscr{L}}{\partial \phi} - \partial_\mu \left(\frac{\partial \mathscr{L}}{\partial(\partial_\mu \phi)}\right) + \partial_\nu \partial_\mu \left(\frac{\partial \mathscr{L}}{\partial(\partial_\nu \partial_\mu \phi)}\right) = 0$$

Field theory generalises the Lagrangian to a Lagrange density, and we found this necessary to deal with spatial dimensions. We can classify the Lagrange density as **local** or **non-local**; it is local if it is dependent on only a *single* spacetime point, with the field or its nth-order derivatives being *evaluated at that point*, While a non-local Lagrange density will depend on *more* than one point in spacetime. Non-local Lagrange densities aren't ideal in a physical theory, since they can be tricky to deal with.

An instanton is a local solution to the field equations, that is, it is a local extrema of the action. In classical mechanics, we expect global and local minima to correspond to unique solutions to the equations of motion, that is, to have the same dynamics. However, in quantum field theory, the different solutions all have different contributions to the probability of an event and, therefore, differences between global and local extrema.

Chapter summary

- Continuous distributions called **fields** replace discrete coordinates $q(t) \rightarrow \phi(x)$, since space and time are treated on equal footing as smooth parameters.
- The action is an integral of the Lagrangian $\mathscr{L}$ with respect to time; but the Lagrangian is written as an integral over a **Lagrange density** with respect to space.
- We define the **4-vector** x^μ with three space coordinates and time, the **Lagrangian field equation** is given by

$$\frac{\partial \mathscr{L}}{\partial \phi} - \partial_\mu \left(\frac{\partial \mathscr{L}}{\partial (\partial_\mu \phi)} \right) = 0$$

the solution to which is called an **instanton**.

Exercise 25.1 As a brief example, we could consider a real, massless scalar field given by the Lagrange density $\mathcal{L} = \frac{1}{2}(\partial_\mu \phi)^2$. A scalar is a function that maps to the reals $\phi : \mathbb{R}^n \rightarrow \mathbb{R}$, taking $x \mapsto \phi(x)$. The field equation is then

$$\partial_\mu \partial^\mu \phi = \frac{\partial^2 \phi}{\partial t^2} - \nabla^2 \phi = 0$$

To include mass in the field, we make the Lagrange density a function of both the change in field over spacetime $(\partial_\mu \phi)$ and its magnitude (ϕ) and we include a mass term m in the potential density:

$$\mathcal{L} = \frac{1}{2}(\partial_\mu \phi)^2 - \frac{1}{2}m^2\phi^2$$

generating the field equation

$$(\partial_\mu \partial^\mu + m^2)\phi(\boldsymbol{x}, t) = (\hat{\Box} + m^2)\phi = 0$$

which is known as the **Klein-Gordon equation**, a homogenous PDE, in which the little square is the d'Alembert operator: $\hat{\Box} = \partial_\mu \partial^\mu = \frac{\partial^2}{\partial t^2} - \nabla^2$. So, we have our real-massive-scalar field! You can think of the Klein-Gordon equation as the field theoretic version of the harmonic oscillator. We can then allow our massive-scalar field interact with an external potential $-J(x)\phi(x)$, where $J(x)$ is called a **source** of the field. The Lagrange density for a sourced, massive-scalar field is

$$\mathcal{L} = \frac{1}{2}(\partial_\mu\phi)^2 - \frac{1}{2}m^2\phi^2 + J\phi$$

with the field equation

$$(\hat{\Box} + m^2)\phi = J$$

which is an inhomogeneous PDE.

We can convert this **position space** Klein-Gordon into a **momentum space** Klein-Gordon equation by using a **Fourier transform**. For any set of spacetime variables, there are also **frequency variables**, more precisely, spatial frequency $\boldsymbol{k}$ and temporal frequency ω, which together constitute a 4-vector $(\omega, \boldsymbol{k})$ along with $(t, \boldsymbol{x})$. To interconvert between the two representations, we utilise the four-dimensional Fourier transform $\tilde{f}(x)$ of a function of spacetime $f(x)$:

$$\tilde{f}(k) = \int d^4x\, e^{ik\cdot x} f(x)$$

With the inverse being

$$f(x) = \int \frac{d^4x}{(2\pi)^4} e^{-ik\cdot x} \tilde{f}(k)$$

There is a factor of 2π for each integral. So, getting back to the Klein-Gordon equation, we define a conjugate function for $\phi(\boldsymbol{x}, t)$ and call it the momentum space field $\tilde{\phi}(\boldsymbol{p}, t)$. The two fields are related via the Fourier relationship:

$$\phi(\boldsymbol{x}, t) = \int \frac{d^3p}{(2\pi)^3} e^{ip\cdot x} \tilde{\phi}(\boldsymbol{p}, t)$$

This is a very useful formula for general field theory and so it should be well learnt; there are many times that we switch representation in condensed matter science. Entering this expression into the Klein-Gordon equation we obtain

$$\int \frac{d^3p}{(2\pi)^3} e^{ip\cdot x} \left(\frac{\partial^2}{\partial t^2} - \nabla^2 + m^2 \right) \tilde{\phi}(\boldsymbol{p}, t) = \int \frac{d^3p}{(2\pi)^3} e^{ip\cdot x} \left(\frac{\partial^2}{\partial t^2} - \|\boldsymbol{p}\|^2 + m^2 \right) \tilde{\phi}(\boldsymbol{p}, t) = 0$$

Since the exponents are linearly independent in x for different p, every term inside the integral is zero; if you can't spot this straight off, then multiply both sides by $\exp(-ip'\cdot x)$ and integrate over the spatial dimensions. We thus obtain

$$\int d^3x \frac{d^3p}{(2\pi)^3} e^{i(p-p')\cdot x} \left(\frac{\partial^2}{\partial t^2} - \|\boldsymbol{p}\|^2 + m^2 \right) \tilde{\phi}(\boldsymbol{p}, t) = 0$$

Then, using the general result $\int d^3x\, e^{-i(p-p')\cdot x} = \delta(p - p')$:

$$\int \frac{d^3p}{(2\pi)^3} \delta(p - p') \left(\frac{\partial^2}{\partial t^2} - \|\boldsymbol{p}\|^2 + m^2 \right) \tilde{\phi}(\boldsymbol{p}, t) = 0$$

Hence, as predicted, each term vanishes, independently giving us the momentum space Klein-Gordon equation:

$$\left(\frac{\partial^2}{\partial t^2} - \|\boldsymbol{p}\|^2 + m^2 \right) \tilde{\phi}(\boldsymbol{p}, t) = 0$$

Exercise 25.2 Here, we will consider a crystal as a regular lattice of particles connected by springs that can be perturbed from equilibrium positions by $\eta_i(\boldsymbol{r})$. On the macroscopic scale, such a system can be modelled by a continuous elastic medium using **infinitesimal strain theory**. The Lagrangian for the system can be written by considering the rotational and translational invariance of the system at a microscopic level and hence must be comprised of symmetric derivatives:

$$\mathcal{L} = \rho(\partial_t \eta_i)^2 - 2\mu \eta_{ij} \eta_{ij} - \lambda \eta_{ii}^2, \qquad \eta_{ij} = \frac{1}{2}(\partial_i \eta_j + \partial_j \eta_i) \qquad (25.0.17)$$

We have introduced the mass density ρ and the first and second **Lamé parameters** λ and μ to describe the isotropic media (note that repeated indices are summed over). Find the equations of motion for the system and show that the solution is a plane wave.

We begin by computing the derivatives $\partial_k \eta_{ij} = \partial_k \partial_i \eta_j + \partial_k \partial_j \eta_i$ and $\partial_k \eta_{ii} = \partial_k \partial_i \eta_i$, from which we build the Euler-Lagrange equation:

$$\rho \partial_t^2 \eta_i = (\mu + \lambda) \partial_i \partial_j \eta_j + \mu \partial_j \partial_j \eta_i \qquad (25.0.18)$$

It is then clear that $\eta_i = \xi_i \exp\{i(\boldsymbol{k} \cdot \boldsymbol{r} - \omega t)\}$ is a solution.

Exercise 25.3 Show that the **Korteweg-de Vries equation** that describes waves on shallow water surfaces or phonons (acoustic waves) in crystals is the field equation for a scalar field ψ whose Lagrange density is given by $\mathcal{L}$:

$$\mathcal{L} = \frac{1}{2} \partial_x \psi \partial_t \psi + (\partial_x \psi)^3 - \frac{1}{2}(\partial_x^2 \psi)^2 \qquad (25.0.19)$$

Define the quantity $\phi = \partial_x \psi$ to simplify the resulting equation of motion.

26
Hamiltonian Field Theory

Despite deriving our field theoretic model with the one-dimensional rod of oscillators, we can see that the Euler-Lagrange equation derived in this setting applies to any continuous field system; the rod was just a nice introduction. With this in mind, we can develop a Hamiltonian field model via a **Hamiltonian density** $\mathcal{H}$ in a similar way to that used for the discrete scenario.

If we now take our generalised distributed coordinate $\phi = \phi(x^\mu)$ to be the continuous analogue of the localised coordinate $q = q(t)$, we can define the **4-momentum density** π_μ according to equation 26.0.1:

$$\pi_\mu = \frac{\partial \mathcal{L}}{\partial(\partial_\mu \phi)} \tag{26.0.1}$$

You may consider this to be the 4-vector generalisation of a momentum density (e.g. we might use the zeroth component $\pi_t = \partial \mathcal{L}/\partial(\partial_t \phi)$, or derivatives with respect to space, too $\pi_x = \partial \mathcal{L}/\partial(\partial_x \phi)$). The Hamiltonian density $\mathcal{H}$ also has the familiar expression from the Legendre transform of the Lagrangian density, assuming no singularities in the Hessian:

$$\mathcal{H}(\phi, \partial_x \phi, \pi, x^\mu) = \pi_\mu \partial^\mu \phi - \mathcal{L} \tag{26.0.2}$$

The Hamiltonian density of the linear chain of oscillators is now readily found to be:

$$\mathcal{H} = \frac{\pi_x^2}{2\mu} + \frac{\mathcal{Y}}{2}(\partial_x \phi)^2$$

The interconversion between Lagrangian and Hamiltonian frameworks is governed by a familiar-looking Hessian condition that each field must satisfy.

$$\frac{\partial^2 \mathcal{L}}{\partial(\partial_\mu \phi^i)\partial(\partial_\nu \phi^j)} \neq 0 \quad \forall i, j$$

The Hamiltonian H of the system is related to the Hamiltonian density $\mathcal{H}$ by the integral over all the *spatial* dimensions of the field but not time. You may wonder why we went to the trouble of defining a 4-momentum if we only define the Hamiltonian

Lagrangian & Hamiltonian Dynamics. Peter Mann, Oxford University Press (2018).
© Peter Mann. DOI: 10.1093/oso/9780198822370.001.0001

with the zeroth component ($\dot{\phi}$); you may have expected to write the Hamiltonian as the $\int d^4x$ over the density $\mathcal{H} = \pi_\mu \partial^\mu \phi - \mathcal{L}$. The attractiveness of this equation is derived from the fact that it is inherently defining a **covariant phase space formulation**. However, despite the Lagrange density being Lorentz invariant (and, indeed, the action integral, too), the Hamiltonian formalism is not! This is because the Hamiltonian is the *generator* of time translations, and writing it this way, as the zeroth component, connects to much more of our previous discrete theory. The covariant Hamiltonian de Donder-Weyl formalism is an interesting subject for advanced readers, and our initial definition will hopefully start you off in the correct direction!

With such a definition in mind, it becomes useful to consider only the zeroth component of our definition of the canonical momenta and field derivatives, (e.g. only with respect to time $H : \mathbb{R}^3 \to \mathbb{R}$).

$$H(\mathcal{H}) = \int_\mathcal{R} \mathcal{H} d^3x \tag{26.0.3}$$

If we compute some derivatives and, considering equations 25.0.11, 26.0.1 and 26.0.2, rearrange for $\partial \mathcal{L}/\partial \phi$ again, remembering we use the zeroth component, we obtain the first of Hamilton's equations:

$$\frac{\partial \mathcal{H}}{\partial \pi_i} = \frac{\partial \phi^i}{\partial t}$$

The second, is slightly more complicated:

$$\frac{\partial \mathcal{H}}{\partial \phi^i} = -\frac{\partial \mathcal{L}}{\partial \phi^i} = -\partial_\mu \left(\frac{\partial \mathcal{L}}{\partial(\partial_\mu \phi)} \right) = -\frac{\partial}{\partial t} \left(\frac{\partial \mathcal{L}}{\partial(\partial_t \phi)} \right) + \frac{\partial}{\partial x} \left(\frac{\partial \mathcal{H}}{\partial(\partial_x \phi)} \right)$$

where $\partial \mathcal{H}/\partial(\partial_x \phi) = -\partial \mathcal{L}/\partial(\partial_x \phi)$. The apparent loss in symmetry of the canonical field equations can be regained if we define the **functional derivative** operator δ as follows:

$$\frac{\delta}{\delta u_i} = \frac{\partial}{\partial u_i} - \frac{\partial}{\partial x_i} \left(\frac{\partial}{\partial(\partial_x u_i)} \right) \tag{26.0.4}$$

If we let $u_i = \phi_i$ and $u_i = \pi_i$ then we generate, respectively, the field equations (note that $\partial_x \pi = 0$, since it is the zeroth component)[1]:

$$\frac{\partial \phi^i}{\partial t} = \frac{\delta \mathcal{H}}{\delta \pi_i}, \qquad \frac{\partial \pi_i}{\partial t} = -\frac{\delta \mathcal{H}}{\delta \phi^i} \tag{26.0.5}$$

In field theory, we like to treat space and time on equal footing but the canonical formalism insists we treat time as being special, and therefore, by definition, it is not covariant!

---

[1] If you are struggling deriving these, work backwards from $\delta \mathcal{H}/\delta u$.

Consequently, most field theoretic models are built on Lagrangian formulations. A covariant canonical formalism is the subject of the **de Donder-Weyl formalism** and the covariant Hamilton's equations are given by

$$\partial_\mu \phi^i = \frac{\partial \mathcal{H}}{\partial \pi_i^\mu}, \qquad \partial_\mu \pi_i^\mu = -\frac{\partial \mathcal{H}}{\partial \phi^i}$$

which are generated from the following variational principle:

$$\delta \int d^4 x (\pi_i^\mu \partial_\mu \phi^i - \mathcal{H}) = 0$$

We do not treat the de Donder formalism in any more depth, due to its in-depth requirements. For further reading please see Paufler and Römer's (2002) article, de Donder-Weyl equations and multisymplectic geometry.

We can also consider a generalised Poisson bracket in the continuous form for two local smooth functionals of phase space, $\mathcal{F}[\phi, \pi]$ and $\mathcal{G}[\phi, \pi]$; it is itself a local functional in the set of smooth functionals on the phase space:

$$\{\mathcal{F}, \mathcal{G}\} := \int d^3 x \left(\frac{\delta \mathcal{F}}{\delta \phi^i} \frac{\delta \mathcal{G}}{\delta \pi_i} - \frac{\delta \mathcal{F}}{\delta \pi_i} \frac{\delta \mathcal{G}}{\delta \phi^i} \right) \qquad (26.0.6)$$

Any observable expressed as a smooth function $F(x)$ can be converted to a smooth functional by using a test function $\omega(x)$:

$$\mathcal{F}[\omega] = \int d^n x F(x) \omega(x) \qquad (26.0.7)$$

It is straightforward to derive familiar relations in this formalism, for instance $\dot{\pi} = \{\pi, \mathcal{H}\} = -\delta \mathcal{H} / \delta \phi$:

$$\{\mathcal{H}, \pi(z)\} = \int d^3 x \frac{\delta \mathcal{H}}{\delta \phi(z')} \frac{\delta \pi(z)}{\delta \pi(z')} - \frac{\delta \mathcal{H}}{\delta \pi(z')} \frac{\delta \pi(z)}{\delta \phi(z')} = \int d^3 x \frac{\delta \mathcal{H}}{\delta \phi(z')} \delta(z - z') = \frac{\delta \mathcal{H}}{\delta \phi(z)}$$

We mention that, the Poisson bracket for a covariant formulation is given by the **Peierls bracket**, over the off-shell phase space.

We can also define canonical transformations $(\phi, \pi) \mapsto (\psi, \mathcal{P})$ by the invariance of the field Cartan form up to some spatial divergence that vanishes at the boundary points:

$$\int_\mathcal{R} \left(\pi_i \dot{\phi}^i - \mathcal{H}(\phi, \pi) \right) d^3 x = \int_\mathcal{R} \left(\mathcal{P}_i \dot{\psi}^i - \mathcal{H}'(\psi, \mathcal{P}) \right) d^3 x + \int_\mathcal{R} \frac{\partial \mathcal{K}}{\partial x^\nu} d^3 x$$

Last, we can consider the Hamilton-Jacobi field equation:

$$H\left[\phi, \frac{\delta S}{\delta \phi}\right] + \frac{\partial S}{\partial t} = 0 \tag{26.0.8}$$

with Hamilton's principal functional $S = S[\phi, \alpha, t]$, based on the canonical transformation from (ϕ, π) to $(\psi, \mathcal{P})$.

Exercise 26.1 Continuing our example of the real unsourced, massive scalar field Lagrangian, we can compute the Hamiltonian with ease:

$$\mathcal{L} = \frac{1}{2}((\partial_\mu \phi)^2 - m^2 \phi^2) = \frac{1}{2}(\dot{\phi}^2 - (\nabla \phi)^2 - m^2 \phi^2)$$

You then compute the conjugate momentum $\pi = \dot{\phi}$, to give the Hamiltonian density:

$$\mathcal{H} = \frac{1}{2}(\pi^2 + (\nabla \phi)^2 + m^2 \phi^2)$$

The Hamilton-Jacobi field equation would then take the form:

$$\frac{1}{2}\int_{\mathcal{R}}\left[\left(\frac{\delta S}{\delta \phi}\right)^2 + (\nabla \phi)^2 + m^2 \phi^2\right] d^3 x = 0$$

while the canonical field equations are given by

$$\dot{\phi} = \frac{\partial \mathcal{H}}{\partial \pi}, \qquad \dot{\pi} = -m^2 \phi$$

27

Classical Electromagnetism

Fig. 27.1: James
Clerk Maxwell
(1831–1879).

In this chapter, we will look at the electromagnetic fields $\boldsymbol{B}$ and $\boldsymbol{E}$ that are the focus of **electrodynamics**; although some background would probably help, it will not be necessary for our journey. If we were interested in using electromagnetism at a formal level, we would use tensorial notation to ease computation; however for our discussion we will follow a more intuitive approach (for a first reading) and complexify later. Alternatively, you can formulate electrodynamics using differential forms and exterior algebra, as for mechanics on a manifold.

We recall from chapter 25 that field equations can be obtained from the extremals of a Lagrange density integrated over space and time. We will now investigate a special type of field called the **electromagnetic field**, whose equations of motion are known as the four **Maxwell equations**. (See figure 27.1 for a picture of James Clerk Maxwell.) Until pioneers such as Faraday and Maxwell, electric vector fields $\boldsymbol{E}$ and magnetic vector fields $\boldsymbol{B}$ were regarded as separate phenomena entirely and it was only in the late 19th century that scientists saw them as components of a larger concept, the electromagnetic field. If the magnetic field is zero and both $\boldsymbol{E}$ and $\boldsymbol{B}$ are independent of time, then the electromagnetic field is called an electrostatic field and, conversely, with the opposite case we have magnetostatic fields.

The first task is to decide on the generalised coordinates for the field ϕ^i. We choose to use the **scalar potential** φ and the **vector potential** $\boldsymbol{A}$ with components A_x, A_y and A_z; you may also think of the scalar potential as the zeroth component of a 4-vector potential A_μ, with $\varphi = A_0$. The electric and magnetic fields are defined to be

$$\boldsymbol{E} = -\vec{\nabla}\varphi - \frac{\partial \boldsymbol{A}}{\partial t} = \left(-E_x - \frac{\partial A_x}{\partial t}\right)\hat{\boldsymbol{x}} + \left(-E_y - \frac{\partial A_y}{\partial t}\right)\hat{\boldsymbol{y}} + \left(-E_z - \frac{\partial A_z}{\partial t}\right)\hat{\boldsymbol{z}} \quad (27.0.1)$$

and

Lagrangian & Hamiltonian Dynamics. Peter Mann, Oxford University Press (2018).
© Peter Mann. DOI: 10.1093/oso/9780198822370.001.0001

$$\boldsymbol{B} = \vec{\nabla} \times \boldsymbol{A}$$

$$= \det \begin{pmatrix} \hat{\boldsymbol{x}} & \hat{\boldsymbol{y}} & \hat{\boldsymbol{z}} \\ \partial_x & \partial_y & \partial_z \\ A_x & A_y & A_z \end{pmatrix}$$

$$= \underbrace{\left(\frac{\partial A_z}{\partial y} - \frac{\partial A_y}{\partial z} \right)}_{B_x} \hat{\boldsymbol{x}} + \underbrace{\left(\frac{\partial A_x}{\partial z} - \frac{\partial A_z}{\partial x} \right)}_{B_y} \hat{\boldsymbol{y}} + \underbrace{\left(\frac{\partial A_y}{\partial x} - \frac{\partial A_x}{\partial y} \right)}_{B_z} \hat{\boldsymbol{z}} \qquad (27.0.2)$$

with $\partial \varphi / \partial x = E_x$ and so on. The second task is to construct the Lagrange density for the field. We are going to cut right to the answer since, in order to write it, we must have previous intuition about the equations of motion of the field, which is what we are trying to derive (although, historically that's the way the Lagrange density was written) or we could proceed from the fundamental structure of differential geometry in the same way that Hamilton's equations can be derived from the Poincaré-Cartan form:

$$\mathcal{L} := \frac{1}{2} \left(\epsilon_0 E^2 - \frac{1}{\mu_0} B^2 \right) - \rho_{el} \varphi + \boldsymbol{J} \cdot \boldsymbol{A} \qquad (27.0.3)$$

where $\boldsymbol{J}$ is the **current density** (electric current per unit area of cross section), ϵ_0 and μ_0 are constants called, respectively, the **permittivity** and **permeability** of free space[1] and ρ_{el} is the **electric charge density** (the electric charge per unit volume of space)[2].

With this information, the field equations can be constructed. We have four Euler-Lagrange equations of motion - one for each of the potentials φ, A_x, A_y and A_z; we will solve φ and A_x, since the other two are essentially the same as A_x. For the scalar potential, we write

$$\frac{d}{dt} \left(\frac{\partial \mathcal{L}}{\partial (\partial_t \varphi)} \right) + \frac{d}{dx} \left(\frac{\partial \mathcal{L}}{\partial (\partial_x \varphi)} \right) + \frac{d}{dy} \left(\frac{\partial \mathcal{L}}{\partial (\partial_y \varphi)} \right) + \frac{d}{dz} \left(\frac{\partial \mathcal{L}}{\partial (\partial_z \varphi)} \right) - \frac{\partial \mathcal{L}}{\partial \varphi} = 0 \quad (27.0.4)$$

The Lagrangian does not depend on $\dot{\varphi} = \partial_t \varphi$, so we can cross the first term off; we also note that neither B^2 nor $\boldsymbol{J} \cdot \boldsymbol{A}$ depends on φ so we can cross them both out of the Lagrangian, too. However, the electric field $\boldsymbol{E}$ does depend on the spatial derivatives

---

[1] They just characterise the response of the media (in this case, a vacuum) to perturbations in fields; in material science, we would use *relative* values.

[2] As a side note, the Lagrangian we are familiar with from previous examples throughout the book for charged particles in electric or magnetic fields describes the evolution of a charged particle in a field (i.e it describes the interaction of the field with a charge):

$$\mathscr{L}(\boldsymbol{r}, \dot{\boldsymbol{r}}, t) = \frac{1}{2} m \dot{\boldsymbol{r}} \cdot \dot{\boldsymbol{r}} + e \dot{\boldsymbol{r}} \cdot \boldsymbol{A} - e\varphi$$

This Lagrangian is describing the Lorentz force law.

of φ and the electric charge density ρ_{el} depends on φ, allowing us to reformulate the problem as

$$\frac{d}{dx}\left(\frac{\partial \mathcal{L}}{\partial (E_x)}\right) + \frac{d}{dy}\left(\frac{\partial \mathcal{L}}{\partial (E_y)}\right) + \frac{d}{dz}\left(\frac{\partial \mathcal{L}}{\partial (E_z)}\right) - \frac{\partial \mathcal{L}}{\partial \varphi} = 0 \tag{27.0.5}$$

Let's break this up and look at the very first term. The only part of the Lagrange density that is a function of E_x is $\frac{1}{2}\epsilon_0 E^2 = \frac{1}{2}\epsilon_0 (E_x^2 + E_y^2 + E_z^2)$; ignoring the $\partial_t A$ in the definition of $\boldsymbol{E}$, we have

$$\frac{\partial \mathcal{L}}{\partial E_x} = \frac{\partial (\frac{1}{2}\epsilon_0 E^2)}{\partial E_x} = \epsilon_0 E_x \tag{27.0.6}$$

Doing this for all three terms gives us $-\epsilon_0 \vec{\nabla} \cdot \boldsymbol{E}$, so we obtain

$$\frac{d}{dx}(\epsilon_0 E_x) + \frac{d}{dy}(\epsilon_0 E_y) + \frac{d}{dz}(\epsilon_0 E_z) = \epsilon_0 \vec{\nabla} \cdot (E_x \hat{\boldsymbol{x}} + E_y \hat{\boldsymbol{y}} + E_z \hat{\boldsymbol{z}}) \tag{27.0.7}$$

Comparing this with the definition of the scalar potential (the origin of that minus sign), we conclude that the Euler-Lagrange equation can be written

$$-\epsilon_0 \vec{\nabla} \cdot \boldsymbol{E} - \frac{\partial \mathcal{L}}{\partial \varphi} = 0 \tag{27.0.8}$$

The last term is pleasingly straightforward to obtain; looking at the Lagrange density, we obtain

$$\vec{\nabla} \cdot \boldsymbol{E} = \rho_{el}/\epsilon_0 \tag{27.0.9}$$

This is the first equation of motion of the electromagnetic field; it is more commonly known as **Gauss's law**. We will save a discussion of its meaning until the end of this chapter, for now, let's derive the other three. The Lagrange equation for the second coordinate A_x is given by

$$\frac{d}{dt}\left(\frac{\partial \mathcal{L}}{\partial (\partial_t A_x)}\right) + \frac{d}{dx}\left(\frac{\partial \mathcal{L}}{\partial (\partial_x A_x)}\right) + \frac{d}{dy}\left(\frac{\partial \mathcal{L}}{\partial (\partial_y A_x)}\right)$$
$$+ \frac{d}{dz}\left(\frac{\partial \mathcal{L}}{\partial (\partial_z A_x)}\right) - \frac{\partial \mathcal{L}}{\partial A_x} = 0 \tag{27.0.10}$$

Inspection of the Lagrange density and potentials shows that $\frac{1}{2}\epsilon_0 E^2$ is a function of $\dot{A}_x$, while $-B^2/2\mu_0$ is a function of $\partial_{x,y,z} A_x$, and finally, $\boldsymbol{J} \cdot \boldsymbol{A}$ is a function of A_x:

$$\frac{d}{dt}\frac{\partial(\frac{1}{2}\epsilon_0 E^2)}{\partial \dot{A}_x} - \frac{d}{dx}\left(\frac{\partial(B^2/2\mu_0)}{\partial(\partial A_x/\partial x)}\right) - \frac{d}{dy}\left(\frac{\partial(B^2/2\mu_0)}{\partial(\partial A_x/\partial y)}\right) -$$
$$\frac{d}{dz}\left(\frac{\partial(B^2/2\mu_0)}{\partial(\partial A_x/\partial z)}\right) - \frac{\partial}{\partial A_x}(J_x A_x + A_y J_y + A_z J_z) = 0 \tag{27.0.11}$$

The first step is to pull the constants outside to clear things up. We can then chain rule the first terms, treating $f(B) = B^2$ and $f(E) = E^2$.

$$\frac{\epsilon_0}{2}\frac{d}{dt}\left(\frac{\partial E^2}{\partial E}\frac{\partial E}{\partial \dot{A}_x}\right) - \frac{1}{2\mu_0}\left[\frac{d}{dx}\left(\frac{\partial B^2}{\partial B}\frac{\partial B}{\partial(\partial A_x/\partial x)}\right)\right.$$
$$\left. + \frac{d}{dy}\left(\frac{\partial B_z^2}{\partial B_z}\frac{\partial B_z}{\partial(\partial A_x/\partial y)}\right) + \frac{d}{dz}\left(\frac{\partial B_y^2}{\partial B_y}\frac{\partial B_y}{\partial(\partial A_x/\partial z)}\right)\right] \quad (27.0.12)$$

Substituting for the expressions for E and B from equations 27.0.1 and 27.0.2 directly, we obtain

$$-\epsilon_0\frac{dE_x}{dt} - \frac{1}{2\mu_0}\left[\frac{d}{dx}(0) + \frac{d}{dy}(2B_z \cdot -1) + \frac{d}{dz}(2B_y \cdot 1)\right]$$
$$= -\epsilon_0\frac{dE_x}{dt} - \frac{1}{\mu_0}\left[\frac{d}{dy}(-B_z) + \frac{d}{dz}(B_y)\right] \quad (27.0.13)$$

The sharp-eyed reader will recognise the last term as the $\hat{x}$ component of the curl of $\boldsymbol{B}$:

$$\vec{\nabla} \times \boldsymbol{B} = \det\begin{pmatrix} \vec{x} & \hat{y} & \hat{z} \\ \partial_x & \partial_y & \partial_z \\ B_x & B_y & B_z \end{pmatrix} = \left(\frac{\partial B_z}{\partial y} - \frac{\partial B_y}{\partial z}\right)\hat{x} + \left(\frac{\partial B_x}{\partial z} - \frac{\partial B_z}{\partial x}\right)\hat{y} + \left(\frac{\partial B_y}{\partial x} - \frac{\partial B_x}{\partial y}\right)\hat{z}$$

Therefore, we may write,

$$-\epsilon_0\frac{\partial E_x}{\partial t} + \frac{1}{\mu_0}(\vec{\nabla} \times \boldsymbol{B})\hat{x} - J_x = 0 \quad (27.0.14)$$

where we have returned to equation 27.0.11 for the last part of the Lagrange equation to complete the expression for the equation of motion of A_x. Repeating the procedure for A_y and A_z gives you exactly the same result (well, obviously with different components, but the same form), allowing us to write the above in vectorial notation; we recognise this as the second Maxwell equation; which is also known as the **Ampére circuit law**:

$$-\epsilon_0\frac{\partial \boldsymbol{E}}{\partial t} + \frac{1}{\mu_0}(\vec{\nabla} \times \boldsymbol{B}) - \boldsymbol{J} = 0 \quad (27.0.15)$$

The remaining two of Maxwell's equations can be derived from the two that we have here, which is lucky, since we had run out of Euler-Lagrange equations! Thankfully, they are much easier to derive than the first two and follow on from the laws of vector calculus:

$$\vec{\nabla} \cdot \boldsymbol{B} = \vec{\nabla}(\vec{\nabla} \times \boldsymbol{A}) = 0 \quad (27.0.16)$$

and

$$\vec{\nabla} \times \boldsymbol{E} = -\vec{\nabla} \times \vec{\nabla}\varphi - \frac{\partial}{\partial t}(\vec{\nabla} \times \boldsymbol{A}) = -\frac{\partial \boldsymbol{B}}{\partial t} \qquad (27.0.17)$$

which are the **Gauss's law for magnetic fields** and **Faraday's law**, respectively. Therefore, the field components φ and $\boldsymbol{A}$ that make the density integral stationary or critical are exactly those that satisfy Maxwell's equations. The forms of Maxwell's equations are generally the same but they do change, depending on the system being studied, for example, if its in media, if charges or magnets are present, and so on. This derivation represents Maxwell's equations for the simplest case possible - a vacuum.

Faraday's law relates the curl of the electric field at a point to the dynamical rate of change of the magnetic field at that point. Ampére's law relates the curl of the magnetic field to the current density and the rate of change of the electric field at that point. Therefore, a **spatially varying** electric field induces a **time-dependent** magnetic field, and vice versa; therefore, the electric field propagation is always accompanied by a magnetic field propagation - they are **coupled**.

Asking how these fields propagate leads to the important result of Maxwell's equations, the **wave equation**. We start by taking the curl of both sides of Faraday's law:

$$\vec{\nabla} \times (\vec{\nabla} \times \boldsymbol{E}) = \vec{\nabla} \times \left(-\frac{\partial \boldsymbol{B}}{\partial t} \right) = -\frac{\partial(\vec{\nabla} \times \boldsymbol{B})}{\partial t} \qquad (27.0.18)$$

We use the relation from vector calculus $\vec{\nabla} \times (\vec{\nabla} \times \boldsymbol{X}) = \vec{\nabla}(\vec{\nabla} \cdot \boldsymbol{X}) - \vec{\nabla}^2 \boldsymbol{X}$ on the left-hand side,

$$\vec{\nabla}(\vec{\nabla} \cdot \boldsymbol{E}) - \vec{\nabla}^2 \boldsymbol{E} = -\frac{\partial(\vec{\nabla} \times \boldsymbol{B})}{\partial t} \qquad (27.0.19)$$

We then use Ampere's law for the curl of $\boldsymbol{B}$, and Gauss's law for the divergence of $\boldsymbol{E}$:

$$\vec{\nabla}\left(\frac{\rho_{el}}{\epsilon_0}\right) - \vec{\nabla}^2 \boldsymbol{E} = -\frac{\partial}{\partial t}\left[\mu_0 \boldsymbol{J} + \mu_0 \epsilon_0 \frac{\partial \boldsymbol{E}}{\partial t} \right] \qquad (27.0.20)$$

Collecting terms and setting the charge and current densities to zero (meaning we consider an isolated field in a vacuum), we obtain, following a similar analysis of the curl of the $\boldsymbol{B}$ field in Ampere's law:

$$\vec{\nabla}^2 \boldsymbol{E} = \mu_0 \epsilon_0 \frac{\partial^2 \boldsymbol{E}}{\partial t^2}, \qquad \vec{\nabla}^2 \boldsymbol{B} = \mu_0 \epsilon_0 \frac{\partial^2 \boldsymbol{B}}{\partial t^2} \qquad (27.0.21)$$

We identify these fascinating results as second order PDEs called **wave equations**[3]. We treated the wave equation in detail in chapter 5, where we considered the separation

---

[3]We can use the d'Alembertian operator $\hat{\Box} = \vec{\nabla}^2 - \frac{1}{c^2}\frac{\partial^2}{\partial t^2}$ and hence identify c, the speed of propagation, as $c = \frac{1}{\sqrt{\mu_0 \epsilon_0}}$.

procedure for a wave on a sphere. The above wave equations describe how the electric and magnetic fields propagate as coupled waves in a vacuum. Any independent solution to either of these wave equations must also satisfy Maxwell's relations, as well as the superposition principle; thus, the two fields are found to be **orthogonal** to each other and the direction of propagation (in a vacuum) and hence are called *transverse waves*. There are two important solutions: **plane waves** and **spherical waves**. The plane wave solutions to the two wave equations are given by

$$\boldsymbol{E} = \boldsymbol{E}_0 \cos(\omega t - \boldsymbol{k} \cdot \boldsymbol{r} + \phi_E), \qquad \boldsymbol{B} = \boldsymbol{B}_0 \cos(\omega t - \boldsymbol{k} \cdot \boldsymbol{r} + \phi_B) \qquad (27.0.22)$$

where $\boldsymbol{E}_0$ and $\boldsymbol{B}_0$ are amplitudes such that $\boldsymbol{E}/\boldsymbol{B} = \boldsymbol{E}_0/\boldsymbol{B}_0 = c$, where c is a constant equal to the speed of propagation of the wave, $\phi_{E/B}$ are phase constants, $\boldsymbol{k}$ is a wave vector and ω is an angular frequency. Direct application of Maxwell's equations shows that the dot products of the electric field, the magnetic field and the wave vector are all mutually orthogonal to each other and that the phase factor is zero (so peaks and troughs match up) in vacuum. This, however, changes when they are in materials, due to conductivity effects and, hence, we see a difference in phase, along with amplitude effects and other dispersion phenomena.

A **static charge** distribution is a situation where $\boldsymbol{J} = 0$ and the electric field is time independent ($\partial \boldsymbol{E}/\partial t = 0$). Then, by Ampére's law, the curl of the magnetic field vanishes, and, by Gauss's law, we know the divergence is zero. By the **Helmholtz decomposition theorem**[4] the magnetic field also vanishes. So, this is why static charges in classical crystals are assumed not to emit radiation, but classical electron hard spheres moving around Bohr orbits do. Static charges allow us to separate our study into the branches of **electrostatics** and **magnetostatics**. There is a large branch of electrodynamics that investigates the different types of static charge distributions we may encounter (linear chains, spherical distributions, periodic distributions, and so on), and how their motion emits electromagnetic radiation. We will consider the motion of a charged particle in a magnetostatic field at the end of this chapter.

Our last topic, which we can't ignore, is **gauge freedom** in electromagnetism. If we transform the vector potential up to a total divergence term $\vec{\nabla}\Lambda$ where Λ is a sufficiently smooth scalar function and then compute $\boldsymbol{B}$, we notice that the magnetic field remains unchanged, since $\vec{\nabla} \times \vec{\nabla} X = 0$:

$$\boldsymbol{B} = \vec{\nabla} \times (\boldsymbol{A} + \vec{\nabla}\Lambda) = \vec{\nabla} \times \boldsymbol{A} \qquad (27.0.23)$$

However, when we compute $\boldsymbol{E}$, we obtain

---

[4]Helmholtz decomposition theorem says that a sufficiently smooth vector field can be split up into a curl-free vector field and a divergence-free vector field: $\boldsymbol{X} = -\vec{\nabla}\phi + \vec{\nabla} \times A$. If both parts are zero, then $\boldsymbol{X}$ must vanish. In this way, Helmholtz decomposition is saying that, if we know the curl and the divergence of a field, then we know the field itself. This is reassuring for Maxwell's equations, since they are written in these terms.

$$E = -\vec{\nabla}\phi - \frac{\partial A}{\partial t} - \vec{\nabla}\left(\frac{\partial \Lambda}{\partial t}\right) = -\vec{\nabla}[\phi + \partial_t \Lambda] - \partial_t A \qquad (27.0.24)$$

which we note is not invariant. However, if we had also changed the scalar potential to $\phi - \partial_t \Lambda$, then the electric field would remain the same; Maxwell's electrodynamics is **gauge invariant**. That means the same physical fields are described by a whole family of potentials which are only defined up to a scalar function. This is a result of constructing the field theory on potentials; the fields themselves are physical objects, so have a definite value but, since they are defined in terms of the derivatives of potentials, there exists an infinite number of possible expressions that describe the same field. More formally, the theory is invariant under the action of a local continuous group. A **gauge transformation** is understood as a transformation between the potentials that leaves the field invariant.

Rather than treating this as a problem, it should be viewed as a powerful tool to *choose* the potentials so we can simplify our treatment of the system. We can use it to our advantage to select the easiest possible description that we can manage. In practice, we don't simply pick a function Λ; instead, we impose a constraint on the potential. This is where the Dirac bracket comes into play in the choice of **gauge fixing**.

There are two particular gauge conditions we can impose, namely the **Coulomb gauge condition** and the **Lorenz gauge condition** and which one you choose depends on what you want to study. In quantum chemistry and material science, the Coulomb gauge constraint,

$$\vec{\nabla} \cdot A = 0 \qquad (27.0.25)$$

is frequently employed over the Lorentz condition,

$$\vec{\nabla} \cdot A + \frac{1}{c^2}\frac{\partial \varphi}{\partial t} = 0 \qquad (27.0.26)$$

The Lorenz gauge condition is favoured for relativistic mechanics, since it is associated with Lorentz transformations on the potential, and we generate wave equations for the 4-potentials. The Coulomb condition leads to a continuity equation for the scalar potential and so is favourable for computation when there are lots of molecules and charges in solids.

We close our discussion by generalising the theory developed in this chapter to a more mathematically appropriate level. We use the language of 4-vectors $A_\mu = (\varphi, A)$ and the second-rank antisymmetric **electromagnetic tensor** $F_{\mu\nu}$, to describe the electromagnetic field. This allows us to review material we have just covered, and the reader is encouraged to confirm the results!

$$F_{\mu\nu} = \partial_\mu A_\nu - \partial_\nu A_\mu \qquad \textbf{electromagnetic tensor} \qquad (27.0.27)$$

From this object, we can recover both the E and B fields as follows:

$$F^{0i} = \partial^0 A^i - \partial^i A^0 = \partial_0 A^i + \vec{\nabla}^i A^0 = -E^i \tag{27.0.28}$$

and

$$F^{ij} = \partial^i A^J - \partial^j A^i - \epsilon^{ijk} B^k \tag{27.0.29}$$

where we have used $\epsilon^{ijk} = \epsilon^{ijk}\epsilon_{klm} = (\delta^i_l \delta^j_m - \delta^i_m \delta^j_l)$. With this, we note that the Lagrange density

$$\mathcal{L} = -\frac{1}{4\mu_0} F_{\mu\nu} F^{\mu\nu} = \frac{1}{2}(\boldsymbol{E}^2 - \boldsymbol{B}^2) \tag{27.0.30}$$

with the field equation $\partial_\mu F^{\mu\nu} = 0$ gives us unsourced Maxwell equations:

$$\vec{\nabla} \cdot \boldsymbol{E} = 0 \qquad \vec{\nabla} \times \boldsymbol{B} = \partial_0 \boldsymbol{E} \tag{27.0.31}$$

The density $\mathcal{L}$ admits local symmetries $A_\mu \to A_\mu - \partial_\mu \theta$, where $\theta(x)$ is some regular function, and there exists a symmetric, gauge-invariant stress-energy tensor,

$$T^{\mu\nu} = F^{\mu\sigma} F^\nu_\sigma + \frac{1}{4} g^{\mu\nu} F_{\mu\nu} F^{\mu\nu} \tag{27.0.32}$$

where $g^{\mu\nu} = \text{diag}(1, -1, -1, -1)$ is the **Minkowski metric tensor**. The electromagnetic tensor is a 2-form element of $F \in \bigwedge^2 T^*M$, where M is a 4-spacetime manifold with 1-forms dt, dx, dy, and dz:

$$F = \frac{1}{2} F_{\mu\nu} dx^\mu \wedge dx^\nu = (\partial_\mu A_\nu) dx^\mu \wedge dx^\nu \tag{27.0.33}$$

For an n-dimensional space, the **Hodge star** takes a k-form to an $(n-k)$-form. In 4-spacetime, if the Hodge star acts on a 1-form, we generate a 3-form; if we act on a 2-form, we generate another 2-form, and so on. We can write the soured Maxwell equations as

$$dF = 0, \qquad d \star F = \mathcal{J}$$

where d is the exterior derivative, $\mathcal{J}$ is the **current 3-form** $\mathcal{J} = \frac{1}{6} j^\alpha \epsilon_{\alpha\beta\gamma\delta} dx^\beta \wedge dx^\gamma \wedge x^\delta$ and j^α is the component of the current density:

$$\mathcal{J} = \rho dx \wedge dy \wedge dz - j^x dt \wedge dy \wedge dz - j^y dt \wedge dz \wedge dx - j^z dt \wedge dx \wedge dy$$

We also have the continuity equation $d\mathcal{J} = d^2 \star F = 0$.

Chapter summary

- In this chapter, we considered a specific application of field theory: the **electromagnetic field**. The Euler-Lagrange field equations were set up for a vacuum field and solved to yield **Maxwell's equations**.

- Maxwell's equations were used to investigate the detailed nature of the classical wave propagation of light as a **coupled disturbance** in electric and magnetic fields.

Exercise 27.1 Show that the following Lagrangian gives rise to Maxwell's relations and the Lorentz force law:

$$\mathscr{L} = \frac{1}{2}\sum_i m_i \dot{\boldsymbol{r}}_i^2 + \frac{\epsilon_0}{2}\int d^3x [\boldsymbol{E}^2 - c^2\boldsymbol{B}^2] + \sum_i e_i[\dot{\boldsymbol{r}}_i \cdot \boldsymbol{A}(\boldsymbol{r}) - \varphi(\boldsymbol{r})] \qquad (27.0.34)$$

where charge and current densities are given by

$$\rho(\boldsymbol{r}) = \sum_i e_i \delta(\boldsymbol{r} - \boldsymbol{r}_i), \qquad \text{and} \qquad \boldsymbol{J}(\boldsymbol{r}) = \sum_i e_i \dot{\boldsymbol{r}}_i \delta(\boldsymbol{r} - \boldsymbol{r}_i) \qquad (27.0.35)$$

respectively. To answer this question you may choose to solve the Euler-Lagrange equations for $\boldsymbol{r}, \boldsymbol{A}$ and φ as presented throughout the text or be more adventurous and apply Hamilton's principle to the first-order variation in the action.

Exercise 27.2 Here we will illustrate how to compute the trajectory of a charged particle in a constant magnetic field $\boldsymbol{B} = (0,0,B)$ when it was initially at position $\boldsymbol{r}(0) = 0$ with velocity $\dot{\boldsymbol{r}}(0) = (\dot{r}_1, \dot{r}_2, \dot{r}_3)$.

The first step is to find the equations of motion from the Lagrangian for the system:

$$m\ddot{\boldsymbol{r}} = e\frac{d\boldsymbol{r}}{dt} \times \boldsymbol{B} \qquad (27.0.36)$$

Recalling that $\boldsymbol{B}$ is constant, integrate this between $[0,t]$ and substitute in the initial conditions:

$$m\int_{\dot{\boldsymbol{r}}(0)}^{\dot{\boldsymbol{r}}(t)} d\dot{\boldsymbol{r}} = e\left(\int_{\boldsymbol{r}(0)}^{\boldsymbol{r}(t)} d\boldsymbol{r}\right) \times \boldsymbol{B} \qquad (27.0.37)$$

which gives,

$$\dot{\boldsymbol{r}}(t) = (e/m)(\boldsymbol{r}(t) - \boldsymbol{r}(0)) \times \boldsymbol{B} + \dot{\boldsymbol{r}}(0) \qquad (27.0.38)$$

Then we simply split this into components by evaluating the cross product and solve the resulting system of equations:

$$\dot{r}_x = \frac{eB}{m}r_y + \dot{r}_1, \qquad \dot{r}_y = -\frac{eB}{m}r_x + \dot{r}_2, \qquad \dot{r}_z = \dot{r}_3 \qquad (27.0.39)$$

By taking the time derivative of the first two equations and putting the expression generated by the second into the first, we get

$$\ddot{r}_x + \left(\frac{eB}{m}\right)^2 r_x = \left(\frac{eB}{m}\right)\dot{r}_2 \tag{27.0.40}$$

If the initial y component of the velocity is equal to zero ($\dot{r}_2 = 0$) then this equation is simply a harmonic oscillator with frequency $\omega = (eB/m)$. Show that the general solution when $\dot{r}_2 \neq 0$ is given by

$$r_x(t) = \frac{\dot{r}_1}{\omega}(\sin\omega t) + \frac{\dot{r}_2}{\omega}(1 - \cos\omega t)$$

$$r_y(t) = \frac{\dot{r}_1}{\omega}(\cos\omega t - 1) + \frac{\dot{r}_2}{\omega}\sin\omega t$$

$$r_z(t) = \dot{r}_3 t$$

This famous problem shows that the motion of the particle traces a spiral in three dimensions along the B axis. Try this for yourself by plotting the above system of equations with $\dot{r}(0) = (\dot{r}_1, 0, \dot{r}_3)$.

Exercise 27.3 Show that the electromagnetic Lagrange density $\mathcal{L} = F_{\alpha\beta}F^{\alpha\beta}$ gives $\partial_\mu F^{\mu\nu} = 0$ as the corresponding field equations:

$$\frac{\partial \mathcal{L}}{\partial(\partial_\mu A_\nu)} = 2F^{\alpha\beta}\frac{\partial F_{\alpha\beta}}{\partial(\partial_\mu A_\nu)} \tag{27.0.41}$$

With the definition of the electromagnetic tensor, the Levi-Civita symbol and the antisymmetric properties of $F^{\mu\nu} = -F^{\nu\mu}$, we see that

$$2F^{\alpha\beta}\frac{\partial F_{\alpha\beta}}{\partial(\partial_\mu A_\nu)} = 2F^{\alpha\beta}\left(\frac{\partial(\partial_\alpha A_\beta)}{\partial(\partial_\mu A_\nu)} - \frac{\partial(\partial_\beta A_\alpha)}{\partial(\partial_\mu A_\nu)}\right)$$

$$= 2F^{\alpha\beta}(\delta_\alpha^\mu\delta_\beta^\nu - \delta_\beta^\mu\delta_\alpha^\nu)$$

$$= 2F^{\mu\nu} - 2F^{\nu\mu}$$

$$= 4F^{\mu\nu}$$

Hence, we see a factor of $1/4$ in the Lagrangian. The field equation can then be constructed by observing $\partial\mathcal{L}/\partial A_\nu = 0$. As an extension exercise, try coupling the free field to a source $J^\mu = (\rho_{\text{el}}, \boldsymbol{J})$ by writing the Lagrange density as

$$\mathcal{L} = J^\mu A_\mu + \frac{1}{4}F^\mu{}_\nu F_{\mu\nu} \tag{27.0.42}$$

The important distinction is now that $\partial\mathcal{L}/\partial A_\mu = J^\mu$ so the field equation is an inhomogeneous differential equation. As we can see, using tensorial notation is a lot easier than trying to use vector calculus. In order to re-establish the Maxwell equations in their familiar form, simply evaluate the components of $\nu = 0, 1, \dots$:

$$J^0 = \partial_\mu F^{\mu 0} = \partial_i F^{i0} = \partial_i E^i \tag{27.0.43}$$

$$J^1 = \partial_\mu F^{\mu 1} = \partial_0 F^{01} + \partial_2 F^{21} + \partial_3 F^{31} = -\partial_0 E^1 + \partial_2 B^3 - \partial_3 B^2 \tag{27.0.44}$$

Exercise 27.4 Here will derive **molecular multipole moments** by considering a material that is made up of localised, positively charged nuclei and negative electrons and free charges that roam the material. The charge density for N such charges, each of charge e_j, is given by

$$\rho(\mathbf{r}, t) = \sum_{j=1}^{N} e_j \delta[\mathbf{r} - \mathbf{r}_j(t)] \tag{27.0.45}$$

We can split this into contributions from the bound atoms in molecules and the non-localised free charges:

$$\rho(\mathbf{r}) = \underbrace{\sum e_j \delta(\mathbf{r} - \mathbf{r}_j)}_{\text{non-local}} + \underbrace{\sum}_{\text{(molecules)}} \underbrace{\sum}_{\text{(atoms)}} e_j \delta(\mathbf{r} - \mathbf{r}_j) \tag{27.0.46}$$

If we average the charge density of a single molecule, we can then sum up over the number of molecules in the material. To do this, we remember back to chapter 3 and set up a molecular coordinate frame according to figure 27.2, where now the large axis is a laboratory frame, and the origin of the small axis is placed on the center of mass of each molecule. The distance from the origin of the laboratory frame to the center of mass of the nth molecule is $\mathbf{r}_n(t)$, and the distance from the origin of the nth molecular frame to the jth charge in the nth molecule is $\mathbf{r}_{nj}(t)$. The average charge density of the nth molecule is then found for a well-behaved test function f:

$$\langle \rho_n(\mathbf{r}, t) \rangle = \int d\mathbf{r}' f(\mathbf{r}') \delta(\mathbf{r} - \mathbf{r}' - \mathbf{r}_{nj} - n)$$
$$= \sum_{j \in n} e_j f(\mathbf{r} - \mathbf{r}_n - \mathbf{r}_{nj}) \tag{27.0.47}$$

Since the terms in the sum only differ very slightly on an intramolecular scale, we can expand about the intermolecular distance $(\mathbf{r} - \mathbf{r}_n)$:

$$\langle \rho_n(\mathbf{r}, t) \rangle = \sum_{j \in n} e_j \left\{ f(\mathbf{r} - \mathbf{r}_n) - \mathbf{r}_{nj} \cdot \vec{\nabla} f(\mathbf{r} - \mathbf{r}_n) + \frac{1}{2} \sum_{\alpha,\beta} (\mathbf{r}_{nj})_\alpha (\mathbf{r}_{nj})_\beta \frac{\partial^2}{\partial r_\alpha \partial r_\beta} f(\mathbf{r} - \mathbf{r}_n) + \dots \right\}$$

We can then define the molecular **charge** q_n, **dipole moment** $\boldsymbol{\mu}_n$ and **quadrupole moments** Θ_n to be

$$q_n = \sum_{j \in n} e_j, \qquad \boldsymbol{\mu}_n = \sum_{j \in n} e_j \mathbf{r}_{nj}, \qquad \Theta_n = 3 \sum_{j \in n} e_j (\mathbf{r}_{jn})_\alpha (\mathbf{r}_{jn})_\beta \tag{27.0.48}$$

respectively. It is then a matter of totalling these quantities over the material to form macroscopic quantities that can be measured experimentally. Compute the dipole moment in debyes $(1 \text{ D} = 3.336 \times 10^{-30} \text{ C·m})$ for the HCl molecule, assuming that the bond is completely ionic

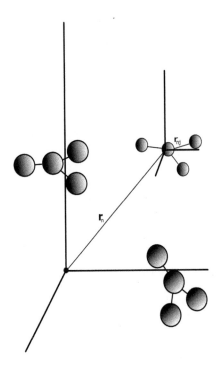

Fig. 27.2: The laboratory frame and the molecular frame.

in character with $r_{\text{HCl}} = 127$ pm. Given that the measured dipole moment is 1.03 D, compute the covalent character of this bond.

28

Noether's Theorem for Fields

We now move on to consider symmetries and conservation laws in field theory, where we see the true importance of Noether's theorem when we recall that *the invariance of the action $\delta S = 0$ under an off-shell infinitesimal set of transformations $\mathfrak{g}$ implies the existence of constants of the motion on-shell.* Remember that a symmetry is any change that leaves the action invariant; a continuous symmetry is a symmetry that can be expressed as an infinitesimal change. This section is founded upon Brading and Brown's (2000) article Noether's theorem and gauge symmetries and A. Zangwill's (2013) book, *Modern Electrodynamics.*

Noether's theorem is derived from a variational *problem.* Unlike the stationary action *principle,* which requires the action to be invariant under arbitrary variations in ϕ with *fixed boundary points,* we study specific variations in ϕ *and* x and look for what conditions are necessary for $\delta S = 0$ to hold. It does *not* impose $\delta\phi = \delta x = 0$ at the boundary points or require a priori that the action is stationary; hence, it should give us a different result than the stationary action principle would. If you recall that the Lagrangian of a system is not unique, since a total divergence term can be added without change in the equations of motion, then Noether's variational problem should really study the vanishing of the action up to a **surface term** in the action; in fact, if you follow a contour integral derivation of Noether's theorem. this term will appear. However, studying gauge invariance of symmetries is a bit beyond this introduction; just be aware of it when you read some of the referenced material.

Under an infinitesimal transformation, the total variation of a general tensor field $\phi(x)$ has two contributions: one from the coordinates and the other from the components of the field (i.e its form). So, we could consider changes in just the fields $\delta\phi$, just the coordinates δx^μ or the coordinates *and* the field $\delta\phi_0$:

$$\delta x^\mu := x'^\mu - x^\mu, \qquad \delta\phi := \phi'(x) - \phi(x), \qquad \delta\phi_0 := \phi'(x') - \phi(x) \qquad (28.0.1)$$

We will consider the variation in the field itself and the functional variation due to changing points in spacetime; however, lets not overcomplicate this introduction and instead crack on with the derivation. To derive Noether's theorem, we start off by considering the change in the action functional under an infinitesimal transformation $\delta S = S' - S$.

Lagrangian & Hamiltonian Dynamics. Peter Mann, Oxford University Press (2018).
© Peter Mann. DOI: 10.1093/oso/9780198822370.001.0001

$$\delta S = \int \mathcal{L}'(\phi', \partial_\mu \phi', x^{\mu'}) dx' - \int \mathcal{L}(\phi, \partial_\mu \phi, x^\mu) dx \qquad (28.0.2)$$

Considering the variation in the Lagrange density and remembering $x' = x + \delta x$, we obtain

$$\mathcal{L}'(\phi', \partial_\mu \phi', x'^\mu) = \mathcal{L}(\phi, \partial_\mu \phi, x^\mu) + \delta \mathcal{L}(\phi, \partial_\mu \phi, x^\mu) \qquad (28.0.3)$$

The integration measure dx' can be written in terms dx by use of the Jacobian of the transition $x \mapsto x'$, such that $dx' = \det(\partial x'/\partial x) \cdot dx$:

Mathematical Background 6

This is a general result of changing coordinates of the volume elements. You can see this in terms of the wedge product volume form defined previously. If we change x_1, x_2 to y_1, y_2, then the relationship between the volume forms is given by,

$$dx_1 \wedge dx_2 = \left(\frac{\partial x_1}{\partial y_1} dy_1 + \frac{\partial x_1}{\partial y_2} dy_2 \right) \wedge \left(\frac{\partial x_2}{\partial y_1} dy_1 + \frac{\partial x_2}{\partial y_2} dy_2 \right)$$

$$= \frac{\partial x_1}{\partial y_2} \frac{\partial x_2}{\partial y_1} dy_2 \wedge dy_1 + \frac{\partial x_1}{\partial y_1} \frac{\partial x_2}{\partial y_2} dy_1 \wedge dy_2$$

$$= \left(\frac{\partial x_1}{\partial y_1} \frac{\partial x_2}{\partial y_2} - \frac{\partial x_1}{\partial y_2} \frac{\partial x_2}{\partial y_1} \right) dy_1 \wedge dy_2 \ .$$

$$= \det \begin{pmatrix} \frac{\partial x_1}{\partial y_1} & \frac{\partial x_1}{\partial y_2} \\ \frac{\partial x_2}{\partial y_1} & \frac{\partial x_2}{\partial y_2} \end{pmatrix} dy_1 \wedge dy_2$$

Taylor expanding the Jacobian to first order, we obtain

$$\det \left(\frac{\partial x'^\mu}{\partial x^\mu} \right) = 1 + \partial_\mu \delta x^\mu + \mathcal{O}(\delta x^2)$$

Now $dx' = (1 + \partial_\mu \delta x^\mu) dx$, so

$$\delta S = \int \left[\mathcal{L}(\phi, \partial_\mu \phi, x^\mu) + \delta \mathcal{L}(\phi, \partial_\mu \phi, x^\mu) \right] \left[1 + \partial_\mu (\delta x^\mu) \right] dx$$

$$- \int \mathcal{L}(\phi, \partial_\mu \phi, x^\mu) dx \qquad (28.0.4)$$

Keeping only first-order terms, we write

$$\delta S = \int \left[\delta \mathcal{L}(\phi, \partial_\mu \phi, x_\mu) + \mathcal{L}(\phi, \partial_\mu \phi, x_\mu) \partial_\mu (\delta x^\mu) \right] dx \qquad (28.0.5)$$

We now Taylor expand the first term under the integral to obtain

$$\delta \mathcal{L}(\phi, \partial_\mu \phi, x_\mu) \Rightarrow \frac{\partial \mathcal{L}}{\partial \phi} \delta \phi_0 + \frac{\partial \mathcal{L}}{\partial (\partial_\mu \phi)} \delta_0 (\partial_\mu \phi) + \partial_\mu \mathcal{L} \delta x^\mu \qquad (28.0.6)$$

And, plugging that back in:

$$\delta S = \int \left(\frac{\partial \mathcal{L}}{\partial \phi} \delta \phi_0 + \frac{\partial \mathcal{L}}{\partial(\partial_\mu \phi)} \delta_0(\partial_\mu \phi) + \underbrace{\partial_\mu \mathcal{L} \delta x^\mu + \mathcal{L} \partial_\mu(\delta x^\mu)}_{\partial_\mu(\mathcal{L}\delta x^\mu)} \right) dx \qquad (28.0.7)$$

Using $\delta_0(\partial_\mu \phi) = \partial_\mu(\delta \phi_0)$, in analogy to the commutativity of the time derivative and the variation in the discrete case we find that

$$\delta S = \int \left(\frac{\partial \mathcal{L}}{\partial \phi} \delta \phi_0 + \frac{\partial \mathcal{L}}{\partial(\partial_\mu \phi)} \partial_\mu(\delta \phi_0) + \partial_\mu(\mathcal{L}\delta x^\mu) \right) dx \qquad (28.0.8)$$

Consider the following product derivative:

$$\partial_\mu \left(\frac{\partial \mathcal{L}}{\partial(\partial_\mu \phi_0)} \delta \phi_0 \right) = \partial_\mu \left(\frac{\partial \mathcal{L}}{\partial(\partial_\mu \phi)} \right) \delta \phi_0 + \frac{\partial \mathcal{L}}{\partial(\partial_\mu \phi)} \partial_\mu(\delta \phi_0)$$

Rearranging this slightly and substituting into the above and collecting like variations gives

$$\delta S = \int \left(\frac{\partial \mathcal{L}}{\partial \phi} \delta \phi_0 + \partial_\mu \left(\frac{\partial \mathcal{L}}{\partial(\partial_\mu \phi)} \delta \phi_0 \right) - \partial_\mu \left(\frac{\partial \mathcal{L}}{\partial(\partial_\mu \phi)} \right) \delta \phi_0 + \partial_\mu(\mathcal{L}\delta x^\mu) \right) dx$$

$$= \int \left[\left\{ \frac{\partial \mathcal{L}}{\partial \phi} - \partial_\mu \left(\frac{\partial \mathcal{L}}{\partial(\partial_\mu \phi)} \right) \right\} \delta \phi_0 + \partial_\mu \left\{ \frac{\partial \mathcal{L}}{\partial(\partial_\mu \phi)} \delta \phi_0 + \mathcal{L}\delta x^\mu \right\} \right] dx \qquad (28.0.9)$$

Therefore, by imposing the condition for a symmetry $\delta S = 0$, we arrive at the general result of Noether's variational problem, called the **Noether condition**:

$$\left\{ \frac{\partial \mathcal{L}}{\partial \phi} - \partial_\mu \left(\frac{\partial \mathcal{L}}{\partial(\partial_\mu \phi)} \right) \right\} \delta \phi_0 + \partial_\mu \left\{ \frac{\partial \mathcal{L}}{\partial(\partial_\mu \phi)} \delta \phi_0 + \mathcal{L}\delta x^\mu \right\} = 0 \qquad (28.0.10)$$

We notice that, on-shell, the first term is exactly zero, since the Euler-Lagrange equations are satisfied. Noether's theorem states that, if the action is invariant under a continuous *global* transformation (explained in a few minutes), then there will be the following conserved **current** j:

$$\partial_\mu \left[\frac{\partial \mathcal{L}}{\partial(\partial_\mu \phi)} \delta \phi_0 + \mathcal{L}\delta x^\mu \right] = \partial_\mu j^\mu = 0 \qquad (28.0.11)$$

We also note that, if we impose the stricter condition that the Lagrangian is invariant up to a total derivative $\delta \mathcal{L} = 0$ then only the first part of the second bracket will appear since $\delta \mathcal{L} = \partial_\mu \mathcal{L} \delta x^\mu$ these are called **quasi-symmetries** while invariance of the action up to a boundary integral are known as just symmetries. If a solution to the field equations do not share all the symmetries that the Lagrange density has then **spontaneous symmetry breaking** occurs.

If you wish to repeat the process using the other type of variation, then you can check the result by setting $\delta x^\mu = 0$, in equation 28.0.10 and swapping $\delta\phi_0 \to \delta\phi$, and you should also check that, for higher-order fields that you generate the Ostrogradsky charge:

$$j^\mu = \frac{\partial\mathcal{L}}{\partial(\partial_\mu\phi)}\delta\phi + \frac{\partial\mathcal{L}}{\partial(\partial_\mu\partial_\nu\phi)}\partial_\nu(\delta\phi) - \partial_\nu\left(\frac{\partial\mathcal{L}}{\partial(\partial_\nu\partial_\mu\phi)}\right)\delta\phi$$

for $\mathcal{L}(\phi, \partial_\mu\phi, \partial_\nu\partial_\mu\phi)$ densities.

We comment that the variations in Hamilton's action principle are the results of the **Lie dragging** of the tensor field under variation; remember, they are **virtual variations** evaluated at the same time $\delta\phi(x) = \phi'(x) - \phi(x)$ (see chapter 13). In Noether's problem, a variation is $\delta\phi_0 = \phi'(x') - \phi(x)$; that is, the new value of the field ϕ' evaluated at the new point in spacetime x' minus the old value of the field ϕ at the old spacetime point x.

Exercise 28.1 In the discrete case, we may consider a spatial translation or a time translation, which are symmetries giving conserved momenta and energies, respectively, but field theory smooshes space and time together in a spacetime manifold. Let's investigate an active infinitesimal transformation on spacetime $x^\nu \mapsto x^\nu - \lambda^\nu$ and hence $\phi(x) \mapsto \phi(x + \lambda) = \phi(x) + \lambda^\nu \partial_\nu\phi(x)$. If the Lagrangian does not depend on the spacetime 4-vectors explicitly (only through ϕ or $\partial_\mu\phi$), then we expect four conserved quantities, one from each of the generators in the 4-vector transformation. From the above formula, we identify $\delta\phi$ with $\lambda^\nu\partial_\nu\phi$, and δx^μ with $-\lambda^\nu$. Pulling out the λ^ν, we only get a non-zero ∂_ν^μ if δ_ν^μ:

$$T_\nu^\mu = \frac{\partial\mathcal{L}}{\partial(\partial_\mu\phi)}\partial_\nu\phi - \delta_\nu^\mu\mathcal{L}$$

We call T_ν^μ the **energy-momentum tensor**, with the continuity equation $\partial_\mu T_\nu^\mu = 0$. It reflects the fact that both energy and momenta are conserved under the spacetime symmetry. It is not unique in that we can add terms up to a spacetime derivative and it will still be divergenceless. This fact can be used to make the tensor symmetric with respect to its indices - a useful property! You will notice that, as we have defined it, we have mixed covariant (downstairs) and contravariant (upstairs) indices but, with help from a **metric tensor** g, we can put them both together, up or down, rewriting the energy-momentum tensor as

$$T^{\mu\nu} = \frac{\partial\mathcal{L}}{\partial(\partial_\mu\phi)}\partial^\nu\phi - g^{\mu\nu}\mathcal{L}$$

We can then prove that it satisfies the continuity condition $\partial_\mu T^{\mu\nu} = 0$:

$$\partial_\mu T^{\mu\nu} = \partial_\mu \left(\frac{\partial \mathcal{L}}{\partial(\partial_\mu \phi)} \partial^\nu \phi - g^{\mu\nu} \mathcal{L} \right)$$

$$= \left[\partial_\mu \frac{\partial \mathcal{L}}{\partial(\partial_\mu \phi)} \partial^\nu \phi + \frac{\partial \mathcal{L}}{\partial(\partial_\mu \phi)} \partial_\mu \partial^\nu \phi \right] - \partial^\nu \mathcal{L}$$

$$= \frac{\partial \mathcal{L}}{\partial \phi} \partial^\nu \phi + \frac{\partial \mathcal{L}}{\partial(\partial_\mu \phi)} \partial_\mu \partial^\nu \phi - \partial^\nu \mathcal{L}$$

$$= \frac{\partial \mathcal{L}}{\partial \phi} \partial^\nu \phi + \frac{\partial \mathcal{L}}{\partial(\partial_\mu \phi)} \partial_\mu \partial^\nu \phi - \frac{\partial \mathcal{L}}{\partial \phi} \partial^\nu \phi - \frac{\partial \mathcal{L}}{\partial(\partial_\mu \phi)} \partial_\mu \partial^\nu \phi$$

$$= 0$$

Thus, we see the tensor is conserved.

Continuous symmetry transformations form a Lie group $\mathcal{G}$ of ρ parameters, the generators of which form a Lie algebra $\mathfrak{g}$. Symmetry transformations form a group, since they can be *inverted* (undone); they are *closed* and *associative* and the identity transformation always provides an element of the set. These properties technically give them an open algebra structure but, for our purpose we can just say they are a Lie group.

We therefore understand a symmetry to be the covariance of an action under a **finite-dimensional** (ρ parameters) Lie group of global transformations. Noether's theorem applies to all global symmetries of a system, be they translational, rotational or time evolutions.

What do we mean by **global symmetry**? What, therefore, is a **local symmetry**? A global symmetry can be viewed as the invariance of the action under a group of continuous transformations depending on a linearly independent *constant* that parametrises the transformation. For instance, an infinitesimal translation is just $x + \delta x$, where δx is just some constant.

A local symmetry can be viewed as the invariance of the action under a group of continuous transformations depending on linearly independent *functions* and their derivatives (normally first order), where $\delta x = f(x) + \partial f(x)$. In this way, the exact nature of the transformation varies from point to point; it is a function of x. The constant is the same no matter where you are on the spacetime manifold (i.e it is globally defined). In contrast, the Lie group of local symmetries is **infinite dimensional**.

A good example of local a symmetry is a **gauge symmetry**. A gauge symmetry is a continuous *local* transformation of coordinates that leaves the Lagrange density invariant (i.e excess mathematical descriptions of the same physical system that leave $\delta \mathcal{L} = 0$). In this way, gauge transformations are slightly different from the symmetries above, in that the state of the system is not changed (i.e a rotation or a translation, by definition, evolves the system, while the gauge does not). *You can therefore understand a gauge theory to be a theory that has a local gauge symmetry induced by the infinite Lie group, and the gauge transformation exchanges the elements of the gauge group* *above a point.* Since local symmetries are very important, can we extend Noether's first theorem to consider them too? This is the context of *Noether's second theorem.*

Returning to the general result of Noether's variational problem, we can take a closer look at the variations. For ease, we will let $\delta x^\mu = 0$, just to consider variations in the dependent variables, the fields, where here $\bar\delta \lambda^r$ represents an infinitesimal of the parameter, not a variation like $\delta\phi$. The index r is bookkeeping for the parametrisation of the symmetry transformation; it runs to ρ. We thus obtain

$$\delta\phi = \sum_r \frac{\partial(\delta\phi)}{\partial(\bar\delta\lambda^r)}\bar\delta\lambda^r$$

Equation 28.0.10 now reads

$$\left[\frac{\partial\mathcal{L}}{\partial\phi} - \partial_\mu\left(\frac{\partial\mathcal{L}}{\partial(\partial_\mu\phi)}\right)\right]\frac{\partial(\delta\phi)}{\partial(\bar\delta\lambda^r)}\bar\delta\lambda^r + \partial_\mu\left(\frac{\partial\mathcal{L}}{\partial(\partial_\mu\phi)}\frac{\partial(\delta\phi)}{\partial(\bar\delta\lambda^r)}\bar\delta\lambda^r\right) = 0 \qquad (28.0.12)$$

If we now have a finite dimensional Lie group of transformations depending on ρ linearly independent parameters λ^r where $(r = 1, 2, \ldots, \rho)$, then we have a global symmetry. For global symmetries, $\bar\delta\lambda^r$ is just a constant and, as for any derivative, we can remove the dark blue term from the bracket and cancel it with the light coloured variation. In this way, we now obtain our ρ constants of the motion on-shell. This is **Noether's first theorem.** If a symmetry requires the equations of motion to be satisfied, it is called a *proper symmetry* of the system; off-shell symmetries are *improper.*

It is the case for local symmetries, however, that $\bar\delta\lambda^r$ is a *function* of the coordinates; not constant. Thus, it cannot be pulled out the derivative, as we could in the global case. Therefore, if the transformations depend on n functions $\lambda^r(x^\mu)$ and their first derivatives $\partial_\mu(\bar\delta\lambda^r(x^\mu))$ that constitute an infinite Lie group, then the infinitesimal transformation of the field $\delta\phi$ can be written as $\delta\phi = \alpha_r(\phi, \partial_\mu\phi, x^\mu)\bar\delta\lambda^r(x^\mu) + \beta_r^\mu(\phi, \partial_\mu\phi, x^\mu)\partial_\mu(\bar\delta\lambda^r(x^\mu))$, with α_r and β_r^μ being functions of those variables for which the symmetry changes. Substituting this expression for $\delta\phi$ under local symmetries, we obtain

$$\left(\frac{\partial\mathcal{L}}{\partial\phi} - \partial_\mu\frac{\partial\mathcal{L}}{\partial(\partial_\mu\phi)}\right)(\alpha_r\bar\delta\lambda^r + \beta_r^\mu\partial_\mu(\bar\delta\lambda^r))$$

$$+ \partial_\mu\left(\frac{\partial\mathcal{L}}{\partial(\partial_\mu\phi)}(\alpha_r\bar\delta\lambda^r + \beta_r^\mu\partial_\mu(\bar\delta\lambda^r))\right) = 0 \qquad (28.0.13)$$

Expanding this and considering the following product derivatives:

$$\partial_\mu\left[\left(\frac{\partial\mathcal{L}}{\partial\phi} - \partial_\mu\frac{\partial\mathcal{L}}{\partial(\partial_\mu\phi)}\right)(\beta_r^\mu\bar\delta\lambda^r)\right] = \partial_\mu\left(\frac{\partial\mathcal{L}}{\partial\phi} - \partial_\mu\frac{\mathcal{L}}{\partial(\partial_\mu\phi)}\right)\beta_r^\mu\bar\delta\lambda^r +$$

$$\left(\frac{\partial\mathcal{L}}{\partial\phi} - \partial\frac{\partial\mathcal{L}}{\partial(\partial_\mu\phi)}\right)\partial_\mu(\beta_r^\mu\bar\delta\lambda^r)$$

Where the coloured term is given by

$$\partial_\mu(\beta_r^\mu \bar{\delta}\lambda^r) = \partial_\mu(\beta_r^\mu)\bar{\delta}\lambda^r + \beta_r^\mu \partial_\mu(\bar{\delta}\lambda^r)$$

We write this as

$$\left\{ \left(\frac{\partial \mathcal{L}}{\partial \phi} - \partial_\mu \frac{\partial \mathcal{L}}{\partial(\partial_\mu \phi)} \right) \alpha_r - \partial_\mu \left[\left(\frac{\partial \mathcal{L}}{\partial \phi} - \partial_\mu \frac{\partial \mathcal{L}}{\partial(\partial_\mu \phi)} \right) \beta_r^\mu \right] \right\} \bar{\delta}\lambda^r + \partial_\mu \Omega = 0 \qquad (28.0.14)$$

where $\partial_\mu \Omega$ is a term that groups variations in the **boundary points**. We thus obtain

$$\partial_\mu \Omega = \partial_\mu \left[\left(\frac{\partial \mathcal{L}}{\partial \phi} - \partial_\mu \frac{\partial \mathcal{L}}{\partial(\partial_\mu \phi)} \right)(\beta_r^\mu \bar{\delta}\lambda^r) + \frac{\partial \mathcal{L}}{\partial(\partial_\mu \phi)}(\alpha_r \bar{\delta}\lambda^r + \beta_r^\mu \partial_\mu(\bar{\delta}\lambda^r)) \right] \qquad (28.0.15)$$

Both λ^r and their derivatives are arbitrary infinitesimal transformations and, without loss of generality, we can choose them such that they vanish at the boundary points. Therefore, $\partial_\mu \Omega$ is set to zero, meaning that the interior variation in the action vanishes *independently* from the boundary point variation in the action, and both terms of equation 28.0.14 vanish. The first gives us

$$\left\{ \frac{\partial \mathcal{L}}{\partial \phi} - \partial_\mu \left(\frac{\partial \mathcal{L}}{\partial(\partial_\mu \phi)} \right) \right\} \alpha_r - \partial_\mu \left\{ \frac{\partial \mathcal{L}}{\partial \phi} - \partial_\mu \left(\frac{\partial \mathcal{L}}{\partial(\partial_\mu \phi)} \right) \right\} \beta_r^\mu = 0 \qquad (28.0.16)$$

This is **Noether's second theorem** for infinitesimal local symmetries. If the action is invariant under the much stronger condition of a infinite dimensional group of transformations (as opposed to a finite Lie group), then there will exist dependancies between the Euler-Lagrange equations and their derivatives. The exact nature depends inherently on the nature of the local transformation itself.

To summarise, Noether's theorem is a variational problem similar to the stationary action principle that seeks the conditions necessary for the condition $\delta S = 0$ to hold. However, it extends the stationary action principle by allowing variation at the boundary points. The first theorem concerns global symmetries, while the second concerns local. It is as applicable to general relativity as it is particle physics and anything in-between.

There is a third theorem that came from Noether's works, which is due to Noether, Klein, Weyl and Utiyama but is often called the **Klein-Utiyama theorem**. Noether's second theorem concerns the vanishing of the interior contribution to the action and we remarked that the vanishing of this is independent of the boundary contribution. The Klein-Utiyama theorem considers the vanishing of the boundary contribution to the Noether condition (i.e the second term in 28.0.14, namely $\partial_\mu \Omega = 0$). We do not explore this theorem but the reader is referred to Brading's (2002) 'Symmetries, conservation laws, and Noether's variational problem' for further information. For further reading on Noether's theorems see Brading and Brown's (2004) article, 'Noether's theorem and gauge symmetries' and Al-Kuwari and Taha's (1991) article, 'Noether's theorem and local gauge invariance', which compactly discusses the implications of Noether's second theorem to gauge theory. In addition a useful paper is Bering's (2011) article, 'Noether's theorem for a fixed region', where quasisymmetries are explored; this covers the differences between a symmetry of the action and a symmetry of the Lagrangian. For all things concerning Noether, see Kosmann-Schwarzbach's (2011) book, *The Noether Theorems*.

An extremely interesting scenario is found by considering Noether's theorem for **discrete symmetries**, which are no longer Lie groups. To start with, we consider an infinite crystal of identical lattice points, each with a finite but constant distance between their neighbours. We expect there to be translational symmetry associated with the model but it is a *discrete* symmetry, since the system is only invariant if we translate a distance of exactly one lattice spacing. It is not well known if such a symmetry could exhibit conserved momenta, but what is certain is that Noether's theorem in its current form does not lead to conserved quantities. There is a fair deal of research into this area; the reader is referred to Caterina and Boghosian's (2006) article, 'A "no-go" theorem for the existence of an action principle for discrete invertible dynamical systems' as well as Hydon and Mansfield's (2011) article, Extensions of Noether's second theorem: From continuous to discrete systems'. Clearly, such a model is useful for both chemistry and crystallography.

Chapter summary

- **Noether's variational problem** was examined for field theories. We generated the following continuity equation

$$\partial_\mu \left[\frac{\partial \mathcal{L}}{\partial(\partial_\mu \phi)} \delta\phi + \mathcal{L}\delta x^\mu \right] = 0$$

It states that there are conserved **currents** for each **continuous symmetry** of the action. We discussed **global** and **local** symmetries and investigated the **first** and **second** theorems.

Exercise 28.2 Consider the following Lagrange density and compute the classical equation of motion. We assume that $\psi(x, p, t)$ is a complex classical Koopman-von Neumann field or "wavefunction" on the phase space, $\hbar$ is an arbitrary constant, $i = \sqrt{-1}$, $U(x, p, t)$ is a potential energy function and the star notation indicates a complex conjugate:

$$\mathcal{L}(\psi, \partial_t \psi, \partial_x \psi, \psi^*, \partial_t \psi^*, \partial_x \psi^*) = i\hbar\psi^* \partial_t \psi - \frac{\hbar^2}{2m} \partial_x \psi^* \partial_x \psi - U\psi^*\psi \qquad (28.0.17)$$

The Euler-Lagrange field equation is then constructed in the usual fashion by taking the relevant partial derivatives with respect to each "coordinate" ψ and ψ^*. Let us consider the Euler-Lagrange equation for ψ^*, the equivalent for ψ being a direct analogy:

$$\frac{\partial \mathcal{L}}{\partial \psi^*} = i\hbar \partial_t \psi - U\psi, \qquad \frac{\partial \mathcal{L}}{\partial (\partial_t \psi^*)} = 0, \qquad \frac{\partial \mathcal{L}}{\partial (\partial_x \psi^*)} = -\frac{\hbar^2}{2m}\frac{\partial \psi}{\partial x} \qquad (28.0.18)$$

Upon constructing the Euler-Lagrange equation of motion, we obtain

$$\frac{\partial \mathcal{L}}{\partial \psi^*} - \frac{\partial}{\partial t}\left(\frac{\partial \mathcal{L}}{\partial \partial_t \psi^*}\right) - \frac{\partial}{\partial x}\left(\frac{\partial \mathcal{L}}{\partial (\partial_x \psi^*)}\right) = 0 \qquad (28.0.19)$$

We obtain the following along with the analogue for the other coordinate:

$$i\hbar\frac{\partial}{\partial t}\psi = -\frac{\hbar^2}{2m}\frac{\partial^2}{\partial x^2}\psi + U\psi, \quad \text{and} \quad i\hbar\frac{\partial}{\partial t}\psi^* = -\frac{\hbar^2}{2m}\frac{\partial^2}{\partial x^2}\psi^* + U\psi^* \qquad (28.0.20)$$

The same result is obtained by following the action variational procedure, which the reader is invited to prove. We can call this the **classical Schrödinger field equation** and, hence, the above Lagrange density is a **Schödinger-Lagrange density**. There exists a whole family of these Schrödinger densities; they all yield the same equations of motion and are related to each other by the addition or subtraction of either $\vec{\nabla} \cdot (\ )$ or d/dt terms. The Schrödinger field equation, however, is not the equation of evolution of the Koopman-von Neumann wave function; we already postulated that the correct equation of motion is the Liouville equation. However, all we have here is the Lagrange equation for this particular Lagrange density.

To reconnect this with familiar theory, we can consider this the result of a canonical transformation from $(\mathcal{S}, \rho) \to (\psi, \psi^*)$, where $\rho(x, p, t)$ is the **Liouville probability density** and $\mathcal{S}(x, p, t)$ is **Hamilton's principal function**. The transformation equations may be considered to be

$$\psi(x, p, t) = \rho^{1/2}e^{i\mathcal{S}/\hbar}, \qquad \psi^*(x, p, t) = \rho^{1/2}e^{-i\mathcal{S}/\hbar} \qquad (28.0.21)$$

where $\rho(x, p, t) = \psi^*\psi$. Upon making the substitution, we can write the above Lagrange density as follows (here are some useful product rule relations to help you verify this):

$$\frac{\partial \psi}{\partial t} = \left[\frac{\partial}{\partial t}\rho^{1/2}e^{i\mathcal{S}/\hbar} + \rho^{1/2}e^{i\mathcal{S}/\hbar}\frac{i}{\hbar}\frac{\partial \mathcal{S}}{\partial t}\right], \qquad \frac{\partial \psi}{\partial x} = \left[\frac{\partial}{\partial x}\rho^{1/2}e^{i\mathcal{S}/\hbar} + \rho^{1/2}e^{i\mathcal{S}/\hbar}\frac{i}{\hbar}\frac{\partial \mathcal{S}}{\partial x}\right] \quad (28.0.22)$$

$$\frac{\partial \psi^*}{\partial t} = \left[\frac{\partial}{\partial t}\rho^{1/2}e^{-i\mathcal{S}/\hbar} - \rho^{1/2}e^{-i\mathcal{S}/\hbar}\frac{i}{\hbar}\frac{\partial \mathcal{S}}{\partial t}\right], \qquad \frac{\partial \psi^*}{\partial x} = \left[\frac{\partial}{\partial x}\rho^{1/2}e^{-i\mathcal{S}/\hbar} - \rho^{1/2}e^{-i\mathcal{S}/\hbar}\frac{i}{\hbar}\frac{\partial \mathcal{S}}{\partial x}\right]$$

Entering these expressions into the Schrödinger-Lagrange density above generates the following expression:

$$\mathcal{L} = -\rho\left[\frac{\partial \mathcal{S}}{\partial t} + \frac{1}{2m}\left(\frac{\partial \mathcal{S}}{\partial x}\right)^2 + U\right] + i\hbar\rho^{1/2}\frac{\partial}{\partial t}\rho^{1/2} - \frac{\hbar^2}{2m}\left(\frac{\partial}{\partial x}\rho^{1/2}\right)^2 \qquad (28.0.23)$$

Taking the classical limit $\hbar \to 0$, we retain only the first portion; we do this because we are interested only in terms that don't contain $\hbar$. In this way, we are using this "arbitrary constant" to keep track of meaningful terms:

$$\mathcal{L}_{\hbar \to 0} = -\rho \left[\frac{\partial \mathcal{S}}{\partial t} + \frac{1}{2m} \left(\frac{\partial \mathcal{S}}{\partial x} \right)^2 + U \right] \tag{28.0.24}$$

We treat $\mathcal{S}$ and ρ in exactly the same way as we treated ψ and ψ^*; they are separate fields and therefore get their own equation of motion. Notice that this new Lagrange density is separable in ρ and $\mathcal{S}$:

$$\frac{\partial \mathcal{L}_{\hbar \to 0}}{\partial \mathcal{S}} - \frac{\partial}{\partial t} \left(\frac{\partial \mathcal{L}_{\hbar \to 0}}{\partial (\partial_t \mathcal{S})} \right) - \frac{\partial}{\partial x} \left(\frac{\mathcal{L}_{\hbar \to 0}}{\partial (\partial_x \mathcal{S})} \right) = 0 \tag{28.0.25}$$

Computing these derivatives we obtain

$$\frac{\partial \mathcal{L}_{\hbar \to 0}}{\partial \mathcal{S}} = 0, \qquad \frac{\partial \mathcal{L}_{\hbar \to 0}}{\partial (\partial_t \mathcal{S})} = -\rho, \qquad \frac{\mathcal{L}_{\hbar \to 0}}{\partial (\partial_x \mathcal{S})} = -\frac{\rho}{m} \frac{\partial \mathcal{S}}{\partial x} \tag{28.0.26}$$

The first field equation is then

$$\frac{\partial \rho}{\partial t} + \frac{\partial}{\partial x} (\rho v) = 0 \tag{28.0.27}$$

with $v = (\partial \mathcal{S} / \partial q)/m$. Putting this result to one side for the moment, we turn to the second equation of motion, proceeding in a similar fashion:

$$\frac{\partial \mathcal{L}_{\hbar \to 0}}{\partial \rho} = - \left[\frac{\partial \mathcal{S}}{\partial t} + \frac{1}{2m} \left(\frac{\partial \mathcal{S}}{\partial x} \right)^2 + U \right], \qquad \frac{\partial \mathcal{L}_{\hbar \to 0}}{\partial (\partial_t \rho)} = 0, \qquad \frac{\mathcal{L}_{\hbar \to 0}}{\partial (\partial_x \rho)} = 0 \tag{28.0.28}$$

The corresponding equation of motion is

$$\frac{\partial \mathcal{S}}{\partial t} + \frac{1}{2m} \left(\frac{\partial \mathcal{S}}{\partial x} \right)^2 + U - 0 \tag{28.0.29}$$

Looking back through the book, you may recognise these two field equations as the Liouville equation of motion and the Hamilton-Jacobi equation of motion for free particles. Furthermore, we can show that the conserved quantity associated with the classical Schrödinger field is ρv, by computing the following current:

$$j = -\frac{i\hbar}{2m} \left(\psi^* \frac{\partial}{\partial x} \psi - \psi \frac{\partial}{\partial x} \psi^* \right) = \frac{\rho}{m} \frac{\partial \mathcal{S}}{\partial x} \tag{28.0.30}$$

Despite being a strong advocate of classical mechanics wherever possible, we cannot claim that this has anything to do with the quantum mechanics of many bodies. This is because of our interpretation of what $\psi(x, p, t)$ represents, which, in this case, is just a complex field depending on canonical variables $\psi : \mathbb{R}^n \times \mathbb{R} \to \mathbb{C}$. To have a quantum mechanical theory, we must endow a rich meaning behind the wavefunction tied in closely to a *quantum* Hilbert space defined over a *non*-commutative algebra of observables. However, as it stands, this is a classical complex field that bears no inference to the uncertainty principle or entanglement. In addition, this isn't Koopman von-Neumann mechanics, since KvN theory uses the Liouville equation of motion as one of the fundamental postulates of the theory!

Similar ideas are used in De Broglie-Bohm's[1] quantum mechanics although the implications

---

[1] I wrote this example to tie in with Koopman-von Neumann theory rather than De Broglie-Bohm mechanics, since this is a classical mechanics text!

behind ψ and ψ^* become a little more serious; in particular we write $\psi(x,t)$, not $\psi(x,p,t)$; $\rho(x,t)$, not $\rho(x,p,t)$; and $S(x,t)$ not $S(x,p,t)$; and we don't take $\hbar \to 0$. In fact, there is more technicality than we have alluded to in this simple pedagogical example, and it must be addressed if one were to be serious about constructing solution curves from a classical limit. However, this is the reason why Bohmian mechanics is very exciting for advocates of classical mechanics. The key difference to remember is that the classical Lagrange density is separable for S and ρ, and the resulting equations of motion are therefore not coupled. When computing the quantum equivalent, we notice that both S and ρ are linked through the quantum potential and are no longer independently chosen; we did this already in chapter 19. This subtlety is the origin of quantum effects in Bohmian mechanics.

The standard reference for this fantastic topic is Holland's (1995) book, *The Quantum Theory of Motion*, where a clear exposition of what is quantum, and what is classical, a discussion of hidden variables and the thoughts of Einstein, Bohm and other great scientists of the twentieth century are provided. A word of caution, however: the probability distribution in the reference is defined on configuration space while, in this example, we have defined a phase space distribution.

Exercise 28.3 Consider the following Lagrange density describing the sound vibrations in an ideal gas:

$$\mathcal{L} = \frac{1}{2}\rho_0\dot{\eta}^2 + P_0\vec{\nabla}\cdot\eta - \frac{1}{2}\gamma P_0(\vec{\nabla}\eta)^2 \qquad (28.0.31)$$

where ρ_0 is the equilibrium density of the gas, P_0 is the equilibrium pressure, γ is the ratio of constant pressure to volume heat capacities and $\eta(\boldsymbol{x},t)$ is a displacement vector. Assuming that γ, ρ_0 and P_0 are not functions of $\boldsymbol{x}$ or t, compute the equation of motion for the displacement field and express it in terms of the Fourier conjugate $\tilde{\eta}(\boldsymbol{p},t)$.

The equation of motion for the η field is given by

$$\rho_0\ddot{\eta} - \gamma P_0\vec{\nabla}(\vec{\nabla}\cdot\eta) = 0 \qquad (28.0.32)$$

The Fourier conjugate to the $\eta(\boldsymbol{x},t)$ field is the momentum space $\tilde{\eta}(\boldsymbol{p},t)$; we thus obtain

$$\eta(\boldsymbol{x},t) = \int \frac{d^3p}{(2\pi)^3}e^{i\boldsymbol{p}\cdot\boldsymbol{x}}\tilde{\eta}(\boldsymbol{p},t) \qquad (28.0.33)$$

This expression is then entered into the equation of motion, which, for convenience, we treat in one dimension for now:

$$\rho_0\frac{d^2}{dt^2}\int \frac{d^3p}{(2\pi)^3}e^{ipx}\tilde{\eta} - \gamma P_0\frac{d}{dx}\left[\frac{d}{dx}\int \frac{d^3p}{(2\pi)^3}e^{ipx}\tilde{\eta}\right] = 0 \qquad (28.0.34)$$

Then, we multiply by $\exp(-ip')$ and, integrating over spatial dimensions $\int d^3x$, we find

$$\rho_0\frac{d^2}{dt^2}\int d^3x\frac{d^3p}{(2\pi)^3}e^{i(p-p')x}\tilde{\eta} - \gamma P_0\frac{d}{dx}\left[\frac{d}{dx}\int d^3x\frac{d^3p}{(2\pi)^3}e^{i(p-p')x}\tilde{\eta}\right] = 0 \qquad (28.0.35)$$

Using $\int d^3x\exp\{i(p-p')x\} = \delta(p-p')$, we obtain

$$\rho_0 \frac{d^2}{dt^2} \int \frac{d^3p}{(2\pi)^3}\delta(p-p')\tilde{\eta} - \gamma P_0 \frac{d}{dx}\left[\frac{d}{dx}\int\frac{d^3p}{(2\pi)^3}\delta(p-p')\tilde{\eta}\right] = 0 \tag{28.0.36}$$

The equation of motion for the Fourier conjugate field is then given by

$$\rho_0 \frac{d^2}{dt^2}\tilde{\eta} - \gamma P_0 \vec{\nabla}(\vec{\nabla}\cdot\tilde{\eta}) = 0 \tag{28.0.37}$$

Exercise 28.4 In material science, we sometimes observe materials that exhibit nanoscale morphology; in particular, rod-like structures can be formed. These "nanorods" or **nanowires** have vast applications in the microelectronics and biosensing industries. Consider the following Lagrangian density describing the displacement field $\eta(x,t)$ when an elastic nanorod is bent in the x-dimension:

$$\mathcal{L} = \frac{1}{2}\left[\rho A\left(\frac{\partial\phi}{\partial t}\right)^2 - \mathcal{Y}I\left(\frac{\partial^2\phi}{\partial x^2}\right)^2\right] \tag{28.0.38}$$

where ρ is the density of the rod (assuming the density is uniform), A is the cross-sectional area, $\mathcal{Y}$ is the Young's modulus and I is the moment of area about the centroid of the nanorod. Noting explicitly that $\mathcal{L} = \mathcal{L}(\partial_t\phi, \partial_{xx}\phi)$ find the equation of motion for the displacement field η:

$$\frac{\partial\mathcal{H}}{\partial t} = -\frac{\partial}{\partial x}\left[\frac{\partial\eta}{\partial t}\frac{\partial}{\partial x}\left(\mathcal{Y}I\frac{\partial^2\eta}{\partial x^2}\right) - \left(\mathcal{Y}I\frac{\partial^2\eta}{\partial x^2}\right)\frac{\partial^2\eta}{\partial t\partial x}\right] \tag{28.0.39}$$

Prove the above conservation equation, holds using the Hamiltonian density for the system.

The important point to remember when solving this problem is the higher-order field dependence, so the usual Lagrange field equation is not to be used here. Instead, we must turn to the second-order Ostrogradsky field equation. In order to prove this result, we can apply the calculus of variations to the Lagrange density and impose Hamilton's principle for the vanishing of the extremals:

$$\delta S = \int_1^2 dx dt\left[\frac{\partial\mathcal{L}}{\partial(\partial_t\eta)}\delta(\partial_t\eta) + \frac{\partial\mathcal{L}}{\partial(\partial_{xx}\eta)}\delta(\partial_{xx}\phi)\right] \tag{28.0.40}$$

The first term is integrated by parts once, according to equation 25.0.9 while the second term, containing second derivatives, requires two partial integrations. The first yields:

$$\int_{x_1}^{x_2}\frac{\partial\mathcal{L}}{\partial(\partial_{xx}\eta)}\delta(\partial_{xx}\eta)dx = \frac{\partial\mathcal{L}}{\partial(\partial_{xx}\eta)}\delta\left(\frac{\partial\eta}{\partial x}\right)\Big|_{x_1}^{x_2} - \int_{x_1}^{x_2}\delta\left(\frac{\partial\eta}{\partial x}\right)\frac{\partial}{\partial x}\left(\frac{\partial\mathcal{L}}{\partial(\partial_{xx}\eta)}\right)dx$$

while the second results in

$$-\int_{x_1}^{x_2}\delta\left(\frac{\partial\eta}{\partial x}\right)\frac{\partial}{\partial x}\left(\frac{\partial\mathcal{L}}{\partial(\partial_{xx}\eta)}\right)dx = -\frac{\partial}{\partial x}\left(\frac{\partial\mathcal{L}}{\partial(\partial_{xx}\eta)}\right)\delta\eta\Big|_{x_1}^{x_2} + \int_{x_1}^{x_2}\delta\eta\frac{\partial^2}{\partial x^2}\left(\frac{\partial\mathcal{L}}{\partial(\partial_{xx}\eta)}\right)dx$$

Using this result and applying Hamilton's principle $\delta S = 0$, we arrive at the equation of motion for the one-dimensional bending of the nanorod:

$$\rho A \frac{\partial^2 \eta}{\partial t^2} + \frac{\partial^2}{\partial x^2}\left(\mathcal{Y}I\frac{\partial^2 \eta}{\partial x^2}\right) = 0 \tag{28.0.41}$$

The Hamilton density $\mathcal{H}$ is found using only the zeroth component of the 4-momentum:

$$\mathcal{H} = \frac{\partial \mathcal{L}}{\partial(\partial_t \eta)}\frac{\partial \eta}{\partial t} - \mathcal{L}$$

$$= \frac{1}{2}\rho A \left(\frac{\partial \eta}{\partial t}\right)^2 + \frac{1}{2}\mathcal{Y}I\left(\frac{\partial^2 \eta}{\partial x^2}\right)^2 \tag{28.0.42}$$

For the next part, we start off by computing the time derivative of the Hamiltonian density before using the equation of motion to rewrite the result (a big clue for this is that we do not see ρ or A in the conserved current):

$$\frac{\partial \mathcal{H}}{\partial t} = \frac{1}{2}\rho A \frac{\partial}{\partial t}\left(\frac{\partial \eta}{\partial t}\right)^2 + \frac{1}{2}\mathcal{Y}I\frac{\partial}{\partial t}\left(\frac{\partial^2 \eta}{\partial x^2}\right)^2$$

$$= \rho A \frac{\partial \eta}{\partial t}\left(\frac{\partial^2 \eta}{\partial t^2}\right) + \mathcal{Y}I\frac{\partial}{\partial t}\left(\frac{\partial^2 \eta}{\partial x^2}\right)\frac{\partial^2 \eta}{\partial x^2} \tag{28.0.43}$$

Use the equation of motion to obtain

$$= -\frac{\partial \eta}{\partial t}\frac{\partial^2}{\partial x^2}\left(\mathcal{Y}I\frac{\partial^2 \eta}{\partial x^2}\right) + \mathcal{Y}I\frac{\partial}{\partial t}\left(\frac{\partial^2 \eta}{\partial x^2}\right)\frac{\partial^2 \eta}{\partial x^2}$$

$$= -\frac{\partial \eta}{\partial t}\frac{\partial^2}{\partial x^2}\left(\mathcal{Y}I\frac{\partial^2 \eta}{\partial x^2}\right) + \mathcal{Y}I\frac{\partial^2 \eta}{\partial x^2}\frac{\partial}{\partial x}\left(\frac{\partial^2 \eta}{\partial t\partial x}\right)$$

$$= -\frac{\partial}{\partial x}\left[\frac{\partial \eta}{\partial t}\frac{\partial}{\partial x}\left(\mathcal{Y}I\frac{\partial^2 \eta}{\partial x^2}\right) - \left(\mathcal{Y}I\frac{\partial^2 \eta}{\partial x^2}\right)\frac{\partial^2 \eta}{\partial t\partial x}\right] \tag{28.0.44}$$

Exercise 28.5 Prove that if the action of a classical field $S[\phi]$ has a symmetry $\phi^i(x) \to \phi'^i(x')$, where $x^\mu \to x'^\mu$ then the equations of motion must also possess the same symmetry. State any assumptions you may choose to make.

This question can be approached in a variety of ways, with some being more intensive than others. Here, we outline a simple yet elegant approach by representing the satisfied ($\approx$) equations of motion by the vanishing of the functional derivative:

$$\frac{\delta S[\phi]}{\delta \phi^i(x)} = \frac{\partial \mathcal{L}}{\partial \phi^i} - \partial_\mu\left(\frac{\partial \mathcal{L}}{\partial(\partial_\mu \phi^i)}\right) \approx 0 \tag{28.0.45}$$

By design, we set the following condition from the definition of a symmetry:

$$\frac{\delta S[\phi]}{\delta \phi^i(x)} = \frac{\delta S[\phi']}{\delta \phi^i(x)} \tag{28.0.46}$$

Next, we chain rule. We assume that the variation is non-global and hence has vanishing boundary points; otherwise we would have surface terms left over:

$$\frac{\delta S[\phi]}{\delta \phi^i(x)} = \int dx' \, \frac{\delta S[\phi']}{\delta \phi'^i(x')} \frac{\delta \phi'^i(x')}{\delta \phi^i(x)}$$

$$= \int dx' \left[\frac{\partial \mathscr{L}'}{\partial \phi'^i} - \partial_\mu \left(\frac{\partial \mathscr{L}'}{\partial(\partial_\mu \phi'^i)} \right) \right] \frac{\delta \phi'^i(x')}{\delta \phi^i(x)} \qquad (28.0.47)$$

And hence the satisfied equations of motion can be written,

$$\frac{\partial \mathscr{L}}{\partial \phi^i} - \partial_\mu \left(\frac{\partial \mathscr{L}}{\partial(\partial_\mu \phi^i)} \right) \approx \frac{\partial \mathscr{L}'}{\partial \phi'^i} - \partial_\mu \left(\frac{\partial \mathscr{L}'}{\partial(\partial_\mu \phi'^i)} \right) \qquad (28.0.48)$$

Exercise 28.6 Show that surface terms in the action that are produced from a total divergence result in the same field equations. Let $\mathcal{L}' = \mathcal{L}(\phi, \partial_\mu \phi) + \partial_\mu \mathcal{K}^\mu(\phi)$, where

$$\partial_\mu \mathcal{K}^\mu(\phi) = \frac{\partial \mathcal{K}^\mu}{\partial \phi^i} \partial_\mu \phi^i$$

To show this, we simply need to evaluate the following on-shell equality:

$$\partial_\mu \left(\frac{\partial \mathcal{L}}{\partial(\partial_\mu \phi^i)} \right) - \frac{\partial \mathcal{L}}{\partial \phi^i} = \partial_\mu \left(\frac{\partial \mathcal{L}'}{\partial(\partial_\mu \phi^i)} \right) - \frac{\partial \mathcal{L}'}{\partial \phi^i} \qquad (28.0.49)$$

Exercise 28.7 Free fields can be excited and their modes computed; however the particles they may correspond to do not interact with each other. Here, we will investigate how to weakly couple fields so that we observe interactions between particles and then we will investigate symmetries. Weak coupling between fields indicates that the new theory is well described by a perturbation potential $U(\phi)$ and a field strength parameter ϵ that modulates the strength of the interaction. Let us first consider a single field:

$$\mathcal{L} = \frac{1}{2}(\partial_\mu \phi) - \frac{1}{2} m^2 \phi^2 - \epsilon U(\phi) \qquad (28.0.50)$$

Generally, the potential is some power of ϕ and can be written

$$U(\phi) = \sum_{n \geq 3} \frac{\epsilon_n}{n!} \phi^n$$

An alternative coupling scheme is to couple multiple *different* scalar fields $\phi_i(x)$ together. The Lagrange density will consist of a potential that couples the fields $U(\phi_1, \phi_2)$ and we are assuming for brevity that the masses are equivalent:

$$\mathcal{L} = \frac{1}{2} \sum_{i=1}^{N} (\partial_\mu \phi_i)^2 - \frac{1}{2} \sum_{i=1}^{N} m^2 \phi_i^2 - \epsilon \left(\sum_{i=1}^{N} \phi_i^2 \right)^2 \qquad (28.0.51)$$

As an example, we can take $U(\phi_1, \phi_2)$ to be a quartic function of ϕ_i^4 with the quadratic cross terms $\phi_i^2 \phi_j^2$:

$$U(\phi_1, \phi_2) = (\phi_1^2 + \phi_2^2)^2$$

The Lagrange density is invariant under a continuous global transformation of the fields $\phi_i(x) \to \phi_i'(x)$ and will exhibit an $\mathcal{SO}(2)$ symmetry or, in the general case, an $\mathcal{SO}(N)$ symmetry. We can understand this as a rotation in the $\phi_i \phi_j$ plane by some constant angle θ about an origin:

$$\begin{pmatrix} \phi_1' \\ \phi_2' \end{pmatrix} = \begin{pmatrix} \cos\theta & -\sin\theta \\ \sin\theta & \cos\theta \end{pmatrix} \begin{pmatrix} \phi_1 \\ \phi_2 \end{pmatrix} \tag{28.0.52}$$

The $\mathcal{SO}(2)$ group is called the **special orthogonal** group and is a set of 2×2 orthogonal matrices with a determinant equal to unity.

Finally, we note that a common extension of a real scalar field ϕ_i and ϕ_j is to define two new fields ψ and $\psi^\dagger$ that constitute a **complex scalar field**:

$$\psi = \frac{1}{\sqrt{2}}(\phi_1 + i\phi_2), \qquad \psi^\dagger = \frac{1}{\sqrt{2}}(\phi_1 - i\phi_2) \tag{28.0.53}$$

The free field Lagrange density for the complex Klein-Gordon equation is then

$$\mathcal{L} = \partial^\mu \psi^\dagger \partial_\mu \psi - m^2 \psi^\dagger \psi \tag{28.0.54}$$

Importantly, we note that we have a $\mathcal{U}(1)$ symmetry, since we can rotate the phase of ψ and $\psi^\dagger$ in the complex plane by

$$\psi \to e^{i\theta} \psi \qquad \text{and} \qquad \psi^\dagger \to e^{-i\theta} \psi^\dagger \tag{28.0.55}$$

without changing the Lagrangian. In the infinitesimal case of θ, we write $\psi \to \psi + \delta\psi$, where $\delta\psi = i\psi$, and similarly for the $\psi^\dagger$ field $\delta\psi^\dagger = -i\psi^\dagger$. The Noether current is then found by application of the Noether condition:

$$\frac{\partial \mathcal{L}}{\partial(\partial_\mu \psi)} \delta\psi + \frac{\partial \mathcal{L}}{\partial(\partial_\mu \psi^\dagger)} \delta\psi^\dagger = i(\psi(\partial^\mu \psi^\dagger) - \psi^\dagger(\partial^\mu \psi)) \tag{28.0.56}$$

You can then write combinations of complex and scalar fields, as well as coupling interactions between them, such as, for instance,

$$\mathcal{L} = \partial^\nu \psi^\dagger \partial_\nu \psi + \frac{1}{2} \partial^\nu \phi \partial_\nu \phi - \mu^2 \psi^\dagger \psi - \frac{1}{2} m^2 \phi^2 - \epsilon \psi^\dagger \psi \phi$$

As an exercise, try to couple different fields and assign their symmetry groups using the Noether condition.

Exercise 28.8 Show that the free complex scalar field is not invariant under a local $\mathcal{U}(1)$ transformation, where $\theta(x)$ is now a function of spacetime rather than being a constant:

$$\psi(x) \to \psi(x) e^{i\theta(x)} \tag{28.0.57}$$

where the Lagrange density is given by $\mathcal{L} = \partial^\mu \psi^\dagger \partial_\mu \psi - m^2 \psi^\dagger \psi$.

The best approach to this question is to simply input the new fields into the Lagrangian and observe if it is changed or not. We will investigate term by term for ease:

$$\partial_\mu \psi(x) \to \partial_\mu \left(\psi(x) e^{i\theta(x)} \right) = e^{i\theta(x)} \partial_\mu \psi(x) + \psi(x) e^{i\theta(x)} i \partial_\mu \theta(x)$$

$$\partial^\mu \psi^\dagger(x) \to \partial^\mu \left(\psi^\dagger(x) e^{-i\theta(x)} \right) = e^{-i\theta(x)} \partial^\mu \psi^\dagger(x) - \psi^\dagger(x) e^{-i\theta(x)} i \partial^\mu \theta(x)$$

$$m^2 \psi(x)^\dagger \psi(x) \to m^2 \psi^\dagger(x) e^{-i\theta(x)} \psi(x) e^{i\theta(x)} = m^2 \psi(x)^\dagger \psi(x)$$

Hence, after reconstructing the Lagrangian, we see that it is quite different from the starting point and therefore not invariant!

29
Classical Path-Integrals

In this short chapter, we develop the **classical path integral** as predominantly developed by E. Gozzi in the 1980s. We will examine both Lagrangian and Hamiltonian formulations before qualitatively discussing some interesting features of gauge fixing. We close the discussion with a peak at operatorial mechanics and revisit the Koopman-von Neumann formalism.

A path integral formulation of classical mechanics is a very convenient approach to studying the differences between quantum and classical systems. Indeed, the solutions of many non-pertubative quantum theories are often very similar to their classical counterparts, and the development of classical Feynman diagrams can be helpful in studying semiclassical systems.

29.1 Configuration Space Integrals

We are aware from quantum theory that the Feynman path integral is an integral over many quantum paths, thereby providing an amplitude between two points. The classical partition function Z must give a weighting of unity to the classical path q_{cl} that satisfies the equations of motion, and zero to all others. We achieve this by using the **functional Dirac delta** $\tilde{\delta}$, which is simply the product of single Dirac deltas:

$$\tilde{\delta}[q - q_{cl}] \equiv \prod_{j=1}^{N} \delta(q^j(t) - q_{cl}^j(t)) \tag{29.1.1}$$

Let us now consider an N-dimensional system with N equations of motion $E^i(\boldsymbol{q}, \dot{\boldsymbol{q}}, \dots) = 0$. The partition function is written

$$Z[J] = \int \prod_{i=1}^{N} dq^i \tilde{\delta}[q - q_{cl}] \exp\left\{ \int dt J_i q^i \right\} \tag{29.1.2}$$

The measure is over the volume $dq^1 \dots dq^N$, and J_i is the source current. The delta can be rewritten, as long as what is inside the square brackets vanishes upon satisfaction of

Lagrangian & Hamiltonian Dynamics. Peter Mann, Oxford University Press (2018).
© Peter Mann. DOI: 10.1093/oso/9780198822370.001.0001

the equations of motion, and we multiply by the Jacobian of the transformation. For instance two possibilities include,

$$\tilde{\delta}[q - q_{\mathrm{cl}}] \equiv \tilde{\delta}[E^i(\boldsymbol{q})] \det\left(\frac{\delta E^i}{\delta q^j}\right) \equiv \tilde{\delta}\left[\frac{\delta S}{\delta q^i}\right] \det\left(\frac{\delta^2 S}{\delta q^i \delta q^j}\right) \qquad (29.1.3)$$

There are two avenues we may traverse from here: the fermionic functional or a Bosonic functional approach. We begin with the fermionic theory and, in order to progress, we now perform two independent steps. The first is to Fourier transform the new delta by the introduction of an additional N variables, λ_i:

$$\tilde{\delta}\left[\frac{\delta S}{\delta q^i}\right] = \int \prod_{i=1}^{N} d\lambda_i \exp\left\{i \int dt \lambda_i \frac{\delta S}{\delta q^i}\right\} \qquad (29.1.4)$$

For the determinant, we turn to a general result in the theory of Berezin integrals or Gaussian functional integrals. Let A be an $N \times N$ matrix and let θ_i and ϕ^i be $2N$ Grassmann variables:

$$\det A = \int d\theta_i d\phi^i \exp\left\{-\theta_i A \phi^i\right\} \qquad (29.1.5)$$

Replacing determinants of square matrices by Gaussian integrals over Grassmann numbers is a useful tool in the quantum field theory of fermions:

$$\det\left[\frac{\delta^2 S}{\delta q^2}\right] = \int \prod_{i=1}^{N} d\theta_i \prod_{i=1}^{N} d\phi^i \exp\left\{\int dt \theta_i \frac{\delta^2 S}{\delta q^i \delta q^j} \phi^i\right\} \qquad (29.1.6)$$

With these changes in place, we can now reconstruct the fermionic partition function:

$$Z[J] = \int \mathcal{D}q^i \mathcal{D}\theta_i \mathcal{D}\phi^i \mathcal{D}\lambda_i \exp\left\{\int dt\left(i\lambda_i \frac{\delta S}{\delta q^i} + \theta_i \frac{\delta^2 S}{\delta q^i \delta q^j} \phi^i + J_i q^i\right)\right\} \qquad (29.1.7)$$

The $\mathcal{D}$ notation indicates the Nth-fold product.

29.2 Phase Space Integrals

We turn now to a phase space description of the classical path integral. The first step is to construct the functional delta for a point in phase space $\varphi^i = \varphi^i(\boldsymbol{q}, \boldsymbol{p})$:

$$\tilde{\delta}[\varphi - \varphi_{\mathrm{cl}}] \equiv \prod_{j=1}^{N} \delta(\varphi^j(t) - \varphi_{\mathrm{cl}}^j(t)) \qquad (29.2.1)$$

We then follow a similar process to the Lagrangian approach by re-expressing the delta in terms of Hamilton's equation $\dot{\varphi}^i = \omega^{ij}\partial_j H(\varphi)$:

$$\tilde{\delta}[\varphi - \varphi_{\mathrm{cl}}] = \tilde{\delta}\left[\dot{\varphi}^i - \omega^{ij}\frac{\partial H}{\partial \varphi^j}\right] \det\left(\frac{\delta}{\delta\varphi^k}\left(\dot{\varphi}^i - \omega^{ij}\frac{\partial H}{\partial\varphi^j}\right)\right)$$

$$= \tilde{\delta}\left[\dot{\varphi}^i - \omega^{ij}\frac{\partial H}{\partial\varphi^j}\right]\det\left(\frac{\partial}{\partial t}\delta^{ik} - \omega^{ij}\frac{\partial^2 H}{\partial\varphi^j\partial\varphi^k}\right) \tag{29.2.2}$$

Unlike the configuration space determinant, the phase space one must be positive definite to ensure that the path is unique. Just as with the Lagrangian case, the next step is to Fourier transform the first part and to exponentiate the determinant:

$$\tilde{\delta}\left[\dot{\varphi}^i - \omega^{ij}\frac{\partial H}{\partial\varphi^j}\right] = \int\prod_{i=1}^{N}d\lambda_i \exp\left\{i\int dt\lambda_i\left[\dot{\varphi}^i - \omega^{ij}\frac{\partial H}{\partial\varphi^j}\right]\right\} \tag{29.2.3}$$

We now have the option to exponentiate the determinant with either fermionic (as previously) or bosonic variables. For variety, we will choose the bosonic pathway and we make use of the following formula, again a well-known result within Gaussian functional integrals:

$$\int \mathcal{D}x\mathcal{D}y\, e^{(ix_i A_{ij} y_j)} \propto \frac{1}{\det[A_{ij}]} \tag{29.2.4}$$

Thankfully, for our case, Gozzi and Regini proved that the determinant is equal to its own inverse (see Gozzi and Regini, 2000): **62**, 067702).

$$\det\left(\frac{\partial}{\partial t}\delta^{ik} - \omega^{ij}\frac{\partial^2 H}{\partial\varphi^j\partial\varphi^k}\right) = \left\{\det\left(\frac{\partial}{\partial t}\delta^{ik} - \omega^{ij}\frac{\partial^2 H}{\partial\varphi^j\partial\varphi^k}\right)\right\}^{-1} \tag{29.2.5}$$

With this in mind, the exponentiated determinant can be written with the new bosonic variables π^i and ξ_i:

$$\det\left(\frac{\partial}{\partial t}\delta^{ik} - \omega^{ij}\frac{\partial^2 H}{\partial\varphi^j\partial\varphi^k}\right) = \int \mathcal{D}\pi\mathcal{D}\xi \exp\left\{i\int dt\pi^k\left(\frac{\partial}{\partial t}\delta^i_k + \omega^{ij}\frac{\partial^2 H}{\partial\varphi^j\partial\varphi^k}\right)\xi_i\right\}$$

We can now construct the partition function for the system by piecing together what we know:

$$Z[J] = \int \mathcal{D}\varphi\mathcal{D}\lambda\mathcal{D}\pi\mathcal{D}\xi \exp\left\{\int dt(i\mathcal{L} + J_i\varphi^i)\right\} \tag{29.2.6}$$

where the Lagrangian $\mathcal{L}$ is given by

$$\mathcal{L} = \lambda_i\left(\dot{\varphi}^i - \omega^{ij}\frac{\partial H}{\partial\varphi^j}\right) + \pi^k\left(\frac{\partial}{\partial t}\delta^i_k + \omega^{ij}\frac{\partial^2 H}{\partial\varphi^j\partial\varphi^k}\right)\xi_i \tag{29.2.7}$$

It is at this point the reader may wish to return to the exponentiation step and progress with Grassmann variables, as we did with the configuration space integral.

An interesting point to note is that, with the introduction of the new variables, we are now dealing with an $8N$-dimensional phase space coordinated by $(\varphi^i, \xi_i, \lambda_i, \pi^i)$. Such a space is known in the business as a **superspace**, and there exists a very rich formulation of both quantum and classical mechanics in this area. As is apparent from the position of the indices, π^i and λ_i are the conjugate momenta to ξ_i and φ^i meaning that the superphase space $\mathcal{M}$ is actually $T^*(T^*M)$, the double cotangent bundle of the original phase space M.

The next appropriate step is to find the Hamiltonian $\mathcal{H}$ for the Lagrangian $\mathcal{L}$. To do this, we can simply take $\mathcal{L}$ and template it over the usual $\mathscr{L} = p\dot{q} - H$ formula, since we only have first-order terms:

$$\mathcal{L} = \lambda_i\dot{\varphi}^i + \pi^i\dot{\xi}_i - \mathcal{H} \tag{29.2.8}$$

Hence, we identify the Hamiltonian as

$$\mathcal{H} = \lambda_i\omega^{ij}\frac{\partial H}{\partial\varphi^j} - \pi^k\omega^{ij}\frac{\partial^2 H}{\partial\varphi^j\partial\varphi^k}\xi_i \tag{29.2.9}$$

Hamilton's equations for (π^i, ξ_i) are given by $\partial_t\pi = -\partial_\xi\mathcal{H}$ and $\partial_t\xi = \partial_\pi\mathcal{H}$:

$$\dot{\xi}_k = -\omega^{ij}\frac{\partial^2 H}{\partial\varphi^j\partial\varphi^k}\xi_i, \qquad \dot{\pi}^i = \omega^{ik}\frac{\partial^2 H}{\partial\varphi^k\partial\varphi^j}\pi^j \tag{29.2.10}$$

while, for λ_i and φ^i we have $\partial_t\lambda = -\partial_\phi\mathcal{H}$ and $\partial_t\varphi = \partial_\lambda\mathcal{H}$:

$$\dot{\lambda}_j = -\lambda_i\omega^{ik}\frac{\partial^2 H}{\partial\varphi^k\partial\varphi^j} + \pi^i\omega^{ik}\frac{\partial^3 H}{\partial\varphi^k\partial\varphi^l\partial\varphi^j}\xi_l, \qquad \dot{\varphi}^i = \omega^{ij}\frac{\partial H}{\partial\varphi^j} \tag{29.2.11}$$

Recall that the variables φ and λ are conjugate to each other. If we reduce the current theory to include only these two (and not the two bosonic conjugate variables ξ and π), we can regenerate a familiar operatorial formulation:

$$\mathcal{L}_{\mathrm{R}} = \lambda_i\dot{\varphi}^i - \lambda_i\omega^{ij}\frac{\partial H}{\partial\varphi^j}, \qquad \mathcal{H}_{\mathrm{R}} = \lambda_i\omega^{ij}\frac{\partial H}{\partial\varphi^j} \tag{29.2.12}$$

Let us now define two multiplicative operators, $\hat{\varphi}^i = \varphi^i$, and two derivative operators, $\hat{\lambda}^i = -i\partial/\partial\varphi^i$. Evaluation of the reduced Hamiltonian shows that, we have the Liouville operator $\mathcal{H}_{\mathrm{R}} \equiv \hat{L}$ from Koopmann-von Neumann theory:

$$\hat{L} = -i\omega^{ij}\frac{\partial H}{\partial\varphi^j}\frac{\partial}{\partial\varphi^i} \tag{29.2.13}$$

It also confirms to us that we cannot build an operatorial approach to classical mechanics only on the $2N$ phase space; we need the additional $2N$ derivative variables obtained from the Fourier transform.

With this in mind, we may now wish to show that the propagator can be expressed in a matrix formulation. More specifically, we wish to show that the transition amplitude between two states, which are denoted $|q_i p_i\rangle$ and $\langle q_f p_f|$, is equal to the classical path integral between the states Z. We will also switch to using qs and ps rather than fiddling with indices of φ:

$$\langle q_f p_f | e^{-i\hat{L}t} | q_i p_i \rangle = \int_1^2 \mathcal{D}q \mathcal{D}p \tilde{\delta}[q^i - q_{cl}] \tilde{\delta}[p_i - p_{cl}] \qquad (29.2.14)$$

To do this, we require the commutation relations from the KvN algebra, which is stated below; this allows us to develop the necessary identity relations in a moment:

$$[\hat{q}, \hat{p}] = [\hat{q}, \hat{\lambda}_p] = [\hat{p}, \hat{\lambda}_q] = 0, \qquad [\hat{q}, \hat{\lambda}_q] = [\hat{p}, \hat{\lambda}_p] = i \qquad (29.2.15)$$

The first stage is to use Trotter's formula to split the exponent operator into N parts:

$$\langle q_f p_f | e^{-i\hat{L}t} | q_i p_i \rangle = \langle q_f p_f | (e^{-i\hat{L}t'})^N | q_i p_i \rangle \qquad (29.2.16)$$

Here, $t' = t/N$. We then insert $(N-1)$ identity relations $1 = \int dq dp |q, p\rangle \langle q, p|$:

$$\int \prod_{i=1}^{N-1} dq^i dp_i \langle q_f p_f | e^{-i\hat{L}t'} | q_{N-1} p_{N-1} \rangle \dots \langle q_1 p_1 | e^{-i\hat{L}t'} | q_i p_i \rangle \qquad (29.2.17)$$

The next step is to take the very first term (in blue) and enter another two identity relations, $\int dp d\lambda_q |p\lambda_q\rangle\langle p\lambda_q|$ and $\int dq d\lambda_p |q\lambda_p\rangle\langle q\lambda_p|$:

$$\int dp d\lambda_q \langle q_1 p_1 | p\lambda_q\rangle\langle p\lambda_x | e^{-i\hat{L}(\hat{p})\lambda_q t'} \int dq d\lambda_p |q\lambda_p\rangle\langle q\lambda_p | e^{-i\hat{L}(\hat{q})\lambda_p t'} | q_i p_i \rangle \qquad (29.2.18)$$

$$\int dp dq d\lambda_p d\lambda_q \langle q_1 p_1 | p\lambda_q\rangle e^{-i\hat{L}(\hat{p})\lambda_q t'} \langle p\lambda_q | q\lambda_p\rangle e^{-i\hat{L}(\hat{q})\lambda_p t'} \langle q\lambda_p | q_i p_i \rangle \qquad (29.2.19)$$

We have taken advantage of the Hermitian nature of the operators and that the Liouville operator can be written as $L(\hat{q}, \hat{p}, \hat{\lambda}_q, \hat{\lambda}_p) = \hat{L}(\hat{p})\lambda_q + \hat{L}(\hat{q})\lambda_p$. The inner products are now computed using the commutator relationships, recalling the anticommutative property of swapping them around that yields a negative sign:

$$\langle q_1 p_1 | p\lambda_q\rangle = \langle p_1 | p\rangle\langle q_1 | \lambda_q\rangle = \delta(p_1 - p)e^{iq_1\lambda_q} \qquad (29.2.20)$$

$$\langle q\lambda_p | q_i p_i\rangle = \langle q | q_i\rangle\langle\lambda_p | p_i\rangle = \delta(q - q_i)e^{-ip_i\lambda_p} \qquad (29.2.21)$$

$$\langle p\lambda_q | q\lambda_p\rangle = e^{ip\lambda_p} e^{-iq\lambda_q} \qquad (29.2.22)$$

Entering these into the equation, integrating over q and p to remove the delta functions and generally tidying up, we are left with the following integral:

$$\int d\lambda_q d\lambda_p \exp\left\{ it'\left[\left(-\hat{L}(\hat{p}) + \frac{q_1 - q}{t'}\right)\lambda_q + \left(-\hat{L}(\hat{q}) + \frac{p - p_i}{t'}\right)\lambda_p\right]\right\} \qquad (29.2.23)$$

Performing this analysis for the full propagator in equation 29.2.17, we are left with the following result:

$$Z = \int \prod_{i=1}^{N-1} dq^i dp_i \int \prod_{i=1}^{N} d\lambda_q^i d\lambda_p^i e^X \qquad (29.2.24)$$

where,

$$X = -it' \sum_{j=1}^{N} \left(\hat{L}_j(\hat{p})\lambda_q^j + \hat{L}_j(\hat{q})\lambda_p^j - \frac{(q_j - q_{j-1})}{t'}\lambda_q^j - \frac{(p_j - p_{j-1})}{t'}\lambda_p^j \right) \qquad (29.2.25)$$

In the limit that $N \to \infty$, we have that $t' \to 0$, and the fractions become normal time derivatives while the sum becomes an integral:

$$Z = \int \mathcal{D}q \mathcal{D}p \mathcal{D}\lambda_q \mathcal{D}\lambda_p \exp\left\{ -i \int dt \left(\hat{L}(\hat{p})\lambda_q + \hat{L}(\hat{q})\lambda_p - \dot{q}\lambda_p - \dot{p}\lambda_q\right)\right\} \qquad (29.2.26)$$

We then collect terms in λ_i, substitute for the Liouville operator and realise that the Lambda integrals are actually Fourier transforms of delta functions:

$$Z = \int \mathcal{D}q \mathcal{D}p \mathcal{D}\lambda_q \mathcal{D}\lambda_p \exp\left\{ -i \int dt \left(\hat{L}(\hat{p}) - \dot{p}\right)\lambda_q + \left(\hat{L}(\hat{q}) - \dot{q}\right)\lambda_p\right\} \qquad (29.2.27)$$

$$= \int \mathcal{D}q \mathcal{D}p \mathcal{D}\lambda_q \mathcal{D}\lambda_p \exp\left\{ -i \int dt \left(\frac{\partial H}{\partial q} + \dot{p}\right)\lambda_q + \left(\frac{\partial H}{\partial p} - \dot{q}\right)\lambda_p\right\} \qquad (29.2.28)$$

$$= \int \mathcal{D}q \mathcal{D}p \tilde{\delta}\left(\frac{\partial H}{\partial q} + \dot{p}\right) \tilde{\delta}\left(\frac{\partial H}{\partial p} - \dot{q}\right) \qquad (29.2.29)$$

Letting $S = p\dot{q} - H(q,p)$ and using partial integration, we finalise the result:

$$Z = \int \mathcal{D}q \mathcal{D}p \tilde{\delta}\left(\frac{\delta S}{\delta q}\right) \tilde{\delta}\left(\frac{\delta S}{\delta p}\right) = \int \mathcal{D}\varphi \tilde{\delta}\left(\frac{\delta S}{\delta \varphi}\right) \qquad (29.2.30)$$

As we can see, the path integral formulation is a very robust theory and allows us to explore many more avenues to probe the nature of classical mechanics. We might try to develop Ward indentites, explore BRST symmetries, develop a Routhian path integral ...the list is almost endless.

We have formulated mechanics on a supermanifold and shown the corresponding matrix formulation is the familiar KvN formalism. We close this section by asking ourselves what the extra superspace variables are, and why they appear when we exponentiate the Jacobian, especially since we only require $4N$ variables to reconstruct KvN theory!

These variables are named **Faddeev-Popov ghosts**. In (non-technical) essence, they arise due to the gauge freedom of the system and the desire to maintain consistency within the path integral formulation. Each distinct physical configuration could have an infinite number of equivalent descriptions, depending on the choice of gauge. Equivalent configurations lie on the same **gauge orbit**, while distinct ones lie on different orbits.

If the system has a gauge symmetry, then the path integral will overcount the configurations corresponding to the same gauge orbit. Hence, the phase space variables are not parallel to the gauge orbits or in line with the gauge constraints. In Abelian gauge theories, we can simply choose the gauge we wish to work in, think back to our foray into electrodynamics with Lorentz gauges. However, non-Abelian gauge theories require a little more work, which is precisely where the Jacobian we introduced earlier comes in. The aim is to pick a single configuration for each orbit to remove the degeneracy in the functional integral so we result in a subset of physical configurations.

We are going to take a step back and have a general look at how these ghosts arise in non-Abelian gauge theory. Taking a general partition function Z, let us fix the gauge according to some local prescription $G[\varphi] = 0$ and make use of the delta function and Jacobian:

$$1 = \int \mathcal{D}\varphi(x)\delta(G[\varphi^i]) \det\left(\frac{\delta G[\varphi^i]}{\delta\varphi(x)}\right) \tag{29.2.31}$$

Motivation of this equation can be seen as the functional generalisation of the following relation:

$$1 = \int \delta(G(\varphi))dG$$
$$= \int d\varphi \delta(G(\varphi))\frac{dG}{d\varphi} \tag{29.2.32}$$

Then take the N-dimensional case and, finally, let $N \to \infty$ to generate the functional unity relation:

$$1 = \left(\prod_{i=1}^{N}\int d\varphi^i\right)\delta^{(N)}(G_i(\varphi)) \det\left(\frac{\partial G_i}{\partial\varphi^j}\right)$$
$$\stackrel{N\to\infty}{=} \int \mathcal{D}\varphi(x)\delta(G[\varphi^i]) \det\left(\frac{\delta G[\varphi^i]}{\delta\varphi(x)}\right) \tag{29.2.33}$$

The path integral is now factored into integrations over the gauge orbits (Haar measure) and the distinct configurations:

$$Z = \int \mathcal{D}\varphi e^{iS}$$
$$= \left(\int \mathcal{D}\varphi(x)\right)\int \mathcal{D}\varphi\delta(G[\varphi^i]) \det\left(\frac{\delta G[\varphi^i]}{\delta\varphi(x)}\right)e^{iS} \tag{29.2.34}$$

When taking expectation values of physical observables (gauge invariant objects), the factored multiplicative gauge measures cancel out on top and bottom, allowing us to compute the desired quantity without an infinite measure:

$$\langle \mathcal{O} \rangle = \frac{1}{Z} \int \mathcal{D}\varphi \mathcal{O}(\varphi) e^{iS} \tag{29.2.35}$$

The ghosts now arise naturally as a result of exponentiating the determinant. The new variables are not physically part of the theory but are just included to express the Jacobian as an integral and cancel out the unphysical degrees of gauge freedom. These variables then neatly place themselves within the Lagrangian and therefore endow themselves with a larger status than they deserve.

Exercise 29.1 In order to understand the ideas presented here, we are going to work through a simple example of the Faddeev-Popov method. The basic idea will be to take any two-dimensional rotationally invariant integral $I(x,y)$ and integrate out the angular dependence to reduce the problem $I(r)$:

$$I = \int_{-\infty}^{\infty} dx \int_{-\infty}^{\infty} dy f(x,y) \qquad \rightarrow \qquad I = \int_0^{\infty} dr g(r) \tag{29.2.36}$$

The problem has now been reduced and, in principle, should be easier to solve.

We can parallel the situation to gauge theory by imagining a set of circles with a Cartesian axis at their center. The gauge transformations are rotations around the origin, and the orbits are circles of fixed radius.

Although not a necessary step for a proof, we are going to give $f(x,y)$ an explicit form for our example:

$$I = \int_{-\infty}^{\infty} dx \int_{-\infty}^{\infty} dy e^{-(x^2+y^2)} \tag{29.2.37}$$

Since the integral is rotationally invariant, we are then going to rotate the axis to a new set $(x,y) \rightarrow (x',y')$ by some angle ϕ. The standard formulas are then given by:

$$x' = x \sin\phi + y \cos\phi, \qquad y' = -\sin\phi + y \cos\phi \tag{29.2.38}$$

We can now simplify things by choosing $\phi = 0$ and fixing $x' = 0$. The unity condition can now be written

$$1 = \int_0^{2\pi} d\phi \delta(x') \det\left(\frac{\partial x'}{\partial \phi}\right) \tag{29.2.39}$$

Evaluate the determinant, recalling the above conditions:

$$\det\left(\frac{\partial x'}{\partial \phi}\right) = \det\left(x + \frac{\partial y}{\partial \phi}\right) \tag{29.2.40}$$

Hence,

$$1 = \int_0^{2\pi} d\phi \delta(y) \det\left(x + \frac{\partial y}{\partial \phi}\right) \tag{29.2.41}$$

The delta will remove the derivative with respect to ϕ, too, since it is constant. Substituting this back into the integral, we obtain,

$$
\begin{aligned}
I &= \int_0^\infty dx \int_0^\infty dy \int_0^{2\pi} d\phi\, e^{-(x^2+y^2)} \delta(y) \det(x) \\
&= \left(\int_0^\infty dx\, e^{-x^2} x \right) \cdot \left(\int_0^\infty dy\, e^{-y^2} \delta(y) \right) \cdot \left(\int_0^{2\pi} d\phi \right) \\
&= 2\pi \int_0^\infty dx\, e^{-x^2} x \\
&= \pi
\end{aligned}
\tag{29.2.42}
$$

The Gaussian integral is easily evaluated by the substitution $u = -x^2$ leaving us with a half multiplying 2π.

Part V

Preliminary Mathematics

30

The (Not So?) Basics

It is often handy to partition a collection of numbers into one package for neatness. This is the idea of a **set**, which is denoted X; it is itself considered to be an object. The numbers that the set contains are its **elements**, which are denoted $x \in X$, which means element x in X. An n-**tuple** is an ordered set whose n elements are listed in a particular order: (cat, dog, rabbit) is not the same as (dog, rabbit, cat). The **Cartesian product** is an operation on sets that generates another set, for example $X \times Y$. We can take the Cartesian product of a set with itself. for example $X \times X \times X \times X = X^4$. Sometimes, it is nice to refer to specific types of numbers such as the set of **real numbers** $\mathbb{R}$ or the set of **rational numbers** $\mathbb{Q}$ or the **integers** $\mathbb{Z}$. A set which doesn't contain any of its boundary points is called an **open set**. A set is **closed** if it contains its boundary points. A **dense set** is a subset of a larger set such that every point in the large set either belongs to the subset or is in the neighbourhood of a point of the subset.

A function is like a little number-crunching box into which you feed in a number $x \in X$ from the **domain** X of the function and you get another number out, $y \in Y$ (element y of the **codomain** Y). It is a mapping from a set of numbers to another set of numbers and there are several ways to write it: $f(x) = y$, where x are the numbers you feed the function; *the argument of f*, and where y is the number you get out; or you can also write it like $f : x \to y$, which means f acts on x and maps to y.

You can, however, get *many* to *one*. **Injective** means that, for each element of the domain, there is a unique member of the codomain corresponding to $f(x)$; the significance of this is that you can reverse the action of the function to go from y to x. This doesn't mean that every member of the codomain has a partner in the domain of the function, since there could be simply more elements of the second set that have nothing to do with the function. If a function is **surjective**, it means that every element of the codomain has a partner in the domain. A **bijective** function is both injective and surjective, meaning there is a one-to-one matching between all the elements of the two sets. The **composition** of two functions is written $(g \circ f)(x)$ or $g(f(x))$; it means that you input a number x to function f and then take that result as the argument for the function g. The **inverse function** is an operation that reverses the action of f on x, taking elements $y \in Y$ and outputting elements $x \in X$. If $f : x \to y$, then $f^{-1} : y \to x$, or $f(x) = y$, then $x = f^{-1}(y) = f^{-1}(f(x))$. An **involutory function** is

Lagrangian & Hamiltonian Dynamics. Peter Mann, Oxford University Press (2018).
© Peter Mann. DOI: 10.1093/oso/9780198822370.001.0001

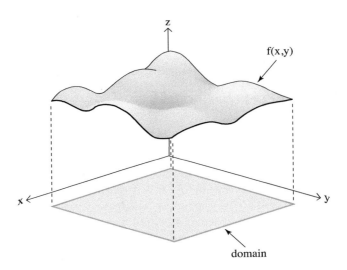

Fig. 30.1: A multivariate function $f(x, y)$ and its domain in the xy-plane.

a function that is its own inverse: $f(f(x)) = x$. An **even function** is a function whose graph for negative x is a reflection in the y axis of its graph of positive x; $f(x)$ is even if $f(-x) = f(x)$. An odd function is $f(-x) = -f(x)$.

A **multivariate function** is a function of more than one argument $f(x_1, x_2, \ldots, x_n)$ (see figure 30.1). A function of a **real variable** is a function whose domain is (a subset of) the real numbers $\mathbb{R}$. A **real-valued function** is a function whose output is a real number $\mathbb{R}$. We could write a real-valued multivariable function as $f : \mathbb{R}^n \to \mathbb{R}$ if it has n real arguments. A line is a one-dimensional space of real numbers $\mathbb{R}$, a plane is a two-dimensional space of real numbers $\mathbb{R} \times \mathbb{R} = \mathbb{R}^n$, a volume is a three-dimensional space of real numbers $\mathbb{R} \times \mathbb{R} \times \mathbb{R} = \mathbb{R}^3$. An element of the line is a number (x). A point on a surface is defined by two coordinates (x, y). A point in a volume is defined by three coordinates (x, y, z). A point in an n-dimensional space $\mathbb{R}^n$ is the n-tuple $\vec{x} = (x_1, x_2, \ldots, x_n)$.

We can use this idea to define some objects. A **curve** is formed as $f : \mathbb{R} \to \mathbb{R}^n$. It is constructed from a collection of functions, each defined over a common variable $r_1(t), r_2(t), \ldots, r_n(t)$, where $r_i : \mathbb{R} \to \mathbb{R}$, we can write this as an n-**tuple** $\vec{r}(t) = [r_1(t), r_2(t), \ldots, r_n(t)]$ with $\vec{r} : \mathbb{R} \to \mathbb{R}^n$.

We call $f : \mathbb{R}^n \to \mathbb{R}$ a **scalar field** since a scalar quantity is the output; it associates a number in $\mathbb{R}$ from a space $\mathbb{R}^n$. If we consider **Euclidean space**, which is what we are familiar with in day-to-day life (x, y, z) then a scalar field would be $f : \mathbb{R}^3 \to \mathbb{R}$. We call $f : \mathbb{R}^m \to \mathbb{R}^n$ a **vector field**, where the output is a vector.

A **functional** A is a box that takes a *function* and gives out a number; it is a function of a function of a variable x. To show this, we use square brackets: $A[f(x)]$; A is a function of f but a functional of x. It is different from the composition because the composition was a function of the *result* of a function, which is just a number.

31
Matrices

We use matrices in many places in the text and it is here that we develop the necessary background to understand how to use them in calculations. When I first saw a 2×2 matrix, I honestly had no idea what on earth was going on because it was so different to anything I had ever seen before. Therefore, in this chapter, I hope to give a guide to someone who has never seen even the simplest of matrices before. We will define the basics of matrix addition, scalar multiplication and how to multiply matrices together before moving on to more specialised topics. We will denote matrices in bold font, and elements in plain font with subscripts.

A matrix is a rectangular array of ordered quantities surrounded by big curly brackets. A general m by n (written $m \times n$) matrix $\boldsymbol{A}$ will have m rows and n columns, with a total of mn elements. We specify a particular element A_{ij} using indexing notation ij, where $i = 1, \ldots, m$ is the row element and $j = 1, \ldots, n$ is the column element. The way to order the elements will usually be obvious from the application at hand. If $m = n$, we have a **square matrix**. If $m = 1$, then we have a one-row matrix; if $n = 1$, the matrix only has one column:

$$\boldsymbol{A} = \begin{pmatrix} A_{11} & A_{12} & \ldots & A_{1n} \\ A_{21} & A_{22} & \ldots & A_{2n} \\ \ldots & \ldots & \ldots & \ldots \\ A_{m1} & A_{m2} & \ldots & A_{mn} \end{pmatrix} \tag{31.0.1}$$

We can now define the basic operations of **matrix algebra**, including matrix addition, subtraction and scalar multiplication, where $\boldsymbol{A}, \boldsymbol{B}$ and $\boldsymbol{C}$ are $m \times n$ matrices. Matrix addition is as follows:

$$\boldsymbol{A} + \boldsymbol{B} = \boldsymbol{B} + \boldsymbol{A} = \boldsymbol{C}, \quad \text{with} \quad C_{ij} = A_{ij} + B_{ij} \tag{31.0.2}$$

Each element of the resultant matrix is the sum of the respective elements of each constituent matrix. Matrix subtraction is

$$\boldsymbol{A} - \boldsymbol{B} = \boldsymbol{C}, \quad \text{with} \quad C_{ij} = A_{ij} - B_{ij} \tag{31.0.3}$$

Multiplication by a scalar a involves multiplying each element by the quantity:

Lagrangian & Hamiltonian Dynamics. Peter Mann, Oxford University Press (2018).
© Peter Mann. DOI: 10.1093/oso/9780198822370.001.0001

$$aA = C, \quad \text{with} \quad C_{ij} = aA_{ij} \tag{31.0.4}$$

Multiplication of two matrices is possible if the number of columns of the first equals the number of rows of the second, so we may multiply an $m \times n$ matrix by a $n \times p$ matrix to give a $m \times p$ matrix:

$$A \cdot B = C, \quad \text{with} \quad C_{ij} = \sum_{j=1}^{n} a_{ij} b_{jk} \tag{31.0.5}$$

For example, consider $A \cdot B = C$, where A is a 3×3 matrix, B is a 3×2 matrix and C is a 3×2 matrix:

$$
\begin{aligned}
A \cdot B &= \begin{pmatrix} A_{11} & A_{12} & A_{13} \\ A_{21} & A_{22} & A_{23} \\ A_{31} & A_{32} & A_{33} \end{pmatrix} \begin{pmatrix} B_{11} & B_{12} \\ B_{21} & B_{22} \\ B_{31} & B_{31} \end{pmatrix} \\
&= \begin{pmatrix} A_{11}B_{11} + A_{12}B_{21} + A_{13}B_{31} & A_{11}B_{12} + A_{12}B_{22} + A_{13}B_{32} \\ A_{21}B_{11} + A_{22}B_{21} + A_{23}B_{31} & A_{21}B_{12} + A_{22}B_{22} + A_{23}B_{32} \\ A_{31}B_{11} + A_{32}B_{21} + A_{33}B_{31} & A_{31}B_{12} + A_{32}B_{22} + A_{33}B_{32} \end{pmatrix} \tag{31.0.6}
\end{aligned}
$$

In general, matrix multiplication is not commutative, so $A \cdot B \neq B \cdot A$. The **transpose** of a matrix is denoted with a superscripted t. It is another matrix that is obtained from the original matrix by swapping the rows for columns, for instance the transpose of A in 31.0.1 is A^t:

$$A^t = \begin{pmatrix} A_{11} & A_{21} & \dots & A_{m1} \\ A_{12} & A_{22} & \dots & A_{m2} \\ \dots & \dots & \dots & \dots \\ A_{1n} & A_{2n} & \dots & A_{mn} \end{pmatrix} \tag{31.0.7}$$

We can see that the elements of the transpose are found by swapping the indices $A^t_{ij} = A_{ji}$. If we replace each element A_{ij} by its complex conjugate, denoted by a star (A^*_{ij}) then the resulting matrix is the **conjugate matrix** to A, which we denote by A^*. The conjugate of the transpose of A is the **adjoint** or the **conjugate transpose**, which is denoted with a dagger symbol: $A^\dagger = A^{t*}$. The elements of the adjoint are $A^\dagger_{ij}$ and are given by the conjugates of the transpose: $A^\dagger_{ij} = A^*_{ji}$. Although, at first, this may seem a really random thing to do, it is very useful in both classical and quantum mechanics. We may consider the above to be three operations that we can perform on a matrix: we can take the transpose, conjugate or conjugate transpose. If the matrix is invariant under a transpose, we say it is a **symmetric matrix**; this is only possible for square matrices, which we look at in a second. If the matrix is invariant under taking the conjugate, then it is said to be a **real** matrix. If the matrix is invariant under taking the conjugate transpose, it is said to be **Hermitian**.

A **square matrix** is a $n \times n$ matrix. A square matrix is said to be **symmetric** if $A_{ij} = A_{ji}$ under transpose, and **antisymmetric** if $A_{ij} = -A_{ji}$. A square matrix is

diagonal if all its off-diagonal elements are zero and the elements on the diagonal A_{ij} with $i = j$ are non-zero:

$$\boldsymbol{A} = \begin{pmatrix} A_{11} & 0 & \dots & 0 \\ 0 & A_{22} & \dots & 0 \\ \dots & \dots & \dots & \dots \\ 0 & 0 & \dots & A_{nn} \end{pmatrix} \tag{31.0.8}$$

Physically, diagonal matrices correspond to a better choice of variables to simplify the problem at hand; matrix diagonalisation (the process of making the matrix diagonal) is therefore similar to a coordinate transformation. A special case is when all diagonal elements are equal to 1; this matrix is called the **identity matrix** $\mathbb{I}$, whose elements comprise the Kronecker delta:

$$\mathbb{I} = \begin{pmatrix} 1 & 0 & \dots & 0 \\ 0 & 1 & \dots & 0 \\ \dots & \dots & \dots & \dots \\ 0 & 0 & \dots & 1 \end{pmatrix} \quad \text{with} \quad \mathbb{I}_{ij} = \delta_{ij} \begin{cases} 1 & i = j \\ 0 & i \neq j \end{cases} \tag{31.0.9}$$

A square matrix $\boldsymbol{A}^{-1}$ that satisfies the following relation is called the **inverse matrix** of $\boldsymbol{A}$, which is an **invertible** matrix:

$$\boldsymbol{A}^{-1} \cdot \boldsymbol{A} = \boldsymbol{A} \cdot \boldsymbol{A}^{-1} = \mathbb{I} \tag{31.0.10}$$

If no matrix can be found that fulfils these properties, so that $\boldsymbol{A}$ has no inverse, then we say that it is **singular**. The **trace** of a matrix is the sum of its diagonal elements, it is denoted $Tr\ \boldsymbol{A}$. An **orthogonal** matrix is a matrix whose transpose is its own inverse: $\boldsymbol{A}^t \cdot \boldsymbol{A} = \mathbb{I}$. A **unitary matrix** is one whose adjoint is its inverse: $\boldsymbol{A}^\dagger \cdot \boldsymbol{A} = \mathbb{I}$.

The **determinant** of a *square* matrix is a mathematical object that we can associate with the matrix; it is often denoted by two straight lines or a "det" in front of the matrix. The determinants of a 2×2 and a 3×3 matrix are the most important for our discussions and so are listed below:

$$\det \begin{pmatrix} A_{11} & A_{12} \\ A_{21} & A_{22} \end{pmatrix} = \begin{vmatrix} A_{11} & A_{12} \\ A_{21} & A_{22} \end{vmatrix} = A_{11}A_{22} - A_{12}A_{21} \tag{31.0.11}$$

$$\det \begin{pmatrix} A_{11} & A_{12} & A_{13} \\ A_{21} & A_{22} & A_{23} \\ A_{31} & A_{32} & A_{33} \end{pmatrix} = A_{11}A_{22}A_{33} - A_{11}A_{23}A_{32} - A_{12}A_{21}A_{33} + A_{12}A_{23}A_{31}$$

$$+ A_{13}A_{21}A_{32} - A_{13}A_{22}A_{31} \tag{31.0.12}$$

Mathematical Background 7

In general, the determinant of an $n \times n$ square matrix is defined to be

$$\det \boldsymbol{A} = \det \begin{pmatrix} A_{11} & A_{12} & \dots & A_{1n} \\ A_{21} & A_{22} & \dots & A_{2n} \\ \dots & \dots & \dots & \dots \\ A_{n1} & A_{n2} & \dots & A_{nn} \end{pmatrix} = \sum_{ij\dots k} \epsilon_{ij\dots k} A_{i1} A_{j2} \dots A_{kn} \qquad (31.0.13)$$

where $\epsilon_{ij\dots k}$ is the **Levi-Civita** symbol or nth order (n indices), building on mathematical background 2:

$$\epsilon_{ij\dots k} \begin{cases} +1 & ij\dots k \text{ is even} \\ -1 & ij\dots k \text{ is odd} \\ 0 & \text{if any index is repeated} \end{cases} \qquad (31.0.14)$$

Importantly, it is **antisymmetric**, meaning that we stick a minus sign up front when we swap any two indices, so $\epsilon_{ijk} = -\epsilon_{ikj}$. By 'even' and 'odd' we mean the number of permutations of the indices to get back to the particular order we define as being equivalent to 1. So, for a third-order Levi-Civita symbol, we define $\epsilon_{123} \equiv 1$; then, consider swapping two indices so at least one of the three indices retains its original place ($\epsilon_{132} = \epsilon_{321} = \epsilon_{123}$). Since we have only performed one swapping, all of these permutations are -1. If we decide to perform two swappings ($\epsilon_{231} = \epsilon_{312} = \epsilon_{123}$), since two is even, all these take the value of $+1$. This all follows on from the antisymmetry property. So, although it may look complicated for any set of non-repeated indices, it is really just raising (-1) by positive or negative powers.

There are n^n permutations in the summation, of which $n!$ are non-zero distinct combinations of the indices. Therefore, for a 2×2 matrix, we may write the determinant as

$$\det \begin{pmatrix} A_{11} & A_{12} \\ A_{21} & A_{22} \end{pmatrix} = \epsilon_{11} A_{11} A_{12} + \epsilon_{12} A_{11} A_{22} + \epsilon_{21} A_{21} A_{12} + \epsilon_{22} A_{21} A_{22}$$

$$= A_{11} A_{22} - A_{21} A_{12}$$

Note there are 2^2 total terms and $2!$ non-zero terms; both ϵ_{11} and ϵ_{22} are 0, since there are repeated indices, while we define ϵ_{12} as being equal to $+1$ exactly and, hence, a single permutation is odd and so it is evaluated to -1.

We will return to the Levi-Civita permutation symbol later when we discuss vector calculus in chapter 34. There we will assume the reader is familiar with the rules of the permutation symbol and its antisymmetric properties. The antisymmetric property ϵ_{ijk} is the reason why repeated indices vanish. Consider $\epsilon_{i_1 j i_2} = -\epsilon_{i_2 j i_1}$. Since both is are the exact same, we realise that what we have is $a = -a$. The only number for which this relation holds is $a = 0$; hence, repeated indices vanish.

We can perform operations with the Levi-Civita symbol; for instance, taking the product $\epsilon_{ijk}\epsilon_{lmn}$, we state without proof that it is the determinant of the following matrix of Kronecker deltas:

$$\epsilon_{ijk}\epsilon_{lmn} = \det \begin{pmatrix} \delta_{il} & \delta_{im} & \delta_{in} \\ \delta_{jl} & \delta_{jm} & \delta_{jn} \\ \delta_{kl} & \delta_{km} & \delta_{kn} \end{pmatrix} \tag{31.0.15}$$

As a special case of this product, we note the following result, which we will use in future chapters:

$$\epsilon_{ijk}\epsilon_{imn} = \delta_{jm}\delta_{kn} - \delta_{jn}\delta_{km} \tag{31.0.16}$$

By **Laplace expansion**, we can express the determinant of an $n \times n$ matrix as the weighted sum of determinants of $(n-1) \times (n-1)$-order determinants. For instance, we can write the determinant of a 3×3 matrix as

$$\det \boldsymbol{A} = \begin{vmatrix} A_{11} & A_{12} & A_{13} \\ A_{21} & A_{22} & A_{23} \\ A_{31} & A_{32} & A_{33} \end{vmatrix} = A_{11}\begin{vmatrix} A_{22} & A_{23} \\ A_{32} & A_{33} \end{vmatrix} - A_{12}\begin{vmatrix} A_{21} & A_{23} \\ A_{31} & A_{33} \end{vmatrix} + A_{13}\begin{vmatrix} A_{21} & A_{22} \\ A_{31} & A_{32} \end{vmatrix} \tag{31.0.17}$$

Qualitatively, the procedure involves picking an element A_{ij} and removing its row and column from the 3×3 determinant; we call what is left the **minor** M_{ij} of A_{ij}. For instance, if we choose A_{11} then cancel the top row and first column, we are left with the following minor:

$$M_{11} = \begin{vmatrix} A_{22} & A_{23} \\ A_{32} & A_{33} \end{vmatrix}$$

We then define the **cofactor** C_{ij} of element A_{ij} to be the signed minor:

$$C_{ij} = (-1)^{i+j} M_{ij} \tag{31.0.18}$$

We now say that the determinant is given by the sum of each element of a particular row (above, we chose the first), weighted by its cofactor. It doesn't matter which row you use; in fact, we can even use columns and get the same answer but, for simplicity, we will stick to using the first row. The Laplace expansion of an $n \times n$ matrix determinant along the ith row is given by

$$\det \boldsymbol{A} = \sum_{j=1}^{n} A_{ij}C_{ij} = \sum_{j=1}^{n} (-1)^{i+j} A_{ij} M_{ij} \tag{31.0.19}$$

We can form a matrix of all the cofactors of the elements A_{ij} of $\boldsymbol{A}$; this is called the **cofactor matrix**, and is denoted $\boldsymbol{C}$:

$$\boldsymbol{C} = \begin{pmatrix} C_{11} & C_{12} & \dots & C_{1n} \\ C_{21} & C_{22} & \dots & C_{2n} \\ \dots & \dots & \dots & \dots \\ C_{n1} & C_{n2} & \dots & C_{nn} \end{pmatrix} \tag{31.0.20}$$

The inverse matrix for $\boldsymbol{A}$ is given by A^{-1}. The inverse of $\boldsymbol{A}$ is the transpose of the cofactor matrix, divided by the determinant of $\boldsymbol{A}$:

$$\boldsymbol{A}^{-1} = \frac{1}{\det \boldsymbol{A}} \boldsymbol{C}^t \tag{31.0.21}$$

Straight away, we can see that matrices whose determinant is exactly zero don't have an inverse and are singular matrices. In the main text, we will use several matrices and concepts that we develop here. We also used the **eigenvalue problem** in the small-oscillations discussion in chapter 12, however, we developed this quite in detail there and so will not repeat it here.

Last, we will discuss **skew-symmetric** or **antisymmetric** matrices, which are square matrices $\boldsymbol{A}$ whose transpose is also its negative $(-\boldsymbol{A} = \boldsymbol{A}^t)$ and the elements satisfy $A_{ij} = -A_{ji}$. The determinant of an even-dimensional antisymmetric matrix can be written as the square of a non-vanishing polynomial over the elements of the array. The polynomial is called the **Pfaffian** of the matrix, symbolised by $\mathrm{Pf}(\boldsymbol{A})$. A relevant example of a skew-symmetric matrix is $\boldsymbol{B}$:

$$\boldsymbol{B} = \begin{pmatrix} 0 & B \\ -B & 0 \end{pmatrix} \qquad \mathrm{Pf}(\boldsymbol{B}) = B \tag{31.0.22}$$

For any skew-symmetric $2n \times 2n$ matrix $\boldsymbol{A}$, we can associate an object called a **form** with it (we do not investigate this just now; see chapters 37 and 38 for further details) symbolised by ω:

$$\omega = \sum A_{ij} e^i \wedge e^j \tag{31.0.23}$$

The Pfaffian of the skew-symmetric matrix is then given by

$$\frac{1}{n!} \omega^n = \mathrm{Pf}(\boldsymbol{A}) e^1 \wedge \dots \wedge e^{2n} \tag{31.0.24}$$

This was used a lot in chapter 22!

32
Partial Differentiation

The **derivative** of a smooth single variable function $f(x)$ at a point $x = a$ is the ratio of the difference $\Delta f = f(a + \Delta x) - f(a)$ to a change $\Delta x = (a + \Delta x) - a$ in the limit that the change is tiny or **infinitesimal**:

$$\frac{d}{dx} f(x) = \lim_{\Delta x \to 0} \frac{\Delta f}{\Delta x} = \frac{df}{dx} = d_x f(x) = f'(x) \tag{32.0.1}$$

This is called **Newton's difference quotient**. It is the ratio of the change in the function to the change in the variable of the functions argument in the limit that the change is only really small, the rate of change of f being with respect to x. We can take higher-order derivatives by performing this operation multiple times and, if the first derivative quantifies the rate of change of the function, then the second derivative is the rate of change of the rate of change of the function. The easiest way to picture this is that velocity is the rate of change of the position, while acceleration is the rate of change of the velocity; they are the first and second derivatives, respectively. A **stationary point** or **critical point** of a smooth function is the point where the derivative is zero. A function has a **local minimum** at a x_0 if $f(x) \geq f(x_0)$ for all x in the domain sufficiently close to x_0; if it is for all x in the domain then it is a **global minima**. A function has a local maxima about x_0 if $f(x) \leq f(x_0)$ for x in the neighbourhood of x_0, it is global if it holds for all x in the domain.

There are three main notations used, and the last three equalities in equation 32.0.1 show the Leibniz, Euler and Lagrange notations, respectively. In the particular case of time derivatives (when $x = t$) we also use Newton's notation of a dot above the character: $\dot{f}(t) = df/dt$.

There are three very important rules for differentiation:

- the **product rule**:

$$\frac{d}{dx}(f \cdot g) = \frac{df}{dx} g + f \frac{dg}{dx} \tag{32.0.2}$$

- The **chain rule**:

$$\frac{d}{dx}(f \circ g) = \frac{df}{dg} \frac{dg}{dx} \tag{32.0.3}$$

Lagrangian & Hamiltonian Dynamics. Peter Mann, Oxford University Press (2018).
© Peter Mann. DOI: 10.1093/oso/9780198822370.001.0001

- The **quotient rule:**

$$\frac{d}{dx}\left(\frac{f}{g}\right) = \frac{1}{g^2}\left(\frac{df}{dx}g - f\frac{dg}{dx}\right) \tag{32.0.4}$$

They can all be proven using Newton's quotient formula but, in particular, the product and quotient rules are special cases of the chain rule. The chain rule is the most important one for our discussion. It is the rule for differentiating the composition of functions. The basic rules of differentiation should be learnt and practised; we summarise the most useful here:

Let $f(x) = x^n$:

$$\frac{df}{dx} = nx^{n-1} \tag{32.0.5}$$

Multiply the function by the power and then subtract 1 from the power (e.g $\frac{d}{dx}3x^2 = 6x$). A special case is, if $f(x)$ is a constant (just any number that doesn't have x dependancy), then the derivative is zero, $\frac{df}{dx} = 0$, for all x (we write for all as $\forall\, x$). The second special case is if $f(x) = x$, since anything to the power 0 is 1, which follows from $a^{(b-b)} = (a^b/a^b) = 1$. The derivatives of trigonometric functions are

$$\frac{d}{dx}\sin x = \cos x, \qquad \frac{d}{dx}\cos x = -\sin x \tag{32.0.6}$$

We now turn to consider the calculus of a function of two or more variables; this is the subject of **partial differentiation**.

The domain of a function $f(x, y)$ is the set of points in the xy plane; the domain of a function of n variables is in $\mathbb{R}^n$. The graph of the function $f(x) = y$ is the set of points with coordinates $(x, f(x))$ in the xy plane $\mathbb{R}^2$; it is a curve above a line (axis). The graph of a function of two variables, $f(x, y) = z$, is the set of points in $\mathbb{R}^3$ with coordinates $(x, y, f(x, y))$; we now call this a **surface** over a plane. It can be very tricky to draw these maps all the time; imagine having a geography map that consisted of drawings of the hillsides; it would be quite difficult to produce and read quickly. These geography maps instead use contour lines of equal height and we can quickly build up a three-dimensional surface from a two-dimensional image.

This is called a **topographical map**; it is a reduced dimensional plot. We can represent $f(x, y)$ in a two-dimensional plot in the xy plane by sketching the curves where the function is some constant $f(x, y) = c$, then changing the value of the constant and plotting again, and so on (see figure 32.1). The curves where the function is constant are called **level curves** of the function. Level curves are vertical projections onto the xy plane of the lines where the surface plot intersects horizontal planes of $f(x, y) = c$. We can see the way that the three-dimensional surface can be represented by a two-dimensional topographical map; in general, we can represent an n-dimensional surface by an $(n-1)$-dimensional surface, which we call a **hypersurface**.

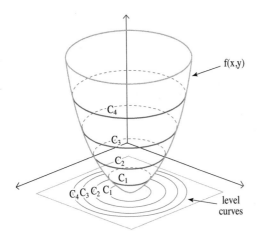

Fig. 32.1: The level curves of a surface in an $(n-1)$-dimensional hypersurface.

The partial derivative of a function $f(x,y)$ with respect to x is the Newtonian quotient of $f(x,y)$ treated as if it were only a function of x, that is, y is held constant. It is the rate of change of the function with respect to the distance in the direction of that particular coordinate axes. We symbolise it with a curly partial sign ∂:

$$\frac{\partial f(x,y)}{\partial x} = \lim_{a \to 0} \frac{f(x+a,y) - f(x,y)}{a}$$

while the partial derivative with respect to y is

$$\frac{\partial f(x,y)}{\partial y} = \lim_{b \to 0} \frac{f(x,y+b) - f(x,y)}{b}$$

Let $f(x,y,z) = 3x^2 - 4xy + 2z^3x^2 + 10zy + 42$ (we can summarise a lot of the rules of partial derivatives using this example) and consider the following quantites:

$$\frac{\partial f(x,y,z)}{\partial x} = 6x - 4y + 4z^3x, \qquad \frac{\partial f}{\partial y} = 10z - 4x, \qquad \frac{\partial f(x,y,z)}{\partial z} = 6z^2x^2 + 10y$$

The partial derivative $\partial f / \partial x$ for $f(x,y)$ is the slope of the curve at the intersection of the surface and a vertical plane from $x = a$ and y fixed at $y = b$. The partial derivative $\partial f / \partial y$ is the slope of the curve at the intersection of the surface and the vertical plane at x fixed at a and a value of $y = b$. The **normal line** through a point p on a surface is a line that is perpendicular to the surface. The **tangent plane** to the surface at p is the plane through p that is perpendicular to the normal line at that point (see figure 32.2). At the point p with coordinates $(a, b, f(a,b))$ on the surface $z = f(x,y)$, the equation of the tangent plane is

$$z = f(a,b) + \frac{\partial f}{\partial x}(a,b)(x-a) + \frac{\partial f}{\partial y}(a,b)(y-b) \qquad (32.0.7)$$

You will note that this is nothing other than a Taylor series or a **Taylor polynomial** of degree 1. **Taylor's theorem**[1] states that a sufficiently smooth function can be

---

[1]Scottish mathematician James Gregory (1638 – 1675), professor of mathematics at the University of St Andrews, was actually the discoverer of Taylor's theorem before Taylor's formal proof 40 years later.

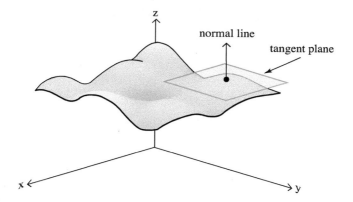

Fig. 32.2: The tangent plane to a point and it's normal line.

expressed as an nth-order polynomial evaluated at the point. Tangent planes therefore approximate a surface near a point; in the neighbourhood of that point, the tangent plane approximates the function; and the Taylor polynomial approximates the function! In one dimension, we can write any degree Taylor polynomial as the following series:

$$f(x) = f(a) + (x-a)\frac{df}{dx}\Big|_{x=a} + \frac{1}{2!}(x-a)^2\frac{d^2f}{dx^2}\Big|_{x=a} + \cdots$$

And, equivalently, if we let $x \to x + \epsilon$ and $a \to x$, as is especially useful for equilibrium problems,

$$f(x+\epsilon) = f(x) + \epsilon\frac{df}{dx} + \frac{1}{2!}\epsilon^2\frac{d^2f}{dx^2} + \cdots$$

If we pull out the $f(x)$, this expression is the power series expression of the exponential function. Hence,

$$f(x+\epsilon) = \exp\left\{\epsilon\frac{d}{dx}\right\}f(x)$$

Higher-order partial derivatives can be found by simply repeating the partial differentiation process: we get both **pure** and **mixed** derivatives. **Clairaut's theorem** states that mixed partial derivatives are commutative:

$$\boxed{\frac{\partial^2 f}{\partial x \partial y} = \frac{\partial^2 f}{\partial y \partial x}} \tag{32.0.8}$$

There are two special examples of the chain rule, and we use both in the text. The first situation concerns a function of two variables that are both functions of a smooth parameter: $z = f(x(t), y(t))$, in which case the derivative of z is

$$\frac{dz}{dt} = \frac{\partial f}{\partial x}\frac{dx}{dt} + \frac{\partial f}{\partial y}\frac{dy}{dt} \tag{32.0.9}$$

The second case where is $z = f(x(s,t), y(st))$ in which case the partial derivatives of z are

$$\frac{\partial z}{\partial s} = \frac{\partial z}{\partial x}\frac{\partial x}{\partial s} + \frac{\partial z}{\partial y}\frac{\partial y}{\partial s}, \qquad \frac{\partial z}{\partial t} = \frac{\partial z}{\partial x}\frac{\partial x}{\partial t} + \frac{\partial z}{\partial y}\frac{\partial y}{\partial t} \tag{32.0.10}$$

It is convenient to express these in a matrix form:

$$\begin{pmatrix} \frac{\partial z}{\partial s} & \frac{\partial z}{\partial t} \end{pmatrix} = \begin{pmatrix} \frac{\partial z}{\partial x} & \frac{\partial z}{\partial y} \end{pmatrix} \begin{pmatrix} \partial x/\partial s & \partial x/\partial t \\ \partial y/\partial s & \partial y/\partial t \end{pmatrix} \tag{32.0.11}$$

The **differential** of $f(x, y, z)$ is the change in $f(x, y, z)$ when all its variables are changed by infinitesimal quantities dx, dy and dz. It is the multivariate version of Newton's quotient rule and can be extended to more variables:

$$df(x, y, z) = \frac{\partial f}{\partial x}dx + \frac{\partial f}{\partial y}dy + \frac{\partial f}{\partial z}dz \tag{32.0.12}$$

We consider df to be a function of twice as many variables as f: both (x, y, z) *and* (dx, dy, dz). The differential is an approximation of the change in a function and, assuming we are treating infinitesimal quantities, the error is assumed to be small $(\Delta f - df \approx 0)$ we thus obtain

$$\Delta f = f(x + dx, y + dy, z + dz) - f(x, y, z) \tag{32.0.13}$$

We now move to generalise this slightly in preparation for our linear algebra discussion. Consider a vector of m functions $(y_1, \ldots, y_m)$ where each function is a function of n variables $y_i = f_i(x_1, \ldots, x_n)$; in other words, each is a function of a point $\vec{x} = (x_1, \ldots, x_n)$ in $\mathbb{R}^n$ and therefore $\vec{y} = (y_1, \ldots, y_m)$ is a point in $\mathbb{R}^m$. We treat this as a **linear transformation** of $\vec{x}$ to $\vec{y} = \vec{f}(\vec{x})$, that is, the domain is in $\mathbb{R}^n$ and the range is in $\mathbb{R}^m$ so $\vec{f} : \mathbb{R}^n \to \mathbb{R}^m$. We may ask the question, what about the rate of change of $\vec{y}$ with respect to $\vec{x}$? To do this, we take the partial derivatives as above: $\partial y_i/\partial x_j$ with $i = 1, \ldots, m$ and $j = 1, \ldots, n$. Just like above, we construct an $m \times n$ matrix of all the possibilities; this is called the **Jacobian matrix** of the transformation $\boldsymbol{D}\vec{f}(\vec{x})$:

$$\boldsymbol{D}\vec{f}(\vec{x}) = \begin{pmatrix} \partial y_1/\partial x_1 & \partial y_1/\partial x_2 & \cdots & \partial y_1/\partial x_n \\ \partial y_2/\partial x_1 & \partial y_2/\partial x_2 & \cdots & \partial y_2/\partial x_n \\ \vdots & \vdots & \ddots & \vdots \\ \partial y_m/\partial x_1 & \partial y_m/\partial x_2 & \cdots & \partial y_m/\partial x_n \end{pmatrix} \tag{32.0.14}$$

The individual partial derivatives are the elements of the matrix, which is only square if $m = n$. The Jacobian matrix of the linear transformation represents the **derivative** of the transformation $\vec{f}(\vec{x})$ (i.e it is a linear approximation of the change). The Jacobian determinant exists for transformations of the form $m : \mathbb{R}^n \to \mathbb{R}^n$ and, for two functions $f(x,y)$ and $g(x,y)$, the Jacobian determinant is written

$$\frac{\partial(f,g)}{\partial(x,y)} = \det \begin{pmatrix} \partial f/\partial x & \partial f/\partial y \\ \partial g/\partial x & \partial g/\partial y \end{pmatrix} = \frac{\partial f}{\partial x}\frac{\partial g}{\partial y} - \frac{\partial f}{\partial y}\frac{\partial g}{\partial x} = \{f,g\} \qquad (32.0.15)$$

The Jacobian determinant of a transformation also measures how much a volume, or area, changes during the transformation (e.g how much it shrinks or expands). It is for this reason that the new volume element (of a phase space, say) under a transformation $x,y \to x',y'$ is $dx'dy'$ but can be written in terms of the old volume element $dxdy$ multiplied by the Jacobian determinant of the transformation:

$$dx'dy' = \frac{\partial(x',y')}{\partial(x,y)}dxdy$$

Moving on, suppose that there is a point $x, z = x_1, \ldots, x_n, z_1, \ldots z_m$ with a set of n smooth functions $y_i = y_i(x_1, \ldots, x_n, z_1, \ldots, z_m)$, where $i = 1, \ldots, n$. If the Jacobian

$$\frac{\partial(y_1, \ldots, y_n)}{\partial(x_1, \ldots, x_n)} \neq 0 \qquad (32.0.16)$$

then, in the neighbourhood of that point, the function $y(x,z)$ can be inverted:

$$x_j = x_j(y_1, \ldots, y_n, z_1, \ldots, z_m), \quad j = 1, \ldots, n \qquad (32.0.17)$$

In this case,

$$y_i(x_1(y,z), \ldots, x_n(y,z), z_1, \ldots, z_m) = y(y,z) \qquad (32.0.18)$$

We can then use the chain rule to generate a useful expression:

$$\frac{\partial y_i}{\partial y_j} = \sum_{k=1}^{n} \frac{\partial y_i}{\partial x_k}\frac{\partial x_k}{\partial y_j} = \delta_{ij} \qquad (32.0.19)$$

This result is called the **inverse function theorem** and it expresses the partial derivative of x with respect to y in terms of the partial derivative of y with respect to x. It says that a function is invertible near a point if the Jacobian determinant is non-zero at that point. The implicit function theorem and the inverse function theorem are important results for point and canonical transformations in mechanics.

We will now consider the **implicit function theorem**. The idea behind implicit functions is that some function of multiple variables (e.g $f(x,y) = 0$) can, in some regions, have one of its variables expressed as a function of the other variables $y(x)$

such that (e.g $f(x, y(x)) = 0$) in some interval, for all x in the interval. This is, of course, providing the functions are sufficiently smooth in all the regions; we now ask, under what conditions can we do this?

We start by implicitly differentiating $f(x, y) = 0$ with respect to x at the point (a, b):

$$\frac{\partial f(x, y)}{\partial x} + \frac{\partial f(x, y)}{\partial y}\frac{dy}{dx} = 0 \qquad (32.0.20)$$

Therefore,

$$\left.\frac{dy}{dx}\right|_{x=a} = -\frac{\partial f(a, b)/\partial x}{\partial f(a, b)/\partial y}, \qquad \text{provided} \qquad \left.\frac{\partial f}{\partial y}\right|_{(a,b)} \neq 0 \qquad (32.0.21)$$

Therefore, if this condition holds, then we can indeed say that, in the neighbourhood of (a, b), there is a function $y(x)$. From elementary theory, it is known that, if $\partial f/\partial y = 0$, we have a vertical tangent line to the level curve. Let's show this for three variable functions before generalising to n and m variables.

Suppose we have $F(x, y, z) = 0$ and some point that satisfies the equation p_0 with coordinates (x_0, y_0, z_0) and some function $z(x, y)$. By implicit differentiation of $f(x, y, z) = 0$ with respect to x and y we obtain

$$\frac{\partial f}{\partial x} + \frac{\partial f}{\partial z}\frac{\partial z}{\partial x} = 0, \qquad \frac{\partial f}{\partial y} + \frac{\partial f}{\partial z}\frac{\partial z}{\partial y} = 0 \qquad (32.0.22)$$

Then, provided $\partial f/\partial z \neq 0$, both $(dz/dx)|_{x_0, y_0}$ and $(dz/dy)_{x_0, y_0}$ exist in that region. We can also perform this analysis with $x(z, y)$ or $y(x, z)$ and obtain similar conditions for the derivatives $\partial f/\partial x \neq 0$ and $\partial f/\partial y \neq 0$. The generalisation of this to a system of equations is known as the implicit function theorem. Given a system of n equations,

$$f_i(x_1, \ldots, x_m, y_1, \ldots, y_n) = 0, \qquad i = 1, \ldots, n \qquad (32.0.23)$$

with some point p_0 and the condition

$$\left.\frac{\partial(f_1, \ldots, f_n)}{\partial(y_1, \ldots, y_n)}\right|_{p_0} \neq 0 \qquad (32.0.24)$$

Then the system of equations can be solved for $y_1, \ldots, y_n$ as functions of $x_1, \ldots, x_m$ in the neighbourhood of the point. In this case, the following equations hold:

$$f_i(x_1, \ldots, x_m, y_1(x_1, \ldots, x_m), \ldots, y_n(x_1, \ldots, x_m)) = 0, \qquad i = 1, \ldots, n \qquad (32.0.25)$$

Importantly (think canonical transformations here), if we have two invertible transformations from $\mathbb{R}^n$ into $\mathbb{R}^n$ $\vec{z} = \vec{g}(\vec{y})$ and $\vec{y} = \vec{f}(\vec{x})$, then, by the chain rule,

$$\frac{\partial(z_1, \ldots, z_n)}{\partial(x_1, \ldots, x_n)} = \frac{\partial(z_1, \ldots, z_n)}{\partial(y_1, \ldots, y_n)}\frac{\partial(y_1, \ldots, y_n)}{\partial(x_1, \ldots, x_n)} \qquad (32.0.26)$$

The partial derivatives above give the rate of change of the function with respect to the distance measured in that direction (i.e x, y or z). However, what if we want to find the derivative in any direction? Is there some form of directional derivative? If we draw an arbitrary line in the xy-plane and then lift it up vertically, then, at some vertical distance, it is going to intersect the surface; the slope of the surface at this point is the **directional derivative** (see figure 32.3). The directional derivative of a point changes as we change direction in the xy-plane. If we imagine a point in the xy-plane, draw lines through this point and then extend them all up to meet the surface, then one of these lines will be the steepest.

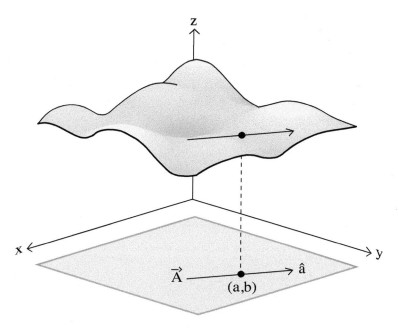

Fig. 32.3: The graphical representation of the directional derivative. A vector $\vec{A}$ in the xy-plane points in the direction of $\hat{a}$. Pulling this up to the surface, we intersect, and the derivative will point in the direction corresponding to $\hat{a}$. As we rotate the direction of $\hat{a}$ in the xy-plane, the steepness of the vector will change.

We now introduce a new operation called the **gradient**. At any sufficiently smooth point of $f(x, y)$, the gradient is defined by

$$\vec{\nabla} f(x, y) = \frac{\partial f}{\partial x}\hat{x} + \frac{\partial f}{\partial y}\hat{y} = \sum_i \frac{\partial}{\partial x_i}\hat{e}_i \qquad (32.0.27)$$

where $\hat{e}_i$ is a unit vector from the origin point $(1, 0)$ or $(0, 1)$. We treat $\vec{\nabla}$ as an operator on f, the result being the gradient. For three dimensions, the gradient operator is

$$\vec{\nabla} = \hat{x}\frac{\partial}{\partial x} + \hat{y}\frac{\partial}{\partial y} + \hat{z}\frac{\partial}{\partial z} \qquad (32.0.28)$$

The gradient at a point is a vector which is located in the xy-plane and points in the direction that the function increases most rapidly, with maximum rate of increase being the magnitude $|\vec{\nabla}f|$. The direction in which the function decreases most rapidly is exactly $-\vec{\nabla}f$, and the maximum rate of decrease is $|\vec{\nabla}f|$. In the above example, it is the directional derivative that returns the steepest line; in other words, the directional derivative is largest when it aligns with the gradient.

We should now take care to note the difference between the surface plot of a function and the hypersurface formed by its level curves. The gradient $\vec{\nabla}f(x,y)$ is normal to the level curves of $f(x,y) = c$, that is to say, it points in the same direction as the normal to a tangent line of a level curve in the xy plane (see figure 32.4). If the gradient is not perpendicular to the level curve, it will have a component in a direction along the level curve, so the function will change along the level curve and so, by definition, is not constant.

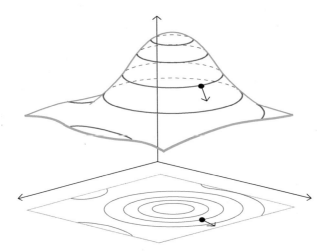

Fig. 32.4: The gradient is perpendicular to the level curves of a surface.

We will return to the gradient operator in chapter 34, as a quick example, let $\phi(x,y,z) = 3x^2 - 4xy + 2z^3x^2 + 10zy + 42$; then the gradient is

$$\vec{\nabla}\phi(x,y,z) = (6x - 4y + 4z^3x)\hat{x} + (10z - 4x)\hat{y} + (6z^2x^2 + 10y)\hat{z}$$

Let us evaluate it at $(1,4,2)$:

$$\vec{\nabla}\phi(1,4,2) = 22\hat{x} + 16\hat{y} + 64\hat{z} \qquad (32.0.29)$$

With the gradient defined, we can express the directional derivative of $f(x, y, z)$, for any old direction $\boldsymbol{A}$ with unit vector $\hat{a}$, as $D_{\boldsymbol{A}}(f)$:

$$D_{\boldsymbol{A}}(f) = \vec{\nabla} \cdot \hat{a} \tag{32.0.30}$$

Th directional derivative of the above scalar field in the direction of $\boldsymbol{A} = \hat{x} + 2\hat{y} + \hat{z}$ is

$$D_{\boldsymbol{A}}\phi = \vec{\nabla}\phi \cdot \frac{\boldsymbol{A}}{|\boldsymbol{A}|}$$

$$= 22\hat{x} + 16\hat{y} + 64\hat{z} \cdot \frac{\hat{x} + 2\hat{y} + \hat{z}}{\sqrt{1+4+1}}$$

$$= 48.3$$

Moving on, there are three situations in which **extrema** of a multivariate function can arise at a point x_0, y_0:

- when the point is a **critical point** of the function, $\vec{\nabla}f(x_0, y_0) = 0$, (if the gradient is anything other than zero, then it is sloped).
- when the point is **singular**, meaning $\vec{\nabla}f(x_0, y_0)$ doesn't exist
- when the point is a **boundary** point

If $\vec{x} = (x_0^1, x_0^2, \ldots, x_0^n)$ is a critical point of $f(\vec{x})$ and sufficiently smooth in the neighbourhood, then the symmetric **Hessian matrix** $\boldsymbol{W}$ is given by

$$\boldsymbol{W}(\vec{x}) = \begin{pmatrix} \frac{\partial^2 f}{\partial x^1 \partial x^1} & \frac{\partial^2 f}{\partial x^1 \partial x^2} & \cdots & \frac{\partial^2 f}{\partial x^1 \partial x^n} \\ \frac{\partial^2 f}{\partial x^2 \partial x^1} & \frac{\partial^2 f}{\partial x^2 \partial x^2} & \cdots & \frac{\partial^2 f}{\partial x^2 \partial x^n} \\ \vdots & \vdots & \ddots & \vdots \\ \frac{\partial^2 f}{\partial x^n \partial x^1} & \frac{\partial^2 f}{\partial x^n \partial x^2} & \cdots & \frac{\partial^2 f}{\partial x^n x^n} \end{pmatrix} \tag{32.0.31}$$

If the Hessian is positive definite, then f has a minimum at $\vec{x}$; if it is negative definite, then we have a maximum; and, if it is indefinite, then we have a saddle point! To find the extremum, we just set $df(x, y) = 0$ and, assuming that x and y are independent variables, we set each partial derivative to 0:

$$df(x, y) = \frac{\partial f}{\partial x}dx + \frac{\partial f}{\partial y}dy = 0 \tag{32.0.32}$$

Thus,

$$\frac{\partial f}{\partial x} = 0, \quad \text{and} \quad \frac{\partial f}{\partial y} = 0 \tag{32.0.33}$$

which can then be used to solve the problem. The key part is assuming x and y are independent, so each term vanishes independently.

Finding the extrema of functions subject to a **constraint** complicates the problem, since we are restricting the domain of the function. A constraint is a relationship between the function's variables which means they are no longer independent but are linked through the constraint equation. There are three main ways to tackle constrained extrema problems: **elimination** of the constraint function, **implicit differentiation** and the method of **Lagrange multipliers**.

We must find the extrema of a function $f(x,y)$ where both x and y are related through another function $\phi(x,y) = c$, where c is a constant. We therefore know that, at extrema, $df = 0$ and, since ϕ is constant, $d\phi = 0$:

$$df = \frac{\partial f}{\partial x}dx + \frac{\partial f}{\partial y}dy = 0, \qquad d\phi = \frac{\partial \phi}{\partial x}dx + \frac{\partial \phi}{\partial y}dy = 0 \qquad (32.0.34)$$

We multiply $d\phi$ by some multiplier λ and add it to (or subtract it from, as we do in mechanics) df:

$$df + \lambda(d\phi) = \left(\frac{\partial f}{\partial x} + \lambda\frac{\partial \phi}{\partial x}\right)dx + \left(\frac{\partial f}{\partial y} + \lambda\frac{\partial \phi}{\partial y}\right)dy = 0 \qquad (32.0.35)$$

We now choose λ such that one of the brackets is 0 (e.g $\lambda = -(\partial f/\partial y)/(\partial \phi/\partial y)$) then,

$$\frac{\partial f}{\partial y} + \lambda\frac{\partial \phi}{\partial y} = 0 \qquad (32.0.36)$$

Therefore,

$$\frac{\partial f}{\partial x} + \lambda\frac{\partial \phi}{\partial x} = 0 \qquad (32.0.37)$$

We can use these two equations together with the constraint equation to solve for x, y and λ. This method is entirely equivalent to constructing a function $F(x,y) = f(x,y) + \lambda\phi(x,y)$; equivalently some authors define $F(x,y) = f(x,y) + \lambda(\phi(x,y) - c)$, explicitly putting in the constant; but, in most cases, we take $c = 0$, so it doesn't matter. Extending this to higher dimensions is straightforward; we will consider $f(x,y,z)$ subject to $\phi(x,y,z) = c$ and, hence, $d\phi = 0$. We set $df = 0$ to search for extrema and write

$$df = \frac{\partial f}{\partial x}dx + \frac{\partial f}{\partial y}dy + \frac{\partial f}{\partial z}dz = 0 \qquad (32.0.38)$$

$$d\phi = \frac{\partial \phi}{\partial x}dx + \frac{\partial \phi}{\partial y}dy + \frac{\partial \phi}{\partial z}dz = 0 \qquad (32.0.39)$$

We then construct $F(x,y,\lambda) = f(x,y) + \lambda\phi(x,y)$

$$dF = df + \lambda(d\phi) = \left(\frac{\partial f}{\partial x} + \lambda\frac{\partial \phi}{\partial x}\right)dx + \left(\frac{\partial f}{\partial y} + \lambda\frac{\partial \phi}{\partial y}\right)dy + \left(\frac{\partial f}{\partial z} + \lambda\frac{\partial \phi}{\partial z}\right)dz \quad (32.0.40)$$

We select λ such that

$$\frac{\partial f}{\partial z} + \lambda\frac{\partial \phi}{\partial z} = 0 \qquad (32.0.41)$$

If $dy = 0$, then

$$\frac{\partial f}{\partial x} + \lambda\frac{\partial \phi}{\partial x} = 0 \qquad (32.0.42)$$

while, for $dx = 0$,

$$\frac{\partial f}{\partial y} + \lambda\frac{\partial \phi}{\partial y} = 0 \qquad (32.0.43)$$

Then, to find the critical points of $f(x, y, z)$ subject to $\phi = c$, we set the three partial derivatives of F to 0 and solve the results for x, y, z and λ. While the maxima we find may not be the true unconstrained maxima of $f(x, y, z)$, it is the maxima in the direction of any neighbouring point where $\phi = c$ (see figure 32.5). In this case, the gradients $\vec{\nabla}f$ and $\vec{\nabla}\phi$ are parallel and we assume $\vec{\nabla}\phi \neq 0$; therefore, there will exist some λ such that $\vec{\nabla}f = -\lambda\vec{\nabla}\phi$ (the minus sign would be a plus sign if we subtract the multiplier instead). As demonstrated in the following example, the extrema of the function occur when the constraint curve is tangent to a level curve of the function, or, in other words, at the point where their gradients align.

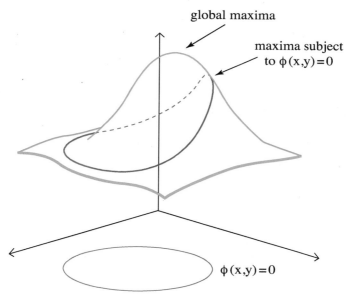

Fig. 32.5: The constrained maxima and the unconstrained maxima may give quite dissimilar results upon optimisation.

Exercise 32.1 We consider the integral $I[x]$ as a functional of two functions $x_1(t)$ and $x_2(t)$ that maps to the reals:

$$I[x] = \int_1^2 f(x_i(t), \dot{x}_i(t), t)dt \tag{32.0.44}$$

The **functional derivative** (see 36) of I with respect to x_i gives the rate of change of I with respect to a change in x_i and is given by

$$\frac{\delta I}{\delta x_i} \equiv \frac{\partial f}{\partial x_i} - \frac{d}{dt}\left(\frac{\partial f}{\partial \dot{x}_i}\right), \quad i = 1, 2 \tag{32.0.45}$$

We construct a gradient similar to the vector gradient, $\vec{\nabla} I[x]$, that points in the direction normal to the level curves of I, where the rate of change of I in the space of $x_1(t)$ and $x_2(t)$ is largest. We construct the same for the constraint functional $\phi[x_1, x_2] = 0$, and we say that $\vec{\nabla} \phi[x]$ points normal to the constraint curve. Both functions x_1 and x_2 are chosen such that $\vec{\nabla} \phi[x]$ is parallel to $\vec{\nabla} I[x]$; then up to some magnitude factored by λ, we can write $\vec{\nabla} I[x] = \lambda \vec{\nabla} \phi[x]$. If we explicitly write the gradients as

$$\vec{\nabla}[x] := \left[\frac{\delta I}{\delta x_1}, \frac{\delta I}{\delta x_2}\right], \qquad \vec{\nabla}\phi[x] := \left[\frac{\delta \phi}{\delta x_1}, \frac{\delta \phi}{\delta x_2}\right] = \left[\frac{\partial \phi}{\partial x_1}, \frac{\partial \phi}{\partial x_2}\right] \tag{32.0.46}$$

we find that

$$\frac{\partial f}{\partial x_i} - \frac{d}{dt}\left(\frac{\partial f}{\partial \dot{x}_i}\right) = \lambda \frac{\partial \phi}{\partial x_i} \tag{32.0.47}$$

The solutions x_1, x_2 at that point are both extrema to $I[x]$ and consistent with the constraint. In other words, $\vec{\nabla} I[x]$ is perpendicular to the surface of $\phi = 0$. If we have m constraints, then the intersection of their hypersurfaces will be an $(N - m)$-dimensional surface where the motion is allowed. Then $\vec{\nabla} I[x]$ is some linear combination of the m normal $\vec{\nabla} \phi$ gradients: $\vec{\nabla} I[x] = \sum \lambda \vec{\nabla} \phi[x]$.

33

Legendre Transforms

The Legendre transform is a mathematical tool that we can us to convert the variables of a function through the methods of partial differentiation in a one-to-one fashion. Developed by Adrien-Marie Legendre in the 19th century, it is central to converting between the Lagrange and Hamiltonian frameworks or between action principles, generating functions and thermodynamic potentials. The general result of the Legendre transform is that a smooth function $f(x, y)$ is transformed into another function of $g(x, p)$, with $p = \partial f / \partial y$ being the **conjugate variable** to y, and the function g defined as

$$g(x, p) = f(x, y) - py \tag{33.0.1}$$

An important condition for this is that $p = \partial f / \partial y$ exists and is non-zero; thereby, the implicit function theorem guarantees us that the inverse function exists. We show this by considering the total differential of a function $f(x, y)$,

$$df = \frac{\partial f}{\partial x} dx + \frac{\partial f}{\partial y} dy = u dx + p dy \tag{33.0.2}$$

where we have re-symbolised the coloured terms for ease. Subtract from df the quantity $d(py)$:

$$d(f - py) = u dx + p dy - p dy - y dp = u dx - y dp \tag{33.0.3}$$

We now define the function $g := f - py$:

$$dg = u dx - y dp \tag{33.0.4}$$

Observing this function and recalling the rules of total differentiation, we deduce that $g(x, p)$ and we see that we have replaced $y \to p$. There is no reason why we couldn't perform the Legendre transform with respect to other variable (e.g $h(u, y)$) or even both at the same time (e.g. $k(up)$). This means that there are four ways that we can express two variables, using the idea of *conjugate pairs*, and it just depends on what differential quantity you subtract. Variables that we do not consider in the transformation are

Lagrangian & Hamiltonian Dynamics. Peter Mann, Oxford University Press (2018).
© Peter Mann. DOI: 10.1093/oso/9780198822370.001.0001

called passive variables, while the important ones are the active variables. Taking the total derivative of $g(x, p)$ we obtain

$$dg = \frac{\partial g}{\partial x}dx + \frac{\partial g}{\partial p}dp \qquad (33.0.5)$$

Notice, however, that we already have an expression for dg just above in equation 33.0.4, equating these we obtain

$$udx - ydp = \frac{\partial g}{\partial x}dx + \frac{\partial g}{\partial p}dp \qquad (33.0.6)$$

Collecting terms in like infinitesimals:

$$(u - \frac{\partial g}{\partial x})dx - (y + \frac{\partial g}{\partial p})dp = 0 \qquad (33.0.7)$$

There is no good reason why varying one independent variable and then varying another should cancel out in a trivial case. Therefore, in order for equality to hold in the general case, we must realise that the terms inside each bracket are exactly 0.

$$u = \frac{\partial g}{\partial x}, \qquad y = -\frac{\partial g}{\partial p} \qquad (33.0.8)$$

The idea of comparing the differentials of a function is an important one that we use repeatedly in the text.

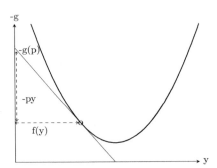

Fig. 33.1: The Legendre transform.

The Legendre transform also has a graphical interpretation, as in figure 33.1 for a graph of $g(x, p)$ versus y on the horizontal axis. A function is convex when the Hessian with y is everywhere positive definite. For a convex function $f(x, y)$, the Legendre transform takes a tangent at a point and looks for where it intersects the vertical axis, giving us the unique value of $g(x, p)$. This ensures that it is a one-to-one transformation. The Legendre transform is usually taken as the negative of the vertical axis value where the tangent cuts and, otherwise, $g = f - py$ as above, but it's just convention. If the function is non-convex, we must extend the definition of the supremum limit over the domain of definition to select the lowest function value of all possible slopes when scanned over the domain of y:

$$g(x, p) = \sup\{py - f\} \tag{33.0.9}$$

The supremum is similar to a maximum yet allows generalisations to non-convex functions to be made.

34

Vector Calculus

In this chapter we introduce only the necessary vector calculus required to understand some of the operations we perform in the text and perhaps support self-learning in more advanced topics. This analysis will not be definitive, as a full exposition will take up too much space, so we will not linger too long on this topic. We will consider the axioms of vector algebra, vector multiplication and vector differentiation, after which we close our foray by tackling the gradient, divergence, curl and other elements of vector integration. In this chapter, we denote vectors in bold font, or in arrowed notation where appropriate, and scalars in normal text.

A **vector** is a quantity characterised by both magnitude and direction, while a **scalar** is only a magnitude. Imagine a vector v in $\mathbb{R}^3$, as is our everyday experience. We known that any vector can be described by specifying its **components** in three directions: x, y and z, relative to a coordinate frame. There is something special about vectors that point exactly down the axes, however, as they only have components in that one direction. For instance, the vector pointing down the x axis can be expressed by x-axis components only: $(x, 0, 0)$. A mathematical way of saying this is that vectors that point down an axis are **linearly independent** of the other axes; we don't need any other vector axis to specify them. Any vector that is not pointing down the axis can be written as a **linear combination** of vectors pointing down the axis.

To illustrate this, suppose we had a coordinate frame with two axes labelled *north* and *west*, respectively. For an arbitrary vector pointing in some direction between the two axes, we can say that it is pointing 3 units north and 2 units west. We express it as a linear combination of vectors along the axes. The vector in the arbitrary direction is not linearly independent of the two axes, since we require both to describe it. The vectors that comprise a linearly independent set are called **basis vectors** and are denoted e_i. Any vector can be written as a linear combination of basis vectors, each weighted by a coefficient[1].

$$A = \sum_i = a_i e_i \qquad (34.0.1)$$

[1]Later, we raise the index of the component, since the basis is a subscript. Basis and component indices are always opposites, but we won't worry about this for now.

Lagrangian & Hamiltonian Dynamics. Peter Mann, Oxford University Press (2018).
© Peter Mann. DOI: 10.1093/oso/9780198822370.001.0001

If we wish to describe basis vectors themselves, then the weighting coefficients are set to zero in every direction other than the direction of that axis. This idea is quite obvious from a practical point of view; in order to describe a generic vector, you need something down each axis and a way of specifying the contribution of each of them; therefore, there will be as many basis vectors as there are dimensions for the space we are describing.

A basis may not be unique for a given space and there may be more than one linearly independent set that spans the space. Another way to think of it that is perhaps more intuitive for chemists is that, in order to describe molecular orbitals, we require a basis set of atomic orbitals; they provide a basis for any combination we may make. Linear independence among a set of vectors is essentially a fancy way of saying that we can't write any member as a linear combination of others in the set. You can't construct a west-pointing vector from any linear combination of north-pointing vectors no matter how many you add up. This is expressed as

$$\sum_i c_i \boldsymbol{A}_i = 0 \tag{34.0.2}$$

If we **normalise** the basis vectors in each dimension so they all have a length of 1, then we say they are **unit vectors** $\hat{\boldsymbol{u}}$. We can actually normalise any vector but it is common to normalise the basis set. This involves dividing a vector $\boldsymbol{u}$ by its length $\|\boldsymbol{u}\|$:

$$\hat{\boldsymbol{u}} = \frac{\boldsymbol{u}}{\|\boldsymbol{u}\|} \tag{34.0.3}$$

The unit vectors of cartesian space are denoted $\hat{\boldsymbol{x}}$, $\hat{\boldsymbol{y}}$ and $\hat{\boldsymbol{z}}$ or sometimes by $\hat{\boldsymbol{i}}$, $\hat{\boldsymbol{j}}$ and $\hat{\boldsymbol{k}}$ or just $\boldsymbol{i}$, $\boldsymbol{j}$, and $\boldsymbol{k}$, but this may cause confusion with indices, so the former are preferred in this text. They are simply normed basis vectors along each axis and we may use these to describe any vector $\boldsymbol{r}(x, y, z)$ in $\mathbb{R}^3$:

$$\boldsymbol{r} = x\hat{\boldsymbol{x}} + y\hat{\boldsymbol{y}} + z\hat{\boldsymbol{z}} \tag{34.0.4}$$

Note how this is nothing more than a linear combination of components and basis vectors! Vectors can be added together to give a resultant vector and they can be multiplied by a scalar. There are certain axioms for vector addition and scalar multiplication:

- $(\boldsymbol{A} + \boldsymbol{B}) + \boldsymbol{C} = (\boldsymbol{A} + \boldsymbol{B}) + \boldsymbol{C}$ (associative addition)
- $\boldsymbol{A} + \boldsymbol{B} = \boldsymbol{B} + \boldsymbol{A}$ (commutative addition)
- $a(\boldsymbol{A} + \boldsymbol{B}) = a\boldsymbol{A} + a\boldsymbol{A}$ (Distributive scalar multiplication)
- $a(b\boldsymbol{A}) = (ab)\boldsymbol{A}$ (associative scalar multiplication)

Vector addition and scalar multiplication follows on from the axioms

$$\begin{aligned} \boldsymbol{A} + \boldsymbol{B} &= (a_1\hat{\boldsymbol{x}} + a_2\hat{\boldsymbol{y}} + a_3\hat{\boldsymbol{z}}) + (b_1\hat{\boldsymbol{x}} + b_2\hat{\boldsymbol{y}} + b_3\hat{\boldsymbol{z}}) \\ &= (a_1 + b_1)\hat{\boldsymbol{x}} + (a_2 + b_2)\hat{\boldsymbol{y}} + (a_3 + b_3)\hat{\boldsymbol{z}} \end{aligned} \tag{34.0.5}$$

$$bA = b(a_1\hat{x} + a_2\hat{y} + a_3\hat{z})$$
$$= (ba_1)\hat{x} + (ba_2)\hat{y} + (ba_3)\hat{z} \tag{34.0.6}$$

A **scalar field** ϕ assigns a number $\mathbb{R}$ to each point in the region of space $\phi(x, y, z)$ over which it is defined. A **vector field** V assigns a vector to each point in the region of space over which it is defined: $V(x, y, z)$. The **dot product** of two vectors A and B is denoted $A \cdot B$; it is the product of the magnitudes of the two vectors and the cosine of the angle between them, and it is a scalar quantity:

$$A \cdot B = \|A\|\|B\| \cos\theta, \qquad 0 \leq \theta \leq \pi \tag{34.0.7}$$

The dot product is also called the **scalar product**; this emphasises the fact that the dot product returns a scalar not another vector. If the dot product is zero for two non-zero vectors, then they are perpendicular to each other. The dot product is commutative $(A \cdot B = B \cdot A)$, it is distributive $(A \cdot (B+C) = A \cdot B + A \cdot C)$ and scalar multiplication is defined: $a(A \cdot B) = (aA) \cdot B = A \cdot (aB)$. Importantly, we have the orthonormality relations $\hat{x} \cdot \hat{x} = \hat{y} \cdot \hat{y} = \hat{z} \cdot \hat{z} = 1$, while the dot product with any other unit vector is $\hat{x} \cdot \hat{y} = \hat{y} \cdot \hat{z} = \hat{z} \cdot \hat{x} = 0$:

$$A \cdot B = (a_1\hat{x} + a_2\hat{y} + a_3\hat{z}) \cdot (b_1\hat{x} + b_2\hat{y} + b_3\hat{z}) = a_1b_1 + a_2b_2 + a_3b_3 \tag{34.0.8}$$

We can use the dot product to define the Euclidean length of a vector $\|A\| = \sqrt{A \cdot A}$; this follows from inserting the value $\cos 0 = 1$ in equation 34.0.7. In addition, we can also write **orthonormality relations** between a set of linearly independent vectors:

$$A \cdot B = \pm\delta_{ij} \tag{34.0.9}$$

That is to say, the dot product between any of the vectors in the set is zero unless we take the dot product of the same vector. The set, for example, could be the basis set.

The **cross product** of two vectors A and B is another vector, denoted $A \times B$. The magnitude of the cross product is the product of the magnitudes of $\|A\|$, and $\|B\|$ and the sine of the angle between them:

$$A \times B = \|A\|\|B\| \sin\theta u \tag{34.0.10}$$

where u is a unit vector in the direction perpendicular to the plane of A and B. The cross product is antisymmetric $(A \times B = -(B \times A))$ and distributive $(A \times (B+C) = A \times B + A \times C)$, with scalar multiplication $(a(A \times B) = (aA) \times B = A \times (aB))$. Importantly, $\hat{x} \times \hat{x} = \hat{y} \times \hat{y} = \hat{z} \times \hat{z} = 0$; $\hat{x} \times \hat{y} = \hat{z}$; $\hat{y} \times \hat{z} = \hat{x}$; and $\hat{z} \times \hat{x} = \hat{y}$. It should be noticed that the cross product is actually the determinant of the following square matrix for two vectors $A = a_1\hat{x} + a_2\hat{y} + a_3\hat{z}$ and $B = b_1\hat{x} + b_2\hat{y} + b_3\hat{z}$:

$$\boldsymbol{A} \times \boldsymbol{B} = \det \begin{pmatrix} \hat{\boldsymbol{x}} & \hat{\boldsymbol{x}} & \hat{\boldsymbol{x}} \\ a_1 & a_2 & a_3 \\ b_1 & b_2 & b_3 \end{pmatrix} = \det \begin{pmatrix} a_2 & a_3 \\ b_2 & b_3 \end{pmatrix} \hat{\boldsymbol{x}} - \det \begin{pmatrix} a_1 & a_3 \\ b_1 & b_3 \end{pmatrix} \hat{\boldsymbol{y}} + \det \begin{pmatrix} a_1 & a_2 \\ b_1 & b_2 \end{pmatrix} \hat{\boldsymbol{z}}$$

We can also combine both the dot and cross product into a **triple product** between three vectors, for instance $\boldsymbol{A} \times (\boldsymbol{B} \times \boldsymbol{C})$ or $\boldsymbol{A} \cdot (\boldsymbol{B} \times \boldsymbol{C})$.

As previously discussed, we can differentiate any sufficiently smooth vectors and scalars that depend on a smooth scalar variable t, with the Newtonian quotient being a vector in the tangent line:

$$\frac{d\boldsymbol{A}(t)}{dt} = \lim_{\Delta t \to 0} \frac{\boldsymbol{A}(t + \Delta t) - \boldsymbol{A}(t)}{\Delta t} \tag{34.0.11}$$

Exercise 34.1 For a **position vector** $r(t)$ or $\vec{r}(t)$ in $\mathbb{R}^3$ we can write $\vec{r}(t) = x(t)\hat{\boldsymbol{x}} + y(t)\hat{\boldsymbol{y}} + z(t)\hat{\boldsymbol{z}}$, and the derivative as:

$$\frac{d\vec{r}}{dt} = \frac{dx}{dt}\hat{\boldsymbol{x}} + \frac{dy}{dt}\hat{\boldsymbol{y}} + \frac{dz}{dt}\hat{\boldsymbol{z}} \tag{34.0.12}$$

This follows on from the Newtonian quotient above:

$$\frac{d\vec{r}}{dt} = \lim_{\Delta t \to 0} \frac{[x(t + \Delta t)\hat{\boldsymbol{x}} + y(t + \Delta t)\hat{\boldsymbol{y}} + z(t + \Delta t)\hat{\boldsymbol{z}}] - [x(t)\hat{\boldsymbol{x}} + y(t)\hat{\boldsymbol{y}} + z(t)\hat{\boldsymbol{z}}]}{\Delta t}$$

$$= \lim_{\Delta t \to 0} \frac{x(t + \Delta t) - x(t)}{\Delta t}\hat{\boldsymbol{x}} + \frac{y(t + \Delta t) - y(t)}{\Delta t}\hat{\boldsymbol{y}} + \frac{z(t + \Delta t) - z(t)}{\Delta t}\hat{\boldsymbol{z}} \tag{34.0.13}$$

For the most part, we assume that this result is so familiar that we end up writing just the components, as follows:

$$\vec{v} = \frac{d\vec{r}}{dt} = \left(\frac{dx}{dt}, \frac{dy}{dt}, \frac{dz}{dt} \right) := (\dot{x}, \dot{y}, \dot{z}) = (\dot{x}^i) \quad i = 1, 2, 3$$

For partial derivatives, the dot and cross product, obey the product rule:

$$\frac{\partial}{\partial x}(\boldsymbol{A} \cdot \boldsymbol{B}) = \boldsymbol{A} \cdot \frac{\partial \boldsymbol{B}}{\partial x} + \frac{\partial \boldsymbol{A}}{\partial x} \cdot \boldsymbol{B}, \qquad \frac{\partial}{\partial x}(\boldsymbol{A} \times \boldsymbol{B}) = \boldsymbol{A} \times \frac{\partial \boldsymbol{B}}{\partial x} + \frac{\partial \boldsymbol{A}}{\partial x} \times \boldsymbol{B}$$

The gradient a scalar field is written

$$\vec{\nabla}\phi = \left(\frac{\partial}{\partial x}\hat{\boldsymbol{x}} + \frac{\partial}{\partial y}\hat{\boldsymbol{y}} + \frac{\partial}{\partial z}\hat{\boldsymbol{z}} \right)\phi = \frac{\partial \phi}{\partial x}\hat{\boldsymbol{x}} + \frac{\partial \phi}{\partial y}\hat{\boldsymbol{y}} + \frac{\partial \phi}{\partial z}\hat{\boldsymbol{z}}$$

The **divergence** of a vector field sufficiently smooth at each point is the dot product (hence returning a scalar) of the gradient operator and the vector field:

$$\vec{\nabla} \cdot V = \left(\frac{\partial}{\partial x} \hat{x} + \frac{\partial}{\partial y} \hat{y} + \frac{\partial}{\partial z} \hat{z} \right) \cdot (v_1 \hat{x} + v_2 \hat{y} + v_3 \hat{z}) = \frac{\partial v_1}{\partial x} + \frac{\partial v_2}{\partial y} + \frac{\partial v_3}{\partial z} \qquad (34.0.14)$$

The **curl** of a vector field V is the cross product of the gradient operator and the vector field:

$$\vec{\nabla} \times V = \left(\frac{\partial}{\partial x} \hat{x} + \frac{\partial}{\partial y} \hat{y} + \frac{\partial}{\partial z} \hat{z} \right) \times (v_1 \hat{x} + v_2 \hat{y} + v_3 \hat{z})$$

$$= \left(\frac{\partial v_3}{\partial y} - \frac{\partial v_2}{\partial z} \right) \hat{x} + \left(\frac{\partial v_1}{\partial z} - \frac{\partial v_3}{\partial x} \right) \hat{y} + \left(\frac{\partial v_2}{\partial x} - \frac{\partial v_1}{\partial y} \right) \hat{z} \qquad (34.0.15)$$

This is conveniently expressed as the determinant of the following 3×3 matrix:

$$\vec{\nabla} \times \boldsymbol{V} = \det \begin{pmatrix} \hat{x} & \hat{y} & \hat{z} \\ \partial_x & \partial_y & \partial_z \\ v_1 & v_2 & v_3 \end{pmatrix} \qquad (34.0.16)$$

Exercise 34.2 In the text we make use of the following three relations (i) $\vec{\nabla} \times (\vec{\nabla}\phi) = 0$, when defining the force as the negative of a potential gradient, and (ii) $\vec{\nabla} \cdot (\vec{\nabla} \times \boldsymbol{A}) = 0$ and (iii) $\vec{\nabla} \times (\vec{\nabla} \times \boldsymbol{A}) = \vec{\nabla}(\vec{\nabla} \cdot \boldsymbol{A}) - \nabla^2 \boldsymbol{A}$, both in our discussion of electrodynamics.

The first is the most straightforward of the three relations and only assumes that Clairaut's theorem holds:

$$\vec{\nabla} \times (\vec{\nabla}\phi) = \left(\frac{\partial}{\partial x} \hat{x} + \frac{\partial}{\partial y} \hat{y} + \frac{\partial}{\partial z} \hat{z} \right) \times \left(\frac{\partial \phi}{\partial x} \hat{x} + \frac{\partial \phi}{\partial y} \hat{y} + \frac{\partial \phi}{\partial z} \hat{z} \right)$$

$$= \left(\frac{\partial^2 \phi}{\partial y \partial z} - \frac{\partial^2 \phi}{\partial z \partial y} \right) \hat{x} + \left(\frac{\partial^2 \phi}{\partial z \partial x} - \frac{\partial^2 \phi}{\partial x \partial z} \right) \hat{y} + \left(\frac{\partial^2 \phi}{\partial x \partial y} - \frac{\partial^2 \phi}{\partial y \partial x} \right) \hat{z}$$

$$= 0$$

The second relation is proved in much the same way:

$$\vec{\nabla} \cdot (\vec{\nabla} \times \boldsymbol{A}) = \vec{\nabla} \cdot \left[\left(\frac{\partial a_3}{\partial y} - \frac{\partial a_2}{\partial z} \right) \hat{x} + \left(\frac{\partial a_1}{\partial z} - \frac{\partial a_3}{\partial x} \right) \hat{y} + \left(\frac{\partial a_2}{\partial x} - \frac{\partial a_1}{\partial y} \right) \hat{z} \right]$$

$$= \frac{\partial}{\partial x} \left(\frac{\partial a_3}{\partial y} - \frac{\partial a_2}{\partial z} \right) + \frac{\partial}{\partial y} \left(\frac{\partial a_1}{\partial z} - \frac{\partial a_3}{\partial x} \right) + \frac{\partial}{\partial z} \left(\frac{\partial a_2}{\partial x} - \frac{\partial a_1}{\partial y} \right)$$

$$= \frac{\partial^2 a_3}{\partial x \partial y} - \frac{\partial^2 a_2}{\partial x \partial z} + \frac{\partial^2 a_1}{\partial y \partial z} - \frac{\partial^2 a_3}{\partial y \partial x} + \frac{\partial^2 a_2}{\partial z \partial x} - \frac{\partial^2 a_1}{\partial z \partial y}$$

$$= 0 \qquad (34.0.17)$$

The third relation is a little more involved:

$$\vec{\nabla} \times (\vec{\nabla} \times \boldsymbol{A}) = \vec{\nabla} \times \left[\left(\frac{\partial a_3}{\partial y} - \frac{\partial a_2}{\partial z} \right) \hat{\boldsymbol{x}} + \left(\frac{\partial a_1}{\partial z} - \frac{\partial a_3}{\partial x} \right) \hat{\boldsymbol{y}} + \left(\frac{\partial a_2}{\partial x} - \frac{\partial a_1}{\partial y} \right) \hat{\boldsymbol{z}} \right]$$

$$= \left[\frac{\partial}{\partial y} \left(\frac{\partial a_2}{\partial x} - \frac{\partial a_1}{\partial y} \right) - \frac{\partial}{\partial z} \left(\frac{\partial a_1}{\partial z} - \frac{\partial a_3}{\partial x} \right) \right] \hat{\boldsymbol{x}} + \left[\frac{\partial}{\partial z} \left(\frac{\partial a_3}{\partial y} - \frac{\partial a_2}{\partial z} \right) \right.$$

$$\left. - \frac{\partial}{\partial x} \left(\frac{\partial a_2}{\partial x} - \frac{\partial a_1}{\partial y} \right) \right] \hat{\boldsymbol{y}} + \left[\frac{\partial}{\partial x} \left(\frac{\partial a_1}{\partial z} - \frac{\partial a_3}{\partial x} \right) - \frac{\partial}{\partial y} \left(\frac{\partial a_3}{\partial y} - \frac{\partial a_2}{\partial z} \right) \right] \hat{\boldsymbol{z}}$$

We then collect mixed and partial derivatives:

$$= \left(-\frac{\partial^2 a_1}{\partial y^2} - \frac{\partial^2 a_1}{\partial z^2} \right) \hat{\boldsymbol{x}} + \left(-\frac{\partial^2 a_2}{\partial z^2} - \frac{\partial^2 a_2}{\partial x^2} \right) \hat{\boldsymbol{y}} + \left(-\frac{\partial^2 a_3}{\partial x^2} - \frac{\partial^2 a_3}{\partial y^2} \right) \hat{\boldsymbol{z}}$$

$$+ \left(\frac{\partial^2 a_2}{\partial y \partial x} + \frac{\partial^2 a_3}{\partial z \partial x} \right) \hat{\boldsymbol{x}} + \left(\frac{\partial^2 a_3}{\partial z \partial y} + \frac{\partial^2 a_1}{\partial x \partial y} \right) \hat{\boldsymbol{y}} + \left(\frac{\partial^2 a_1}{\partial x \partial z} + \frac{\partial a_2}{\partial y \partial z} \right) \hat{\boldsymbol{z}}$$

Next fill in the missing derivatives by adding and subtracting the relevant relations:

$$= \left(-\frac{\partial^2 a_1}{\partial x^2} - \frac{\partial^2 a_1}{\partial y^2} - \frac{\partial^2 a_1}{\partial z^2} \right) \hat{\boldsymbol{x}} + \left(-\frac{\partial^2 a_2}{\partial z^2} - \frac{\partial^2 a_2}{\partial x^2} - \frac{\partial^2 a_2}{\partial y^2} \right) \hat{\boldsymbol{y}}$$

$$+ \left(-\frac{\partial^2 a_3}{\partial x^2} - \frac{\partial^2 a_3}{\partial y^2} - \frac{\partial^2 a_3}{\partial z^2} \right) \hat{\boldsymbol{z}} + \left(\frac{\partial^2 a_2}{\partial y \partial x} + \frac{\partial^2 a_3}{\partial z \partial x} + \frac{\partial^2 a_1}{\partial x^2} \right) \hat{\boldsymbol{x}}$$

$$+ \left(\frac{\partial^2 a_3}{\partial z \partial y} + \frac{\partial^2 a_1}{\partial x \partial y} + \frac{\partial^2 a_2}{\partial y^2} \right) \hat{\boldsymbol{y}} + \left(\frac{\partial^2 a_1}{\partial x \partial z} + \frac{\partial a_2}{\partial y \partial z} + \frac{\partial^2 a_3}{\partial z^2} \right) \hat{\boldsymbol{z}}$$

$$= -\left(\frac{\partial^2}{\partial x^2} + \frac{\partial^2}{\partial y^2} + \frac{\partial^2}{\partial z^2} \right) (a_1 \hat{\boldsymbol{x}} + a_2 \hat{\boldsymbol{y}} + a_3 \hat{\boldsymbol{z}}) + \frac{\partial}{\partial x} \left(\frac{\partial a_1}{\partial x} + \frac{\partial a_2}{\partial y} + \frac{\partial a_3}{\partial z} \right) \hat{\boldsymbol{x}}$$

$$+ \frac{\partial}{\partial y} \left(\frac{\partial a_1}{\partial x} + \frac{\partial a_2}{\partial y} + \frac{\partial a_3}{\partial z} \right) \hat{\boldsymbol{y}} + \frac{\partial}{\partial z} \left(\frac{\partial a_1}{\partial x} + \frac{\partial a_2}{\partial y} + \frac{\partial a_2}{\partial z} \right) \hat{\boldsymbol{z}}$$

$$= -\vec{\nabla}^2 \boldsymbol{A} + \vec{\nabla} \left(\frac{\partial a_1}{\partial x} + \frac{\partial a_2}{\partial y} + \frac{\partial a_3}{\partial z} \right)$$

$$= -\vec{\nabla}^2 \boldsymbol{A} + \vec{\nabla} (\vec{\nabla} \cdot \boldsymbol{A}) \tag{34.0.18}$$

We will now look at the fundamentals of vector integration and culminate our discussion with the divergence theorem and Stokes' theorem. If we have a vector $\boldsymbol{A}(t)$ in $\mathbb{R}^3$ that is a function of a scalar variable t, then the **definite integral** is written between two limits $t = 1$ and $t = 2$:

$$\int_1^2 \boldsymbol{A}(t) dt = \hat{\boldsymbol{x}} \int_1^2 a_1(t) dt + \hat{\boldsymbol{y}} \int_1^2 a_2(t) dt + \hat{\boldsymbol{z}} \int_1^2 a_3(t) dt \tag{34.0.19}$$

The integral of $y = f(x)$ can be interpreted as the area in the xy plane bounded by the graph of the function (area under the graph). There are different ways to add up this area and the three main methods are the (i) **Riemann integral**, (ii) **Darboux integral** and the (iii) **Lebesgue integral**. The first two involve the sum of tiny vertical rectangles of base $\Delta x_i = x_i + dx_i$ (they may or may not be of regular width) times the height of the function at that point, while the latter involves the sum of tiny horizontal rectangles of width $\Delta y_i = y_i + dy_i$.

Exercise 34.3 Try integrating the vector $\boldsymbol{A}(t) = 2t^4\hat{\boldsymbol{x}} + t^3\hat{\boldsymbol{y}} - 3\hat{\boldsymbol{z}}$ and then evaluating it between 2 and 1:

$$\int \boldsymbol{A}(t)dt = \hat{\boldsymbol{x}}\int 2t^3\,dt + \hat{\boldsymbol{y}}\int t^2\,dt + \hat{\boldsymbol{z}}\int -3dt$$

$$= \left(\frac{t^4}{2} + c_1\right)\hat{\boldsymbol{x}} + \left(\frac{t^3}{3} + c_2\right)\hat{\boldsymbol{y}} + \left(-3t + c_3\right)\hat{\boldsymbol{z}}$$

$$= \frac{t^4}{2}\hat{\boldsymbol{x}} + \frac{t^3}{3}\hat{\boldsymbol{y}} - 3t\hat{\boldsymbol{z}} + \boldsymbol{c} \qquad (34.0.20)$$

with $\boldsymbol{c} = c_1\hat{\boldsymbol{x}} + c_2\hat{\boldsymbol{y}} + c_3\hat{\boldsymbol{z}}$ being a constant vector. Evaluating this at 2 and 1 gives

$$\int_1^2 \boldsymbol{A}(t)dt = (15/2)\hat{\boldsymbol{x}} + (7/3)\hat{\boldsymbol{y}} - 3\hat{\boldsymbol{z}}$$

A **contour integral** is an integral of a function that is evaluated along a curve $\mathcal{C}$ between two points 1 and 2, and hence gives the area under a curve over a surface. To illustrate the contour integral consider a surface in three dimensions as in figure 34.1.

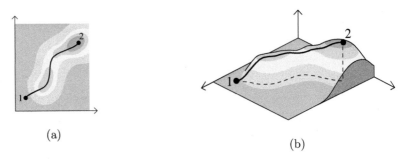

(a)

(b)

Fig. 34.1: A two-dimensional contour map (a) with a function between two points, and the same map realised in three dimensions (b).

The aim of the contour integral is to evaluate the area underneath this curve which is the blue region. We can imagine the curve as a ruffled curtain and in figure 34.2 we are straightening it out.

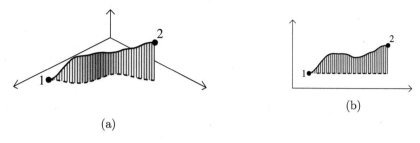

(a)

(b)

Fig. 34.2: The area under the curve is identified in (a) and then evaluated in (b).

In this case, we have little rectangles composed of tiny shifts in arc length of the curve ds, multiplied by the height of the function $f(x,y)$, which when added together become

$$\int_1^2 f(x,y)ds = \int_C f(x,y)ds \tag{34.0.21}$$

Extending this to the case where the integration volume is a vector parametrised by a scalar variable $r(t) = x(t)\hat{x} + y(t)\hat{y} + z(t)\hat{z}$, we can consider three cases: a scalar field:

$$\int_C \phi(r)dr = \hat{x}\int_C \phi dx + \hat{y}\int_C \phi dy + \hat{z}\int_C \phi dz \tag{34.0.22}$$

a vector field dot product:

$$\int_C V \cdot dr = \int_C v_1 dx + \int_C v_2 dy + \int_C v_3 dz \tag{34.0.23}$$

and a vector field cross product:

$$\int_C V(r) \times dr = \hat{x}\int_C (v_2 dz - v_3 dy) + \hat{y}\int_C (v_1 dz - v_3 dx)\hat{y} + \hat{z}\int_C (v_1 dy - v_2 dx) \tag{34.0.24}$$

where $\phi(r)$ is a scalar field, and $V = v_1\hat{x} + v_2\hat{y} + v_3\hat{z}$ is a vector field, both of which are defined along the curve. Our interests will be with the middle relation, the dot product.

Mathematical Background 8

As a special case, if we suppose that the curve is a **Jordan curve** or a simple closed curve (so that it only crosses itself at the boundary points (e.g a circle), then we may write (assuming all components of V and their first partial derivatives are continuous) the integral around the curve:

$$\oint_C \boldsymbol{V} \cdot d\boldsymbol{r} \qquad (34.0.25)$$

The circle notation indicates that the curve is closed and we now call the entire integral the **circulation** of the vector field. In addition, if the vector field is the gradient of a *single-valued* scalar field $\boldsymbol{V} = \vec{\nabla}\phi$ everywhere in the region, then

$$\int_1^2 \boldsymbol{V} \cdot d\boldsymbol{r} \qquad (34.0.26)$$

Importantly, it is only a function of the end points 1 and 2, meaning it is independent of the path between the two points; hence, the integral has the same value for all paths between these points. The proof is as follows:

$$\int_1^2 \boldsymbol{V} \cdot d\boldsymbol{r} = \int_1^2 \vec{\nabla}\phi \cdot d\boldsymbol{r}$$

$$= \int_1^2 \left(\frac{\partial \phi}{\partial x}\hat{\boldsymbol{x}} + \frac{\partial \phi}{\partial y}\hat{\boldsymbol{y}} + \frac{\partial \phi}{\partial z}\hat{\boldsymbol{z}} \right) \cdot (dx\hat{\boldsymbol{x}} + dy\hat{\boldsymbol{y}} + dz\hat{\boldsymbol{z}})$$

$$= \int_1^2 \left(\frac{\partial \phi}{\partial x}dx + \frac{\partial \phi}{\partial y}dy + \frac{\partial \phi}{\partial z}dz \right)$$

$$= \int_1^2 d\phi$$

$$= \phi(2) - \phi(1) \qquad (34.0.27)$$

meaning that the line integral is an **exact differential**. A second consequence of this is that the circulation for a path-independent line integral of a simple closed curve is zero:

$$\oint_C \boldsymbol{V} \cdot d\boldsymbol{r} = 0 \qquad (34.0.28)$$

When these two conditions hold, we say that $\boldsymbol{V}$ is a **conservative field** and that ϕ is its scalar potential. The proof of the second statement is a result of the previous assumption that the line integral is path independent.

$$\oint_C \boldsymbol{V} \cdot d\boldsymbol{r} = \int_1^2 \boldsymbol{V} \cdot d\boldsymbol{r} + \int_2^1 \boldsymbol{V} \cdot d\boldsymbol{r} = \int_1^2 \boldsymbol{V} \cdot d\boldsymbol{r} - \int_1^2 \boldsymbol{V} \cdot d\boldsymbol{r} = 0 \qquad (34.0.29)$$

where in the last step we have used $\int_1^2 f(x)dx = -\int_2^1 f(x)dx$. In this way, a closed integral can be thought of as the sum of two piecewise smooth paths between P_1

and P_2. Now, we recall from above that our line integral is path independent and hence the two integrals are equivalent, meaning the right hand side vanishes.

Exercise 34.4 To illustrate the above, we will consider a contour integral three ways. Let $\phi = 2xy^2z^2$, and $\vec{r} = x\hat{x} + y\hat{y} + z\hat{z}$, $\boldsymbol{F} = x\hat{x} + xy\hat{y} + zx^2\hat{z}$ along a curve $\mathcal{C}$ such that $x = t^2$, $y = 2t$ and $z = t^3$ over the interval $[0, 1]$. With this data, evaluate

$$\text{(i)} \ \int_{\mathcal{C}} \phi \, d\boldsymbol{r}, \qquad \text{(ii)} \ \int_{\mathcal{C}} \boldsymbol{F} \cdot d\boldsymbol{r}, \qquad \text{(iii)} \ \int_{\mathcal{C}} \boldsymbol{F} \times d\boldsymbol{r} \qquad\qquad (34.0.30)$$

We start by entering the expression for the curve into the position vector and compute the volume element: $d\boldsymbol{r} = (2t\hat{x} + 2\hat{y} + 3t^2\hat{z})dt$. Then the integrals follow:

(i)

$$\int_{\mathcal{C}} \phi \, d\boldsymbol{r} = \int_0^1 [(2)(t^2)(2t)^2(t^3)(2t\hat{x} + 2\hat{y} + 3t^2\hat{z})dt]$$

$$= \hat{x} \int_0^1 16t^{11} \, dt + \hat{y} \int_0^1 16t^{10} \, dt + \hat{z} \int_0^1 24t^{12} \, dt$$

$$= \frac{4}{3}\hat{x} + \frac{16}{11}\hat{y} + \frac{24}{13}\hat{z} \qquad\qquad (34.0.31)$$

(ii)

$$\int_{\mathcal{C}} \boldsymbol{F} \cdot d\boldsymbol{r} = \int_0^1 [(t^2\hat{x} + 2t^3\hat{y} + t^7\hat{z}) \cdot (2t\hat{x} + 2\hat{y} + 3t^2\hat{z})] dt$$

$$= \left[\frac{3}{2}t^4 + \frac{3}{10}t^{10} + c \right]_0^1 = 9/5 \qquad\qquad (34.0.32)$$

(iii)

$$\int_{\mathcal{C}} \boldsymbol{F} \times d\boldsymbol{r} = \int_0^1 [(t^2\hat{x} + 2t^3\hat{y} + t^7\hat{z}) \times (2t\hat{x} + 2\hat{y} + 3t^2\hat{z})] dt$$

$$= \int_0^1 \det \begin{pmatrix} \hat{x} & \hat{y} & \hat{z} \\ t^2 & 2t^3 & t^7 \\ 2t & 2 & 3t^2 \end{pmatrix} dt$$

$$= \hat{x} \int_0^1 (6t^5 - 2t^7) dt + \hat{y} \int_0^1 (2t^8 - 3t^5) dt + \hat{z} \int_0^1 (2t^2 - 4t^4) dt$$

$$= \frac{3}{4}\hat{x} - \frac{5}{18}\hat{y} - \frac{2}{15}\hat{z} \qquad\qquad (34.0.33)$$

Suppose that there is a surface $\mathcal{S}$ above a region in the xy plane and we partition it into lots of little segments, each of infinitesimal area $d\mathcal{S}$; if we associate with each area a direction outwards (positive **orientation**), the unit normal n to the point, then we have a vector $d\boldsymbol{S} = \boldsymbol{n}dS$ of magnitude dS and direction $\boldsymbol{n}$:

$$d\boldsymbol{S} = \hat{\boldsymbol{x}}dS_x + \hat{\boldsymbol{y}}dS_y + \hat{\boldsymbol{z}}dS_z \qquad (34.0.34)$$

with $dS_x = \pm dydz$, $dS_y = \pm dzdx$ and $dS_z = \pm dxdy$. We may define three types of **surface integral** or flux integral, in a similar way as for the line integrals. They are termed flux integrals because they indicate the amount of something flowing out of the surface in the direction of the normal vector:

$$\iint_S \phi(\boldsymbol{r})d\boldsymbol{S} = \hat{\boldsymbol{x}} \iint_S \phi dS_x + \hat{\boldsymbol{y}} \iint_S \phi dS_y + \hat{\boldsymbol{z}} \iint_S \phi dS_z$$

$$\iint_S \boldsymbol{V}(\boldsymbol{r}) \cdot d\boldsymbol{S} = \iint_S v_1 dS_x + \iint_S v_2 dS_y + \iint_S v_3 dS_z$$

$$\iint_S \boldsymbol{V}(\boldsymbol{r}) \times d\boldsymbol{S} = \hat{\boldsymbol{x}} \iint_S (v_2 dS_z - v_3 dS_y)\hat{\boldsymbol{y}} \iint_S (v_1 dS_z - v_3 dS_x)\hat{\boldsymbol{y}} + \hat{\boldsymbol{z}} \iint_S (v_1 dS_y - v_2 dS_x)$$

The middle integral, for instance, is a scalar quantity; it can be thought of as the flux of $\boldsymbol{V}$ across S in the direction of $\boldsymbol{n}$. For closed surfaces, that is, surfaces that don't have a boundary (e.g a torus), we may use the symbol $\oiint_S$. We could then study parametric surfaces, which are surfaces parametrised by two variables. In general, the specification of the surface will give us two integration elements and hence a double integral projected onto a coordinate plane, but we will not discuss this.

If we use two integrals to integrate over a surface, then the obvious extension is to use triple integrals to integrate over volumes. If we have some volume V, then we may define

$$\iiint_V \boldsymbol{A}dV \quad \text{and} \quad \iiint_V \phi dV \qquad (34.0.35)$$

The **fundamental theorem of calculus**, as proved in general by Newton, has the following implication for definite integrals: if we integrate the derivative of a function then

$$\int_a^b \frac{d}{dx}f(x)dx = f(b) - f(a) \qquad (34.0.36)$$

That is to say, the definite integral of a sufficiently smooth function $F(x) = df/dx$ in a closed interval $[a, b]$ is the difference in the antiderivative $f(x)$ of the function evaluated at the boundaries (a, b). Curiously we are not inputting anything in-between the boundary points. This can be extended to higher dimensions and is the subject of the subsequent discussion to close this chapter. Although we do not use all of the ideas below, we may as well cover them while we are here. The pioneers of the discussion to follow are Gauss, Ostrogradsky, Stokes and Green and their work is used all over theoretical physics.

There is a similar relationship between a surface integral over a region and a closed line integral over the boundary of the region; there is a relationship between the triple integral of a volume and the surface integral of that volume.

The **divergence theorem** of Gauss relates the outward (positive orientation) flow of a vector field V through a closed surface $\mathcal{S}$ (surface integral) to the volume integral of the divergence of the field over the region/volume V enclosed by the surface:

$$\iiint_V \vec{\nabla} \cdot V \, dV = \oiint_\mathcal{S} V \cdot d\mathcal{S} \tag{34.0.37}$$

We can take V to be some region within $\mathbb{R}^3$, and $\mathcal{S}$ as a piecewise smooth boundary, although the divergence theorem applies equally to subsets of $\mathbb{R}^n$.

The **Kelvin-Stokes theorem** is a special case of **Stokes' theorem** (see mathematical background 5). It states that the surface integral of the curl of a vector field V on $\mathbb{R}^3$, defined over a surface $\mathcal{S}$, is equal to the line integral of the boundary curve $\mathcal{C}$ taken to be a simple closed curve:

$$\iint_\mathcal{S} (\vec{\nabla} \times V) \cdot d\mathcal{S} = \oint_\mathcal{C} V \cdot d\boldsymbol{r} \tag{34.0.38}$$

The last famous theorem is **Green's theorem**, which is the two-dimensional case of the Kelvin-Stokes theorem. Green's theorem relates the line integral of a simple closed curve $\mathcal{C}$ to the double integral over the region $\mathcal{R}$ it encloses in the xy-plane. If there are two functions $L(x, y)$ and $M(x, y)$ that are sufficiently smooth in the region $\mathbb{R}$, then

$$\oint_\mathcal{C} (L \, dx + M \, dy) = \iint_\mathcal{R} \left(\frac{\partial M}{\partial x} - \frac{\partial L}{\partial y} \right) dx \, dy \tag{34.0.39}$$

where the curve is travelled in the positive direction. Importantly, we can see that the Green's theorem is equivalent to Gauss' theorem restricted to the plane. The derivation of these three theorems is slightly involved and is beyond this introduction to vector analysis.

Mathematical Background 9

Consider a volume of fluid V that is enclosed by a surface $\mathcal{S}$ of density $\rho(x, y, z, t)$ and has velocity $\boldsymbol{v}(x, y, z, t)$ and mass (at any instant) M. If there are no sources or sinks in the fluid, then the net rate of mass flow out of the volume must equal the rate of decrease of the mass of fluid within the volume. We use the divergence theorem to develop the **continuity equation**, which describes the evolution of a conserved quantity. The mass of fluid in the volume is given by the integral of the density over a little volume element dV:

$$M = \iiint_V \rho dV \qquad (34.0.40)$$

The rate of increase of the mass or the mass gain is

$$\frac{\partial M}{\partial t} = \iiint_V \frac{\partial \rho}{\partial t} dV \qquad (34.0.41)$$

The mass of fluid leaving per unit time is given by the following surface integral:

$$\iint_S \rho \boldsymbol{v} \cdot \boldsymbol{n} dS \qquad (34.0.42)$$

meaning that the rate of increase in the mass is given by the negative of this; we then use Stokes' theorem to convert this to a volume integral:

$$-\iint_S \rho \boldsymbol{v} \cdot \boldsymbol{n} dS = -\iiint_V \vec{\nabla} \cdot (\rho \boldsymbol{v}) dV \qquad (34.0.43)$$

All that remains is to equate the two relations for the mass gain and, since dV is arbitrary, the term under the integral must vanish:

$$\iiint_V \left(\vec{\nabla} \cdot (\rho \boldsymbol{v}) + \frac{\partial \rho}{\partial t} \right) dV = 0 \qquad (34.0.44)$$

Therefore, we obtain

$$\vec{\nabla} \cdot \boldsymbol{j} + \frac{\partial \rho}{\partial t} = \partial_\mu j^\mu = 0, \quad \text{or, in integral form,} \quad \frac{dM}{dt} + \oiint_S \boldsymbol{j} \cdot d\boldsymbol{S} = 0$$

This result is the continuity equation, with $j = \rho \boldsymbol{v}$. Although we derived it for fluid mechanics it appears all over physics and chemistry, in classical mechanics, quantum mechanics, general relativity, electromagnetism, thermodynamics and molecular dynamics — anywhere where there is a conserved quantity, there will be a continuity equation.

The notation developed in the chapter is useful for simple procedures and, in most calculations, it is very difficult to work with. We now introduce the Kronecker delta δ_{ij} and the Levi-Civita notation ϵ_{ijk} for vector calculus, as well as the summation convention. We can write the dot and cross products as follows for $\boldsymbol{A} = a_i e_i$ and $\boldsymbol{B} = b_i \boldsymbol{\theta}_i$, with $\boldsymbol{A}, \boldsymbol{B} \in \mathbb{R}^3$:

$$\boldsymbol{A} \cdot \boldsymbol{B} = \sum_i a_i b_i = \sum_i \sum_j \delta_{ij} a_i b_j = \delta_{ij} a_i b_j \qquad (34.0.45)$$

and

$$\boldsymbol{A} \times \boldsymbol{B} = \det \begin{pmatrix} \hat{\boldsymbol{x}} & \hat{\boldsymbol{y}} & \hat{\boldsymbol{z}} \\ a^1 & a^2 & a^3 \\ b^1 & b^2 & b^3 \end{pmatrix} = \sum_i \sum_j \sum_k \epsilon_{ijk} \hat{\boldsymbol{e}}_i a_j b_k \qquad (34.0.46)$$

Component wise, we may write $(\boldsymbol{A} \times \boldsymbol{B})_i = \epsilon_{ijk} a_j b_k$ such that $(\boldsymbol{A} \times \boldsymbol{B})_1 = a_2 b_3 - a_3 b_2$, and so on:

$$(\boldsymbol{A} \times \boldsymbol{B})_1 = \epsilon_{111} a_1 b_1 + \epsilon_{112} a_1 b_2 + \epsilon_{113} a_1 b_3$$
$$+ \epsilon_{121} a_2 b_1 + \epsilon_{122} a_2 b_2 + \epsilon_{123} a_2 b_3$$
$$+ \epsilon_{131} a_3 b_1 + \epsilon_{132} a_3 b_2 + \epsilon_{133} a_3 b_3$$

Despite it being amazingly simple, we don't often use this notation unless strictly necessary in this book, because it can be a bit off-putting upon first glance; but it makes calculations so much easier once a certain familiarity is reached. The gradient of a scalar field becomes $(\vec{\nabla}\phi)_i = \partial_i \phi$. The divergence of a vector field simply becomes $\vec{\nabla} \cdot \boldsymbol{A} = \partial_i a_i$, while $(\vec{\nabla} \times \boldsymbol{A})_i = \epsilon_{ijk} \partial_j a_k$.

You may recall in equation 34.0.17 that we set about proving $\vec{\nabla} \cdot (\vec{\nabla} \times \boldsymbol{A}) = 0$. Although not particularly difficult, it was long and it was easy to make mistakes. In the Levi-Civita notation, this is reduced to two lines:

$$\vec{\nabla} \cdot (\vec{\nabla} \times \boldsymbol{A}) = \partial_i (\epsilon_{ijk} \partial_j a_k) = \epsilon_{ijk} \partial_i \partial_j a_k = 0 \qquad (34.0.47)$$

But why does the last term vanish? Mixed partial derivatives commute; therefore, $\partial_i \partial_j$ is symmetric under the swapping of i and j, such that $\partial_i \partial_j = \partial_j \partial_i$. We know full well that the Levi-Civita permutation symbol is antisymmetric under a swapping of i and j, $\epsilon_{ijk} = -\epsilon_{jik}$. It is a general rule that the product of symmetric and antisymmetric objects is zero. This can be seen by letting $a_{ij} = -a_{ji}$ be an antisymmetric object and $b_{ij} = b_{ji}$ be a symmetric object. Their product is

$$a_{ij} b_{ij} = a_{11} b_{11} + a_{12} b_{12} + a_{21} b_{21} + a_{22} b_{22} = a_{12} b_{12} - a_{12} b_{12} = 0$$

And there we have the first vector identity we proved longhand earlier. The second identity in equation 34.0.18 was more tricky but again reduces to a few lines:

$$[\vec{\nabla} \times (\vec{\nabla} \times \boldsymbol{A})]_i = \epsilon_{ijk} \partial_j (\vec{\nabla} \times \boldsymbol{A})_k$$
$$= \epsilon_{ijk} \partial_j (\epsilon_{klm} \partial_l a_m)$$
$$= \epsilon_{ijk} \epsilon_{klm} \partial_j (\partial_l a_m) \qquad (34.0.48)$$

We then use the following identity from mathematical background 8 for the product of two permutation symbols: $\epsilon_{ijk} \epsilon_{klm} = \delta_{il} \delta_{jm} - \delta_{im} \delta_{jl}$. We thus obtain,

$$[\vec{\nabla} \times (\vec{\nabla} \times \boldsymbol{A})]_i = (\delta_{il}\delta_{jm} - \delta_{im}\delta_{jl})\partial_j\partial_l a_m$$
$$= \partial_i\partial_j a_j - \partial_j\partial_j a_i$$
$$= \partial_i(\vec{\nabla} \cdot \boldsymbol{A}) - \nabla^2 a_i \tag{34.0.49}$$

which is the same result as 34.0.18, as expected, but using a lot less space! It could be considered a nice exercise to compute all our workings in the book in terms of this notation, with particular reference to Maxwell's electrodynamics.

Exercise 34.5 One important result is the calculation of the following derivative of a quadruple product:

$$\frac{d}{d\boldsymbol{r}}\left[(\boldsymbol{\omega} \times \boldsymbol{r}) \cdot (\boldsymbol{\omega} \times \boldsymbol{r})\right] = -2\boldsymbol{\omega} \times (\boldsymbol{\omega} \times \boldsymbol{r}) \tag{34.0.50}$$

In the preceding notation, this is a bit tricky, but our new notation should sort it with comparative ease. The first step is to write the problem out in component form:

$$\frac{d}{dr_m}\left[(\epsilon_{ijk}\omega_j r_k)(\epsilon_{inl}\omega_n r_l)\right] = (\epsilon_{ijk}\omega_j \frac{dr_k}{dr_m})(\epsilon_{inl}\omega_n r_l) + (\epsilon_{ijk}\omega_j r_k)(\epsilon_{inl}\omega_n \frac{dr_l}{dr_m})$$
$$= (\epsilon_{ijk}\omega_j \delta_{km})(\epsilon_{inl}\omega_n r_l) + (\epsilon_{ijk}\omega_j r_k)(\epsilon_{inl}\omega_n \delta_{lm})$$
$$= (\epsilon_{ijm}\omega_j)(\epsilon_{inl}\omega_n r_l) + (\epsilon_{ijk}\omega_j r_k)(\epsilon_{inm}\omega_n) \tag{34.0.51}$$

It is a general observation that any index that is repeated is a dummy index and may be relabelled at will (of course, maintaining consistency with others). Here ,we note that k, l, n and j are all repeated; therefore, if we choose them carefully enough, we can greatly simplify the result:

$$= (\epsilon_{ijm}\omega_j)(\epsilon_{inl}\omega_n r_l) + (\epsilon_{inl}\omega_n r_l)(\epsilon_{ijm}\omega_j)$$
$$= 2(\epsilon_{ijm}\omega_j)(\epsilon_{inl}\omega_n r_l) \tag{34.0.52}$$

This corresponds to changing the indices in the second half of the expression, with $k \to l$ and $n \to j$. Converting the second portion back into the familiar vector notation, we have

$$= 2(\epsilon_{ijm}\omega_j)(\boldsymbol{\omega} \times \boldsymbol{r})_i \tag{34.0.53}$$

If we permute the m and i indices in $\epsilon_{ijm} \to -\epsilon_{mji}$, then the rules of the Levi-Civita symbol give us a negative sign, since it is an odd permutation:

$$= -2\epsilon_{mji}\omega_j(\boldsymbol{\omega} \times \boldsymbol{r})_i \tag{34.0.54}$$

If we treat the bracketed term as a single entity, then we can write this entire term as another cross product. This is why we did the permutation, so that we can identify it with the above cross product formula:

$$= -2\boldsymbol{\omega} \times (\boldsymbol{\omega} \times \boldsymbol{r}) \tag{34.0.55}$$

35

Differential Equations

It is a common trend in physics and nature, or perhaps just the way numbers and calculus come together to describe the evolution of things, that most theories admit a **differential equation** of low order. A differential equation is a relationship between derivatives of a function. The **order** of a differential equation is the highest derivative it contains. A differential equation is called an ODE if it contains only one variable; if it depends on more than one variable, then it is a PDE.

We may use this to characterise equations, for instance, Newton's second law is a ODE of second-order:

$$m\frac{d^2 x}{dt^2} = F(x) \tag{35.0.1}$$

The wave equation is a second-order PDE because there are derivatives with respect to x and t:

$$\frac{1}{c^2}\frac{\partial^2 \phi}{\partial t^2} = \frac{\partial^2 \phi}{\partial x^2} \tag{35.0.2}$$

while Hamilton's equations are coupled (because the variables x and p appear in both equations) first-order ODEs:

$$\frac{dp}{dt} = -\frac{\partial H}{\partial x}, \qquad \frac{dx}{dt} = \frac{\partial H}{\partial p} \tag{35.0.3}$$

The Euler-Lagrange equation is a second-order ODE:

$$\frac{d}{dt}\left(\frac{\partial \mathscr{L}}{\partial \dot{x}}\right) - \frac{\partial \mathscr{L}}{\partial x} = 0 \tag{35.0.4}$$

(to see this, expand the time derivative of $\partial \mathscr{L}/\partial \dot{x}$). The Lagrange equation for fields is a PDE of second-order, since it depends on the field and its space *and* time derivatives:

$$\partial_\mu \left(\frac{\partial \mathcal{L}}{\partial (\partial_\mu \phi)}\right) - \frac{\partial \mathcal{L}}{\partial \phi} = 0$$

The Hamilton-Jacobi equation is a PDE, as equations of motion often involve both $\partial S/\partial x$ and $\partial S/\partial t$.

Lagrangian & Hamiltonian Dynamics. Peter Mann, Oxford University Press (2018).
© Peter Mann. DOI: 10.1093/oso/9780198822370.001.0001

$$H\left(x, \frac{\partial S}{\partial x}\right) + \frac{\partial S}{\partial t} = 0 \qquad (35.0.5)$$

In addition, we can distinguish between **linear** and **non-linear** differential equations, be they partial or ordinary. A differential equation can be written as a differential operator $\hat{\mathcal{O}}$ acting on an unknown function $y(x)$, with the right-hand side being a function of x and called the **source term** $f(x)$; note that it is independent of unknown function $y(x)$. The aim is to find the unknown function $y(x)$ through a process called **integrating the differential equation**, although it is more complicated than simple integration. If the source term is exactly zero, we say that the differential equation is **homogeneous**, if it is non-zero, we say it is **non-homogeneous**:

$$\hat{\mathcal{O}}[y(x)] = f(x) \qquad (35.0.6)$$

If $\hat{\mathcal{O}}$ is a linear operator, then the differential equation is a linear equation, if not it is non-linear. A linear operator satisfies

$$\hat{\mathcal{O}}[c_1 y_1(x) + c_2 y_2(x)] = c_1 \hat{\mathcal{O}}[y_1(x)] + c_2 \hat{\mathcal{O}}[y_2(x)] \qquad (35.0.7)$$

A linear differential equation may therefore be written in the following form (note that both y and it's derivatives only appear to the power of 1; we can't write product terms):

$$a_0 y + a_1 y' + a_2 y'' + \cdots = b \qquad (35.0.8)$$

where a and b are either constants or functions of x. Importantly, some non-linear differential equations can be approximated by a linear differential equation in some region about an **equilibrium**; we say that it has been **linearised**. The benefit of this is that linear ODEs are, in general, far easier to solve than their non-linear counterparts. We spent chapter 12 linearising the harmonic oscillator ODE.

We may now therefore classify what differential equation we are dealing with, by specifying its order and whether it is partial or ordinary, homogeneous or non-homogeneous, and linear or non-linear. Non-linear mechanics is also called chaotic dynamics and is in itself a vastly complex and almost entirely separate branch of mechanics. That is not to say that some of the equations of motion in this book aren't non-linear, consider the Hamilton-Jacobi equation with a Hamiltonian containing squared momentum terms, it will produce equations of motion with $(\partial S/\partial q)^2$ factors in them. It is a lucky feature, however, of the mathematics of differential equations that we can associate with this a system of first-order ODEs (Hamilton's equations). As a little example of how to classify a differential equation fully, lets consider the wave equation:

$$\frac{1}{c^2} \frac{\partial^2 \phi}{\partial t^2} - \frac{\partial^2 \phi}{\partial x^2} = 0 \qquad (35.0.9)$$

The wave equation is a linear, homogeneous, second-order PDE. It is linear because there are no cross terms; it is homogeneous because all terms involve the wavefunction (or derivatives), meaning there are no source terms; and it is a PDE because it contains derivatives in more than one variable.

Let us now look at the special cases of the simplest differential equations, first of all to introduce some key ideas. Take the following differential equation:

$$\frac{dy}{dx} = f(x, y) \qquad (35.0.10)$$

Imagine as a special case that we could instead write the right-hand side independent of y, then we could solve this for y by simple integration:

$$\frac{dy}{dx} = f(x) \qquad \longrightarrow \qquad y(x) = y(a) + \int_a^x f(x')dx' \qquad (35.0.11)$$

If, as a second special case, we could write the right-hand side as the product of two functions, each of which depends only on one of the variables, then we say we have **separated** the system $f(x, y) = g(x)h(y)$:

$$\frac{dy}{dx} = g(x)h(y) \qquad (35.0.12)$$

With some simple rearranging, we can convert each side as follows and then integrate:

$$\int \frac{dy}{h(y)} = \int f(x)dx \qquad (35.0.13)$$

Consider now the first-order, linear ODE:

$$\left(\frac{d}{dx} + R(x) \right) y = S(x) \qquad (35.0.14)$$

where $R(x)$ and $S(x)$ are smooth functions. If $S(x)$ is zero, the equation is homogeneous; if $S(x) \neq 0$, it is non-homogeneous. The homogeneous system is separable:

$$\frac{dy}{dx} + R(x)y = 0, \qquad \longrightarrow \qquad \int \frac{1}{y}dy = -\int R(x)dx \qquad (35.0.15)$$

We now change this to a definite integral:

$$\int_a^x d\ln y(x') = -\int_a^x p(x')dx' \qquad \longrightarrow \qquad \frac{y(x')}{y(a)} = \exp\left\{ -\int_a^x R(x')dx' \right\} \quad (35.0.16)$$

So we have the solution $y(x')$ and another term $y(a)$, called the **boundary condition** at $x = a$. The boundary conditions are conditions which need to be satisfied by a **particular solution** of the differential equation, if we are dealing with physics, where

x is time, then the boundary condition at $t = 0$ is called an **initial value**. There are different types of boundary conditions we can impose and, in order to have a **complete solution**, we must specify a boundary condition for each order of the equation. Suppose we have an ODE with the domain in the interval $x \in [a, b]$, then we could impose the following conditions:

- **Dirichlet boundary conditions**: a type of boundary condition that specifies the value a solution needs to take on the boundary of the domain (e.g $y(a) = \alpha$ and $y(b) = \beta$).
- **Neumann boundary conditions**: a type of boundary condition that specifies the value that the derivative of a solution is to take on the boundary of the domain (e.g $y'(a) = \alpha$ and $y'(b) = \beta$).
- **Cauchy boundary conditions**: a type of boundary condition that specifies the value of both the function and the derivative, so it is both a Dirichlet and a Neumann condition in one, initial value problems in mechanics correspond to knowing a particle's initial position and initial velocity which constitute a set of Cauchy conditions. Hence, we can associate a Cauchy problem to a dynamical system.

This leads us on to the **general solution** of a differential equation. A general solution of a linear differential equation is a solution to both the homogeneous and non-homogenous equations and will contain a constant for each order of the equation. All the particular solutions can be obtained by letting the constants have particular values. For instance, the Verlet algorithm is a general solution to Newton's second-order ODE:

$$x(t) = x(0) + \dot{x}(0)t + \frac{1}{2}\ddot{x}t^2$$

while, if we *chose* the values of the initial position and velocity (let's impose the Cauchy boundary conditions $x(0) = \dot{x}(0) = 0$), then a particular solution in this instance is $\ddot{x}t^2/2$.

By the **uniqueness theorem**, also known as the **Cauchy-Lipschitz theorem**, the particular solution to a linear ODE is unique provided $f(x, y)$ and y' are continuous over the domain of definition; that is to say, there is only one solution that will satisfy the boundary conditions. The theorem extends to any order ODE; however, we have no such luck with PDEs, which, in general, have their own theorems. For instance, Hamilton's equations require the Hamiltonian H and its first derivative $\nabla_x H \; x = (q, p)$ to be Lipschitz continuous to afford unique solutions; if they weren't unique, then a point in phase space would not be unique, contrary to physicality.

The first-order equation discussed above defines a slope dy/dx at each point (x, y) in the domain of $f(x, y)$, which we call a **slope field**. An **integral curve** to an initial value problem is a particular solution curve that is tangent to the field at each point. Generalising this to higher orders, we have what is known as **flow**. If differential equations cannot be solved explicitly, then computational integrator methods become the easiest way to visualise a solution graphically. There are two broad categories: linear multistep methods and **Runge-Kutta** methods. In the text, we look briefly at symplectic integrators as a method for solving Hamilton's equations. Numerical solutions to dynamical system are usually quick to obtain and are sometimes the only way to proceed.

In general, we write an nth-order ODE using the notation that superscript (n) means the nth derivative of function y, and so on:

$$y_i^{(n)} = \phi_i(x, y, y', \dots, y^{(n-1)}) \tag{35.0.17}$$

In the context of mechanics, we might write a second-order ODE for the acceleration of the system, such as $\ddot{q}^i = f(t, q, \dot{q})$; note we treat $x, y, y', \dots$ as independent variables. It is a mathematical curiosity that such an ODE has an *equivalent* first-order PDE in $(n+1)$ variables, which we can write as,

$$(\partial_x + y'\partial_y + y''\partial_{y'} + \cdots + \phi\partial_{y^{(n-1)}})f = 0 \tag{35.0.18}$$

An important result of nth-order linear ODEs is that we can convert them into a system of n coupled first-order ODEs; this allows us to use methods only available to first-order systems. To do this, we have to introduce an additional n variables.

A **first integral** of the system is a function $I(x, y, y', \dots, y^{(n-1)})$ that is constant along solution curves of the ODE as well as being a solutions to the associated PDE. For our second-order mechanical system, we might write the first integral $I(t, q, \dot{q})$ and show its invariance along solutions by taking the time derivative:

$$\frac{dI}{dt} = I_t + \dot{q}I_q + \ddot{q}I_{\dot{q}} = 0 \tag{35.0.19}$$

One of the most important themes in analytical mechanics is the use of first integrals to reduce the order of the ODE. This can be achieved by inverting the particular first integral to give a function of a constant of the motion, assuming, of course, the Jacobian of the transformation is non-zero: $I_{y^{(n-1)}} \neq 0$. Let $I(x, y, y', \dots, y^{(n-1)}) = I_0$; then, the inverted expression reads $y^{(n-1)} = \phi(x, y, y', \dots, y^{(n-2)}, I_0)$. The scenario where there exist n first integrals means that the general solution to the ODE is now $y(x, I_0^1, \dots, I_0^n)$. In this way, the general solution to the ODE is now dependent only on x and a set of n arbitrary integration constants. Although this is a powerful theorem, there is often difficulty in finding the set of I_0^i, if it even exists!

The more integrals of an ODE we have, the easier it should be to solve. We now consider **Jacobi's last multiplier** method for solving second-order ODEs. This particular theorem has been shown by Nucci and Leach (2008) to be extremely useful to finding the Lagrangian for a mechanical system. Here, we show how a general differential equation can be solved by Jacobi's method. Consider a system of differential equations of the form

$$\frac{dx^i}{dt} = X^i(t, x^1, \ldots, x^n), \quad i-1, \ldots, n \tag{35.0.20}$$

Then, a function $M(t, x^1, \ldots, x^n)$ is called the **last multiplier** of the system if it satisfies

$$\frac{\partial M}{\partial t} + \sum_{k=1}^{n} \frac{\partial(MX^i)}{\partial x^k} = 0 \tag{35.0.21}$$

The Jacobi last multiplier can be used to find Lagrangians as well as first integrals of the differential equation, so proves a valuable tool. We prove in the text that the Euler-Lagrange equation can be written in such a fashion; however, M. C. Nucci and A. G. Choudhury have shown the applicability of this method in general.

Exercise 35.1 Here, we discuss how to solve a linear second-order ODE such as an Euler-Lagrange equation of motion with fixed boundary conditions by using a Fourier transform. It is a general theme to use transforms (Fourier, Laplace, Legendre, etc) to solve ODEs and depending on the problem at hand, some will be more appropriate than others. Consider the following system:

$$a_0 f(t) + a_1 f^{(1)}(t) + \cdots + a_n f^{(n)}(t) = g(t) \tag{35.0.22}$$

where we have coefficients a_i and some function $g(t)$. The Fourier transform $\mathbb{F}$ of a function $f(t)$ is $\tilde{f}(\omega)$ and, together with the inverse operation $\mathbb{F}^{-1}$, generate the following equations:

$$\tilde{f}(\omega) = \mathbb{F}(f(t))(\omega) = \int_{-\infty}^{\infty} dt\, e^{i\omega t} f(t) \qquad f(t) = \mathbb{F}^{-1}(f(\omega))(t) = \frac{1}{2\pi} \int_{-\infty}^{\infty} d\omega\, e^{-i\omega t} f(\omega)$$

We understand that there is a special relationship between t and ω and that they are, in fact, conjugate variables. A rather important property is the effect the Fourier transform has on time derivatives:

$$\mathbb{F}\left(\frac{d}{dt}f(t)\right)(\omega) = \int_{-\infty}^{\infty} dt\, e^{i\omega t} \frac{d}{dt}f(t) = -\int_{-\infty}^{\infty} dt\, \frac{d}{dt} e^{i\omega t} f(t) = -i\omega f(\omega) \tag{35.0.23}$$

In this step, we have integrated by parts and canceled terms evaluated on the boundary. If we apply the Fourier transform to the ODE then we generate a polynomial in ω:

$$a_0 f(\omega) + a_1(-i\omega)f(\omega) + \cdots a_n(-i\omega)^n f(\omega) = g(\omega) \tag{35.0.24}$$

Pulling $f(\omega)$ out of the sum allows us to write this as $A(\omega)f(\omega) = g(\omega)$. Rearranging this for $f(\omega)$ and then applying an inverse transformation, we obtain

$$f(t) = \int_{\infty}^{\infty} \frac{d\omega}{2\pi} e^{-i\omega t} \frac{g(\omega)}{A(\omega)} \tag{35.0.25}$$

This can then be solved by analysis in the complex plane. In this book, we do not use too many transforms in this way, in order to keep the text accessible for those with no prior mathematical background (although one or two crop up when discussing classical field theory).

36
Calculus of Variations

The calculus of variations is an important tool that is used all over modern physics. First developed by Euler to determine the shortest paths between fixed points along a surface, it was applied by Lagrange to mechanical problems, which comprises the essence of analytical mechanics. Here, we will present the general formulation of the calculus of variations as applied to mechanics, relativity and field theories. There are actually many types of variational problems across physics but all will follow the general trend laid out below. The versions in the text have been simplified for ease of understanding upon first introduction; presented here is the general mathematical framework. We develop the Euler-Lagrange equation by first considering variations of the coordinates, then of a function of the coordinates and finally of a contour integral.

Fig. 36.1: Leonhard Euler (1707–1783)

A path is traced out in $\mathbb{R}^3$ between two points as time evolves. It is parametrised by three coordinates that are smooth functions of time: $x(t), y(t), z(t)$. Often, we wish to generalise this to higher-dimensional spaces, so let's say there are now n dimensions, and a path between two points is parametrised by n smooth functions of time $x_i(t)$, with $i = 1, \ldots, n$.

There are an infinite number of paths between two points that we could draw, so let's just pick one arbitrarily and give it a name: the **unvaried path**. The **endpoints** of the path are defined to be at t_1 and t_2:

$$x_i(t_1) \quad \text{and} \quad x_i(t_2) \quad i = 1, \ldots, n \tag{36.0.1}$$

We will now express other curves in terms of this curve by adding a shape function $\eta_i(t)$, each multiplied by a small scaling quantity α:

$$x_i(t, \alpha) = x_i(t) + \alpha \eta_i(t) \tag{36.0.2}$$

For a given curve, α is a constant, that is to say, it is the same at all times. In order that all these curves pass through the same endpoints, we enforce the condition that the

Lagrangian & Hamiltonian Dynamics. Peter Mann, Oxford University Press (2018).
© Peter Mann. DOI: 10.1093/oso/9780198822370.001.0001

shape functions vanish at the endpoints. This is the **Dirichlet boundary condition** for the problem:

$$\eta_i(t_1) = \eta_i(t_2) = 0 \tag{36.0.3}$$

The original path is simply the $\alpha = 0$ path. The basic idea is to compare coordinates on a varied path to those on the unvaried path at the same value of the smooth parameter time. We denote the **variation** $\delta x_i(t)$:

$$\boxed{\delta x_i(t) = x_i(t,\alpha) - x_i(t) = \alpha\eta_i(t)} \tag{36.0.4}$$

Therefore, we may rephase the relation between curves as $x_i(t,\alpha) = x_i(t) + \delta x_i(t)$, again noting the vanishing endpoints. We may also consider the variation of the time derivative of the coordinate, too:

$$\dot{x}_i(t,\alpha) = \frac{dx_i(t,\alpha)}{dt} = \frac{dx_i(t)}{dt} + \alpha\frac{d\eta_i(t)}{dt}$$
$$= \dot{x}_i(t) + \alpha\dot{\eta}_i(t) = \dot{x}_i(t) + \delta\dot{x}_i(t) \qquad i = 1,\dots,n \tag{36.0.5}$$

So, we have two important equations: $x_i(t,\alpha) = x_i(t) + \delta x_i(t)$ and $\dot{x}_i(t,\alpha) = \dot{x}_i(t) + \delta\dot{x}_i(t)$ and the difference between the varied and unvaried path is exactly the variation at a given instant in time.

So far, we have worked out how to vary both the coordinate x and its time derivative $\dot{x}$; this leads us nicely into varying a function $f(x,\dot{x})$. Explicit time dependance isn't going to change the result, since a variation is evaluated at a single time. The variation of a function $f(x,\dot{x})$ is defined in the same fashion as the coordinate variations:

$$\delta f = f(x(t,\alpha),\dot{x}(t,\alpha)) - f(x(t),\dot{x}(t)) \tag{36.0.6}$$

However, we can also Taylor expand the left-hand side. In Lagrangian and Hamiltonian mechanics, we only require **first-order variations** but higher orders can be kept, to characterise the function further:

$$\Delta f = \left(\frac{\partial f}{\partial\alpha}\right)_0 \alpha + \frac{1}{2}\left(\frac{\partial^2 f}{\partial\alpha^2}\right)_0 \alpha^2 + \mathcal{O}(\alpha^3) \tag{36.0.7}$$

The first-order variation is written

$$\delta f(x,\dot{x}) = \left(\frac{\partial f}{\partial\alpha}\right)_0 \alpha = \sum_{i=1}^{n}\left(\frac{\partial f}{\partial x_i}\frac{\partial x_i}{\partial\alpha} + \frac{\partial f}{\partial\dot{x}_i}\frac{\partial\dot{x}_i}{\partial\alpha}\right)\alpha = \sum_{i=1}^{n}\left(\frac{\partial f}{\partial x_i}\delta x_i + \frac{\partial f}{\partial\dot{x}_i}\delta\dot{x}_i\right) \tag{36.0.8}$$

If we now consider a contour integral I over the unvaried curve $\mathcal{C}$, with the function defined along the curve we obtain

$$I(\alpha,[x],[\eta]) = \int_{\mathcal{C}} f(x(t,\alpha),\dot{x}(t,\alpha))dt = \int_{t_1}^{t_2} f(x(t,\alpha),\dot{x}(t,\alpha))dt \tag{36.0.9}$$

Setting $\alpha = 0$ gives us the unvaried path; the variation of I is defined just as before:

$$\delta I = I(\alpha, [x], [\dot{x}]) - I([x], [\dot{x}])$$

$$= \int_{t_1}^{t_2} \{f(x(t, \alpha), \dot{x}(t, \alpha)) - f(x(t), \dot{x}(t))\} dt = \int_{t_1}^{t_2} \delta f \, dt \qquad (36.0.10)$$

Inserting the first-order variation of $f(x, \dot{x})$, we obtain a general formula for the variation of the contour integral:

$$\delta I = \int_{t_1}^{t_2} \sum_{i=1}^{n} \left(\frac{\partial f}{\partial x_i} \delta x_i + \frac{\partial f}{\partial \dot{x}_i} \delta \dot{x}_i \right) dt \qquad (36.0.11)$$

Now, while this is a general result, it is not very useful in its current form. To proceed, we would like to be able to have variations in only one variable. We now use the linearity of the variation to pull the time derivative out:

$$\delta \left(\frac{d}{dt} x_i \right) = \frac{dx_i(t, \alpha)}{dt} - \frac{dx_i(t)}{dt} = \frac{d}{dt} (\delta x_i) \qquad (36.0.12)$$

Then,

$$\delta I = \int_{t_1}^{t_2} \sum_{i=1}^{n} \left(\frac{\partial f}{\partial x_i} \delta x_i + \frac{\partial f}{\partial \dot{x}_i} \frac{d}{dt} \delta x_i \right) dt \qquad (36.0.13)$$

Now consider the product rule $\frac{d}{dt}(AB) = A\frac{dB}{dt} + \frac{dA}{dt}B$ and then rearrange for $A\frac{dB}{dt} = \frac{d}{dt}(AB) - \frac{dA}{dt}B$:

$$\frac{\partial f}{\partial \dot{x}_i} \frac{d}{dt} \delta x_i = \frac{d}{dt} \left(\frac{\partial f}{\partial \dot{x}_i} \delta x_i \right) - \frac{d}{dt} \left(\frac{\partial f}{\partial \dot{x}_i} \right) \delta x_i \qquad (36.0.14)$$

The variation to the contour integral then becomes

$$\delta I = \int_{t_1}^{t_2} \sum_{i=1}^{n} \left(\frac{\partial f}{\partial x_i} \delta x_i + \frac{d}{dt} \left(\frac{\partial f}{\partial \dot{x}_i} \delta x_i \right) - \frac{d}{dt} \left(\frac{\partial f}{\partial \dot{x}_i} \right) \delta x_i \right) dt \qquad (36.0.15)$$

We can now collect the variations into one term:

$$\delta I = \int_{t_1}^{t_2} dt \sum_{i=1}^{n} \left\{ \frac{\partial f}{\partial x_i} - \frac{d}{dt} \left(\frac{\partial f}{\partial \dot{x}_i} \right) \right\} \delta x_i + \int_{t_1}^{t_2} dt \sum_{i=1}^{n} \frac{d}{dt} \left(\frac{\partial f}{\partial \dot{x}_i} \delta x_i \right) \qquad (36.0.16)$$

We cancel the dt factors in the second term and evaluate the exact differential:

$$\int_{t_1}^{t_2} d \left(\frac{\partial f}{\partial \dot{x}_i} \delta x_i \right) = \left(\frac{\partial f}{\partial \dot{x}_i} \delta x_i \right) \Big|_{t_1}^{t_2} \qquad (36.0.17)$$

However, when $\delta x_i(t)$ is evaluated at the endpoints, by definition, it is zero (see above), so this entire term vanishes. This leaves us with the following expression for the variation of the contour integral:

$$\delta I = \int_{t_1}^{t_2} dt \sum_{i=1}^{n} \left\{ \frac{\partial f}{\partial x_i} - \frac{d}{dt}\left(\frac{\partial f}{\partial \dot{x}_i} \right) \right\} \delta x_i \tag{36.0.18}$$

This is the general result of the calculus of variations. However, pretty much every time we use this theory, we ask one specific question: what path is the **extremum path** of the contour integral? In our discussion of partial differentiation, we concluded that the extremum of a function is those regions where the first-order derivative vanishes. In the language of variational calculus, this translates to the vanishing of the first-order variation, specifically $\delta I = 0$.

The variations are arbitrary and have so far not in any way been specified, except that they must vanish at the end points. Therefore, the term in the curly braces must vanish for $\delta I = 0$ to hold in general. This does not happen for any old x and $\dot{x}$, however! Only in the case of those that satisfy the **Euler-Lagrange equation** will this relation hold:

$$\boxed{\frac{\partial f}{\partial x_i} - \frac{d}{dt}\left(\frac{\partial f}{\partial \dot{x}_i} \right) = 0} \tag{36.0.19}$$

Any path whose coordinates satisfy this equation are, by definition, extremum paths of the contour integral. There are n of these equations, one for each coordinate and its time derivative. Multiplying out the time derivative of $f(x, \dot{x})$ we obtain

$$\frac{\partial f}{\partial x_i} - \frac{\partial^2 f}{\partial \dot{x}_i \partial x_j}\frac{dx_j}{dt} - \frac{\partial^2 f}{\partial \dot{x}_i \partial \dot{x}_j}\frac{d\dot{x}_j}{dt} = 0 \tag{36.0.20}$$

It can therefore be seen that the Euler-Lagrange equation is a second-order ODE. We cannot claim that it is linear as that will depend on the nature of $f(x, \dot{x})$. If the Euler-Lagrange equation is satisfied for all $q \in \mathcal{C}^{\infty}([t_1, t_2], \mathbb{R})$ with well-posed boundary conditions $q(t_1) = q_1$ and $q(t_2) = q_2$, then I does not depend on the entire path but only on the boundary points (q_1, t_1) and (q_2, t_2). The integrand function can then be written as $f = \alpha \dot{q} + \beta$ and, hence, $f dt = dg$ is exact (equal to an exact differential), where g is a function of the end points. Hence,

$$\int_{t_1}^{t_2} dg = g(t_1, q_1) - g(t_2, q_2) \tag{36.0.21}$$

Upon satisfaction of the Euler-Lagrange condition, there exists a function $g(x_1, x_2, t_1, t_2)$ such that I is only dependent on the end points, not the entire path!

If the function $f(x, \dot{x}, t)$ does not explicitly depend on a variable, say x_i, then $\partial f / \partial x_i = 0$. We can use this property to show that there will exist a **first integral** of the equation. If there are n first integrals for a $2n$-dimensional canonical system, then the second-order Euler-Lagrange equation can be solved by using these first integrals. This is in contrast to the normal non-canonical system, where we require $2n$ first integrals to solve a $2n$-dimensional system:

$$
\begin{aligned}
\frac{df}{dt} &= \frac{\partial f}{\partial x}\dot{x} + \frac{\partial f}{\partial \dot{x}}\ddot{x} \\
&= \frac{d}{dt}\left(\frac{\partial f}{\partial \dot{x}}\right)\dot{x} + \frac{\partial f}{\partial \dot{x}}\frac{d}{dt}(\dot{x}) \\
&= \frac{d}{dt}\left(\frac{\partial f}{\partial \dot{x}}\dot{x}\right)
\end{aligned}
\tag{36.0.22}
$$

In the second step, we used the Euler-Lagrange equation. This implies that the contents of the bracket below is constant:

$$
\frac{d}{dt}\left(f - \frac{\partial f}{\partial \dot{x}}\dot{x}\right) = 0
\tag{36.0.23}
$$

If the function had depended on more variables $\{x_i\}$, then this condition would become

$$
\frac{d}{dt}\left(f - \sum_i \frac{\partial f}{\partial \dot{x}_i}\dot{x}_i\right) = 0
\tag{36.0.24}
$$

This leads us to neatly gather up and recap the topic of functionals into the idea of the **functional derivative**, which we use for field theory. The functional derivative investigates how a functional I behaves under an infinitely small variation in x by some arbitrary test function ϕ controlled by α. We now imagine taking the continuum limit of the function x_i, where $i = 1, \ldots n$, and let $n \to \infty$. The continuum differential is now

$$
I[x + \alpha\phi] - I[x] = \alpha \cdot DI_x[\phi] + \mathcal{O}(\alpha^2) + \cdots
\tag{36.0.25}
$$

where the linear functional DI_x is

$$
DI_x = \int dt \frac{\delta I[x]}{\delta x(t)}\phi(t), \quad \text{with} \quad \frac{\delta I[x]}{\delta x(t)} := \lim_{\alpha \to 0}\frac{1}{\alpha}(I[x + \alpha\phi] - I[x])
\tag{36.0.26}
$$

The functional derivative is linear and satisfies both product and chain rules. It is the generalisation of the familiar partial derivative, which we recall for a function F, was written as

$$
F(x + \alpha\phi) - F(x) = \alpha \cdot dF_x(\phi) + \mathcal{O}(\alpha^2) + \cdots
\tag{36.0.27}
$$

We then normally take (in partial differentiation) $dF_x(\phi)$ as

$$
dF_x(\phi) = \frac{\partial F(x)}{\partial x(t)}\phi, \quad \text{with} \quad \frac{\partial F(x)}{\partial x(t)} := \lim_{\alpha \to 0}\frac{1}{\alpha}(F(x + \alpha) - F(x))
\tag{36.0.28}
$$

This is similar to what we discussed for partial differentiation in chapter 32. The quotient $\delta I[x]/\delta x(t)$ gives us the rate of change of the functional as the function x changes at the point t. Applying this to the contour integral functional above will generate the Euler-Lagrange equation. If we recall the general result of the calculus of variations from above,

$$\delta I = \int dt \left\{ \frac{\partial f}{\partial x_i} - \frac{d}{dt}\left(\frac{\partial f}{\partial \dot{x}_i}\right) \right\} \delta x_i \qquad (36.0.29)$$

Then we can (with minor rearranging) identify the quotient $\delta I[x]/\delta x(t)$ from the above definition as

$$\boxed{\frac{\delta I}{\delta x_i} = \frac{\partial f}{\partial x_i} - \frac{d}{dt}\left(\frac{\partial f}{\partial \dot{x}_i}\right)} \qquad (36.0.30)$$

which is, as expected, the Euler-Lagrange equation. If we were to consider higher-order (α^2) terms, we would generate a series for the functional derivative:

$$\frac{\delta I}{\delta x_i} = \frac{\partial f}{\partial x_i} - \frac{d}{dt}\left(\frac{\partial f}{\partial \dot{x}_i}\right) + \frac{d^2}{dt^2}\left(\frac{\partial f}{\partial \ddot{x}_i}\right) - \cdots \qquad (36.0.31)$$

This is useful when the function f contains terms dependent on higher-order derivatives (e.g. $f(x, \dot{x}, \ddot{x}, \ldots, x^{(n)})$).

Normally, we define I as an integral that contains, somewhere in its expression $f(x)$; in that case, I is not a function of the independent variable (x), but its functional derivative is. If, however, we evaluate the function at a particular value $x = y$ and let $I = f(y)$, then the functional derivative $\delta f(y)/\delta f(x)$ is zero for all $x \neq y$ and only non-zero for $x = y$. Does this remind us of how the Kronecker delta filters summations out?

The **Dirac delta function** δ is an extension of the Kronecker delta δ_{ij} and has the property that

$$\int_a^b f(x)\delta(x-c)dx = \begin{cases} f(c) & a < c < b \\ 0 & \text{otherwise} \end{cases} \qquad (36.0.32)$$

The Kronecker delta is used in summations as a filtering tool. The Dirac delta is used in integrals for the same purpose. We can use the Dirac delta function to write the functional derivative at a given point. From the above definition of the functional derivative, we realise that we can write

$$\frac{\delta f(y)}{\delta f(x)} = \delta(x-y) \qquad (36.0.33)$$

This can be readily extended to functions instead of variables:

$$\frac{\delta I[f(x)]}{\delta f(x')} = \lim_{\alpha \to 0} \frac{1}{\alpha} I[f(x) + \alpha\delta(x-x')] - I[f(x)] \qquad (36.0.34)$$

Exercise 36.1 As a brief example of the functional derivative, consider the functional derivative of $I[f] = \int_{-1}^{1} f(x)dx$:

$$\frac{\delta I[f]}{\delta f(x')} = \lim_{\alpha \to 0} \frac{1}{\alpha} \left(\int_{-1}^{1} [f(x) + \alpha \delta(x - x')]dx - \int_{-1}^{1} f(x)dx \right)$$

$$= \int_{-1}^{1} \delta(x - x')dx \tag{36.0.35}$$

which is equal to 1 if x' is within the range of the integration limits -1 to 1, and is otherwise zero:

$$\int_{-1}^{1} \delta(x - x')dx = \begin{cases} 1 & -1 \le x' \le 1 \\ 0 & \text{otherwise} \end{cases} \tag{36.0.36}$$

Part VI

Advanced Mathematics

37

Linear Algebra

Since our ultimate goal is to get talking about manifolds and forms, we can't spend too long defining everything from scratch, so there is a certain point at which we must jump in. The best place is by looking at what a **vector space** really is. Informally, a **field** $\mathbb{F}$ is a set of scalars that are endowed with the properties of real numbers: they can be added, subtracted, commutatively multiplied and so on, it is quite a structured object. A vector space V is a set defined over a field such that, if $a \in \mathbb{F}$ and $\boldsymbol{v} \in V$, then

$$
a\boldsymbol{v} \in V, \quad a(\boldsymbol{v} + \boldsymbol{w}) = a\boldsymbol{v} + a\boldsymbol{w}, \quad (a+b)\boldsymbol{v} = a\boldsymbol{v} + b\boldsymbol{v}, \quad a(b\boldsymbol{v}) = (ab)\boldsymbol{v},
$$
$$
\mathbb{I}\boldsymbol{v} = \boldsymbol{v}, \quad \boldsymbol{u} + (\boldsymbol{v} + \boldsymbol{w}) = (\boldsymbol{u} + \boldsymbol{v}) + \boldsymbol{w}, \quad \boldsymbol{0} + \boldsymbol{v} = \boldsymbol{v}, \quad \boldsymbol{u} + \boldsymbol{v} = \boldsymbol{v} + \boldsymbol{u} \quad (37.0.1)
$$

where $\mathbb{I}$ is an identity operator (e.g. multiplication by 1) and $\boldsymbol{0}$ is the zero vector. These are the **axioms** of vector space geometry and they endow the properties of associativity, commutativity, and the scalar distributivity of scalar multiplication with vector addition and multiplication. The vector space can be divided up into **subspaces**, which are also vector spaces. For example, V could be the direct sum of U and W, such as $V = U \bigoplus W$, and the dimensionality of V is the dimensionality of U added to the dimensionality of W.

For two vector spaces V and W, a **linear map** L is defined by $L : V \to W$. If $\boldsymbol{v}_i \in V$ and $a_i \in \mathbb{F}$, then

$$
L\left(\sum_i a_i \boldsymbol{v}_i\right) = \sum_i a_i L\boldsymbol{v}_i \quad (37.0.2)
$$

where $L\boldsymbol{v}_i$ represents the action of L on $\boldsymbol{v}_i$. If L is bijective, then we call it an **isomorphism**, and the vector spaces it maps between are **isomorphic**. If L maps into itself ($L : V \to V$) then it is an **endomorphism**. There may be more than one of these maps acting on a set of vectors and we often consider them as a collective object (i.e. the a set of linear mappings is an object in its own right).

A **linear functional** f on V is a map $f : V \to \mathbb{F}$. The set of all linear functionals is denoted V^* and called the **dual space** to V and it has its own vector space structure. As elements of V are called vectors, we call elements of the dual space **covectors**; if

Lagrangian & Hamiltonian Dynamics. Peter Mann, Oxford University Press (2018).
© Peter Mann. DOI: 10.1093/oso/9780198822370.001.0001

V has a basis e_i, then V^* has a **dual basis** θ_i. In a similar way to vectors, we can expand covectors as a linear combination of components and basis covectors:

$$V^* \ni f = \sum_i f_i \theta_i \tag{37.0.3}$$

where, $f^i \in \mathbb{F}$ are the components with respect to the basis. There exists a **natural pairing** between V and V^* such that $\langle e_i, \theta_i \rangle = \delta_{ij}$.

If we have two basis sets e_i and e'_i of a vector space V, then we can write each of the new basis vectors as a linear combination of the other basis vectors, using a **change of basis matrix**[1]:

$$e'_j = \sum_i e_i A_{ij}$$

This changes the way we describe the vectors in V. We can illustrate this by writing a linear combination in both bases for a vector in V.

$$\sum_i a_i e_i = \sum_i \left(\sum_j A_{ij} a'_j \right) e_i = \sum_{ij} a'_j e_i A_{ij} = \sum_j a'_j e'_j$$

where $a_i = \sum_j A_{ij} a'_j$. When you change the basis on V, you automatically change the basis on the dual space V^*, since we know that we must always preserve the dual pairing. We show this by letting θ_i and θ'_i be the dual bases on V^* (corresponding to e_i and e'_i on V), with the condition that the dual pairing is preserved such that $\langle e_i, \theta_j \rangle = \langle e'_i, \theta'_j \rangle = \delta_{ij}$ and that we can write $\theta'_j = \sum_i \theta_i B_{ij}$, as above. If u is a vector belonging to V^*, then we can change its basis, too, in a very similar way to $v \in V$:

$$u = \sum_i u_i \theta_i = \sum_i \left(\sum_j B_{ij} u'_j \right) \theta_i = \sum_{ij} u'_j \theta_i B_{ij} = \sum_j u'_j \theta'_j = u'$$

where $u_i = \sum_j B_{ij} u'_j$. Since the dual pairing must be preserved, we can develop some relations between B_{ij} and A_{ij} that prove useful:

---

[1]Try not to get too hung up on what a change of basis matrix is; its definition is laid out in both its name and the next equation.

$$\langle e'_i, \theta'_j \rangle = \sum_{hk} \langle e_h A_{hi}, \theta_k B_{kj} \rangle$$

$$= \sum_{hk} A_{hi} B_{kj} \langle e_h, \theta_k \rangle$$

$$= \sum_{hk} A_{hi} B_{kj} \delta_{hk}$$

$$= \sum_{h} A_{hi} B_{hj}$$

$$= \delta_{ij}$$

Therefore, $\boldsymbol{B} = (A^T)^{-1} = (A^{-1})^T$. Using these new relations, we can write our change of basis results in terms of the one matrix, for ease:

$$u'_j = \sum_i (B^{-1})_{ji} u_i = \sum_i (A^T)_{ji} u_i = \sum_i u_i A_{ij}$$

Herein lies the problem: there is no way with the current notation to tell which components belong to the vector and which to the covector. If we hadn't already said that $v \in V$ and $u \in V^*$, with their respective bases, then how would you tell? Therefore, we must change the index notation.

We write the basis of V with a lower index e_i while we use a raised index for the basis covectors θ^i. We then write each of the vector components with the opposite index to its basis, such that, the vectors of either space are written as the sum over an upstairs part and a downstairs part in their linear combinations:

$$v = \sum_i a^i e_i, \qquad u = \sum_i u_i \theta^i$$

Writing them in this way makes identifying the vectors from their duels easy! Note that, if you need to square something or raise it to any power, then you must use brackets, such that $(x^i)^n$ represents the nth power of the ith component of vector x. The **Einstein summation convention** simply says that, whenever we see a sum over the upstairs-downstairs combination, like so, we just ignore the summation notation, because we know exactly what we mean. It is not a mathematical step, instead it is used only for notational ease. One particular type of vector that we use in field physics is the 4-vector x^μ, where the Greek index runs from $0, 1, 2, 3$, that is, a time component 0 and three spatial components. A coordinate transformation $x^\mu \to \bar{x}^\mu$ will transform a vector v^μ as

$$\bar{v} = \left(\frac{\partial \bar{x}^\mu}{\partial x^\nu} \right) v^\nu$$

Note that it is summed over, since there are repeated indices. However, the gradient vector $\partial\phi/\partial x^\mu$ transforms as

$$\frac{\partial\phi}{\partial\bar{x}^\mu} = \left(\frac{\partial x^\nu}{\partial\bar{x}^\mu}\right)\frac{\partial\phi}{\partial x^\nu}$$

We can see that v is a contravariant vector while $\partial_\mu\phi$ is a contravariant vector. In the text, we make use of the Poincaré-Cartan 1-form which is a useful mathematical object that defines time-independent mechanics in the Lagrangian formalism.

$$\theta_{\mathscr{L}} = \frac{\partial\mathscr{L}}{\partial\dot{q}^i}\,dq^i \tag{37.0.4}$$

for some coordinate function $q(t)$. If we change the coordinates $q^i \to Q^i(q)$, we see that the 1-form changes as a contravariant vector:

$$\theta_{\mathscr{L}} = \frac{\partial\mathscr{L}}{\partial\dot{q}^i}\frac{\partial q^i}{\partial Q^j}\,dQ^j \tag{37.0.5}$$

An **inner product** g on V is a map g $: V \times V \to \mathbb{F}$ that satisfies three properties for $x, y, z, w \in V$:

$$g(x + y, z + w) = g(x, z) + g(x, w) + g(y, z) + g(y, w), \tag{37.0.6}$$

and,

$$g(v, w) = \overline{g(w, v)}, \quad g(x, y) = 0 \quad \forall\, y \tag{37.0.7}$$

That is to say, it is **linear**, **Hermitian** and **non-degenerate**, where the bar denotes the complex conjugate. If the field is real ($\mathbb{F} = \mathbb{R}$ (or a subset of)) then, for $a \in \mathbb{R}$, we have $\bar{a} = a$; but, if the field is complex ($\mathbb{F} = \mathbb{C}$ (or a subset of)), then, for $a \in \mathbb{C}$, we have $\bar{a} \neq a$. This has the effect that it is antilinear on the first entry, but linear on the second entry; hence, we term it **sesquilinear**. If $\mathbb{F} = \mathbb{R}$, then the Hermitian condition is a symmetry condition and we say that the form is a **bilinear form**. The non-degeneracy for all y implies that $x = 0$. If V is equipped with g, then it is called an **inner product space**.

If $V = \mathbb{R}^n$, we can equip it with an inner product g that is the scalar product we are familiar with: $v \cdot u$; in this case, we denote $\mathbb{R}^n$ by $\mathbb{E}^n$, to signify it is an inner product space, and call it **Euclidean space**. However, we may choose another inner product and generate a different space (e.g. a **Lorentzian inner product** gives rise to a **Minkowski space**).

Let's move on to consider **tensors**. Tensors are the domain of **multilinear algebra**, which is the generalisation of linear algebra. Its pretty much exactly the same except, instead of just dealing with one vector space at a time, you have any number. Rather than trying to imagine what a tensor is; instead, just accept that tensors are mathematical objects that are more general than vectors. Don't attempt to visualise them just yet; instead, just work with them and see where it goes.

The **tensor product** is the generalisation of the vector product and is denoted by $\otimes$; for any two vectors, we can write their tensor product as $\boldsymbol{v} \otimes \boldsymbol{u}$, which is called a *second-order tensor*. To define a third-order tensor, we write $\boldsymbol{v} \otimes \boldsymbol{u} \otimes \boldsymbol{w}$ and so on; in general, tensor products are not commutative, so the order in which we write them is important. A zeroth-order tensor is called a *scalar* while a first-order tensor is called a *vector*.

The set of all tensors of a given order forms a vector space, so a set of first-order vectors forms a vector space, just like we are familiar with. We denote the set of all tensors of order m by the mth tensor power $\mathcal{T}^m$:

$$\mathcal{T}^m = \underbrace{V \otimes V \otimes \cdots \otimes V}_{m \text{ times}} = V^{\otimes m}$$

The set of all the different orders of vector spaces forms an **algebra**[2] $\mathcal{T} = \bigcup_m T^m$. Just as vector spaces have a basis, so too do tensor spaces. For two vector spaces V and W with the bases $\boldsymbol{e}_i$ and $\boldsymbol{\theta}_i$, respectively, then the tensor product of these two vector spaces[3] is given by $V \otimes W$, and the basis of this new space is given by $\boldsymbol{e}_i \otimes \boldsymbol{\theta}_i$. A tensor in this space is a linear combination of the basis tensors and their components given by T^{ij}, just like any vector can be written. For two vectors in V, we can write their tensor product using this idea:

$$\boldsymbol{v} \otimes \boldsymbol{u} = \sum_{ij} T^{ij} \boldsymbol{e}_i \otimes \boldsymbol{e}_j$$

Using $\boldsymbol{v} = \sum v^i \boldsymbol{e}_i$ you can show that, in this particular case, $T^{ij} = v^i u^j$.

Now that you are comfortable with the above idea of a tensor, we can define a more general tensor, with only a slight addition for the inclusion of the dual space. A **general tensor** is written as $\mathcal{T}^m_n$:

$$\mathcal{T}^m_n := \underbrace{V \otimes V \otimes \cdots \otimes V}_{m \text{ times}} \otimes \underbrace{V^* \otimes \cdots \otimes V^*}_{n \text{ times}} = V^{\otimes m} \otimes (V^*)^{\otimes n}$$

Our previous definition of tensors is just the case where $n = 0$. Tensors may be viewed in two ways: firstly, as elements of a tensor product space, as we are familiar with, and, secondly, as multilinear maps themselves, although we do not explore this here. Furthermore, the components of a tensor can be **symmetric** ($T_{ij} = T_{ji}$) or **antisymmetric** ($T_{ij} = -T_{ji}$). Just like scalar and vector fields take a point and assign it a scalar or a vector, tensor fields assign a tensor to the point!

---

[2]It is common practice to denote the set of all things by the same or similar symbols, so we have to be careful with indices, which are the clues to identifying what we are dealing with.

[3]Yes — we can take the tensor product of the spaces, in addition to the elements in them!

We can define another mathematical object, the **exterior product** or the **wedge product**. This object may seem quite vague at first but it is the staple of an entire branch of theory called **wedge calculus**, which happens to be what most of geometrical physics is based on. Asking why it is relevant is similar to asking why the dot product or the cross product are used to describe physics; the answer is, it works, no matter how abstract it may appear at first viewing. Let $v, u \in V$:

$$v \wedge u := v \otimes u - u \otimes v$$

We call the wedge product of two vectors a 2-vector, and the vector space of all 2-vectors is denoted $\bigwedge^2 V$. The wedge product is **antisymmetric**, such that

$$v \wedge u = -u \wedge v \tag{37.0.8}$$

This indicates that the wedge product of any vector with itself is zero ($v \wedge v = 0$), since only zero is equal to its own negative. The wedge product is also **associative** and **distributive**, meaning

$$v \wedge (u \wedge w) = (v \wedge u) \wedge w, \quad \text{and} \quad v \wedge (u + w) = v \wedge u + v \wedge w \tag{37.0.9}$$

Furthermore, if e_i is a basis for V, then $e_i \wedge e_j$ is a basis for $\bigwedge^2 V$, such that, for two vectors $v, u \in V$, we can write the following:

$$\omega = v \wedge u = \sum_{ij}(v^i e_i \otimes u^j e_j - u^j e_j \otimes v^i e_i) = \sum_{ij} v^i u^j e_i \wedge e_j$$

This shows that we pull coefficients out front. Most of the time when we use the wedge product in the main text, we will always look to collect like terms. This just means that we take a differential and observe what the coefficients are and what it is being "wedged" with. For instance, in the following expression we might say, 'collect the terms in dq':

$$dp \wedge dq - \dot{q}dp \wedge dt - \dot{p}dt \wedge dq$$

The thought process is to look for a dq term, write out its coefficient and then cancel the wedge sign and write in the other side's differential. The first term would give us dp the third term would give us $-\dot{p}dt$. We repeat this for all the wedge products containing dq (making sure that they are the same way around). Then, we collect all the terms we obtained into a bracket (e.g. $(dp - \dot{p}dt)$), and then stick a wedge in-between all the brackets we get via this procedure (e.g. $(dp - \dot{p}dt) \wedge \cdots$) and repeat for dp, and so on.

What about the wedge product of more than two vectors and how about the wedge product of p vectors? The wedge product of p vectors is called a p-**vector**, and the space generated by all the p-vectors is denoted $\bigwedge^p V$. A basis for this space is given by $e_I := e_{i1} \wedge \cdots \wedge e_{ip}$, where the I denotes an **ordered index** from 1 to p that acts as bookkeeping from the original vector space's basis. The dimensionality of the space of p-vectors is given by $\dim \bigwedge^p V = \binom{n}{p}$ where $\dim V = n$. We call the collection of all of these $\bigwedge^p V$ spaces an **exterior algebra** or **Grassmann algebra** and denote it as $\bigwedge V$ for all $p = 0, 1, 2 \ldots$; the exterior algebra can be thought of as a composition of $\wedge^p V$ spaces such that $\bigwedge V = \bigoplus_{i=0}^{\dim V} \bigwedge^i V$. Just as with normal vectors, we can write a p-vector in the space of p-vectors as $\omega \in \bigwedge^p V$, where β^I is the components of ω with respect to e_I:

$$\omega = \sum_I \beta^I e_I$$

We can regenerate any of the previous theory about scalars and vectors by letting p equal 0 and 1, respectively (e.g. $\bigwedge^0 V = \mathbb{F}$, or $\bigwedge^1 V = V$). Some authors will stick a factor of a half in the definition of the wedge product, or, for p vectors $1/p!$, but it doesn't matter too much in the long run, and computations can be altered accordingly with minimal disturbance.

Finally, purely for the enjoyment, we will mention **Grassmann variables**, which crop up when defining classical fermionic fields. A Grassmann variable θ_i commutes with ordinary numbers x but anticommute, with other Grassmann variables:

$$\theta_i \theta_j = -\theta_j \theta_i \qquad \theta_i x = x \theta_i \tag{37.0.10}$$

Interestingly, since anything equal to its own negative (e.g. $\theta_i \theta_i = -\theta_i \theta_i$) must be zero exactly, Grassmann variables are non-zero square roots of zero!

38
Differential Geometry

The subject of differential geometry is vast, and a discussion of the fundamentals starts well before manifolds, but it is hoped that this introduction is satisfactory for a working knowledge. The reality of phase space or a configuration space would not be a very nice place to work if all we knew was linear algebra. Linear algebra requires the spaces to be linear so that we can define our operations consistently. This is not always the case and we need a more general space to work with. An example that will give a heuristic feel to what we are about to do is the following; define some surface that globally we can't work with, due to its surface profile. Then, zoom in really close until it is locally flat. Like a football, globally it is round but, up close, it is like lots of little flat patches. In this way, we can define operations and do calculus between different points on the surface.

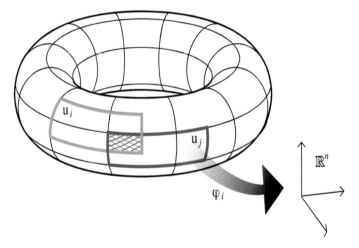

Fig. 38.1: The topological space is depicted here as a torus. On the torus there are coordinate charts, as demonstrated by the blue and red patches. Their intersections are compatible with one another. Locally, each point on the torus is homeomorphic to a Euclidean vector space.

Lagrangian & Hamiltonian Dynamics. Peter Mann, Oxford University Press (2018).
© Peter Mann. DOI: 10.1093/oso/9780198822370.001.0001

Looking at a torus, which we take to be our **topological space** in this example, we can define little patches on it called **open sets**, which are denoted $\{U_i\}$. A topological space is a space endowed with a *topology*. A topology is a collection of open sets. Each of these open sets on the topological space is called a **coordinate patch** or *coordinate neighbourhood*. A *coordinate map* $\{\varphi_i\}$ maps the coordinate patch on the topological space to an $\mathbb{R}^n$ Euclidean vector space denoted $\varphi_i : U \to \mathbb{R}^n$ bijectively (one to one). If the coordinate patch has the *local coordinates* $\{x^i\} = (x^1, \ldots x^n)$, then $(\varphi \circ \varphi^{-1})(x^1, \ldots, x^n) = \varphi(x^1, \ldots, x^n)$, that is to say, when we use $p \in M$ in real-life physics, what we are actually referring to is the result of φ on U where the result is denoted by $\varphi(x^1, \ldots, x^n)$. These maps are subject to two conditions:

- *They must be homeomorphic to a piece of Euclidean vector space.* This means the maps are continuous and invertible, meaning that, if $f : A \to B$ and $g : B \to A$ are smooth, then both A and B are said to be *homeomorphic*.

- In general, we can not always find one open set to cover the entire topological space, so we have to use a patchwork of open sets. If two coordinate patches overlap, then their maps φ_i and φ_j must be *compatible* with each other and we write $\varphi_j \circ \varphi_i^{-1}$, which are called homeomorphic *transition functions*. It ensures that the patches are sort of seamless and that results obtained using one coordinate patch agree with another.

Each patch has a map to $\mathbb{R}^n$ and, together, they make a **chart** (U_i, φ_i). The collection of all charts is an **atlas**. In this way, we can say that the topological space is *locally Euclidean*, meaning that at any point, it can be thought of as a slightly deformed bit of Euclidean space $\mathbb{R}^n$; this allows us to define **local coordinates** that are compatible with other coordinate patches. Therefore, we have entirely covered out topological space with a patchwork of open sets that can be mapped to pieces of Euclidean space that are all compatible with each other[1]. A **smooth manifold** consists of a topological space with a set of coordinate patches that completely cover the surface, together with the set of compatible maps to the local Euclidean regions. That is the definition of a manifold, *a topological space that locally resembles a Euclidean space but globally might be complicated*. The manifold is actually the *pair* of both the topological space and the atlas. When we are talking about points on a manifold, what we really use is $\varphi(x^1, \ldots, x^n)$ to describe the point in the easy-to-use Euclidean space.

This is the most basic type of manifold but there are many different additional structures we can endow our basic manifold with in order to describe different physical things. The most important to physics and our story is the *differentiable* or *smooth* manifold, which is suitably equipped for calculus due to its additional *differentiable*

---

[1]There are subtleties that we have overlooked, such as the requirement that the topological space be *Hausdorff* such that every point *can* be represented by an open set, the idea of closeness between points, that the open sets are *countable*, so that local definition can be extended to global definition and that the manifold is *oriented*. A manifold is orientable if all the transition functions in the atlas have positive Jacobian determinants. But let's assume that these are a given.

structure, and it this type of manifold that forms the basis for the study of differential geometry.

Another class of manifold is the *geometric* manifold, which is equipped with an additional structure called a *geometry*; this allows us to measure the distance and angles between vectors. As we will see in a moment, vectors do not live on the manifold itself; instead, they belong to the *tangent spaces* of the manifold. Therefore, what we need is an inner product on the tangent space, and, to be lavish, we would also like to do calculus with these vectors, meaning we want a *smooth inner product on the tangent space*. We call this smooth inner product a *metric tensor* and, depending on its signature, it gives rise to two different, yet related, types of manifold. If the **signature** of the metric tensor is positive, then we call the manifold a **Riemannian manifold**; if it is not, then we call it a **pseudo-Riemannian manifold**. The metric tensor adds a rather rich structure onto the manifold such that it suddenly becomes stiff and unyielding so that *distance* and *curvature* (the failure to be flat) can be defined.

The manifolds we will be interested in are even-dimensional **symplectic manifolds**[2] and their odd-dimensioned counterparts **contact manifolds**[3]. Before we study these, we have to define a few extra ideas and concepts. On smooth manifolds, we can define a **submanifold**, which is a smooth manifold with the inherited topology from the manifold it is contained in; consider it a subset of the original manifold. There are two types of sumanifolds to be wary of, **immersed submanifolds** and **embedded submanifolds**; but, for our discussion, we need not worry too much: an embedding is a stronger version of an immersion and retains the topology of the parent space, while an immersion is a weaker statement. A **hypersurface** is a smooth submanifold of M of dimension $(n-1)$, so a hypersurface in $\mathbb{R}^3$ would be a 2-dimensional surface, in $\mathbb{R}^n$ it would be an $(n-1)$-dimensional surface. A **foliation** is the decomposition of M into a disjoint union of submanifolds of lesser dimension; we call these **leaves**.

A **tangent vector** X_p at a point $p \in M$ on manifold M is the linear *derivation* at p which can be viewed as a directional derivative operator. The set of all tangent vectors at p is called the **tangent space** (denoted T_pM) where an element of the tangent space is a tangent vector. For a coordinate patch U on a manifold M, we define the coordinates at point $p \in M$ to be $\{x^i\}$, where the index runs to the dimensionality of the manifold: $i = 1, 2, \ldots, n$; therefore, all the tangent spaces of a manifold have the same dimensionality as each other, since the derivation is linear. A basis for the tangent space is given by $\{\partial/\partial x^i|_p\}$, so we can write any vector $X_p \in T_pM$ as follows:

$$X_p = \sum_{i=1}^{n} X^i \frac{\partial}{\partial x^i}\bigg|_p \tag{38.0.1}$$

[2]The domain of classical mechanics.
[3]The domain of thermodynamics.

The scalars $X^i \in \mathbb{R}$ are the components of the tangent vectors with respect to a given basis. A **vector field** X on M is a smooth mapping from $p \mapsto X_p$ such that every point p on a smooth manifold M has an associated tangent vector X_p in the tangent space $T_p M$. We can express a vector field in terms of the local coordinates of $p \in U$; here, we define $X^i(x)$ as a function of the coordinates, *not* a constant coefficient like X^i, while $\{\partial / \partial x^i\}$ is the basis field. For elementary purposes, we determine the local component $X^i(x)$ to be given by $X^i(x) = dx^i/dt$:

$$X = \sum_{i=1}^{n} X^i(x) \frac{\partial}{\partial x^i} \tag{38.0.2}$$

We don't really consider the tangent space as providing additional structure to a manifold, since, by definition, every point has an associated tangent space; it is the choice of a specific vector field and, hence, the specific choice of vector in the tangent space that defines the additional structure a manifold is endowed with. Vector fields are smooth functions which act on a point on a manifold and output vectors in the tangent space; as shown above, they can be expressed in local coordinates and basis fields which can be used to make a patchwork of local vector fields over the manifold.

A **projection** π is a smooth mapping that takes a vector in the tangent space to its point on the manifold $X_p \mapsto p$ where the vector was attached, such that $\pi \circ X = id_M$ (applying a vector field and then a projection above a point) is an identity on M. The set of all tangent vector spaces for all p on M is the **tangent bundle** of M, denoted $TM = \bigcup_{p \in M} T_p M$; it is the *disjoint union*, meaning that there is no overlap between $T_p M$ and $T_q M$ where $p, q \in M$. The vector bundle can itself be treated as a manifold, where the original manifold is called the **base**, and the great thing is that this definition allows us to retain a lot of the concepts from linear algebra, since every point of a tangent bundle is a vector space! The **fibres** of the tangent bundle are the individual tangent spaces and, importantly, the tangent bundle is twice the dimension of the original manifold, since it consists of the pair (p, X), which defines it. Having a chart on the original manifold, (φ_i, U_i), induces a chart on the tangent bundle, too, $(T\varphi_i, TU_i)$, which we call the bundle chart. Any vector bundle is the set of fibre spaces. We can now think of a vector field as mapping a point on the base space to a fibre of the bundle $X : M \to TM$, and the set of all smooth vector fields on M as $\mathcal{X}(M)$ (see figure 38.2) .

The dual vector space of the tangent space is called the **cotangent space** $T_p^* M$. An element of the cotangent space is called a **cotangent vector** ω_p with respect to the basis $\{dx^i|_p\}$ and the coefficients ω_i. The **cotangent bundle** $T^* M$ is the set of all the cotangent spaces over M:

$$\omega_p = \sum_{i=1}^{n} \omega_i dx^i \tag{38.0.3}$$

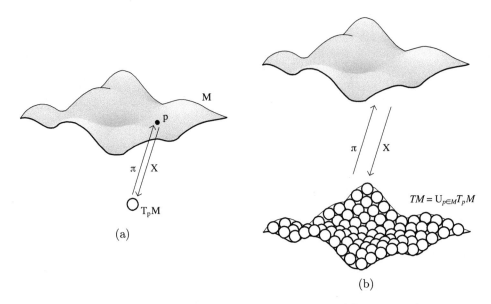

Fig. 38.2: The tangent space above a point on the base space (a), and the disjoint union of the tangent spaces or the tangent bundle (b).

A **smooth covector field** or, for short, a **1-form** ω is a smooth map from $p \mapsto \omega_p$; it maps each point on a manifold to an element of the cotangent space of that point. The 1-form can be expanded as a linear combination of smooth functions on the coordinate patch $\omega_i(x)$ and the basis of the cotangent space:

$$\omega : M \to T^*M \qquad \text{with} \qquad p \mapsto \omega_p \in T_p^*M$$

$$\omega = \sum_{i=1}^{n} \omega_i(x) dx^i \tag{38.0.4}$$

It is well known from linear algebra that the linear mappings from elements of the tangent space to $\mathbb{R}$ span the vector dual space to the tangent space, that is to say, the cotangent space. *Therefore, elements of the cotangent space ω_p are themselves linear maps* (confusingly, also called forms) *that map vectors in the tangent space onto the reals:*

$$\omega_p(X_p) \mapsto \mathbb{R} \tag{38.0.5}$$

Further, since the tangent vectors $X_p \in T_pM$ are generated by the vector field X on a point, we can say that, for $p \in M$, $\omega_p(X_p) \in \mathbb{R}$ can be generalised to a smooth map $\omega(X)$ for $X \in \mathcal{X}(M)$; this is a very important concept, *a 1-form acts on vector fields.*

We can write an expression for the dual pairing between the tangent space and the cotangent space in terms of their bases, note the index is raised on the 1-form basis:

$$\left\langle \frac{\partial}{\partial x^i}, dx^j \right\rangle = \delta_i^j$$

After we discussed tensors in chapter 37, you may recall, we then defined an object called the wedge product and called them p-vectors. We then said that the space of all p-vectors at a point on M was $\bigwedge^p V$, and the set of all these forms an algebra $\bigwedge V$ over M. We now utilise this concept.

A k-**form**[4] $\overset{k}{\omega}$ on U is a smooth map $p \to \omega_p$ where ω_p is an element of the k-fold direct product of the cotangent space $\bigwedge^k(T_p^* M)$ at $p \in U$. As with 1-forms, k-forms map elements of $\bigwedge^k(T_p^* M)$ to the reals. Here, $\omega_p \in \bigwedge^k(T_p^* M)$ maps a tangent vector $X_p \in \bigwedge^k(T_p M)$ to $\mathbb{R}$ or, equivalently, it acts on k vector fields which, yield the elements of the k-th fold tangent product space $\bigwedge^k(T_p M)$, just as in the one-dimensional example above:

$$\omega_p(X_1, \ldots, X_k) \mapsto \mathbb{R} \tag{38.0.6}$$

The set of all k-forms on M is a vector space denoted[5] $\Omega^k(M)$, where $\Omega(M) = \bigoplus_{k=0}^n \Omega^k(M)$ is an algebra. It is important to realise that $0, 1, 2, \ldots$-forms are just special cases of k-forms. Note that, while $\omega_I = \omega_{i1,i2\ldots ik}$ is a *constant* for the given basis in the linear combination, $\omega_I(x) = \omega_{i1,i2\ldots ik}(x)$ is a *function* of the coordinate patch, so it is local to a point. As we know, a 1-form can be written as a linear combination of basis 1-forms dx^i and the k-form can be written in terms of the basis $dx^{i1} \wedge \cdots \wedge dx^{ik}$:

$$\overset{k}{\omega_p} = \sum_I \omega_I dx^I = \sum_{i_1 < i_2 < \cdots < i_k} \omega_{i1,i2\ldots ik} dx^{i1} \wedge \cdots \wedge dx^{ik} \tag{38.0.7}$$

$$\overset{k}{\omega} = \sum_I \omega_I(x) dx^I = \sum_{i_1 < i_2 < \cdots < i_k} \omega_{i1,i2\ldots ik}(x) dx^{i1} \wedge \cdots \wedge dx^{ik} \tag{38.0.8}$$

This is built from the wedge product, which we defined earlier in 37, the wedge product of a k-form and a j-form is a form of degree $(k + j)$. For example, a 2-form is a linear combination of the coefficients ω_{ik} and the basis $dx^i \wedge dx^k$; just by taking the wedge product and pulling out the coefficients. It is a bijective mapping from the direct product space $TM \times TM$ to $\mathbb{R}$:

$$\overset{2}{\omega} = \overset{1}{\omega} \wedge \overset{1}{\omega} = \sum_{i<k=1} \omega_{ik} dx^i \wedge dx^k$$

---

[4]It is defined on U rather than M because the coordinates are defined only in this local neighbourhood; if we have global coordinates, then it is fine to define ω on M.

[5]Some authors denote the set of all k-forms as $\chi^*(M)$, to signify that it is the dual of the set of all vector fields.

We can evaluate a 2-form on two representative vector fields: $X = x^i \partial_i$ and $Y = y^i \partial_i$, given $\overset{2}{\omega} = \alpha \wedge \beta$ and $\alpha = a_i dx^i$ and $\beta = b_i dx^i$, by writing $\overset{2}{\omega}(X, Y)$. In this case, the evaluated product is given by

$$\overset{2}{\omega}(X, Y) = \alpha(X)\beta(Y) - \beta(X)\alpha(Y) \tag{38.0.9}$$

where $\alpha(X) = a_i x^i$, since $\langle dx^i, \partial_j \rangle = \delta^i_j$. Then, it simply becomes a matter of pairing the correct components to evaluate the expression.

The **kernel** of a k-form $\overset{k}{\omega}$ on $\dim(M) = n$ is given by $\ker(\overset{k}{\omega})$:

$$\ker(\overset{k}{\omega}) = \{v \in M \mid \overset{k}{\omega}(v, v_2, \ldots, v_k) = 0, \forall\, v_2, \ldots, v_k \in M\} \tag{38.0.10}$$

If the k-form is decomposable, then the kernel has dimension $(n-k)$. For $\dim(M) = n$, the **Hodge star** takes a k-form to an $(n-k)$-form:

$$\star : \overset{k}{\bigwedge} M \to \overset{n-k}{\bigwedge} M \tag{38.0.11}$$

The action of the star operator on forms α and β satisfies the following:

$$\alpha \wedge \star\beta = \beta \wedge \star\alpha$$
$$\star \star \alpha (-1)^{k(n-k)} \alpha$$
$$\star (\phi_1 \alpha + \phi_2 \beta) = \phi_1(\star\alpha) + (\phi_2(\star\beta))$$
$$\alpha \wedge \star\alpha = 0$$

The **exterior derivative**, or *Cartan derivative*, of a k-form is denoted d and is thought of as the generalisation of the gradient, curl and divergence from vector calculus. The exterior derivative is a linear operator that acts on a local k-form and maps it to a $(k+1)$-form on M; more formally, we write $d : \Omega^k(M) \to \Omega^{k+1}(M)$. It is another way to generate forms, in addition to taking the wedge product of 1-forms for the appropriate number of times. Thus, if d were to operate on the k-form $\overset{k}{\omega}$, we would write the result $d\overset{k}{\omega}$ as follows:

$$d\overset{k}{\omega} = \sum_{i1 < \cdots < ik} d\omega_{i1, \ldots i_k}(x^1, \ldots, x^n) \wedge dx^{i1} \wedge \cdots \wedge dx^{ik}$$
$$= \sum_I \sum_{i=1}^{n} \frac{\partial \omega_I}{\partial x^i} dx^i \wedge dx^I \tag{38.0.12}$$

If the exterior derivative of a k-form is zero ($d\omega = 0$), we call it a **closed form**. A k-form ω is **exact** if there exists a $(k-1)$-form β for which $\omega = d\beta$; by the **Poincaré**

lemma, all closed forms are exact. This follows on from a fundamental observation that applying the exterior derivative twice to something is zero ($d^2\omega = 0$), so, if $\omega = d\beta$, then $d\omega = d(d\beta) = d^2\beta$, which is always zero.

We can now understand the importance of forms with respect to physics. Functions are forms of degree 0; the total derivative is an example of an exterior derivative that converts a 0-form into a 1-form:

$$df = \sum_{i=1}^{n} \frac{\partial f}{\partial x^i} dx^i$$

In Newtonian mechanics, we define a conservative force $\vec{F}(\boldsymbol{r})$ as one that is the negative gradient of a potential energy $U(\boldsymbol{r})$ such that $\vec{F} = -\vec{\nabla} U$. In differential forms, we can say that a 1-form $\phi = \vec{F} \cdot d\boldsymbol{r}$ is exact: $\phi = -dU(\boldsymbol{r})$. Then, the curl of the force should vanish ($\vec{\nabla} \times \vec{F} = 0$) since, by the Poincaré lemma, $d\phi = 0$ is closed.

We can take the exterior product of two forms of degree k and j, according to the rule below, which you can use to derive the product rule for two 0-forms:

$$d(\overset{k}{\omega} \wedge \overset{j}{\omega}) = (d\overset{k}{\omega}) \wedge \overset{j}{\omega} + (-)^k \overset{k}{\omega} \wedge (d\overset{j}{\omega})$$

Exercise 38.1 Prove that the exterior derivative of a k-form $\overset{k}{\omega} = f dx^{i_1} \wedge \cdots dx^{i_k}$ and a j-form $\overset{j}{\omega} = g dx^{i_1} \wedge \cdots dx^{i_j}$ is given by the above formula.

This is done as follows:

$$d(\overset{k}{\omega} \wedge \overset{j}{\omega}) = d(fg dx^{\alpha_1} \wedge \cdots dx^{\alpha_k} \wedge dx^{\beta_1} \wedge \cdots dx^{\beta_j})$$

$$= (g df + f dg) \wedge (dx^{\alpha_1} \wedge \cdots \wedge dx^{\alpha_k} \wedge dx^{\beta_1} \wedge \cdots \wedge dx^{\beta_j}) \qquad (38.0.13)$$

$$= d\overset{k}{\omega} \wedge \overset{j}{\omega} + (-1)^k \overset{k}{\wedge} d\overset{j}{\omega} \qquad (38.0.14)$$

The **interior derivative** (*interior product* or *contraction*) is a linear map from a k-form to a $(k-1)$-form, written $\iota_X : \Omega^k(M) \to \Omega^{k-1}(M)$. If ω is a k-form and there is a set of vector fields $X_1, \ldots, X_k$, we say that the interior product is a *contraction* of a form and a vector field (it is important to note that you can't use 0-forms (functions) for this, since there are no vector field arguments left (you can't have negative forms):

$$\boxed{(\iota_{X_1}\omega)(X_2, \ldots, X_k) := \omega(X_1, X_2, \ldots, X_k)} \qquad (38.0.15)$$

We say that $\iota_X \omega$ is the interior product of X and ω. Remember that, earlier, we said that a 1-form is a function of vector fields; thus, the interior derivative of a 1-form ω is written $\iota_X \omega$ and is equal to $\omega(X)$. Contraction with a vector field is just evaluating the form on the vector field (i.e. determining what the component $X^i(x)$ of the vector field $X = \sum X^i(x)\partial_x$ is in the direction of a particular basis ∂_x). Note that this is really important for later!

If we have a vector field X which is the linear combination of components and basis vectors $X = \sum X^i(x)\partial_x$, then we have the following rule: $\iota_{\partial_j} dx^k = \delta_j^k$. That is to say, the interior product of a vector field and a 1-form is the inner product. This is going way back to the basics of linear algebra; if we had a basis and its duel basis and we take their inner product, then we get the delta function, since one is the contravariant vector and one is the covariant vector.

A **tensor field** on a manifold assigns a smooth tensor to every point on the manifold; it is the generalisation of the vector field and form, in other words, vector fields and form fields are just particular cases of tensor fields. The space generated by the tensor product is the direct analogy of our general tensor defined previously, except, instead of using V and its dual V^*, we use T_pM and T_p^*M, such that the tensor product space is given by $(T_pM)^{\otimes m} \otimes (T_p^*M)^{\otimes n}$ and the basis for this space is formed from the bases of the constituent tangent and cotangent spaces. The **tensor product bundle** of a manifold $TM^{\otimes m} \otimes TM^{*\otimes n}$ is defined to be the fibre bundle $\mathcal{T}_n^m M$ of dimension $(n+m)$, a section of which is a tensor field:

$$\mathcal{T}_n^m M := \underbrace{TM \otimes \cdots \otimes TM}_{m \text{ times}} \otimes \underbrace{T^*M \otimes \cdots \otimes T^*M}_{n \text{ times}}$$

A k-fold direct product space $\bigwedge^k(T_p^*M)$ at $p \in U$ is a sub-bundle of $\mathcal{T}_k^0 M$ (i.e. $\mathcal{T}_1^0 M = T^*M$). We should understand that the components of vector fields are contravariant while the components of forms are covariant with respect to the transformation properties.

Suppose we have two vector bundles TM and T^*M over a common manifold base M with a smooth function f defined on TM that maps to the reals. The **fibre derivative** $\mathbb{L}$ is a fibre-preserving map between the two bundles over the same base point p that takes a fibre in TM to a fibre in T^*M bijectively. The fibre derivative of f at a given point is written $\mathbb{L}_f : TM \mapsto T^*M : X_p \mapsto Df_p(X_p)$. If we assume that the function is regular (i.e no singularities), then $\mathbb{L}_f$ is a diffeomorphism and is valid for all chart domains. Since the coordinate chart of the tangent bundle $\{x^i, \dot{x}^i\}$ has twice as many entries as that of the original manifold $\{x^i\}$, any function on TM will be a function of two arguments. We can therefore take two derivatives of this function: the partials with respect to each coordinate. Therefore, we obtain

$$D_1 f^{(\varphi)} \circ T\varphi^{-1} = \left\{ \frac{\partial f}{\partial x^i} \right\} \tag{38.0.16}$$

$$D_2 f^{(\varphi)} \circ T\varphi^{-1} = \left\{ \frac{\partial f}{\partial \dot{x}^i} \right\} \tag{38.0.17}$$

Since, by the definition of a fibre derivative, the base point is unchanged, we only take the second of these to mean anything and we write the application of the fibre derivative on f as $\mathbb{L}_f(x^i, \dot{x}^i) \mapsto \left(x^i, \frac{\partial f}{\partial \dot{x}^i}\right)$. This is the geometrical version of the Legendre transform.

We now move to one of the more vague ideas in differential geometry: the idea of the *pushforward* and *pullback*. At this point, forms and vector fields may seem similar in that they both take a point on M and output an element of either $\bigwedge^n(T_p^*M)$ or $\bigwedge^m(T_pM)$. Differential forms can be moved from one manifold to another, using a smooth map. If we have two manifolds M and N with a smooth map between them, ϕ, and a function f on N that outputs a real number $\mathbb{R}$, as functions do, then ϕ^*f is called the **pullback** of f by ϕ such that the function f can now take M directly to the result $\mathbb{R}$ (i.e. it is now a function on M rather than N). The pullback, in essence, takes a function and changes its variables from those coordinates on N to those on M and is the basic principle behind changing coordinates in calculus: $(\phi^*f)(p) = f(\phi(p))$. In regard to forms, the pullback just takes a form on N and pulls it back to a form on M.

The **pushforward** is defined in a similar manner yet is of fundamental importance, highlighted by using its other name, the *differential*. The pushforward is a linear map from the tangent space of M at p to the tangent space of N at $\phi(p)$; it pushes tangent vectors on M to tangent vectors on N. We can write the pushforward as a linear map $\phi_* : T_pM \to T_{\phi(p)}N$. That is to say, the differential of ϕ at p is a linear map between the spaces of tangent vectors. If $f : N \to \mathbb{R}$ is a smooth function on N, then we can write the pushforward $(\phi_*X_p)_{\phi(p)}(f) := X_p(\phi^*f)$ for tangent vectors X_p.

Defining the pushforward of a vector field is a little more complicated and is not always possible; it depends on whether ϕ is a *diffeomorphism*[6]. If it is, then defining the pushforward of a vector field is viable and it can be written as $(\phi_*X)_{\phi(p)}(f) := X_p(\phi^*f)$. The pullback of a vector field by a diffeomorphism is the inverse map of the pushforward of a vector field: $\phi^*X := (\phi^{-1})_*X$.

At each point on a manifold, we have a vector field assigning a vector in the tangent space of each point. A curve on a manifold over the interval $I = [t_0, t_1]$ is denoted $\gamma(t) : I \to M$ and is often thought of as a continuous set of points in M. It is more rewarding to think of a curve as a smooth map from $\mathbb{R}^1$ into the manifold so that, therefore, each curve has a point on M, or each point is *parameterised* by a number (we could label this parameter as t, for instance). Every curve also has a tangent vector $\gamma'(t)$. If the curve's tangent vector is also the tangent vector assigned by the vector field over the same point on the manifold, $X_{\gamma(t)}$, $(\gamma'(t) = X_{\gamma(t)})$, as pictured in 38.3, then we call the curve an **integral curve**. For a point on a manifold $p \in M$ with the local coordinates $(x^1, \ldots, x^n)$, we can write the differentiation of the coordinates of the curve by the parameter t in terms of the vector field of that point:

---

[6]If a function is diffeomorphic, both the function and its inverse have smooth differentials; this condition is stronger than that required for a homeomorphism.

$$\frac{dx^i(t)}{dt} = X^i(x^1(t), \dots, x^n(t)) \qquad (38.0.18)$$

This is nothing more than an ODE, which we *integrate*; to find the unique solution curve for which this is true, see *Cauchy-Lipschitz theorem* (see chapter 32).

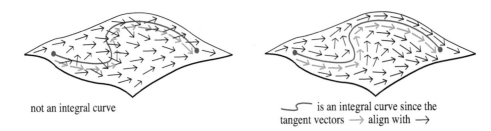

not an integral curve $\smile\!\!\!\frown$ is an integral curve since the
tangent vectors $\longrightarrow$ align with $\longrightarrow$

Fig. 38.3: A curve is an integral curve of a vector field if their tangent vectors coincide.

For all the points on the manifold, the map $\gamma_p(0) \to \gamma_p(t)$ is the **flow** Ψ or *congruence* of M *generated* by X. That is to say, when we advance the parameter t for all of the points on the manifold at the same time, the flow of the curve is the set of tangent vectors generated by the vector field on M. Note that it is a map onto the *same* manifold, but just with the smooth parameter advanced by a small amount Δt. If the set of maps are diffeomorphic for all Δt, then the family of these maps is a *one-parameter Lie group* (see below). If a vector field generates flow all over the manifold, a global flow, then we say it is a *complete vector field*.

The idea of finding the derivative of a vector field or a 1-form on a curved surface is not possible when using the standard notion of differentiation, since, as with any derivative, we require the limit of the quotient of the difference between two points and the *difference* between them, and since these values are, by definition, from different tangent vector spaces, it makes no sense. We remarked earlier that the concept of distance on a manifold is not well defined for this reason and we require extra structure to cope with this, which is called a *connection* and the *covariant derivative*, which are in the context of Riemannian manifolds and so a bit too involved for our discussion. Therefore, we ask, is there a way to define the derivative of any general tensor field on non-geometric manifolds?

Yes, and flow is the basic key to understanding this. For two vectors, one at t and another at $t+\Delta t$ if the flow and the vector field is invariant under the shift in parameter for all shifts we could make (although, in general, they are not) then we say it is **Lie dragged**. If we Lie drag the $(t + \Delta t)$ vector back to the point at t, we define a new vector where the difference between them is unique and we can find the differential.

This is the idea of the **Lie derivative** $\mathcal{L}_X$. For a tensor field T, we can imagine pulling back it to the point p:

$$(\mathcal{L}_X T)_p = \lim_{t \to 0} \left.\frac{\Psi_t^* T - T}{\Delta t}\right|_p \qquad \textbf{Lie derivative} \qquad (38.0.19)$$

So we have used the pullback, the vector field and, most importantly, the flow at a point to define the derivative of a tensor field at that point (i.e. how much the tensor changes as it evolves along the integral curve generated by X); as you can see, it can't always be done if the tensor field is not constant when Lie dragged, and it is a non-local operation, that is to say, it depends on the tensor field in a neighbourhood of a point.

The bilinear **Lie bracket** $[\cdot, \cdot]$ is perhaps more familiar under the name *the commutator* and, for any two vector fields X and W on a smooth manifold M or $X, W \in \mathcal{X}(M)$, their Lie bracket yields a third vector field on M, denoted $[X, W] := (XW - WX)$. For any vector field W, we can write the following relation using the expansion about the $X^i(x)$ functions and the basis $\{\partial/\partial x^i\}$ given previously (note that the Lie derivative of a vector field along the flow generated by another is the Lie bracket):

$$\mathcal{L}_X W = [X, W] = \left[X^k \frac{\partial}{\partial x^k}, W^l \frac{\partial}{\partial x^l} \right] = \left(X^k \frac{\partial W^l}{\partial x^k} - W^k \frac{\partial X^l}{\partial x^k} \right) \frac{\partial}{\partial x^l} \qquad (38.0.20)$$

Further, for $X, W, Z \in \mathcal{X}(M)$, the Lie bracket satisfies the *Jacobi identity* below. The Lie bracket gives or endows the set of vector fields a **Lie algebra** structure:

$$[[X, W], Z] + [[W, Z], X] + [[Z, X], W] = 0 \qquad (38.0.21)$$

A Lie algebra of smooth vector fields on a coordinate patch U on M is a set of vector fields on U; this is a vector space where any linear combination of its elements is another element, and taking the Lie bracket of two elements is again another element. A Lie algebra is a vector space $\mathfrak{g}$ over $\mathbb{R}$ that is equipped with a map $[\ ,\ ] : \mathfrak{g} \times \mathfrak{g} \to \mathfrak{g}$ that satisfies two properties, namely bilinearity (or the Leibniz condition) and skewsymmetry, in addition to the Jacobi identity, where $X, W, Z \in \mathfrak{g}$ and $a, b \in \mathbb{R}$:

$$[aX + bW, Z] = a[X, Z] + b[W, Z] \qquad (38.0.22)$$

$$[X, Z] = -[Z, X] \qquad (38.0.23)$$

We can use these properties of the Lie algebra to show that, for any three vector fields on a manifold, $[\mathcal{L}_X, \mathcal{L}_W]Z = \mathcal{L}_{[X,W]}Z$ holds. A famous theorem, the *Frobenius'-theorem*, states that, if E is a subbundle of TM, then it is *involutive* if the Lie bracket of two vector fields in E is also a vector field in E. The subbundle E is *integrable* if there is a submanifold on M whose tangent bundle equals E restricted to the submanifold. These can then be extended to get a connected immersed submanifold. The collection of these is called a *foliation* of M. So, the subbundle is integrable if and only if it is a regular foliation of M.

We can take the Lie derivative of a k-form $\omega \in \Omega^k(M)$ with the vector field $X \in \mathcal{X}(M)$ to give **Cartan's formula**:

$$\mathcal{L}_X \omega = (i_X d + d i_X)\omega \tag{38.0.24}$$

It is the **anticommutator** of the interior and exterior derivatives; remember that the interior product is just a map from k-forms to $(k-1)$-forms, while the exterior derivative is a map from a k to $(k+1)$-form, such that the Lie derivative is a derivation of degree 0. Importantly, Lie derivatives can be used to prove invariance of a tensor field under a transformation generated by a vector field. Both the tensor field and the vector field will be physically important in this case; if Cartan's equation equals zero then the k-form is conserved.

Groups G are collections of symmetry transformations on a set. More formally, a group of a set of elements possess a binary product $G \times G \to G$ such that it is associative, an identity exists and an inverse operation exists. In chemistry, we are used to these as rotations and reflections that leave a molecule seemingly unchanged when the transformations act on the molecule. Groups act on sets, and the **group action** is the result. Here, the action of the group is a homeomorphic mapping to the symmetric group or an automorphism. Classical mechanics can be abstracted to groups acting on an algebra of commuting observables. A specific example would be the three dimensional **special Euclidean group** $SE(3)$ on a set of position vectors $\boldsymbol{r} \in \mathbb{R}^3$. The group action can be written as $(O, \boldsymbol{r}_0)\boldsymbol{r} = O\boldsymbol{r} + \boldsymbol{r}_0$ and can be thought of as consisting of a rotational portion from $O \in SO(3)$, where $O^t = O^{-1}$(the group of three-dimensional rotations, called a **special orthogonal group**, for which the transpose is equal to the inverse) and vector translations $\boldsymbol{r}_0 \in \mathbb{R}^3$, although the $SE(3)$ group is more complicated than a simple direct product of two subgroups, due to conditions on the inverse of certain actions. An important property of groups is that they can be multiplied, for instance $(\tilde{O}, \tilde{\boldsymbol{r}}_0)(O, \boldsymbol{r}_0) = (\tilde{O}O, \tilde{O}\boldsymbol{r}_0 + \tilde{\boldsymbol{r}}_0)$. Translations $\mathbb{R}^3 \subset SE(3)$ act by vector addition:

$$\mathbb{R}^3 \times \mathbb{R}^3 \to \mathbb{R}^3 : (I, \tilde{\boldsymbol{r}}_0)(I, \boldsymbol{r}_0) = (I, \boldsymbol{r}_0 + \tilde{\boldsymbol{r}}_0) \tag{38.0.25}$$

while rotations $SO(3)$ act by composition:

$$SO(3) \times SO(3) \to SO(3) : (\tilde{O}, \boldsymbol{0})(O, \boldsymbol{0}) = (\tilde{O}O, \boldsymbol{0}) \tag{38.0.26}$$

where I and $\boldsymbol{0}$ are the identity and null vectors, respectively. Any element $(O, \boldsymbol{r}_0) \in SE(3)$ can be written as the composition of a translation acting on the left of a rotation $(O, \boldsymbol{r}_0) = (I, \boldsymbol{r}_0)(O, \boldsymbol{0})$. In the main text, we don't make much reference to groups apart from the $U(1)$ and $SO(N)$ groups; however, this abstraction can always be performed and the entirety of mechanics can be formulated in a group theoretic language.

A **Lie group** $\mathcal{G}$ is a group of infinitesimal transformations (smooth/continuous, such as when rotating a linear diatomic around the internuclear axis) that may also be regarded as a smooth manifold. Lie groups may be finite- or infinite-dimensional and the properties of each type are quite distinct. In addition, $\mathcal{G}$ has an associated vector space and a dual vector space: the Lie algebra $\mathfrak{g}$ and its dual $\mathfrak{g}^*$. The Lie algebra may be regarded as the infinitesimal of the Lie group. By infinitesimal (best to think in manifold terms), the Lie algebra is defined on the tangent space of the group at the identity. Another way to picture this scenario is if we imagine $\mathcal{G}$ as a base manifold (remember, we said that $\mathcal{G}$ is also a manifold) and then take tangent and cotangent bundles $T\mathcal{G}$ and $T^*\mathcal{G}$, but then restricted their fibres to the identity element I of $\mathcal{G}$. This allows us to write $T_I\mathcal{G} \equiv \mathfrak{g}$ and $T_I\mathcal{G} \equiv \mathfrak{g}^*$. The Lie algebra and the dual are therefore vector spaces (or, if you prefer, any vector space with a Lie bracket satisfying the above listed properties is a Lie algebra). The basis for a Lie algebra are called the **generators** of the group. We can use a tool called the *exponential map* to reconstruct the Lie group from its algebra; however, difficulties arise if more than one Lie group has the same Lie algebra. The dual of the Lie algebra carries a **Lie-Poisson bracket** for $\kappa \in \mathfrak{g}^*$ and $F, G \in \mathcal{C}^\infty(\mathfrak{g}^*)$:

$$\{F, G\}_\pm(\kappa) = \pm\left\langle \kappa, \left[\frac{\delta F}{\delta \kappa}, \frac{\delta G}{\delta \kappa}\right]\right\rangle \tag{38.0.27}$$

where $\langle \cdot, \cdot \rangle : \mathfrak{g}^* \times \mathfrak{g} \to \mathbb{R}$ is a pairing between the Lie algebra and the dual, and $[\cdot, \cdot] : \mathfrak{g} : \mathfrak{g} \to \mathfrak{g}$ is the Lie bracket. If we were to take the Lie bracket of two of the basis vectors of the Lie algebra $e_i e_j \in \mathfrak{g}$, we obtain a linear superposition of basis vectors whose coefficients c_{ij}^k are known as **structure constants**:

$$[e_i, e_j] = c_{ij}^k e_k \tag{38.0.28}$$

The structure constants obey the Jacobi identity and are antisymmetric with respect to interchange of $i \to j$. It is an interesting and extremely useful property that the Lie algebra can be obtained from the structure constants.

Algebroids and groupoids are slight generalisations of groups and algebras; their definitions in traditional group theoretic language are a little too complex for our needs, so we will instead switch to a differential geometry definition. A **Lie algebroid** is a vector bundle over a base manifold with a Lie bracket on its space of vector fields as well as a bundle morphism (called an **anchor**) from the vector bundle to the tangent bundle of the base manifold.

A vector bundle $\pi : E \to M$ over $\dim(M) = n$ is a Lie algebroid if its space of sections has a Lie bracket $[\cdot, \cdot]$; if there is an anchor $\rho : E \to TM$ inducing a Lie algebra homomorphism from vector fields on E to vector fields on M, together they satisfy the Leibniz rule:

$$[X, FY] = \rho(X)F \cdot Y + F[X, Y] \tag{38.0.29}$$

where $F \in \mathcal{C}^\infty(E)$ and $X, Y \in \chi(E)$. A **principal bundle** is a vector bundle P with a fibre-preserving Lie group acting freely on the right: $P \times \mathcal{G} \to P$. The quotient space $P/\mathcal{G}$ is also a manifold with a surjection $\pi : P \to M$. The fibres of P are isomorphic to $\mathcal{G}$ and hence inherit the group structure locally.

We will not need to know too much about Lie groups for our discussion, but they are closely related to an awful lot of ideas in physics! This is because the relationship between smooth symmetries and their generators is exactly described by Lie theory, and symmetries are universal to physics and chemistry, from gauge bosons to general relativity, to crystals, and everything in-between!

We are now going to summarise some basic results in the differential geometry of **jets**. We denote (E, π, M) to be a fibre bundle over base space M with projection π such that $\pi : E \to M$ and, as we know, a fibre bundle can be treated as a fibered manifold (this is more general than specifying E as the tangent or cotangent bundles). The tangent bundle to E is denoted TE. The tangent bundle can actually be considered as a direct sum of two subbundles called the *horizontal* and *vertical* bundles, respectively ($TE = \mathcal{V}E \oplus \mathcal{H}E$), and, for a point $u \in E$, the tangent space decomposes as $T_u E = \mathcal{V}_u E + \mathcal{H}_u E$ such that any vector in the tangent space at u consists of vertical and horizontal components. In this way, a full variation can be split up into the form variation and the coordinate variation — the vertical and horizontal subbundles! Consider this over a trivial bundle $E \equiv M \times \mathbb{R}$ where $\mathbb{R}$ is a smooth time line and M is a configuration space.

Vector fields are solutions to differential equations, so jet theory is focused around vector fields, which we have previously called X in notation; we might also choose to generalise the word vector field to a *smooth section* and denote it by $X \equiv \phi$. In brief, jets are sets of **equivalence classes** of vector fields at a point coinciding up to their rth-order derivative. So, if two vector fields X, at a point p are equal to each other, and their first derivatives are also equal to each other, then we say we have a **first-order jet** at that point. We denote this first-order jet space as $j_p^1 X$ at point $p \in M$. The disjoint union of all the first-order jet spaces makes a first-order jet manifold $J^1 E$ above fibre bundle E. Jets play a role similar to the tangent and cotangent bundles we defined in the discrete case.

More quantitatively, given two local sections $X, Y \in \Gamma_p E$, we define an *equivalence relation* in the set of all sections above a point $\Gamma_p E$ if and only if their values at p coincide such that $X(p) = Y(p)$ and $T_p X = T_p Y$; we are not saying anything about other points in M. This means that, for X and Y, we can write not only *their* equality but their derivatives too; the order of the derivative that they are equivalent to is given by r:

$$X^\alpha(p) = Y^\alpha(p), \qquad \left.\frac{\partial^r X^\alpha}{\partial (x^i)^r}\right|_p = \left.\frac{\partial^r Y^\alpha}{\partial (x^i)^r}\right|_p \qquad (38.0.30)$$

If $r = 1$, then we say we have a *first-order jet of X at p*; we denote $j_p^1 X$. The set of first-order jets, the disjoint union of the spaces $J^1E = \cup_{p \in M} j_p^1 X$, is a first-order *jet manifold* of a fibre bundle[7]. This concept generalises to rth order jets above a point.

We define projections π_1 and $\pi_{1,0}$, called the **source** and **target** projections, respectively, such that $\pi_1 : J^1E \mapsto M$ taking $j_p^1 X \mapsto p \in M$, and $\pi_{1,0} : J^1E \mapsto E$, taking $j_p^1 X \mapsto X(p) \in E$, where $X(p)$ is the result of the vector field (i.e. section along the vector field) at $p \in M$, which is a point in $u \in E$. So far, we have a sort of triple layer of spaces consisting of the base manifold, the vector bundle and the jet manifold.

The *rth jet prolongation* of a section $X \in \Gamma E$ is the map $j^r X : M \mapsto J^r E$; it is itself a section of $\pi_{1,0}$ over every point $p \in M$:

$$(j^1 X)(p) := j_p^1 X \tag{38.0.31}$$

The best way to understand this is that $j^1 X$ inputs a point on M and spits out a point in J^1E; we say it is a section along $\pi_{1,0}$ because it is the exact opposite of this operation. If two functions had the same nth order prolongation (and so map to the same point in J^1E), then they have the same nth-order Taylor polynomial at $p \in M$ (i.e. they are equal up to their nth derivative). We would say the two functions have nth-order **contact**, and the jet space is the set of nth-order equivalence classes.

Just like the situation with the configuration manifold and the tangent bundle, the spacetime manifold induces coordinates on both E and J^1E such that the jet manifold describes the partial derivatives of the coordinates on E over a point on M. If M has the local coordinates (x^μ), E has (x^μ, ϕ^α), and the first-order jet manifold has $(x^\mu, \phi^\alpha, \partial_\mu \phi^\alpha)$; higher-order jets (equivalence classes) will lead to the appearance of higher-order derivatives in the jet's local coordinates. To regenerate particle mechanics, observe the following structures and local coordinates: $M = \mathbb{R}$ with (t), $E = Q$ (a configuration manifold) spanned by (q, t), and the first-order jet bundle is $J^1E = TQ$, with coordinates $(t, q, \dot{q})$.

A general vector field on E with coordinates x^i, u^α is given by the sum of horizontal and vertical terms, each written in terms of a basis and components:

$$X = X^i \frac{\partial}{\partial x^i} + X^\alpha \frac{\partial}{\partial u^\alpha}$$

A horizontal vector field is one where the vertical coefficients vanish ($X^\alpha = 0$) while a vertical vector field is one where the horizontal coefficients vanish. A vertical vector field will generate a vertical flow and, vice versa.

The *vertical differential* of a section $X : M \mapsto E$ of π through the point $X(p) \mapsto u \in E$ maps $T_u E \mapsto \mathcal{V}_u E$. It is the projection of tangent vectors onto their vertical

---

[7]Some authors use $J^1\pi$ in place of J^1E, if you decide to read around, note that π is the projection of E. The same is true with the set of all sections $\Gamma_p E$ or $\Gamma_p \pi$.

components. In the local coordinates of E (x^i, u^α), we write the vertical differential as follows:

$$d_u^v X = \left(du^\alpha - \frac{\partial X^\alpha}{\partial x^i} dx^i \right) \otimes \frac{\partial}{\partial u^\alpha} \tag{38.0.32}$$

We can use the vertical differential to construct a 1-form θ on $J^1 E$, where the values of this 1-form are defined in the vertical bundle $\mathcal{V}E$. A form is called a **contact form** if it is annihilated when the pullback of the prolongation of a section $X : M \mapsto J^1 E$ is $(j^1 X)^* \theta = 0$. A **semi-basic contact form** is one in which the local expression contains differentials of x^μ and ϕ^α only, with none in $\partial_\mu \phi^\alpha$. In local coordinates, every contact 1-form can be written as a linear combination of basis forms θ_j^α that constitute a basis for the space of contact forms on $J^1 E$; this is called the *contact ideal* over the jet manifold. In the local coordinates of the jet manifold $(x^i, u^\alpha, u_i^\alpha)$, we write the 1-form as follows:

$$\theta = \theta_j^\alpha \otimes \frac{\partial}{\partial u^\alpha} = (du_j^\alpha - u_i^\alpha dx^i) \otimes \frac{\partial}{\partial u^\alpha}$$

If we have a natural coordinate system $(x^i, u^\alpha, u_i^\alpha)$ on $J^1 E$, then $X(p)$ is $(x^i, g^\alpha(x^i), f_i^\alpha(x^i))$:

$$(j^1 X)^* \theta = \left(\frac{\partial g^\alpha}{\partial x^i} dx^i - f_i^\alpha dx^i \right) \otimes \frac{\partial}{\partial u^\alpha} = 0$$

Then, $f_i^\alpha = \partial g^\alpha / \partial x^i$. With the definition of a contact form in place, we end our discussion of jets, having more than enough information to cope with the concepts in time dependent-mechanics.

Part VII

Exam-Style Questions

- (Q) Describe both mathematically and qualitatively how time evolution of a canonical system can be viewed as the continuous unfolding of infinitesimal canonical transformations. (7)
- (A) Consider an infinitesimal shift in the coordinates by δq^i and δp_i such that the new canonical variables are given by

$$Q^i = q^i + \delta q^i \qquad P_i = p_i + \delta p_i \quad (1)$$

The generating function of the transformation is a function on the phase space and can be written

$$\delta q^i = \epsilon \frac{\partial G(q,p,t)}{\partial p_i} \qquad \delta p_i = -\epsilon \frac{\partial G(q,p,t)}{\partial q^i} \quad (1)$$

If the small parameter represents a small elapse of time $\epsilon = dt$ (1) and the generator is the Hamiltonian $G = H$ (1) then

$$\delta q^i = dt \frac{\partial H}{\partial p_i} = dt\dot{q}^i = dq^i \qquad \delta p_i = -dt \frac{\partial H}{\partial q^i} = dt\dot{p}_i = dp_i \quad (1)$$

Qualitatively, we may say that the Hamiltonian is the generator of infinitesimal canonical transformations and causes a physical shift (1) in the coordinates during a time interval dt. (1)

- (Q) Prove Noether's theorem in the canonical formalism by considering the time derivative of a function $F(\boldsymbol{q}, \boldsymbol{p})$ and show that an off-shell symmetry leads to a conserved quantity. (10)
- (A) We first assume that the following condition holds off-shell:

$$\{F, H\} = 0 \quad (1)$$

Then consider an infinitesimal canonical transformation generated by F and parametrised by ϵ:

$$\delta q^i = \{q^i, F\}\epsilon = \frac{\partial F}{\partial p_i}\epsilon \qquad \delta p_i = \{p_i, F\}\epsilon = -\frac{\partial F}{\partial q^i}\epsilon \quad (2)$$

We know that the action S can be written using the Poincaré-Cartan invariant:

$$S[\boldsymbol{q}, \boldsymbol{p}] = \int dt(p_i \dot{q}^i - H(\boldsymbol{q}, \boldsymbol{p}, t)). \quad (1)$$

Suppose that the phase space Lagrangian is invariant $\delta\mathscr{L} = 0$ (1) up to a total time derivative for the off-shell variation (i.e. we have a quasisymmetry):

$$\delta\mathscr{L}' = \dot{q}^i \delta p_i - \dot{p}_i \delta q^i - \delta H + \frac{d}{dt}(p_i \delta q^i) \quad (1)$$

Using the quantities above and $\delta H = \{H, F\}\epsilon$, we arrive at,

$$\delta\mathscr{L}' = -\dot{q}^i \frac{\partial F}{\partial q^i}\epsilon - \dot{p}_i \frac{\partial F}{\partial p_i}\epsilon - \{H, F\}\epsilon + \frac{d}{dt}\left(p_i \frac{\partial F}{\partial p_i}\epsilon\right)$$

$$= -\frac{dF}{dt}\epsilon + \frac{d}{dt}\left(p_i \frac{\partial F}{\partial p_i}\epsilon\right) \quad (2)$$

We recall that $\{H, F\} = -\{F, H\}$, which we agreed was zero at the start. Re-symbolising (for our convenience) the term in brackets to F', we can write

$$\delta\mathscr{L}' = \epsilon \frac{d}{dt}(F' - F)$$

$$= \epsilon \frac{d}{dt} f$$

where $f = F' - F$. If we assume that there is a quasisymmetry, then this expression vanishes and, hence, on-shell, the total time derivative of F vanishes, indicating that the symmetry condition $\{F, H\} = 0$ gives rise to a conserved charge! In other words, an off-shell symmetry gave rise to a conserved quantity and we are safe to write

$$\frac{d}{dt}F = \{F, H\} = 0 \quad (2)$$

The important part to note is that we did not assume during any derivation work that the equations of motion were satisfied to arrive at this result.

- (Q) Given a two-dimensional first-order system:

$$\dot{x} = \alpha(x, y), \qquad \dot{y} = \beta(x, y)$$

Show that locally, away from singularities, the system is superintegrable. (10)
- (A) We first create a 1-form: $\theta = f\,dy - g\,dx$ (1). Locally, there exists an exact differential:

$$\kappa\theta = dH \qquad (1)$$

where $\kappa(x, y)$ is a non-zero integrating factor and $H(x, y)$ is a Hamiltonian defined on a manifold, M. Given this, we have a Poisson structure:

$$\dot{x} = \{x, y\} = -\{y, x\} = \frac{1}{\kappa}, \qquad \{x, x\} = \{y, y\} = 0 \qquad (3)$$

Thus, we have a Hamiltonian system locally defined in the neighbourhood of a point in M:

$$\dot{x} = \frac{1}{\kappa}\frac{\partial H}{\partial y}, \qquad \dot{y} = -\frac{1}{\kappa}\frac{\partial H}{\partial x}$$

In this neighbourhood, we assume the existence of two smooth, functionally independent (one can't be written as a function of the other), Poisson-commuting functions f_1 and f_2 such that $\{f_1, f_2\} = 0$ and the following volume is nowhere vanishing:

$$df_1 \wedge df_2 \neq 0 \qquad (2)$$

By the Carathédory-Jacobi-Lie theorem, on an open neighbourhood of this chart, we have a symplectic 2-form ω and an additional two smooth functions g_1 and g_2 such that

$$\omega = f_1 \wedge g_1 + f_2 \wedge g_2 \qquad (3)$$

Therefore, local maximal superintegrability about a point on M is a generic property for the symplectic manifold (M, ω).

- (Q) What is the significance of the Poisson bracket $\{f, g\}$ for two functions f and g on the phase space? (5)
- (A) The Poisson bracket $\{f, g\}$ evaluates the derivative of g with respect to the action of a one-parameter group of canonical transformations generated by f. (1)

  If g is invariant under the set of transformations generated by f, then we can write $\{f, g\} = 0$ and, by the asymmetry of the Poisson bracket, we can also write $\{g, f\} = 0$, therefore meaning that f is invariant under the action of the one-parameter family of canonical transformations generated by f. (2)

  Noether's theorem is then stated as follows: g is invariant under the action of a one-parameter set of canonical transformations generated by f if and only if f is invariant under the action of the one-parameter group of canonical transformations generated by g. (2)

- (Q) The passage from the Lagrangian description to the Hamiltonian formalism can proceed via the Legendre transform if the Lagrangian is non-degenerate in the generalised velocities:

$$\det \left| \frac{\partial^2 L}{\partial \dot{q}_i \partial \dot{q}_j} \right| \neq 0$$

  Show that this condition is independent of the choice of generalised coordinates, and therefore is a property of the dynamics, by performing a point transformation $\{q^i\} \to \{Q^i\}$ given the invertible transformation equations $Q^i = Q^i(q^1, \ldots, q^n, t)$ and $q^i = q^i(Q^1, \ldots, Q^n)$, where the Jacobian of the transformation does not vanish:

$$\det \left| \frac{\partial Q^i}{\partial q^j} \right| \neq 0 \qquad (5)$$

- (A) Begin by using the chain rule to find the momentum:

$$\frac{\partial \mathscr{L}}{\partial \dot{q}^i} = \sum_{k=1}^{n} \frac{\partial \mathscr{L}}{\partial \dot{Q}^k} \frac{\partial \dot{Q}^k}{\partial \dot{q}^i} . \qquad (1)$$

  before constructing the Hessian condition:

$$\frac{\partial^2 \mathscr{L}}{\partial \dot{q}^j \partial \dot{q}^i} = \frac{\partial}{\partial \dot{q}^j}\left(\sum_{k=1}^{n} \frac{\partial \mathscr{L}}{\partial \dot{Q}^k} \frac{\partial \dot{Q}^k}{\partial \dot{q}^i} \right) \quad (1)$$

$$= \sum_{l=1}^{n} \frac{\partial}{\partial \dot{Q}^l}\left(\sum_{k=1}^{n} \frac{\partial \mathscr{L}}{\partial \dot{Q}^k} \frac{\partial \dot{Q}^k}{\partial \dot{q}^i} \right) \frac{\partial \dot{Q}^l}{\partial q^j} \quad (1)$$

$$= \sum_{k,l=1}^{n} \frac{\partial^2 \mathscr{L}}{\partial \dot{Q}^l \partial \dot{Q}^k} \frac{\partial \dot{Q}^k}{\partial \dot{q}^i} \frac{\partial \dot{Q}^l}{\partial \dot{q}^j} \quad (1)$$

where we have again used the chain rule. We then cancel the dot (see appendix C) to obtain

$$\frac{\partial^2 \mathscr{L}}{\partial \dot{q}^j \partial \dot{q}^i} = \sum_{k,l=1}^{n} \frac{\partial^2 \mathscr{L}}{\partial \dot{Q}^l \partial \dot{Q}^k} \frac{\partial Q^k}{\partial q^i} \frac{\partial Q^l}{\partial q^j} \quad (1)$$

It is clear therefore, that the Hessian condition is invariant under a point transformation:

$$\det\left| \frac{\partial^2 \mathscr{L}}{\partial \dot{q}^j \partial \dot{q}^i} \right| \neq 0 \rightarrow \det\left| \frac{\partial^2 \mathscr{L}}{\partial \dot{Q}^j \partial \dot{Q}^i} \right| \neq 0$$

1. In the Lagrangian formalism what is the difference between the following equations?

-
$$\frac{\partial \mathscr{L}}{\partial q^i} - \frac{d}{dt}\left(\frac{\partial \mathscr{L}}{\partial \dot{q}^i} \right) - \sum_{j=1}^{m} \lambda_j \frac{\partial \phi_j}{\partial q^i} = 0$$

-
$$\frac{d}{dt}\left(\frac{\partial \mathscr{L}}{\partial \dot{q}^i} \right) - \frac{\partial \mathscr{L}}{\partial q^i} = 0$$

-
$$\frac{d}{dt}\left(\frac{\partial \mathscr{L}}{\partial \dot{q}^j} \right) - \frac{\partial \mathscr{L}}{\partial q^j} = \mathcal{F}_j$$

-
$$\frac{\partial \mathcal{L}}{\partial \phi} - \partial_\mu \left(\frac{\partial \mathcal{L}}{\partial (\partial_\mu \phi)} \right) = 0$$

For a given Lagrangian do they lead to identical equations of motion under all circumstances? If not, are the equations of motion equivalent in terms of their solutions? In this case, should we talk about reference frames in analytical mechanics? Comment on the coordinates for each equation and the structure of the configuration space. What happens if we add a total time derivative to each of these equations? (10)

2. What is the difference between the Jacobi energy function $J(q, \dot{q}, t)$ and the Hamiltonian $H(q, p, t)$? Are they defined on the same space? Are they invariant under point and canonical transformations, respectively? (3)

3. Derive Hamilton's principle from d'Alembert's and state any assumptions you may make. (10)

4. How does the Noether variational problem differ from Hamilton's principle? What is the significance of the following variations?

$$\delta t = t' - t, \qquad \delta q^i = q'^i(t) - q^i(t), \qquad \delta q_0^i = q'^i(t') - q^i(t)$$

Repeat the derivation of the Noether condition from appendix A with δq^i instead of δq_0^i to obtain the following result: (10)

$$\delta A = \left[\frac{\partial \mathscr{L}}{\partial q^i} - \frac{d}{dt}\left(\frac{\partial \mathscr{L}}{\partial \dot{q}^i} \right) \right] \delta q^i + \frac{d}{dt}\left[\frac{\partial \mathscr{L}}{\partial \dot{q}^i} \delta q^i \right]$$

5. What is a Noether symmetry of a Lagrangian and how does it lead to conserved quantities? What is the difference between a Noether symmetry and a gauge symmetry? (10)

6. What are non-Noether symmetries and, with the use of an example, show how solution-equivalent equations of motion can arise from different Lagrangians not related via a total time derivative. Discuss parameter invariance and point transformations within this context and comment on key differences. (15)

7. How does the Suslov form generalise Hamilton's principle? (1)

8. Describe the vakonomic approach to constrained mechanics and contrast it with the principles of Lagrange and d'Alembert. (5)

9. Write an essay on how constraints are treated in mechanics. Include in your answer what types of constraints we might encounter and the various tools created to describe them. Derive the Suslov form of the constrained Hamiltonian dependent on a non-holonomic constraint equation. (10)

10. Derive the Gibbs-Appell equations from Gauss's principle. (5)

11. Can we have invariants of the phase space without specifying a metric on the configuration space? If so, what is the geometrical structure of classical mechanics? (2)

12. Show that Poissonian mechanics gives rise to higher-order Nambu structures by writing the binary Poisson bracket as an n-fold Nambu bracket: (15)

$$\eta^{ij} \frac{\partial}{\partial \xi^i}(p) \wedge \frac{\partial}{\partial \xi^j}(dF, dG) \qquad \rightarrow \qquad \eta^{i_1 \cdots i_n}(p) \frac{\partial}{\partial \xi^{i_1}} \wedge \cdots \wedge \frac{\partial}{\partial \xi^{i_n}}$$

13. What is the Routhian function R and why is it useful in solving problems? (6)

14. Describe in detail the process of finding the normal modes of a system of coupled oscillations. (14)

15. What is the difference between the various quantities we call "action" in classical physics? Can you show the equivalence of some of them? (6)

16. Write an essay on the structure of phase space. Include in your answer a detailed account of symplectic tori, symplectic forms, level sets and frequencies. Describe time as an axis in extended phase space and comment on how the Poincaré-Cartan 1-form gives rise to the Hamilton-Jacobi problem. (30)

17. Show that time-dependent mechanics has a contact structure. (5)

18. Describe how the Dirac bracket can be used to investigate constrained dynamics and comment on the different classification of constraints. (7)

19. Consider the 1-form $L = \mathscr{L}(q^i, \dot{q}^i, t)dt$. Let θ be the variational form

$$\theta = \frac{\partial \mathscr{L}}{\partial \dot{q}^i} \delta q^i$$

Evaluate the Euler-Lagrange form $\mathrm{E} = \delta L + d\theta$ using wedge calculus to derive the Euler-Lagrange equation without reference to integration. (5)

20. Describe the Hamilton-Jacobi method using an example, and comment on the relationship between a first-order non-linear PDE and a set of coupled ODEs. (12)

21. Discuss the canonical perturbation theory of an integrable system with an example. (5)

22. What are degenerate Lagrangians? Give a condition for degeneracy. Comment on how KAM theory treats tori associated with degeneracies. (10)

23. Comment on the similarities between the Lagrangian field equation and the discrete Lagrange equation? (4)

24. What is the significance of Liouville's theorem? (6)

25. Koopman-von Neumann theory is an operator approach to a Hilbert-space-based classical mechanics. Can you outline a set of axioms to abstract the theory from our previous discussion? The best method is to observe structures in quantum mechanics and consider what we expect for classical results. For instance, we expect an algebra of commuting observables so the commutator should be exactly zero by design: $[\hat{q}, \hat{p}] = 0$. Is there a superposition principle in classical mechanics? Do we observe phase-amplitude effects in classical mechanics? Comment on the comparative differences you find with quantum mechanics. Prove that the axioms defined lead to the equation of motion being the first-order Liouville equation rather than a second-order Schrödinger equation. (30)

26. Consider the phase space $\mathbb{R}^2$. Show that the following transformation $(q, p) \rightarrow (Q, P)$ is canonical.

$$\begin{bmatrix} Q \\ P \end{bmatrix} = \Sigma \begin{bmatrix} q \\ p \end{bmatrix}, \qquad \Sigma := \begin{bmatrix} -2 & 1 \\ 3 & -2 \end{bmatrix}$$

Find the generating function for this transformation and hence, find the associated symplectic flow. Hint: the last part is a trick question! (7)

27. Prove that the following transformation $(q, p) \rightarrow (Q, P)$, is canonical, where $P = q^{-1}$ and $Q = pq^2$. (2)

28. Use Liouville's theorem to show that the phase space probability density function ρ behaves like an incompressible fluid (1). Classical mechanics is based over a commutable algebra of observables. Using a non-commutative bracket, show that the corresponding quantum phase space density function (the Wigner function) behaves like a compressible fluid (9).

29. How can we formulate a classical density functional theory in terms of a one-particle density function from the Liouville phase space density function? (5)

30. Review the path integral formulation of mechanics and demonstrate how the classical propagator gives rise to the Lagrangian formulation using the Koopman-von Neumann formulation (as was shown for Hamilton's mechanics in the text, using the phase space integral). (20)

31. The central force problem is an important result in classical mechanics. Here, a particle moves in a central potential $U(r)$ that is only dependent on the radius. Using spherical coordinates, the Lagrangian for this system is

$$\mathscr{L} = \frac{m}{2}(\dot{r}^2 + r^2\dot{\theta}^2 + r^2\sin^2\theta\dot{\phi}^2) - U(r)$$

- Construct (3) the Euler-Lagrange equations for each coordinate.
- Show that the momenta conjugate to r, θ and ϕ are

$$p_r = m\dot{r}, \qquad p_\theta = mr^2\dot{\theta}, \qquad p_\phi = mr^2\sin^2\theta\dot{\phi} \qquad (12)$$

- Show that the Hamiltonian can be written as

$$H = \frac{p_r^2}{2m} + \frac{p_\theta^2}{2mr^2} + \frac{p_\phi^2}{2mr^2\sin^2\theta} + U(r)$$

- Construct the Hamilton-Jacobi equation for this system:

$$H\left(r, \theta, \phi, \frac{\partial S}{\partial r}, \frac{\partial S}{\partial \theta}, \frac{\partial S}{\partial \phi}, t\right) + \frac{\partial S}{\partial t} = 0$$

- The Hamilton-Jacobi equation you obtain is separable. Using $W = S^1(\phi) + S^2(r) + S^3(\theta)$, rewrite the Hamilton-Jacobi equation as

$$\frac{1}{2m}\left(\frac{\partial S^2}{\partial r}\right)^2 + \frac{1}{2mr^2}\left(\frac{\partial S^3}{\partial \theta}\right)^2 + \frac{L_z^2}{2mr^2\sin^2\theta} + U(r) = E$$

- The radial equation is

$$\frac{1}{2m}\left(\frac{dS^2}{dr}\right)^2 + \frac{L^2}{2mr^2} + U(r) = E$$

Integrate this result and, using $t = dS^2/dE$, show that:

$$t = \int \frac{mdr}{\sqrt{2mE - 2mU(r) - \frac{L^2}{r^2}}}$$

- Show that the separation function for θ can be written as

$$S^3(\theta) = \int \left(L^2 - \frac{L_z^2}{\sin^2 \theta}\right) d\theta$$

- Perform a Legendre transform to obtain ϕ from the expression for $S^3(\theta)$ as $\phi = \partial S^3 / \partial L_z$. Hence, obtain:

$$\phi = \int \frac{L_z d\theta}{\sqrt{L^2 = \frac{L_z^2}{\sin^2 \theta} \sin^2 \theta}}$$

- Solve that the integral for ϕ to obtain

$$\phi = -\tan^{-1} \frac{L_z \cos \theta}{\sqrt{L^2 \sin^2 \theta - L_z^2}} + \phi_0$$

Hint: start by using a substitution $x = \cos \theta$, $\sin^2 \theta = 1 - x^2$. Then use:

$$x = \sqrt{\frac{L^2 - L_z^2}{L^2}} \sin \phi'$$

The resulting integral can then be solved by letting $y = \tan \phi'$.

Appendix A
Noether's Theorem Explored

In this appendix, we will derive **Noether's theorem** with more rigour and, along the way, prepare for our discussion of field theory. (For a picture of Amalie Emmy Noether, see figure H.8.)

We can consider three types of variations in classical mechanics: just time, just coordinates or both together:

$$\delta t := t' - t, \qquad \delta q^i := q'^i(t) - q^i(t), \qquad \delta_0 q^i := q'^i(t') - q^i(t)$$

The variations used in Hamilton's action principle are called **form variation**; they are the difference between the new variable q', evaluated at a point in time t, and the old variable q evaluated at that same point: $\delta q(t) = q'(t) - q(t)$. We then imposed boundary conditions on these variations to derive the Lagrange equation.

Fig. A.1: Amalie Noether (1882–1935).

In Noether's problem, we consider a variation that arises from both varying the dependent *and* independent variables; that is, $q \to q'$ and $t \to t'$, respectively. Note, however, that the integration volumes are no longer the same (i.e. $dt \to dt'$, see mathematical background 6; so we multiply the old volume element, dt, by the Taylor expansion of the **Jacobian** of the transformation; this is just a mathematical way of expressing the new volume element in terms of the old one: $dt' = (\text{Jacobian})dt$. We then Taylor expand to keep track of the order of the variation:

$$\delta_0 A = A[\boldsymbol{q} + \boldsymbol{\delta_0 q}] - A[\boldsymbol{q}]$$

$$= \int \left[\mathscr{L}(\boldsymbol{q}, \dot{\boldsymbol{q}}, t) + \delta_0 \mathscr{L}(\boldsymbol{q}, \dot{\boldsymbol{q}}, t) \right] dt' - \int \mathscr{L}(\boldsymbol{q}, \dot{\boldsymbol{q}}, t) dt$$

$$= \int \left[\mathscr{L}(\boldsymbol{q}, \dot{\boldsymbol{q}}, t) + \delta \mathscr{L}(\boldsymbol{q}, \dot{\boldsymbol{q}}, t) \right] \left[1 + \frac{\partial}{\partial t} \delta t \right] dt - \int \mathscr{L}(\boldsymbol{q}, \dot{\boldsymbol{q}}, t) dt \qquad (A.0.1)$$

You now multiply the brackets out and only keep terms with **first-order variations** cancelling $\mathcal{O}\delta^2$.

$$= \int \left[\mathscr{L} \frac{\partial}{\partial t} \delta t + \delta \mathscr{L} \right] dt$$

$$= \int \left[\mathscr{L} \frac{\partial}{\partial t} \delta t + \frac{\partial \mathscr{L}}{\partial q^i} \delta_0 q^i + \frac{\partial \mathscr{L}}{\partial \dot{q}^i} \delta_0 \dot{q}^i + \frac{\partial \mathscr{L}}{\partial t} \delta t \right] dt \qquad \text{(A.0.2)}$$

Now take the first and last terms:

$$\frac{d}{dt} (\mathscr{L} \delta t) = \frac{\partial \mathscr{L}}{\partial t} \delta t + \mathscr{L} \frac{\partial}{\partial t} \delta t$$

which give us

$$= \int \left[\frac{\partial \mathscr{L}}{\partial q^i} \delta_0 q^i + \frac{\partial \mathscr{L}}{\partial \dot{q}^i} \delta_0 \dot{q}^i + \frac{d}{dt} (\mathscr{L} \delta t) \right] dt \qquad \text{(A.0.3)}$$

We now focus on the second term and the variation of $\delta_0 \dot{q}^i$, realising that the time derivative commutes with the variation $\delta_0 q = q'(t') - q(t)$:

$$\delta \left(\frac{d}{dt} q^i \right) = \frac{dq'}{dt} - \frac{dq}{dt} = \frac{d}{dt} (\delta_0 q^i)$$

Therefore,

$$= \int \left[\frac{\partial \mathscr{L}}{\partial q^i} \delta_0 q^i + \frac{\partial \mathscr{L}}{\partial \dot{q}^i} \frac{d}{dt} (\delta_0 q^i) + \frac{d}{dt} (\mathscr{L} \delta t) \right] dt \qquad \text{(A.0.4)}$$

We can use the product rule to show that $\frac{d}{dt}(AB) = \dot{A}B + A\dot{B}$ and then rearrange for $A\dot{B} = \frac{d}{dt}(AB) - \dot{A}B$:

$$= \int \left[\frac{\partial \mathscr{L}}{\partial q^i} \delta_0 q^i + \frac{d}{dt} \left(\frac{\partial \mathscr{L}}{\partial \dot{q}^i} \delta_0 q^i \right) - \frac{d}{dt} \left(\frac{\partial \mathscr{L}}{\partial \dot{q}^i} \right) \delta_0 q^i + \frac{d}{dt} (\mathscr{L} \delta t) \right] dt \qquad \text{(A.0.5)}$$

Collect variations in $\delta_0 q^i$ and terms in d/dt to obtain

$$\delta A = \int \left\{ \left[\frac{\partial \mathscr{L}}{\partial q^i} - \frac{d}{dt} \left(\frac{\partial \mathscr{L}}{\partial \dot{q}^i} \right) \right] \delta_0 q^i + \frac{d}{dt} \left[\frac{\partial \mathscr{L}}{\partial \dot{q}^i} \delta_0 q^i + \mathscr{L} \delta t \right] \right\} dt \qquad \text{(A.0.6)}$$

The first term is nothing other than the Lagrange equation of motion, which we know vanishes on-shell. In Hamilton's principle, the second term is an exact differential, since the two dt factors cancel out, meaning the expression is evaluated at the boundaries. Given the imposed boundary conditions, this would vanish, but in Noether's problem, considered more general, we do not impose these vanishing conditions.

Now, let us assume that the variation *is* a Noether symmetry, such that the action vanishes ($\delta_0 A = 0$); allowing us to write the following:

$$\left[\frac{\partial \mathscr{L}}{\partial q^i} - \frac{d}{dt} \left(\frac{\partial \mathscr{L}}{\partial \dot{q}^i} \right) \right] \delta_0 q^i + \frac{d}{dt} \left[\frac{\partial \mathscr{L}}{\partial \dot{q}^i} \delta_0 q^i + \mathscr{L} \delta t \right] = 0 \qquad \text{(A.0.7)}$$

This is called the **Noether condition**; it is a relationship between the Euler-Lagrange equation and a total time derivative. It shows that the term in the second brackets is conserved on-shell when there is a symmetry; if there is no symmetry, then the action will not vanish. The action will still vanish on-shell if we impose boundary conditions on the variations; in that case we regenerate the exact differential. Therefore, off-shell symmetries give on-shell conservation laws; that is the very heart of the Noether condition! The apparent discrepancy between the term we derive in chapter 11 and the one we derive in this appendix is because, in the latter, we allowed variations in time too. If you cross off the δt terms in the Jacobian and the variation in the Lagrangian, then the results tie up.

The **homogeneity of time** and the conservation of energy can be proven by letting $\delta t = \epsilon$ be just an infinitesimal constant; if the Lagrangian is not an explicit function of time, then $\partial \mathscr{L}/\partial t = 0$ in this case, $\delta_0 q^i = -\dot{q}^i \delta t = -\epsilon \dot{q}^i$, and the conserved quantity is the Jacobi energy function:

$$\frac{d}{dt} J = -\epsilon \frac{d}{dt} \left[\frac{\partial \mathscr{L}}{\partial \dot{q}^i} \dot{q}^i - \mathscr{L} \right] = 0 \qquad (A.0.8)$$

The **homogeneity of space** leads to the conservation of linear momentum. For a shift in the coordinate reference frame by some constant vector $\delta q^i = c$ with $\delta \dot{q}^i = 0$ and $\delta t = 0$ (since we are not altering the time), the Noether condition becomes,

$$\frac{d}{dt} p_i = \frac{d}{dt} \left(\frac{\partial \mathscr{L}}{\partial \dot{q}^i} c \right) \qquad (A.0.9)$$

Since the choice of c is arbitrary, the conjugate momenta must be conserved for equality to hold. There are lots of other transformations we could make, such as Galilean transformations (transformation between two inertial reference frames, conserving centre of mass in a system of particles), Lorentz transformations (transformations between two reference frames, one at rest to another non-accelerating frame; importantly, they leave the line element dS invariant), and so on. Some are shifts in the coordinate frames, while others are true motions. The general theme is to group transformations together that have the same symmetry properties. We call these **symmetry groups**. Most important to spacetime symmetries is the **Poincaré group** and it includes translations, rotations and boosts (transformations between two uniformly moving bodies). For instance, application of the Galilean group to a free particle induces $\dot{q}(t) \to \dot{q}(t) + v$ with $v \in \mathbb{R}$ and $\mathscr{L} = \frac{1}{2} m\dot{q}$. This gives a total time derivative of the form

$$\mathscr{L}(\boldsymbol{q}, \dot{\boldsymbol{q}}, t) \to \mathscr{L}(\boldsymbol{q}, \dot{\boldsymbol{q}}, t) + \frac{d}{dt} \left(\frac{1}{2} mv^2 t + mvq \right) = EL + \frac{1}{2} mv^2 (t_2 - t_1) + mvq(t_2) - mvq(t_1)$$

Another important coordinate transformation is the **Lorentz transformation**. Einstein's special relativity tells us that the speed of light in all inertial frames is constant.

The Cartesian coordinates of two inertial frames at rest to one another are related via the following set of transformations in space and time:

$$\bar{t} = \gamma(t - \frac{vx}{c}), \qquad \bar{x} = \gamma(x - vt), \qquad \bar{y} = y, \qquad \bar{z} = z \qquad \text{(A.0.10)}$$

with $\gamma = (1 - (v/c)^2)^{-1/2}$ and c being the speed of light. If the equations of motion are sensibly behaved under the element of a Lorentz group, then they are said to be **Lorentz covariant**, while quantities that are invariant under the transformation are **Lorentz invariant**.

Appendix B
The Action Principle Explored

B.1 Geodesics

On a Riemannian configuration manifold, the correct choice of **metric** g_{ij} converts the physical path to a **geodesic**, which is the *shortest* path between two points across the curved configuration manifold's surface. Therefore, we expect that the Euler-Lagrange equation in the context of a Riemannian manifold will give us the **geodesic equation**, which is a well known mathematical result. We aren't going to discuss *connections* and *parallel transport*, due to their involved nature, but instead motivate this result in a simple way.

We assume that the configuration space is Euclidean $\mathbb{E}^n$. The velocity of a free particle in cartesian coordinates is a vector:

$$\vec{v} = \frac{d\vec{r}}{dt} = \left(\frac{dx}{dt}, \frac{dy}{dt}, \frac{dz}{dt} \right) := (\dot{x}, \dot{y}, \dot{z}) = (\dot{x}^i) \quad i = 1, 2, 3 \tag{B.1.1}$$

The speed of the particle is the scalar v:

$$v := \|\vec{v}\| = \sqrt{(\dot{x})^2 + (\dot{y})^2 + (\dot{z})^2} \tag{B.1.2}$$

which takes the form $\vec{v} = (\dot{x}^i)$ and $v = \sqrt{\dot{x}^i \cdot \dot{x}^j}$ in components. A metric on a smooth configuration manifold $\mathbb{E}^n$ induces an **inner product** on the tangent space above every point $p \in \mathbb{E}^n$; this is written $g_p(\boldsymbol{v}, \boldsymbol{w})$ for two vectors $\boldsymbol{v}, \boldsymbol{w} \in T_p\mathbb{E}^n$. Since we have defined the speed in terms of the scalar product, we write[1] $v = \sqrt{g_{ij} \dot{v}^i \dot{w}^j}$. In local coordinates, the Lagrangian for a free particle[2]:

---

[1]We assume that $\langle \cdot, \cdot \rangle$ is non-degenerate, symmetric and bilinear on $T_p\mathbb{E}^n$. We can write the **Riemannian line element** ds^2 as follows, choosing an invertible set of coordinates $x_i(y_j)$ and therefore $dx_i = (\partial x_i / \partial y_j) dy_j$,

$$ds^2 = dx_i dx_i = \frac{\partial x_i}{\partial y_j} \frac{\partial x_i}{\partial y_k} dy_j dy_k = g_{jk} dy_j dy_k$$

Then construct a kinetic energy by taking the time derivative $(ds/dt)^2$.

[2]Sometimes the Lagrangian is defined $\mathscr{L} = \sqrt{-g_{ij}(q)\dot{q}^i \dot{q}^j}$ they both lead to geodesic equations but the latter Lagrangian leads to *affinely parametrised* geodesics. If you are reading into general relativity, this is most useful!

$$\mathscr{L} = \frac{1}{2} \sum_{ij} g_{ij} \dot{q}^i \dot{q}^j \tag{B.1.3}$$

The mass has disappeared because we associate $g_{ij} \equiv M_{ij}$, the mass matrix from the small oscillations discussion in chapter 12, you will note that it is symmetric and bilinear. The Euler-Lagrange equation is now

$$\frac{d}{dt} \left(\sum_j g_{ij} \dot{q}^j \right) - \frac{1}{2} \sum_{jk} \frac{\partial g_{ij}}{\partial q^i} \dot{q}^j \dot{q}^k = 0 \tag{B.1.4}$$

Expanding the time derivative and chain ruling the metric, assuming it is a function of q, we obtain

$$\frac{d}{dt} \left(\sum_j g_{ij} \dot{q}^j \right) = \sum_j \frac{\partial g_{ij}}{\partial q^k} \dot{q}^j \dot{q}^k + \sum_{ij} g_{ij} \ddot{q}^j \tag{B.1.5}$$

The metric tensor satisfies the relationship $g^{is} g_{sj} = \delta^i_j$, meaning that we can isolate the acceleration by multiplying by g^{si}:

$$\ddot{q}^s + \sum_{ijk} g^{si} \left(\frac{\partial g_{ij}}{\partial q^k} - \frac{1}{2} \frac{\partial g_{ik}}{\partial q^i} \right) \dot{q}^j \dot{q}^k \tag{B.1.6}$$

We can make use of the symmetry of the metric tensor by adding and subtracting the same term:

$$\ddot{q}^s + \frac{1}{2} \sum_{ijk} g^{si} \left(\frac{\partial g_{ij}}{\partial q^k} + \frac{\partial g_{ik}}{\partial q^j} - \frac{\partial g_{jk}}{\partial q^i} \right) \dot{q}^j \dot{q}^k = \ddot{q}^s + \sum_{jk} \Gamma^s_{jk} \dot{q}^j \dot{q}^k = 0 \tag{B.1.7}$$

In the last step, we have substituted in the symmetric **Christoffel symbol** Γ^s_{jk}. This proves that, when a suitable metric is chosen for the configuration manifold, classical curves are exactly geodesics:

$$\Gamma^s_{jk} = \frac{1}{2} \sum_i g^{si} \left(\frac{\partial g_{ij}}{\partial q^k} + \frac{\partial g_{ik}}{\partial q^j} - \frac{\partial g_{jk}}{\partial q^i} \right) \tag{B.1.8}$$

The geodesic deviation can be found by considering two neighbouring geodesics, $q^s(t)$ and $\tilde{q}^s(t)$, which are joined by some deviation vector $\delta q(t)$ satisfying $\tilde{q}^s = q^s + \delta q^s$ and $\tilde{\Gamma}^s_{jk} = \Gamma^s_{jk} + \partial_l \Gamma^s_{jk} x^l$. The two equations of motion are

$$\ddot{\tilde{q}}^s + \tilde{\Gamma}^s_{jk} \dot{\tilde{q}}^j \dot{\tilde{q}}^k = 0 \quad \text{and} \quad \ddot{q}^s + \Gamma^s_{jk} \dot{q}^j \dot{q}^k = 0 \tag{B.1.9}$$

We take the left-hand side from the right to obtain

$$\delta\ddot{q}^s + \tilde{\Gamma}^s_{jk}\dot{\tilde{q}}^j\dot{\tilde{q}}^k - \Gamma^s_{jk}\dot{q}^j\dot{q}^k = 0 \tag{B.1.10}$$

with $\delta\ddot{q}^s = \ddot{\tilde{q}}^s - \ddot{q}^s$. Finally, with the above two definitions for $\tilde{q}^s$ and $\tilde{\Gamma}^S_{jk}$ this becomes

$$\delta\ddot{q}^s + \Gamma^s_{jk}\dot{q}^j\delta\dot{q}^k + \Gamma^s_{jk}\delta\dot{q}^j\dot{q}^k + \partial_l\Gamma^s_{jk}\delta q^l\dot{q}^j\dot{q}^k + \partial_l\Gamma^s_{jk}\delta q^l\dot{q}^j\delta\dot{q}^k + \partial_l\Gamma^s_{jk}\delta q^l\delta\dot{q}^j\dot{q}^k = 0 \tag{B.1.11}$$

The final two terms are truncated if we consider only first-order deviations, thus leaving us with the Jacobi equation for geodesics:

$$\boxed{\delta\ddot{q}^s + 2\Gamma^s_{jk}\dot{q}^j\delta\dot{q}^k + \partial_l\Gamma^s_{jk}\delta q^l\dot{q}^j\dot{q}^k = 0} \tag{B.1.12}$$

For further reading on geodesics and introductory Riemannian geometry, the reader is recommended to see Lovett's (2010) book, *Differential Geometry of Manifolds*; to continue any further, we would need to define the covariant derivative and curvature tensors. The reader more at home with all the mathematics is referred to Jóźwikowski's fascinating (2013) paper, 'Jacobi vector fields for Lagrangian systems on algebroids', where you will find a proof of the Jacobi equation and how it relates to our derivation. For detail on the principle of virtual work on a Riemannian manifold, please see V. G. Ivancevic and T. T. Ivancevic's (2007) book, *Applied Differential Geometry: A Modern Introduction*.

Appendix C
Useful Relations

In this short appendix, we consider a collection of useful relations that we use at various points. In particular, we have the *cancelling of the dot, swapping derivatives* and the *generalised force*. In all cases, we assume $\phi(q, t)$ and blue-coloured terms vanish, in this case not indicating equality.

- **Cancelling the dot**

$$\frac{\partial \dot{\phi}}{\partial \dot{q}^i} = \frac{\partial}{\partial \dot{q}^i} \left[\frac{d\phi}{dt} \right]$$

$$= \frac{\partial}{\partial \dot{q}^i} \left[\frac{\partial \phi}{\partial q^j} \dot{q}^j + \frac{\partial \phi}{\partial t} \right]$$

$$= \frac{\partial}{\partial \dot{q}^i} \left(\frac{\partial \phi}{\partial q^j} \dot{q}^j \right) + \frac{\partial}{\partial \dot{q}^i} \left(\frac{\partial \phi}{\partial t} \right)$$

$$= \frac{\partial}{\partial \dot{q}^i} \left[\frac{\partial \phi}{\partial q^j} \right] \dot{q}^j + \frac{\partial \phi}{\partial q^j} \frac{\partial \dot{q}^j}{\partial \dot{q}^i}$$

The first terms vanish, since ϕ is not a function of $\dot{q}$, and the last highlighted term is the delta function leaving only the black term.

- **Swapping derivatives**

$$\frac{d}{dt} \left(\frac{\partial \phi}{\partial q^i} \right) = \frac{\partial}{\partial q^j} \left(\frac{\partial \phi}{\partial q^i} \right) \dot{q}^j + \frac{\partial}{\partial t} \left(\frac{\partial \phi}{\partial q^i} \right)$$

$$= \frac{\partial}{\partial q^i} \left(\frac{\partial \phi}{\partial t} \right) + \frac{\partial}{\partial q^i} \left(\frac{\partial \phi}{\partial q^j} \right) \dot{q}^j$$

$$= \frac{\partial}{\partial q^i} \left(\frac{d\phi}{dt} \right) \qquad (C.0.1)$$

We can therefore swap the above two derivatives. However, we cannot swap the following:

$$\frac{d}{dt} \left(\frac{\partial \phi}{\partial \dot{q}^i} \right) \neq \frac{\partial}{\partial \dot{q}^i} \left(\frac{d\phi}{dt} \right) \qquad (C.0.2)$$

the left-hand side is zero since ϕ is not a function of $\dot{q}$, while the second term is as above:

- **Generalised force** The generalised force appears many times in part II. Here, we show how it is the negative gradient of the potential with respect to the generalised coordinates.

$$
\begin{aligned}
\mathcal{F}_i &= -\vec{\nabla}_i U \cdot \frac{\partial \vec{r}_i}{\partial q^j} \\
&= -\left[\frac{\partial}{\partial x_i} \hat{\boldsymbol{x}} + \frac{\partial}{\partial y_i} \hat{\boldsymbol{y}} + \frac{\partial}{\partial x_i} \hat{\boldsymbol{z}} \right] U \cdot \frac{\partial}{\partial q^j} [x_i \hat{\boldsymbol{x}} + y_i \hat{\boldsymbol{y}} + z_i \hat{\boldsymbol{z}}] \\
&= -\frac{\partial U}{\partial q^j}
\end{aligned}
\tag{C.0.3}
$$

Appendix D
Poisson & Nambu Brackets Explored

In this appendix, we will explore new notations and offer a proof of the basic axiomatic rules of the Poisson bracket and their generalisation to higher-order structures. We first show that the Poisson brackets can be written using directional notation for derivatives. If a differential operator is placed in-between two functions, an arrow may be placed on the top to indicate which function it acts on. For instance,

$$F \overleftarrow{\partial}_{q_i} G = \frac{\partial F}{\partial q^i} G, \qquad F \overrightarrow{\partial}_{p_i} G = F \frac{\partial G}{\partial p_i} \tag{D.0.1}$$

Using this notation, the Poisson bracket may be written as

$$\{F, G\} = F(\overleftarrow{\partial}_{q_i} \overrightarrow{\partial}_{p_i} - \overleftarrow{\partial}_{p_i} \overrightarrow{\partial}_{q_i}) G \tag{D.0.2}$$

The axiomatic Poisson bracket rules that define the Lie algebra structure are now developed, using local coordinates. The **antisymmetric rule** is

$$\begin{aligned}
\{F, G\} &= \frac{\partial F}{\partial q} \frac{\partial G}{\partial p} - \frac{\partial F}{\partial p} \frac{\partial G}{\partial q} \\
&= -\left(\frac{\partial G}{\partial q} \frac{\partial F}{\partial p} - \frac{\partial G}{\partial p} \frac{\partial F}{\partial q} \right) \\
&= -\{G, F\}
\end{aligned}$$

Leibniz's rule is

$$\begin{aligned}
\{FG, I\} &= \frac{\partial (FG)}{\partial q} \frac{\partial I}{\partial p} - \frac{\partial (FG)}{\partial p} \frac{\partial I}{\partial q} \\
&= F \frac{\partial G}{\partial q} \frac{\partial I}{\partial p} + G \frac{\partial F}{\partial q} \frac{\partial I}{\partial p} - F \frac{\partial G}{\partial p} \frac{\partial I}{\partial q} - G \frac{\partial F}{\partial p} \frac{\partial I}{\partial q} \\
&= F\{G, I\} + G\{F, I\}
\end{aligned}$$

Jacobi's identity is

$$\{F, G\} = F_q G_p - F_p G_q$$

$$\{\{F, G\}, I\} = \frac{\partial \{F, G\}}{\partial q} I_p - \frac{\partial \{F, G\}}{\partial p} I_q \tag{D.0.3}$$

$$= (F_{qq}G_p + F_q G_{pq} - F_{pq}G_q - F_p G_{qq})I_p - \\ (F_{qp}G_p + F_q G_{pp} - F_{pp}G_q - F_p G_{qp})I_q \tag{D.0.4}$$

where we have used the product rule for the partial derivatives, and eased notation by letting $\partial y / \partial x = y_x$. If we do the same for the other two relations, we find

$$\{\{G, I\}, F\} = (G_{qq}I_p + G_q I_{pq} - G_{pq}I_q G_p I_{qq})F_p \\ -(G_{qp}I_p + G_q I_{pp} - G_{pp}I_q - G_p I_{qp})F_q$$

$$\{\{I, F\}, G\} = (I_{qq}F_p + I_q F_{pq} - I_{pq}F_q - I_p F_{qq})G_p \\ -(I_{qp}F_p + I_q F_{pp} - I_{pp}F_q - I_p F_{qp})G_q$$

You then add them all together, remember that the order of derivatives doesn't matter, and you will arrive at Jacobi's identity.

Linearity is

$$\{F_1 + F_2, G\} = \sum_j \left(\frac{\partial (F_1 + F_2)}{\partial q_j} \frac{\partial G}{\partial p_j} - \frac{\partial (F_1 + F_2)}{\partial p_j} \frac{\partial G}{\partial q_j} \right) \tag{D.0.5}$$

$$= \sum_j \left(\frac{\partial F_1}{\partial q_j} \frac{\partial G}{\partial p_j} - \frac{\partial F_1}{\partial p_j} \frac{\partial G}{\partial q_j} \right) + \sum_j \left(\frac{\partial F_2}{\partial q_j} \frac{\partial G}{\partial p_j} - \frac{\partial F_2}{\partial p_j} \frac{\partial G}{\partial q_j} \right) \tag{D.0.6}$$

$$= \{F_1, G\} + \{F_2, G\} \tag{D.0.7}$$

We take this opportunity to really discuss the Poisson bracket and show its power. This topic should most definitely be **skipped** upon first reading of canonical mechanics.

D.1 Symplectic Notation & Nambu Brackets

In the usual case, we study Poisson manifolds $\mathcal{P}$ whose Poisson structure is defined by the map $\{\cdot, \cdot\}$ on the algebra of commutative observables $\mathcal{C}^\infty(\mathcal{P})$ such that $\{\cdot, \cdot\}$: $\mathcal{C}^\infty(\mathcal{P}) \otimes \mathcal{C}^\infty(\mathcal{P}) \mapsto \mathcal{C}^\infty(\mathcal{P})$. The Poisson bracket is the determinant of the **Jacobian matrix** for two functions F and G:

$$\frac{\partial (F, G)}{\partial (q, p)} = \det \begin{pmatrix} \frac{\partial F}{\partial q} & \frac{\partial F}{\partial p} \\ \frac{\partial G}{\partial q} & \frac{\partial G}{\partial p} \end{pmatrix} = \frac{\partial F}{\partial q} \frac{\partial G}{\partial p} - \frac{\partial F}{\partial p} \frac{\partial G}{\partial q} \tag{D.1.1}$$

If one of the functions is a Hamiltonian, then the Poisson bracket is a time derivative and we can understand canonical transformations from $(q, p \to Q, P)$ by simple chain ruling:

$$\dot{F} = \frac{\partial(F,H)}{\partial(q,p)} = \frac{\partial(F',H')}{\partial(Q,P)}\frac{\partial(Q,P)}{\partial(q,p)} = \frac{\partial(F',H')}{\partial(Q,P)}\{Q,P\} \tag{D.1.2}$$

where $H' = H(q(Q,P), p(Q,P))$. We know that if this transformation is canonical, then the highlighted term is unity and we can clearly see that we have another binary bracket. We then extended our definition using the symplectic 2-form and the geometry of vector fields and forms:

$$\{F,G\} = \omega(X_F, X_G) = \omega^{-1}(dF, dG)$$

As an example, consider $\{H, \cdot\} = \omega^{-1}(dH, \cdot) = X_H$. We can write this in coordinates using $2n \times 2n$ **symplectic matrices** (see chapter 22). In this notation, if we let a point in phase space be written $\xi^{\mu} = (q^i, p_j)$, with Greek indices running over $2n$, then $\omega^{\mu\nu} = -\omega_{\mu\nu}$, with $\omega^{\mu\nu} = -\omega^{\nu\mu}$ and $\omega^{\mu\nu}\omega_{\nu\eta} = \delta^{\mu}_{\eta}$ (prove that this last relation gives a 2×2 matrix with a diagonal of 1's which is the identity matrix).

$$\omega_{\mu\nu} = \begin{pmatrix} 0 & 1 \\ -1 & 0 \end{pmatrix}$$

The time derivative is given by $\dot{\xi}^{\mu} = (\dot{q}^i, \dot{p}_j)$ and Hamilton's equations are constructed as follows:

$$\dot{\xi}^{\mu} = \omega_{\mu\nu}\frac{\partial H}{\partial \xi^{\mu}} = -\omega^{\mu\nu}\frac{\partial H}{\partial \xi^{\mu}} = -\begin{pmatrix} 0 & -1 \\ 1 & 0 \end{pmatrix}\begin{pmatrix} \partial H/\partial q^i \\ \partial H/\partial p_i \end{pmatrix} = \begin{pmatrix} \partial H/\partial p_i \\ -\partial H/\partial q^i \end{pmatrix} \tag{D.1.3}$$

You will note that we could also write this in terms of the Hamiltonian gradient dH, which was defined when discussing *the structure of phase space* in chapter 23, hence, we have defined **symplectic notation**. The Poisson bracket is now constructed as follows:

$$\{F,G\} = \omega_{\mu\nu}\frac{\partial F}{\partial \xi^{\mu}}\frac{\partial G}{\partial \xi^{\nu}} = -\omega^{\mu\nu}\frac{\partial F}{\partial \xi^{\mu}}\frac{\partial G}{\partial \xi^{\nu}} \tag{D.1.4}$$

When reading around, be careful of this definition, as some sources swap the minus sign with the other set of indices. This goes back to our definition of the 2-form as the *negative* of the exterior derivative of the Poincaré-Cartan 1-form $\omega = -d\theta_H$; some authors define $\omega = d\theta_H$, and the swapping of the wedge would absorb the minus sign, too. Whatever convention you prefer, the theory is the same. We can easily verify this result:

$$\{F,G\} = \omega_{\mu\nu}\frac{\partial F}{\partial \xi^{\mu}}\frac{\partial G}{\partial \xi^{\nu}} = \begin{pmatrix} \partial F/\partial q^i \\ \partial F/\partial p_i \end{pmatrix}\begin{pmatrix} 0 & 1 \\ -1 & 0 \end{pmatrix}\begin{pmatrix} \partial G/\partial q^i \\ \partial G/\partial p_i \end{pmatrix} = \begin{pmatrix} \partial F/\partial q^i \\ \partial F/\partial p_i \end{pmatrix}\begin{pmatrix} \partial G/\partial p_i \\ -\partial G/\partial q^i \end{pmatrix}$$

We therefore recover the canonical form of the Poisson bracket from the symplectic matrices:

$$\{F, G\} = \frac{\partial F}{\partial q^i} \frac{\partial G}{\partial p_i} - \frac{\partial F}{\partial p_i} \frac{\partial G}{\partial q^i}$$

It is customary to re-label the inverse matrix ω^{-1} as the **Poisson tensor** or Poisson bi-vector field $\eta \in \chi(\bigwedge^2 T\mathcal{P})$, where $T\mathcal{P}$ is the tangent bundle of the Poisson manifold. The Poisson tensor is contravariant, and the Poisson bracket could be written

$$\{F, G\} = \omega^{-1}(dF, dG) = \eta(dF, dG), \qquad \text{where} \qquad \eta = \sum_{i,j} \eta^{ij}(p) \frac{\partial}{\partial \xi^i} \wedge \frac{\partial}{\partial \xi^j} \quad \text{(D.1.5)}$$

We can also represent canonical transformations using symplectic notation; consider $q, p \to Q, P$ and let $\xi = (q, p)$ and $\chi(Q, P)$. The transformation equations can be written $\chi^\mu(\xi)$ and $\xi^\mu(\chi)$. Chain ruling both sides of Hamilton's equation we obtain

$$\frac{\partial \xi^\mu}{\partial \chi^\nu} \frac{d\chi^\nu}{dt} = \omega_{\mu\nu} \frac{\partial H}{\partial \chi^\sigma} \frac{\partial \chi^\sigma}{\partial \xi^\nu} \tag{D.1.6}$$

Multiplying by the Jacobian of the transformation $\partial \chi^\varsigma / \partial \xi^\mu$ we find that

$$\frac{\partial \chi^\varsigma}{\partial \xi^\mu} \frac{\partial \xi^\mu}{\partial \chi^\nu} \frac{d\chi^\nu}{dt} = \frac{\partial \chi^\varsigma}{\partial \xi^\mu} \omega_{\mu\nu} \frac{\partial H}{\partial \chi^\sigma} \frac{\partial \chi^\sigma}{\partial \xi^\nu} \tag{D.1.7}$$

If we then cancel the $\partial \xi^\mu$ on the left-hand side, we have the delta function $\delta^{\varsigma\nu}$:

$$\frac{d\chi^\varsigma}{dt} = \frac{\partial \chi^\varsigma}{\partial \xi^\mu} \omega_{\mu\nu} \frac{\partial H}{\partial \chi^\sigma} \frac{\partial \chi^\sigma}{\partial \xi^\nu} \tag{D.1.8}$$

Let

$$\Omega_{\mu\nu} := \frac{\partial \chi^\varsigma}{\partial \xi^\mu} \omega_{\mu\nu} \frac{\partial \chi^\sigma}{\partial \xi^\nu} \tag{D.1.9}$$

We therefore regenerate Hamilton's equations for χ:

$$\frac{d\chi^\varsigma}{dt} = \Omega_{\mu\nu} \frac{\partial H}{\partial \chi^\sigma} \tag{D.1.10}$$

Hamiltonian dynamics posits that the motion can be defined by a single scalar function $H : \mathbb{R}^{2n} \to \mathbb{R}$ acting on an even-dimensional differentiable manifold. We now ask, is there a generalisation of this, perhaps to those manifolds that are not even dimensional, or to a hypothetical situation where we require more scalar functions to describe the dynamics?

Indeed, it was Y. Nambu who asked this question in 1971 and we now discuss the **Hamilton-Nambu formalism** of classical mechanics. Hamilton-Nambu dynamics involves replacing the binary Poisson operation with a higher-order one, one that has more input Hamiltonians and operates over more observables. The key idea behind the

Nambu formalism is the introduction of a nth-order bracket, the **Nambu bracket**, on the nth tensor product space of observables on a manifold $\mathcal{P}$; the bracket satisfies the antisymmetry, Leibniz and Jacobi identities (whose generalisation is called the *fundamental identity*) such that $\{\cdot, \ldots, \cdot\}_{\mathrm{NB}} : \mathcal{C}^\infty(\mathcal{P})^{\otimes^n} \mapsto \mathcal{C}^\infty(\mathcal{P})$ defines an n-Lie algebra structure. A manifold with a Nambu bracket defined on its observables is a **Nambu-Poisson manifold** and it may be odd dimensional.

We do not at the moment specify the extra variables, only saying that their presence is required to describe the system. If we take n phase space functions $F_i(x^1, \ldots, x^n)$ with coordinates $\{x^i\}$, with $i = 1, \ldots n$, then the Nambu-Poisson bracket $\{F_1, \ldots, F_n\}_{\mathrm{NB}}$ is given by the nth-order Jacobian determinant:

$$\{F_1, \ldots, F_n\}_{\mathrm{NB}} = \frac{\partial(F_1, \ldots, F_n)}{\partial(x^1, \ldots, x^n)} = \eta(dF_1, \ldots, dF_n) \qquad \textbf{Nambu bracket} \quad \text{(D.1.11)}$$

The bracket is an n-ary operation that defines a higher-order algebraic structure on the phase space, such that a second-order Nambu bracket is just a Poisson bracket! The **Nambu tensor** $\eta \in \Gamma\big(\bigwedge^n T\mathcal{P}\big)$ is given in local coordinates by

$$\eta = \eta^{i_1 \ldots i_n}(p) \frac{\partial}{\partial x^{i_1}} \wedge \cdots \wedge \frac{\partial}{\partial x^{i_n}}$$

In analogy to the symplectic 2-form on a differentiable manifold defined a symplectic manifold, the Nambu-Poisson tensor defines the Nambu-Poisson manifold $(\mathcal{P}, \eta)$; for any Hamiltonian vector field, its Lie derivative is zero ($\mathcal{L}_{X_H}\eta = 0$) and it is closed and non-degenerate. To describe the time evolution of an n-dimensional system, we require $(n-1)$ Hamiltonians:

$$\frac{dF}{dt} = \{F, H_1, \ldots, H_{n-1}\}$$

while the **Nambu equations of motion** are expressed using the dim(N) Levi-Civita symbol:

$$\frac{dx_i}{dt} = \sum_{i_1, \ldots, i_{N-1}}^{N} \epsilon_{i_1 \ldots i_{N-1}} \frac{\partial H_1}{\partial x_{i_1}} \cdots \frac{\partial H_{N-1}}{\partial x_{i_{N-1}}} \qquad \text{(D.1.12)}$$

The Nambu formalism finds application in constrained mechanics and in **superintegrable** systems. It is not possible to have more than n functionally independent first integrals in an n-dimensional system; if any others exist, they will be linked to the others (so not independent and not in involution). Superintegrable systems are integrable systems, that have $(2n-1)$m globally defined, functionally independent first integrals. In this case, the extra integrals of motion are used like the other Hamiltonians.

For more information on Nambu structures, please see Takhtajan's (1994) article, *On foundation of the generalised Nambu mechanics*, Horikoshi and Kawamura's (2013) article, Hidden Nambu mechanics: A variant formulation of Hamiltonian systems, and, of course, Nambu's (1973) paper. The second of these papers is focused on exploiting the natural Nambu structures in constrained mechanics thereby indicating that the Nambu formalism is another method for studying constrained Hamiltonian dynamics. The authors construct a variational procedure for deriving the Nambu equations as well as a partition function.

Appendix E
Canonical Transformations Explored

This appendix considers the relationships between the remaining two generator functions by use of the Legendre transform before exploring the invariance of the Poisson bracket and phase space volume under canonical transformation. From the text, we know that

$$F_2(q, P, t) = F_1(q, Q, t) + QP \qquad \text{(E.0.1)}$$

We can compute the differentials of the left side

$$dF_2 = \frac{\partial F_2}{\partial q} dq + \frac{\partial F_2}{\partial P} dP + \frac{\partial F_2}{\partial t} dt \qquad \text{(E.0.2)}$$

and the right side

$$dF_2 = (K - H)dt + pdq - PdQ + PdQ + QdP \qquad \text{(E.0.3)}$$
$$= (K - H)dt + pdq + QdP \qquad \text{(E.0.4)}$$

Now, match the terms with the same differential and realise that, for independent variations, each term must vanish:

$$\frac{\partial F_2}{\partial q} dq + \frac{\partial F_2}{\partial P} dP + \frac{\partial F_2}{\partial t} dt = (K - H)dt + pdq + QdP \qquad \text{(E.0.5)}$$

$$\left(\frac{\partial F_2}{\partial q} - p \right) dq + \left(\frac{\partial F_2}{\partial P} - Q \right) dP = \left(K - H - \frac{\partial F_2}{\partial t} \right) dt \qquad \text{(E.0.6)}$$

And we have the F_2 relations in each of the brackets. To get to F_3 from F_1, we want to switch $q \to p$ and, to do this, we take away the following term:

$$F_3(p, Q, t) = F_1(q, Q, t) - qp \qquad \text{(E.0.7)}$$

where the left side is

$$dF_3 = \frac{\partial F_3}{\partial p} dp + \frac{\partial F_3}{\partial Q} dQ + \frac{\partial F_3}{\partial t} dt \qquad \text{(E.0.8)}$$

and the right is

$$dF_3 = \left(K - H\right)dt + pdq - PdQ - qdp - pdq \tag{E.0.9}$$
$$= \left(K - H\right)dt - PdQ - qdp \tag{E.0.10}$$

Equating these we obtain

$$\frac{\partial F_3}{\partial p}dp + \frac{\partial F_3}{\partial Q}dQ + \frac{\partial F_3}{\partial t}dt = \left(K - H\right)dt - PdQ - qdp \tag{E.0.11}$$

and upon collecting like variations we arrive at the final result:

$$\left(\frac{\partial F_3}{\partial p} + q\right)dp + \left(\frac{\partial F_3}{\partial Q} + P\right)dQ = \left(K - H - \frac{\partial F_3}{\partial t}\right)dt \tag{E.0.12}$$

For the last generating function, we need to change both variables in F_1 namely $q, Q \rightarrow p, P$; so, naturally, we have to perform the Legendre transform on both of them, with the terms established above for each transformation:

$$F_4(p, P, t) = F_1(q, Q, t) - qp + QP \tag{E.0.13}$$

Expanding the left-hand side we obtain

$$dF_4 = \frac{\partial F_4}{\partial p}dp + \frac{\partial F_4}{\partial P}dP + \frac{\partial F_4}{\partial t}dt \tag{E.0.14}$$

and expanding the right-hand side

$$dF_4 = \left(K - H\right)dt + pdq - PdQ - qdp - pdq + QdP + PdQ \tag{E.0.15}$$
$$= \left(K - H\right)dt - qdp + QdP \tag{E.0.16}$$

we then equate the two sides

$$\frac{\partial F_4}{\partial p}dp + \frac{\partial F_4}{\partial P}dP + \frac{\partial F_4}{\partial t}dt = \left(K - H\right)dt - qdp + QdP \tag{E.0.17}$$

and collect variations

$$\left(\frac{\partial F_4}{\partial p} + q\right)dp + \left(\frac{\partial F_4}{\partial P} - Q\right)dP\left(K - H - \frac{\partial F_4}{\partial t}\right)dt \tag{E.0.18}$$

With these results in place, let us now turn the general proof of the invariance of the Poisson bracket with respect to a canonical transformation such that $\{F, G\}_{Q,P} = \{F, G\}_{q,p}$:

$$\{F,G\}_{Q,P} = \left(\frac{\partial F}{\partial Q^i}\frac{\partial G}{\partial P_i} - \frac{\partial F}{\partial P_i}\frac{\partial G}{\partial Q^i} \right)$$

$$= \left(\frac{\partial F}{\partial q^j}\frac{\partial q^j}{\partial Q^i} + \frac{\partial F}{\partial p_j}\frac{\partial p_j}{\partial Q^i} \right)\left(\frac{\partial G}{\partial q^k}\frac{\partial q^k}{\partial P_i} + \frac{\partial G}{\partial p_k}\frac{\partial p_k}{\partial P_i} \right)$$

$$- \left(\frac{\partial F}{\partial q^j}\frac{\partial q^j}{\partial P_i} + \frac{\partial F}{\partial p_j}\frac{\partial p_j}{\partial P_i} \right)\left(\frac{\partial G}{\partial q^k}\frac{\partial q^k}{\partial Q^i} + \frac{\partial G}{\partial p_k}\frac{\partial p_k}{\partial Q^i} \right)$$

$$= \frac{\partial F}{\partial q^j}\frac{\partial G}{\partial q^k}\{q^j,q^k\}_{Q,P} + \frac{\partial F}{\partial q^j}\frac{\partial G}{\partial p_k}\{q^j,p_k\}_{Q,P} + \frac{\partial F}{\partial p_j}\frac{\partial G}{\partial q^k}\{p_j,q^k\}_{Q,P}$$

$$+ \frac{\partial F}{\partial p_j}\frac{\partial G}{\partial p_k}\{p_j,p_k\}_{Q,P}$$

$$= \frac{\partial F}{\partial q^j}\frac{\partial G}{\partial p_k}\delta^j_k - \frac{\partial F}{\partial p_j}\frac{\partial G}{\partial q^k}\delta^k_j$$

We used the chain rule to factor in the dependancy of the original phase space in the first step and then multiplied everything out in the second and grouped terms accordingly. In the next line, we used the axiomatic rules that Poisson brackets obey, including the fundamental brackets and the antisymmetric rule, to swap the order to get the negative sign; finally, the deltas are equal to unity when $j = k$ and we arrive at the Poisson bracket, with all the same subscripts.

We can prove Liouville's theorem for the invariance of the phase space volume by using the following theory regarding volume elements in mathematics; from here, we will see that Liouville's theorem gives rise to canonical transformations, or perhaps vice versa, depending on your viewpoint. For a change of coordinates $(q,p) \to (Q,P)$, we say that the new volume element is given in terms of the old one multiplied by the Jacobian determinant of the transformation (we prove that when we discuss field theory using wedge calculus in mathematical background 6):

$$dQdP = \det \begin{pmatrix} \frac{\partial Q}{\partial q} & \frac{\partial Q}{\partial p} \\ \frac{\partial P}{\partial q} & \frac{\partial P}{\partial p} \end{pmatrix} dqdp \tag{E.0.19}$$

We may also recall (see appendix D) that the Poisson bracket is the determinant of such a Jacobian matrix:

$$\det \begin{pmatrix} \frac{\partial Q}{\partial q} & \frac{\partial Q}{\partial p} \\ \frac{\partial P}{\partial q} & \frac{\partial P}{\partial p} \end{pmatrix} = \frac{\partial Q}{\partial q}\frac{\partial P}{\partial p} - \frac{\partial Q}{\partial p}\frac{\partial P}{\partial q} = \{Q,P\}_{q,p} = 1 \tag{E.0.20}$$

Therefore, the volume of phase space is conserved and Liouville's theorem is proved! We may view the phase space transformations in two different ways: either as the moving of a phase space region $\mathcal{R} \to \mathcal{R}'$ in a fixed axis system or, the phase points do not change but instead the coordinate axes are curvilinear. The area of phase space $\mathcal{R}$ can be enclosed by a curve γ and, by using the Kelvin-Stokes theorem, be written as a contour integral:

$$\iint_{\mathcal{R}} dq dp = \oint_{\gamma} p \, dq = \oint_{\gamma} P \, dQ \tag{E.0.21}$$

Let us pick (q, Q) as a pair of independent variables such that we have $p(q, Q)$ and $P(q, Q)$. Then,

$$\oint_{\gamma} [p(q, Q) dq - P(q, Q) dQ] = 0 \tag{E.0.22}$$

Since this is valid for any closed curve γ, the integrand is equal to a perfect differential of some function $F_1(q, Q)$:

$$\oint_{\gamma} (p \, dq - P \, dQ) = \oint dF_1(q, Q) = \oint_{\gamma} \frac{\partial F_1}{\partial q} dq + \oint_{\gamma} \frac{\partial F_1}{\partial Q} dQ \tag{E.0.23}$$

Comparing like differentials gives us the following:

$$p = \frac{\partial F_1}{\partial q}, \qquad P = -\frac{\partial F_1}{\partial Q} \tag{E.0.24}$$

Of course, from here, we regenerate the theory of canonical transformations entirely from Liouville's preservation of phase space. However, the time-dependent transformations we developed earlier are more general.

Appendix F
Action-Angle Variables Explored

This appendix looks in detail at action-angle coordinates and how to find them, treating the harmonic oscillator as an example. Action-angle variables can be useful tools to exploit the physics of a phase space description of mechanics, but the question is, how can we find these variables if, indeed the system has any?

The key point about action-angle variables is that all the Q^is are cyclic, meaning that the Hamiltonian is only a function of the P_is (i.e. is a function of only n coordinates $H = H(\boldsymbol{P})$). When this is the case, the flow curves in phase space are trivialised to a straight line. We would say that the flow is **rectified**. In these coordinates, Hamilton's equations read

$$\dot{Q}^i = \frac{\partial H(\boldsymbol{P})}{\partial P_i} = \text{constant} := \boldsymbol{\omega}_i(P), \qquad \dot{P}_i = -\frac{\partial H}{\partial Q^i} = 0 \qquad \text{(F.0.1)}$$

This means that there are n conjugate momenta that are constants of the motion, while the $\dot{Q}^i$s are also constants, meaning that the Q^is are linear in time and can be integrated up to some additive constant $Q^i(t) = \boldsymbol{\omega}_i(P)t + Q^i(t_0)$, while the action variable doesn't change: $P_i(t) = P_i(t_0)$.

The important criteria for these special systems is that they are **periodic** in some coordinate. There are two important examples of periodicity in physical systems. The first is a *closed path in phase space*; this is periodic in both the p_is and the q^is. The second example is when p_i is periodic in q^i, meaning that, for one period (whatever it is) of the variable q^i, we arrive back at the same q^i, but the Q^is take a value defined by whatever one period later is. There are two familiar systems for which these two conditions apply: the harmonic oscillator and the rigid rotor; in both of these systems, we can rotate about a fixed axis periodically, leaving q^i periodic and with Q^i advancing by a period each rotation.

Okay, so how can we generate these transformations then? As you would expect, we can form a generating function that will give the action-angle variables from the original set. In this new description P_i is constant and Q^i is cyclic with a period [1] of

---

[1] The 2π is convention ($0 \leq Q^i \leq 2\pi$) but you can redefine this to 1. In general, when integrating a variable over a cycle, it will be broken into positive and negative parts. The values are just the turning points of the closed path in phase space.

2π. If we take the generating function to be $F_1 = F_1(\boldsymbol{q}, \boldsymbol{Q})$ of one degree of freedom to start with, we can define two relations:

$$p_i = \frac{\partial F_1}{\partial q^i}, \qquad P_i = -\frac{\partial F_1}{\partial Q^i} \qquad \text{(F.0.2)}$$

For an F_1 generating function that is a function of Q and q, we can take its total differential:

$$dF_1 = \frac{\partial F_1}{\partial q^i} dq^i + \frac{\partial F_1}{\partial Q^i} dQ^i$$
$$= p_i dq^i - P_i dQ^i$$

Assuming P_i is constant and Q^i is 2π periodic, integration over one period is given by,

$$\oint dF_1 = \oint p^i_i - P_i \oint dQ^i$$

If q^i and Q^i are periodic, then so too is F_1, so by the very definition of the operation over a closed path $\oint F_1 = 0$:

$$0 = \oint p_i dq^i - P_i 2\pi$$

We thus obtain

$$P_i = \frac{1}{2\pi} \oint p_i dq^i \qquad \text{(F.0.3)}$$

where P_i is a constant for these special systems. This is in agreement with Liouville's theorem, too, since $\oint p_i dq^i = \iint dp_i dq^i$, using **Stokes' theorem** in an area in phase space. If we integrate the differential of the generating function, we can show that

$$F_1(\boldsymbol{q}, \boldsymbol{Q}) = \int p_i dq^i - P_i Q^i \qquad \text{(F.0.4)}$$

We can simplify this by using an F_2 generating function using a Legendre transform:

$$F_2(\boldsymbol{q}, \boldsymbol{P}) = F_1(\boldsymbol{q}, \boldsymbol{Q}) + Q^i P_i = \int p_i(\boldsymbol{q}, \boldsymbol{P}) dq^i$$

From here, just use the relations of the F_2 function and substitute the above:

$$Q^i = \frac{\partial F_2}{\partial P_i} = \frac{\partial}{\partial P_i} \int p_i dq^i \qquad \text{(F.0.5)}$$

Thus, we have two expressions for the action and angle coordinates. If this integral is too hard to do in practice, we could just integrate equation F.0.1 to get an expression for Q^i with respect to t; this is what we do below. Let's call the constant $\boldsymbol{\omega}_i(P)$ the **frequency**, for historical reasons, and, hence, $\int \dot{Q}^i dt = \boldsymbol{\omega}_i(P)t + Q^i_0$, where Q^i_0 is the constant of integration (sometimes called ϕ). Since F.0.1 depends only on the value of P_i, so too does $\boldsymbol{\omega}_i = \boldsymbol{\omega}_i(P)$ and, since P_i is constant in time, then so too is $\boldsymbol{\omega}_i$.

Exercise F.1 In exercise 18.1, we plucked a coordinate change for the harmonic oscillator out of seemingly thin air; well, we can, in fact, derive it by using action-angle variable theory. The Hamiltonian for the simple harmonic oscillator is given by

$$H = \frac{1}{2m}\{p^2 + m^2\omega^2 q^2\}$$

Rearrange for p:

$$p = \pm\sqrt{2mE - m^2\omega^2 q^2}$$

Use equation F.0.3:

$$P = \frac{1}{2\pi}\oint \sqrt{2mE - m^2\omega^2 q^2}\ dq$$

Unfortunately, this integral is a bit too hard to evaluate analytically since it changes from positive to negative over the path, so we will have to work on the highlighted term a little. This is actually really quite simple: pull out $2mE$ and note the form of what we have in the brackets, $1 - something$. From elementary maths, we know that, if we can say this something is $\sin^2\theta$, then we can write the entire bracketed term as $\cos^2\theta$, since $\cos^2\theta + \sin^2\theta = 1$:

$$2mE - m^2\omega^2 q^2 = 2mE\left(1 - \frac{m\omega^2}{2E}q^2\right)$$

Let

$$\frac{m\omega^2}{2E}q^2 = \sin^2 Q$$

The sine function, as we know, is periodic in 2π and, hence, so is Q; this is indeed the substitution for q that we suggested earlier! If we use this to change variables of the integration to Q and integrate over one period from 0 to 2π we obtain

$$P = \frac{E}{\pi\omega}\oint_{2\pi}\cos^2 Q\, dQ$$

This integral is easy to evaluate and, with these limits, gives us exactly π; therefore, $P = E/\omega$, as expected:

$$dq = \sqrt{\frac{2E}{m\omega^2}}\cos Q\ dQ$$

$$P = \frac{1}{2\pi}\oint_0^{2\pi}\sqrt{2mE\cos^2 Q}\sqrt{\frac{2E}{m\omega^2}}\cos Q\ dQ$$

$$= \frac{1}{2\pi}\oint_0^{2\pi}\cos^2 Q\sqrt{2mE}\cdot\sqrt{\frac{2E}{m\omega^2}}dQ$$

$$= \frac{1}{2\pi}\oint_0^{2\pi}\cos^2 Q\ dQ\cdot\frac{2E}{\omega}$$

$$= \frac{1}{2\pi}\pi\cdot\frac{2E}{\omega}$$

This now gives us our result: $P = \frac{E}{\omega}$. Therefore, using action-angle variable theory, we can derive the canonical transformation just by the rules of differential and integral calculus, rather than trying to guess, as we suggested earlier!

We can calculate the change in Q^i over a period of motion, using the F_2 generator, by noting that

$$\frac{\partial Q^i}{\partial q^i} = \frac{\partial}{\partial P_i}\left(\frac{\partial F_2}{\partial q^i}\right)$$

just by switching the order of partial differentiation. Using this, we find that, over one period, the change in the angle variable ΔQ is

$$\Delta Q^i = \oint dQ^i$$

$$= \oint \frac{\partial Q^i}{\partial q^i}dq^i = \frac{\partial}{\partial P_i}\oint \frac{\partial F_2}{\partial q^i}dq^i = \frac{\partial}{\partial P_i}\oint p_i dq^i = \frac{\partial}{\partial P_i}(2\pi P_i)$$

$$= 2\pi$$

You can see how, whatever we define the period to be, the value we chose would just replace the 2π. Further, we can define the time for a period T:

$$2\pi = Q^i(t+T) - Q^i(t)$$
$$= \left[\omega_i(t+T) + \phi\right] - \left[\omega_i t + \phi\right]$$
$$= \omega_i T$$

By definition, the frequency of a period of the angle variable is the **angular frequency** $\omega_i = 2\pi/T$. This is an alternative way of identifying the frequency of the periodic motion in Q^i without solving oscillatory equations and using only our knowledge of the Hamiltonian and action-angle theory.

Exercise F.2 Synchronisation in chemical and biological oscillators is described using the **Kuramoto model**. For N uncoupled oscillatorsm the natural frequency of the motion is given by $\dot{\theta}_i = \omega_i, \forall\, i \in N$. For coupled oscillators, we can write the time derivative of the angle variable via an **coupling function** Γ_{ij}:

$$\dot{\theta}_i = \omega_i \sum_{j=1}^{N}\Gamma_{ij}(\theta_j - \theta_i) \tag{F.0.6}$$

In the simple Kuramoto model, we choose $\Gamma_{ij} = (k/N)\sin(\theta_j - \theta_i)$ to give a system of coupled limit-cycle oscillators. For certain values of the parameter k, the system may exhibit synchronisation between the phase of the oscillators. In other words, imagine each oscillator spinning around on a unit circle. In some situations, the motions will clump together (synchronise) and in others, at any point in time, they could be found anywhere around the circle, with no apparent coupling. To investigate this, we switch to polar variables by introducing the *phase coherence* r and the *average phase* ϕ. Each point is now a complex number:

$$re^{i\phi} = \frac{1}{N}\sum_{j=1}^{N} e^{i\theta_j} \tag{F.0.7}$$

When r is small, there is no real synchronisation between the phases and, hence, ϕ is scattered across the cycle. When r is large, the phases are clumped together and the system is said to be in synchronisation. Multiplying the above by $e^{-i\theta_i}$ and taking the imaginary part, we obtain

$$r\sin(\phi - \theta_i) = \frac{1}{N}\sum_{j=1}^{N} \sin(\theta_j - \theta_i) \tag{F.0.8}$$

Since the parameters r and ϕ are averages over the system of oscillators, the model becomes a mean field model. Hence, each phase θ_i approaches the mean phase ϕ, equality being reached at equilibrium. The strength of this approach is given by the r parameter. There is a critical value of coupling k beyond which synchronisation occurs.

The Hamiltonian of this model is given by

$$H(\boldsymbol{q},\boldsymbol{p},t) = \sum_{i=1}^{N} \frac{1}{2}\omega_i(q_i^2 + p_i^2) + \frac{k}{4N}\sum_{i,j=1}^{N}(q_i p_j - q_j p_i)(q_j^2 + p_j^2 - q_i^2 - p_i^2) \tag{F.0.9}$$

This Hamiltonian admits an action-angle variable description by a change of variables $(q,p) \to (Q,P)$:

$$P_i = \frac{(q_i^2 + p_i^2)}{2}, \qquad Q_i = \tan^{-1}\left(\frac{q_i}{p_i}\right) \tag{F.0.10}$$

Find the Hamiltonian for these new variables and Hamilton's equations of motion. For further reading on this subject please see the article by Acebrón and Bonilla (2005).

Appendix G
Statistical Mechanics Explored

This appendix is an exploration of how probability theory and classical mechanics can couple together and lead to some very powerful ideas central to modern theoretical chemistry.

G.1 The Boltzmann Factor

The Boltzmann factor is the probability distribution for the canonical ensemble and, if we recall our in-text discussion, such a system is allowed to exchange heat with the environment. We have not mentioned *how* it exchanges heat and we have to assume that, whatever the mechanism is, it does not change the state of the environment (e.g. we can take heat out of the environment and not affect it). The equilibrium canonical system is a **closed isothermal system**, meaning that the total number of particles is constant and so is the temperature.

There are quite probably tens of different ways to derive the Boltzmann factor, some more straight forward than others, and most chemistry undergraduate courses will follow similar pathways of probability and counting microstates. To be a little different, we can instead show that the Boltzmann factor is the unique solution to a differential equation.

Consider a system composed of two non-interacting subsystems A and B that are both in contact with the same heat bath. Their individual probability distributions are $\rho_A(\boldsymbol{\xi}_A)$ and $\rho_B(\boldsymbol{\xi}_B)$ with Hamiltonians $H_A(\boldsymbol{\xi}_A)$ and $H_B(\boldsymbol{\xi}_B)$ and the phase space vectors $\boldsymbol{\xi}$. The overall system has the Hamiltonian $H_{AB}(\boldsymbol{\xi}_{AB}) = H_A(\boldsymbol{\xi}_A) + H_B(\boldsymbol{\xi}_B)$ and the probability distribution $\rho_{AB}(H_{AB}) = \rho_{AB}(H_A + H_B) = \rho_{AB}(H_A)\rho_B(H_B)$, where, in the first equality, we substituted for H_{AB} and, in the last step, we realised that the system is separable, since the subsystems do not interact and the probability distribution is a function of a single variable.

Mathematical Background 10

For any function of a single variable, $f(x)$, that satisfies such a relation where $f(x + y) = f(x)f(y)$, the y variable is superfluous and we may treat it to be a small shift in x such that $y = \delta x$. Setting $x = y = 0$, we have $f(0+0) = f(0) \cdot f(0)$

and, hence, require a solution which is its own square, meaning $f(0)$ is either 1 or 0. Setting $f(0) = 0$ is an arbitrary solution, so we let $f(0) = 1$. Next, we use the definition of the derivative as the Newtonian quotient to form an ODE and, letting $\beta = -f'(0)$ we obtain

$$\frac{df(x)}{dx} = -\beta f(x) \tag{G.1.1}$$

Any function equal to its own derivative is given by the exponential function (up to a multiplying scalar) and we are guaranteed uniqueness by the Cauchy-Lipschitz theorem. Therefore, we find that $f(x) = e^{-\beta x}$.

Returning to thermodynamics, we identify the probability distribution of the canonical ensemble with the following exponential, which is known as the Boltzmann factor:

$$\rho(\boldsymbol{\xi}, t) = \exp\{-\beta H(\boldsymbol{\xi}, t)\} \tag{G.1.2}$$

The Boltzmann factor tells us that the probability of occupying a particular state is proportional to the negative exponent of the energy multiplied by a variable β. We call this the **Boltzmann distribution**. At large β, the probability falls rapidly so the system is dominated by lower energy states and the average energy decreases; in the limit of small β, all states are contributing to the average.

G.2   Fluctuations

The average energy in the canonical ensemble is given by the ensemble average and was found in the text to be:

$$-\frac{1}{Z}\frac{\partial Z}{\partial \beta} = -\frac{\partial \log Z}{\partial \beta} = \langle E(\beta) \rangle \tag{G.2.1}$$

Since we are dealing with average energies, it is fruitful to investigate the spread about the mean of a distribution. This is known as the **fluctuation** about a probability distribution and is given by the variance or uncertainty. For the energy, we consider the fluctuation given by

$$(\Delta E)^2 = \langle E^2 \rangle - \langle E \rangle^2 = \frac{1}{Z}\frac{\partial^2 Z}{\partial \beta^2} - \frac{1}{Z^2}\left(\frac{\partial Z}{\partial \beta}\right)^2 \tag{G.2.2}$$

With a little algebra, this can be re-configured:

$$(\Delta E)^2 = \frac{1}{Z}\frac{\partial^2 Z}{\partial \beta^2} + \frac{\partial}{\partial \beta}\left(\frac{1}{Z}\right)\frac{\partial Z}{\partial \beta} = \frac{\partial}{\partial \beta}\left(\frac{1}{Z}\frac{\partial Z}{\partial \beta}\right)$$

where $\frac{\partial}{\partial \beta}\left(\frac{1}{Z}\right) = -\frac{1}{Z^2}\frac{\partial Z}{\partial \beta}$ is nothing but the chain rule in disguise, and the last step is the reverse of a product ruling. The term in brackets is the average energy:

$$(\Delta E)^2 = -\frac{\partial}{\partial \beta}\langle E \rangle \tag{G.2.3}$$

Should we be all that interested in fluctuation theory? How much can a little displacement *really* affect the results? To illustrate this, we will follow a very basic yet intuitive example of an N-particle system that has a set of N independent variables E_i (take this as the energy of the particle), each of which can take one of two possible states, $+1$ and -1, with equal probability. The total energy E of the system is defined as the sum of the states over all of the particles:

$$E = \sum_{i=1}^{N} E_i$$

Since the two states are equally probable, then the expectation value of E is zero:

$$\langle E \rangle = \sum_{i=1}^{N}\langle E_i \rangle = \sum_{i=1}^{N}\left(\frac{1}{2}(+1) + \frac{1}{2}(-1)\right) = 0$$

The fluctuation about the expectation value is given by $(\Delta E)^2 = \langle E^2 \rangle - \langle E \rangle^2$, of which only the first term is important, since zero squared is still zero. Squaring E_i and accounting for the different choices of particles we could take, we obtain

$$E^2 = \sum_{i=1}^{N} E_i^2 + 2\sum_{i=1}^{N}\sum_{j=i+1}^{N} E_i E_j$$

Then the expectation value $\langle E^2 \rangle$ is

$$\langle E^2 \rangle = \sum_{i=1}^{N}\langle E_i^2 \rangle + 2\sum_{i=1}^{N}\sum_{j=i+1}^{N}\langle E_i E_j \rangle = \sum_{i=1}^{N}\left(\frac{1}{2}(+1)^2 + \frac{1}{2}(-1)^2\right) + 2\sum_{i=1}^{N}\sum_{j=i+1}^{N}\langle 0 \cdot 0 \rangle = N$$

The fluctuation is then not arbitrary and increases with the number of particles in the system; this is quite different from the naive result of zero:

$$(\Delta E)^2 = \langle E^2 \rangle - \langle E \rangle^2 = N$$

Obviously, this is a simplistic model but the idea is to show that, in some cases, the fluctuation can be extremely important.

Appendix H
Biographies

There is a rich and wonderful history behind classical mechanics, and the founders can be considered the pioneers of the modern-day scientific method. While researching mechanics, I have become very interested in the lives and workings of many of the great figures in the history of science, and below is a short account of the most famous and influential to our journey. For more information, please see the University of St Andrews History of Mathematicians website (http://www-history.mcs.st-and.ac.uk/). The images have been purchased from either Corbis^{TM}or wikicommons^{TM}.

H.1 Sir Isaac Newton

Fig. H.1: Sir Isaac Newton (1643–1727).

Sir Isaac Newton (1643–1727) was born near Grantham in Lincolnshire. (For a picture of Newton, see figure H.1.) His father, also named Isaac, died 3 months before his son was born. Newton's father was an uneducated farmer who could not even write his own name; however, he was wealthy for the time in that he owned property and numerous livestock. When Newton was 3 years old, his mother remarried and Barnabas Smith became his stepfather, with whom a great rift developed. Newton attended the Free Grammar School in Grantham and he soon excelled. Despite living nearby, he lodged with families nearer to the school. He ended up lodging with the school's headmaster who, it is thought, prepared him for university studies.

In June 1661, Newton entered Trinity College Cambridge. It is during this period of study that he wrote *Quaestiones quaedam philosophicae* (*Certain philosophical questions*). He soon became interested in mathematics; reading works by Euclid and others kindled his interest in the subject. However, in this desire to learn about geometry and algebra, he also devised his own theorems, those necessary for studying dynamics.

In 1665, Newton was forced to return home to Lincolnshire, as the university was closed due to a plague outbreak. During this time away from Trinity College, he began to formulate calculations for differential and integral calculus, which he called the 'method of fluxions'. This was several years before Gottfried Leibniz, another eminent natural philosopher of the day, came up with his own thoughts on calculus. In 1669, having already returned to Trinity College and obtained his masters degree, Newton was appointed to the Lucasian Chair of Mathematics at Cambridge, officially established by King Charles II in January 1664. Upon taking over this role, Newton produced a reflecting telescope and also showed that white light could be split up into constituent coloured components, which he thought were particle beams, as opposed to waves. This belief brought him into conflict with other scientists, who argued that light was a wave, and also with Jesuits over his colour theory. It is at this time that his mother also died, a fundamental blow to Newton. Following the loss of his mother and the criticism of his theories, Newton appears to have suffered from a nervous breakdown; he withdrew from all but essential contacts.

However, despite his almost irrational fear of criticism, Newton went on to publish one of the most famous books in the history of science: the *Principia* (1687). This book formulated the laws of motion and universal gravitation and held up for over 200 years until Albert Einstein came along with the theory of general relativity. His famous *standing on the shoulders of giants* quote perhaps refers to others in Oxford and Cambridge, such as Merton (of Merton College fame), who were among the first in history to write quantitative dynamical laws of motion. Up until this time, the ancient Greek paradigm of physical science had been to ask *why* rather than *how*. Medieval natural philosophy had been restricted by religion and superstition; Newton's work was the first real evidence of the modern quantitative scientific method. Although *change* had been studied for millennia, thanks to the ancient Greeks and Mesopotamians, they spoke of *violent changes* and *natural changes* (e.g forces or decay), as one subject. Newtonian physics is the first real example of what we now call *dynamics* as the rate of change of position. Even today, we are no closer to the why's of science but, thanks to the paradigm shifts of Merton, Oresme, Galileo and Newton, we are nowadays well prepared to explain the how's.

Newton's works brought him into conflict with many religious groups, who were reluctant to believe in a heliocentric solar system or that the heavens, the realm of the divine, were subject to rules or explanation, just as men were. Newton himself often told the story that he was inspired to develop his theory of gravitation by watching an apple falling to the ground. He was honoured with a monument in Westminster Abbey and also had his image placed upon a Bank of England note, both fitting tributes to a gifted scientist and a Master of the Royal Mint.

H.2 Leonhard Euler

Leonhard Euler (1707–1783) was a Swiss-born natural philosopher who is considered to be one of the most influential scientists in history. (For a picture of Euler, see figure H.2.) Euler enrolled at the University of Basel in his early teens and showed a greatness for mathematics. While there, he met with Johann Bernoulli, an eminent professor who nurtured his talent and directed him towards new topics. His first graduate placement was at the newly founded St Petersburg Academy of Sciences in 1727 at the age of 19 before being made professor of physics in 1730 and senior chair of mathematics in 1733.

His increasing prominence and financial stability allowed him to support a family and he married Katharina Gsell in 1734 to become a doting family man. It is at this time that he presented his two-volume *Mechanica*, which is an influential text on the mathematical foundations of analytical mechanics, and developed the calculus of variations. He then received an offer of a position at the Berlin Academy from Frederick the Great; with the resulting great freedom of research he described himself as *the happiest man in the world*. He became good friends with Maupertuis, who was another eminent developer of analytical mechanics.

Fig. H.2: Leonhard Euler (1707–1783).

In Berlin, he met d'Alembert, the French mathematician and another of the founders of analytical mechanics. The two began an academic rivalry which, along with personal differences with Voltaire, caused Euler to return to Russia in 1776. He then lost his wife in 1773. In addition, he suffered from loss of eyesight (attributed to his intricate drawings for his research) and was totally blind upon returning to Russia. He continued to output mathematical greatness despite this, being helped by his sons and academic friends to write down his thoughts. He died aged 76 while discussing celestial mechanics with friends at lunch in 1783, leaving the field of pure and applied mathematics forever in his debt.

H.3   Jean d'Alembert

Jean-Baptiste le Rond d'Alembert (1717–1783) was a French natural philosopher and mathematician. (For a picture of d'Alembert, see figure H.3.) D'Alembert was an illegitimate child of an artillery officer and the writer Claudine Geérin de Tencin. He was left just a few days old on the steps of the Saint-Jean-le-Rond de Paris church, after which he was named, and sent to an orphanage. His father, upon his return to Paris, took him out of the orphanage and arranged for him to live with the wife of a glazier, Madame Rousseau, with whom he lived with until middle age. He considered his foster mother his real family and was very happy living there. His father also secretly arranged education at a private school in Paris and, upon his death left the 9 year-old d'Alembert a small sum to assure his security and continue his education.

Due to the nature of his birth, he could not take his father's family name and instead settled on d'Alembert. He originally studied law and theology at Jansenist Collége des Quatre Nations but soon found a love for mathematics and would devote his spare time to reading and studying lectures, despite working towards a law course. He produced his first paper in a communication to the Académie des Sciences, which was the centre of mathematical development at the time. He applied for a position at the Paris Academy of Science and was unsuccessful on three occasions before admission.

While at the academy, he explained natural phenomena such as refraction, studied Latin texts, and published his famous *Traité de dynamique*. At the academy, he developed rivalries with other eminent natural philosophers, in particular Clairaut. He developed a unyielding self confidence and dismissed thoughts that his work may be incorrect. This attitude led to conflict with his colleagues and resulted in arguments with almost all he encountered. By his early thirties, he had managed to trade criticism and insults with Euler, Bernoulli, Clairaut, König and Voltaire. The affair has been termed the "ugliest of all scientific disputes"[1].

Fig. H.3: Jean D'Alembert (1717–1783).

Despite this, however, his mathematical work was on point and he made foundational contributions to PDEs and physical sciences. He also developed the d'Alembert ratio test to check for series convergence and made changes to the way we think about limits as quotients. D'Alembert was invited in 1764 to join the Berlin academy by Frederick II but declined, perhaps due to the rivalry with Euler. Towards the end of his life, d'Alembert's focus switched to philosophy, literature and culture. He worked on the Encyclopédie alongside Diderot, produced musical theories and, as part of the French enlightenment, attended many Parisian salons. He is associated with feminism, liberalism and cultural decadence.

H.4   Joseph-Louis Lagrange

Joseph-Louis Lagrange (1736–1813) was an Italian mathematician responsible for a wide range of mathematical concepts we hold dear today, including number theory, analysis, acoustics, fluid dynamics and, of course, classical mechanics. (For a picture of Lagrange, see figure H.4.) Lagrange is considered as influential as Euler in his contributions to mathematics and astronomy. His notable students include Fourier and Poisson, among others.

In his early life, Lagrange did not show any enthusiasm for mathematics. He was born in Turin to a family of French origin and christened Giuseppe Lodovico Lagrangia. He identified strongly with his French ancestry, signing his surname name LaGrange. When he was 17 he obtained a copy of a paper by Halley and it sparked his interests in mathematics. He threw himself into study and soon sent letters to eminent mathematicians, including Euler. His early works were considered a little amateur but, such criticism only spurred him to study further and, by the age of 19 he had established

---

[1]See http://www.encyclopedia.com/science/dictionaries-thesauruses-pictures-and-press-releases/koenig-konig-johann-samuel

himself as a leading figure in enlightenment mathematics. In 1755, he was appointed assistant professor at an artillery and ballistics academy. He was among the very first to teach calculus to engineers, now a common practise in any modern degree program, much to the dismay of first-year engineers!

It was Lagrange's work on the calculus of variations that gave him international recognition, particularly among Euler and Maupertuis. Due to this and his work in dynamics, he was invited by Euler to the Berlin Academy, but refused, due perhaps to an unwillingness to be esteemed and an inherent shyness. He instead founded the Turin Academy of Sciences in 1758. He developed much of the modern understanding of the propagation of sound and vibrating strings during this time and even pointed out some mistakes by Newton in the general treatment of acoustics. It was at this time that he developed some of the most beautiful and masterfully written pieces on dynamics and the calculus of variations and, together with Euler, defined the Euler-Lagrange equation.

Fig. H.4:
Joseph-Louis
Lagrange
(1736–1813).

It was finally due to letters by d'Alembert and King Frederick of Prussia that Lagrange accepted a position in the Berlin Academy in 1766. He accepted this only because he knew that Euler was leaving for St Petersburg. While most were very pleased by this move, there were some elders who perhaps felt threatened by a young upstart, notably J. Castillon (1704-1791). Lagrange married his cousin Vittoria Conti but the couple did not have children. Both man and wife were subject to poor health and Vittoria died in 1783, casting Lagrange into depression.

Despite offers enticing him back to Italy, Spain and Naples, he moved to Paris under the offer of Louis XVI to the French Academy of Sciences. In 1788 he published *Méchanique analytique* under the council of Laplace and Legendre. Due to his depression, however, it is said the printed copy remained in his drawer for 2 years, untouched.

His book transformed mechanics from Newton's theory to that of analytical mechanics and paved the way for the firm setting of the dynamics as a branch of mathematics in its own right. As he says in the preface to the book,

"No figures will be found in this work. The methods I present require neither constructions, nor geometrical or mechanical arguments, but solely algebraic operations subject to a regular and uniform procedure. (Lagrange, [1811] 1997, p. 7)"

Lagrange formulated his mechanics from the principle of virtual work, from which he derives the entirety of dynamics. The overriding feel of the book is to find one unifying principle which encapsulates everything. This work was the first to introduce generalised coordinates rather than being bound to analysis in a given frame. In addition, he demonstrates the method of Lagrange multipliers, so this book is one of the first instances of their publication.

During this time, there was the unrest in Paris, indeed, in the entirety of France under the Revolution. Under the "Reign of Terror" which saw A. Lavoisier thrown out of the Academy, Lagrange experienced a different treatment, due to his Italian nationality. The same unrest saw Lavoisier executed by guillotine. Despite this, the temporary government and even Napoleon sought to indulge Lagrange with honours and awards to the Legion of Honour and Count of the Empire in 1808. In his later years, he worked on the development of units of measurement and lecturing in École Polytechnique. He died in 1813 and is buried in the Panthéon in Paris.

H.5   Carl Gustav Jacobi

Carl Gustav Jacobi (1804–1851) was a German mathematician and theoretical physicist remembered for astounding contributions to dynamics and number theory. (For a picture of Jacobi, see figure H.5.) Jacobi was born in Potsdam to a Jewish banker, along with three other siblings, although the family would later become Christian. He was well educated and rapidly progressed in his studies, partly because of his uncle's teaching and his own talent for language, history and, above all, mathematics. He was promoted to final year before he had completed his first year of schooling and applied at just 12 to the University of Berlin. Jacobi, however, was unsuccessful in gaining a place due to age restrictions so he remained stationary until he was 16. Despite this, he was not dissuaded and spent time reading philosophy, languages and quite advanced mathematics texts, including some of Euler's analysis books, as well as works by Lagrange and Laplace.

Fig. H.5: Carl Gustab Jacobi (1804–1851).

He entered Berlin in 1821 but mainly studied philosophy as the mathematics education was poor at this time. By 1825, the 21-year-old Jacobi had a doctorate and was lecturing on the theory of curves and surfaces at the University of Berlin. Surprisingly, one of his early papers was not published, as referees criticised it as essentially being nothing special. They must have soon realised, however, that this was just a blip, as the output and quality of Jacobi's subsequent works was outstanding. Over the next few years, he impressed the likes of Legendre, Neumann, Gauss, Bessel, Fourier and Poisson. Jacobi's work on analysis and curves placed him in leading position to become professor at the University of Könisberg, where he remained for some time. Reports from his students and colleagues speak only of his talent, enthusiasm and warm personality.

Despite his success, he struggled in later life with money and health problems, including diabetes, and required financial aid from Dirichlet and von Humboldt. His health suffered quite a bit during this time and prevented him working and publishing. In later life, Jacobi rejected the views of Lagrange on analytical mechanics that it is an axiomatically defined theory; instead, on philosophical grounds, he believed that the physical reasonings behind mechanics should dominate the progress of the field. He died from small pox in 1851 following political unrest in Prussia and stress over money and health. He had also suffered from a breakdown prior to this, due to overworking. We remember him for integrating the Euler-Lagrange equation to give the Jacobi energy function, Jacobian determinants in transformation theory, the Hamilton-Jacobi method and his differential equation solved by Jacobi fields.

H.6   Sir William Hamilton

Fig. H.6: Sir William Hamilton (1805–1865).

Sir William Rowan Hamilton (1805–1865) was an Irish physicist and mathematician born in Dublin. (For a picture of Sir Hamilton, see figure H.6.) At a young age, he was very adept at learning languages, at least by the standards of a young child. He was taught by his uncle Rev. James Hamilton, with whom William lived. In 1813, an American called Zerah Colburn, dubbed the "mental calculator" because he could perform amazing mental calculations visited Dublin. Hamilton was a year Colburn's junior and the two were pitted against each other in a competition. Hamilton was defeated and it is thought that this defeat spurred him to study mathematics with a passion. His ability in languages allowed him to study foreign books of the great French mathematicians. He studied the works of Clairaut, Newton and Laplace, even spotting an error in Laplace's *Mécanique céleste*. He wrote his results to Dr J. Brinkley, Astronomer Royal, who encouraged his talents.

He entered Trinity College Dublin in 1823 aged 18 and became professor in 1827 prior to graduating. He submitted his first paper in 1824 entitled "On caustics" although it was initially rejected. In 1827 he presented his "Theory of systems of rays" introducing the principal and characteristic functions supporting the wave theory of light. The study of optics and its application to mechanics led to the publication of two celebrated papers on dynamics. The papers (edited by David R. Wilkins) can be found online and are a must read for all interested in the history of dynamics. In the papers, Hamilton constantly refers to "varying the action". This is not in the same way as that of Euler or Lagrange but, instead, in the manner we have discussed in the text. Hamilton's variation of the action leads to a more formal structure of dynamics being uncovered, hidden by the former variational procedures. It was Hamilton that first termed the Lagrangian function $\mathscr{L} = T - U$ and his reformulation allows action to be a quantity used across physics.

Before his graduation, he accepted an astronomy position in Trinity College; however, the practicalities of this role were always of secondary interest to the theoretical mathematics he enjoyed. He also wrote poetry and struck a friendship with the poet Wordsworth, although the poet suggested to Hamilton that his talents lay in science instead! Hamilton had one great love in his life: Catherine Disney. He met her during his undergraduate years but, unfortunately for him, she was married off to an older but rich clergyman named Barlow. The distraught Hamilton's work suffered and it is reported he considered suicide.

He had gained worldwide fame after his general treatment of dynamics and was corresponding with St Petersburg Academy of Science and the United States National Academy of Sciences. However overworking was taking its toll on him and he became unwell. He visited Catherine during a vacation to Armagh with his student and was so nervous that he (as the Royal Astronomer of Ireland) broke the eyepiece in the telescope while showing her constellations. After the visit, he slipped further into melancholy. Without Catherine, he is thought not to care to *whom* it was that he married, and he settled with Helen Bayly. Their marriage was not born out of true love and he spent their honeymoon working on a third treatment of optics.

While out walking with Helen, clearly attentively listening to her, he happened upon the correct way to multiply quaternions. He took his penknife and carved $i^2 = j^2 = k^2 = ijk = -1$ into the stone of the Broom Bridge over the Royal Canal in Dublin. He spent the rest of his professional career investigating quaternion calculus, publishing *Lectures on Quaternions* in 1853.

Catherine contacted him and they began correspondence. It clearly adversely affected him and he experienced a period of alcoholism, even losing control of his faculties at a society dinner. In this correspondence, both he and Catherine became quite personal, leading to guilt on Catherine's part. She told Barlow of her correspondence and attempted suicide unsuccessfully. Catherine went to live with her family away from her husband and the pair continued their letters. She died after he finished his first work on quaternions in 1853. The extreme depression Hamilton suffered drove him to argue with Helen and engross himself further into the subject of quaternions until his death in 1865.

H.7 Siméon Denis Poisson

The son of a soldier, Siméon Denis Poisson (1781–1840) was born into a working-class family in revolutionary France. (For a picture of Poisson, see figure H.7.) His family, who had lost previous children, devoted themselves to securing his education and nurtured his early years. At this time, France was undergoing the social and political revolution against the establishment and ruling monarchy of King Louis XVI, who had spent most of the country's wealth in the French-Indian war and the American Revolution. Due to this unrest, Poisson's father took advantage of the situation and became president of their local district. Poisson's father decided that a medical career would be a suitable choice for his son and sent him to study medicine. Fortunately for natural science, Poisson was awful at the finer intricacies of surgery and enlisted instead in École Centrale in a more theoretical discipline. Despite his lowly background

Fig. H.7: Siméon Denis Poisson (1781–1840).

and lack of formal education, he was allowed to sit entrance exams and was soon top of the class, attracting the attention of his professors, Lagrange, Legendre and Laplace. He was allowed to assist in certain teaching roles while still a student.

Over the next few years, in his late teens and early twenties, Poisson soon gained attention among the mathematical elite and quickly became professor. He took up a position in the chair of classical mechanics in the Faculté des Sciences and, in 1811, published two volumes on *Traité de mécanique*. Poisson is also one of the founding fathers of the study of electromagnetism and charged bodies and he revolutionised the teaching of physics and mathematics in 19th century France. He married in 1817 and had four children. For our story, he is the first to popularise the definition of the conjugate momentum to a generalised coordinate and, of course, the developer of the Poisson bracket and the subsequent Poisson algebra of commuting observables.

H.8 Amalie Emmy Noether

Amalie Emmy Noether (1882–1935) was a German mathematician remembered for vast contributions to both pure mathematics and theoretical physics. (For a picture of Noether see figure H.8.) She came from a well educated Jewish family, with her father and two of her brothers being professors of mathematics and chemistry. Noether never hinted at her future greatness at school and revealed instead a passion for languages and their teaching. Thankfully for physics, she decided not to pursue this avenue, instead sitting in on lectures at the University of Erlangen at the turn of the twentieth century. This was a bold move for a women during these times, as women's rights were only just surfacing across Western Europe. She became a fully matriculated student in 1904 and earned a doctorate 3 years later under Paul Gordan, a student of Jacobi.

She worked for 7 years without pay at Erlangen's Mathematical Institute and looked after her father, who had grown ill. During this time, she was guided by Ernst Fischer towards the work of David Hilbert and abstract algebra. Her fame grew over the next few years and she took on doctorate students. David Hilbert and Felix Klein were two giants in the mathematical world in Noether's time; they revolutionised mathematics beyond recognition. Both saw Noether as a budding talent and tried their best to recruit her to work with them. Unfortunately, their efforts were unsuccessful, as prejudice against women in science was the norm. Despite this, she remained with Hilbert and helped him with the physics aspect of his work. It was then in 1918 that she developed what are now called the *Noether theorems*, the significance of which was stressed by Albert Einstein. Noether lectured under Hilbert's name at the University of Göttingen, again without pay.

Fig. H.8: Amalie E. Noether (1882–1935).

In 1918 Germany underwent a revolution that lead to the establishment of the Weimar Republic. With this came a surge of equal rights protesters who successfully turned Germany into a modern society before the reversal and establishment of traditional values that the Nazi era brought. This brief freedom allowed Emmy Noether to pursue her *habilitation* and become a tenured lecturer.

Over the next 10 years, she output some of the most mathematically inspired papers in the field of abstract algebra and eventually she was paid a small salary in 1923. The lack of money definitely shaped her life and she is said to have lived very modestly, only on essentials. Noether was truly passionate about mathematics and physics and there are some great anecdotes about her wild conversations after lectures or at a dinner where she was so engrossed that she spilled her dinner on her jersey but barely noticed. She took on many students while at Göttingen and often allowed others to publish her ideas to further their own careers at her own expense. Van der Waerden described her as "completely unegotistical and free of vanity, she never claimed anything for herself, but promoted the works of her students above all" (van der Waerden 1935, p. 98).

With the Nazi party gaining political prowess in the 1930s, Noether was forced to leave her position as a lecturer; instead, she gathered students at her home to continue educating. Despite her political sharpness, Weyl suggested at her funeral memorial that she was unaware of the true evil of the Nazi regime and did not see the hatred or violence churning around her[2]. After a 1-year lecturing tour of America, she worked

---

[2]See http://www-history.mcs.st-and.ac.uk/Extras/Weyl_Noether.html.

in Bryn Mawr, Pennsylvania in the USA. It is here that she died unexpectedly, 4 days after an operation on a tumour.

Despite a lifetime of being discouraged and unpaid, her resilience meant that Emmy Noether ended her career being one of the most respected mathematicians in history. Her works are considered to have fuelled and inspired an entire generation of mathematical research and she was praised most highly by the likes of Hilbert, Klein, Weyl, Curie and Einstein. To echo a quote from Einstein to the editor of the *New York Times* in 1935, her unselfish dedication to mathematics and theoretical physics was met with dismissal by the authorities and cost her the means of maintaining her simple life and her studies.

H.9   Ludwig Eduard Boltzmann

Ludwig Boltzmann (1884–1906) was born in Vienna, Austria, at, arguably, the golden era of classical mechanics. (For a picture of Boltzmann, see figure H.9.) His farther, Ludwig Georg Boltzmann was a revenue official and the young Boltzmann was initially educated at home, having a private tutor, before attending high school in Linz, Austria. Boltzmann then went on to study physics in 1863 at the University of Vienna. In 1866, Boltzmann completed his doctorate degree on the kinetic theory of gases. Boltzmann worked under the guidance and teaching of Joseph Stefan and it was through Stefan that Boltzmann was introduced to Maxwell's workings.

Boltzmann had a varied academic career, starting off at the university of Graz, where he had been appointed to a professorship in mathematical physics. He also had placements at Heidelberg and Berlin and was appointed as professor of mathematics at the University of Vienna in 1873. In 1876, he married Henrietta con Aigentler, who was a teacher of mathematics and physics in Graz. They had five children together.

Fig. H.9: Ludwig Boltamann (1884–1906).

Boltzmann returned to the University of Graz, where he taught experimental physics. It is here he developed his theories of classical statistical mechanics and developed the foundations of statistical physics, as we are familiar with today. In 1890, he was appointed to the chair of theoretical physics at the University of Munich. In 1893, Boltzmann became the professor of theoretical physics at the University of Vienna, taking over the post from his former mentor Stefan.

During his life, Boltzmann spent considerable effort defending his theories against several of his colleagues and, in 1903, Boltzmann founded the Austrian Society of Mathematics. He found that, while many physicists were doubtful of his mechanics, mathematicians of the day had a more favourable outlook. Unfortunately, during his life, Boltzmann was subject to severe bouts of depression and also had attempted suicide. In September 1906, he hanged himself during a family holiday to Italy. It is speculated that it was bipolar disorder and the lack of appreciation of his work by his peers that drove him to suicide. He is buried in Vienna and his gravestone bears the inscription of his entropy formula, $S = K \log W$. For details on his works, please see C. Cercignani's (1998) book, *The Man Who Trusted Atoms*.

H.10 Edward Routh

Edward Routh (1831–1907) was a Canadian-English mathematician remembered for his skill in teaching and lecturing at the University of Cambridge as well as for developing Routhian mechanics. (For a picture of Routh, see figure H.10.) Routh was born in the British colony of Quebec, where his father was posted as a commissary-general, himself a veteran of the Battle of Waterloo and the defeat of Napoleon. Routh spent his childhood in Canada until aged, when the family moved to London, where he enrolled at University College School and then gained a scholarship at University College London in 1847 aged 16. He was tutored by De Morgan (himself a fascinating scientific character) under whose study he was guided towards a mathematical career.

Fig. H.10: Edward Routh (1831–1907).

Routh entered Peterhouse Cambridge in 1850 at the same time as James Clerk Maxwell, who later transferred to Trinity. He obtained his M. A. from London in 1853 and was awarded the Gold Medal for Mathematics. Upon graduation from Cambridge, he was Senior Wrangler in Mathematics (Maxwell coming second) and he was soon appointed as a lecturer. At this time, he met his future wife Hilda Airy, who was the daughter of the famous Lucasian professor of mathematics and Astronomer Royal George Airy. With Hilda he had five children and became famous for teaching mathematics at Cambridge. His fame grew due to the sheer number of his students that became senior wranglers themselves, beating by far any other tutor to date. A great anecdote to his skill was later recounted in his obituary in the Times newspaper:

> The case of a student of hydrodynamics was alleged as typical of the trials to which [his patience] was exposed. The troubled undergraduate's primary difficulty lay in conceiving how anything could float. This was so completely removed by Dr Routh's lucid explanation that he went away sorely perplexed as to how anything could sink! [3]

Routh seasoned his lecture notes with mathematical jokes and was extremely well thought of by his colleagues and pupils; he was said to be shy with strangers but kindly

[3] See http://www.educ.fc.ul.pt/icm/icm2003/icm14/Routh.htm.'

and a conversationalist with friends. He is remembered for quite a vast contribution to theoretical mechanics and, for our journey, is a pioneer in reduction procedures.

H.11   Hendrika van Leeuwen

Fig. H.11: The Dutch physicist Hendrika van Leeuwen (1887–1974).

Hendrika van Leeuwen (1887–1974) was a Dutch physicist present at the downfall of classical mechanics. (For a picture of van Leeuwen, see figure H.11.) Van Leeuwen studied at Leiden University under Hendrik Lorentz who was responsible for Lorentz transformations and the force law. Van Leeuwen completed her dissertation in 1919 *Problems de la Théorie Électronique du Magnétism*, which was published in *Journal de Physique et le Radium* **2** (12) in which she uses the Routhian method to consider classical collisions of charged particles and, in doing so, proves that classical mechanics cannot, in general, provide a molecular interpretation of magnetism. This conclusion went against the results of others in the field such as Langevin and, at the dawn of women's rights in post-great-war Europe, this shows van Leeuwen had a strong and courageous enthusiasm for her work.

Until 1947, she worked as a lab assistant researching magnetic materials and was then promoted to a reader in theoretical and applied physics at Delft University of Technology. The Bohr van Leeuwen theorem is a key result in the classical statistical mechanics of molecules and solids and is, in my opinion, one of the clearest motivators for a quantum theory.

Bibliography

Abraham, R., and Marsden, J. E. (1987). *Foundations of Mechanics* (2nd edition). Redwood City: Addison-Wesley.

Abraham, R., Marsden, J. E., and Ratiu, T. (1993). *Manifolds, Tensor Analysis, and Applications* (2nd edition). New York: Springer.

Acebrón, J. A., Bonilla, L. L., Pérez Vicente, C. J., Ritort, F., and Spigler, R. (2005). The Kuramoto model: A simple paradigm for synchronization phenomena. *Reviews of Modern Physics*, 77(1), 137–85.

Aldaya, V., and de Azcárraga, J. A. (1978). Variational principles on rth order jets of fibre bundles in field theory. *Journal of Mathematical Physics*, 19(9), 1869–75.

Aldaya, V., and de Azcarraga, J. A. (1980). Geometric formulation of classical mechanics and field theory. *La Rivista del Nuovo Cimento* (1978–1999), 3(10), 1.

Al-Kuwari, H. A., and Taha, M. O. (1991). Noethers theorem and local gauge invariance. *American Journal of Physics*, 59(4), 363–5.

Arnold, V. I. (1989). *Mathematical Methods of Classical Mechanics* (2nd edition). New York: Springer.

Arnold, V. I., Kozlov, V. V., and Neishtadt, A. I. (1997). *Mathematical Aspects of Classical and Celestial Mechanics* (2nd edition). New York: Springer.

Atkins, P. W., and Friedmann, R. S. (1997). *Molecular Quantum Mechanics* (3rd edition). Oxford: Oxford University Press.

Balseiro, P., Marrero, J. C., de Diego, D. M., and Padrón, E. (2010). A unified framework for mechanics: Hamilton–Jacobi equation and applications. *Nonlinearity*, 23(8), 1887.

Bering, K. (2011). Noethers theorem for a fixed region. *Archivum Mathematicum*, 47(5), 337–56.

Boas, M. L. (2005). *Mathematical Methods in the Physical Sciences* (3rd edition). New York: Wiley.

Bohm, D. (1993). *Wholeness and the Implicate Order*. London: Routledge and Kegan Paul.

Born, M. (1927). *The Mechanics of the Atom* (translated by J. W. Fisher and revised by D. R Hartree). London: G. Bell and Sons.

Brading, K. (2002). *Symmetries, Conservation Laws, and Noethers Variational Problem*. PhD thesis, University of Oxford.

Brading, K., and Brown, H. R. (2000). Noethers theorems and gauge symmetries. *arXiv preprint hep-th/0009058*.

Brading, K., and Brown, H. R. (2004). Gauge symmetry transformations observed? *The British Journal for the Philosophy of Science*, 55(4), 645–65.

Brading, K., and Castellani, E. (2003). *Symmetries in Physics: Philosophical Reflections*. Cambridge: Cambridge University Press.

Brown, H. R. (2007). *Physical Relativity*. Oxford: Oxford University Press.

Brumer, P., and Gong, J. (2006). Born rule in quantum and classical mechanics. *Physical Review A.*, 73(5), 052109.

Cariñena, J. F., and Fernández-Núñez, J. (2010). A geometric approach to the Gibbs–Appell equations in Lagrangian mechanics. *Journal of Physics. A, Mathematical and Theoretical*, 43(50), 505205.

Cariñena, J. F., Gracia, X., Marmo, G., Martínez, E., Muñoz-Lecanda, M. C., and Román-Roy, N. (2006). Geometrical Hamilton–Jacobi theory. *International Journal of Geometric Methods in Modern Physics*, 3(7), 1417.

Carta, P., Gozzi, E., and Mauro, D. (2006). Koopmanvon Neumann formulation of classical YangMills theories: I. *Annalen der Physik*, 15(3), 177–215.

Caterina, G., and Boghosian, B. (2006). A no-go theorem for the existence of an action principle for discrete invertible dynamical systems. *arXiv preprint nlin/0611058*.

Cendra, H., Marsden, J. E., and Ratiu, T. S. (2001). *Geometric Mechanics, Lagrangian Reduction, and Nonholonomic Systems*. Berlin: Springer-Verlag.

Cercignani, C. (1998). Ludwig Boltzmann: *The Man Who Trusted Atoms*. Oxford: Oxford University Press.

Christodoulou, D., and Kaelin, I. (2013). On the mechanics of crystalline solids with a continuous distribution of dislocations. *Advances in Theoretical and Mathematical Physics*, 17(2), 399–477.

Cisneros-Parra, J. U. (2012). On singular Lagrangians and Diracs method. *Revista mexicana de física*, 58(1), 61–8.

Coddington, E. A., and Levison, N. (1984). *Theory of Ordinary Differential Equations*. Malabar: Robert E. Krieger.

Courant, R., and Hilbert, D. (1962). *Methods of Mathematical Physics* (Vol. II). New York: Wiley.

Courant, T. J. (1990). Dirac manifolds. *Transactions of the American Mathematical Society*, 319(2), 631–61.

Craig, D. P., and Thirunamachandran, T. (1998). *Molecular Quantum Electrodynamics*. London: Academic Press Incorporated.

Currie, D. G., and Saletan, E. J. (1966). q-Equivalent particle Hamiltonians. I. The classical one-dimensional case. *Journal of Mathematical Physics*, 7(6), 967–74.

de León, M., Salgado, M., and Vilariño, S. (2014). Methods of differential geometry in classical field theories: k-symplectic and k-cosymplectic approaches. *arXiv preprint arXiv:1409.5604*.

Dedecker, P. (1957). A property of differential forms in the calculus of variations. *Pacific Journal of Mathematics*, 7(4), 1545–9.

Delphenich, D. (2012). The role of integrability in a large class of physical systems. *arXiv preprint arXiv:1210.4976*.

Dirac, P. A. M. (1958). *The Principles of Quantum Mechanics* (4th edition). Oxford: Oxford University Press.

Dirac, P. A. M. (1964). *Lectures on Quantum Mechanics*. New York: Yeshiva University.

Dugas, R. (1988). *A History of Mechanics* (Editions du Griffon, Neuchâtel, Switzerland, 1955) (translated by J. R. Maddox). New York: Dover Publications.

Evans, R., Oettel, M., Roth, R., and Kahl, G. (2016). New developments in classical density functional theory. *Journal of Physics: Condensed Matter*, 28(24), 240401.

Faddeev, L. D. (1969). The Feynman integral for singular Lagrangians. *Theoretical and Mathematical Physics*, 1(1), 1–13.

Faddeev, L., and Jackiw, R. (1988). Hamiltonian reduction of unconstrained and constrained systems. *Physical Review Letters*, 60(17), 1692–94.

Fasano, A., and Marmi, S. (2006). *Analytical Mechanics*. Oxford: Oxford University Press.

Fetter, A. L., and Walecka, J. D. (1980). *Theoretical Mechanics of Particles and Continua*. New York: McGraw-Hill.

Goldstein, H., Poole, C., and Safko, J. (2002). *Classical Mechanics* (2nd edition). San Francisco: Addison Wesley.

Goldstin, H. H. (1980). *A History of the Calculus of Variations From the 17th through the 19th Century*. New York: Springer.

Gonzalez, O. (1999). Mechanical systems subject to holonomic constraints: Dierential-algebraic formulations and conservative integration. *Physica D. Nonlinear Phenomena*, 132(1–2), 165–74.

Gozzi, E., and Pagani, C. (2010). Universal local symmetries and nonsuperposition in classical mechanics. *Physical Review Letters*, 105(15), 150604.

Gozzi, E., and Regini, M. (2000). Addenda and corrections to work done on the path-integral approach to classical mechanics. *Physical Review D: Particles and Fields*, 62(6), 067702.

Grabowska, K., and Grabowski, J. (2011). Dirac algebroids in Lagrangian and Hamiltonian mechanics. *Journal of Geometry and Physics*, 61(11), 2233–53.

Gray, C. G., Karl, G., and Novikov, V. A. (2004). Progress in classical and quantum variational principles. *Reports on Progress in Physics*, 67(2), 159–208.

Gray, C. G., and Taylor, E. F. (2007). When action is not least. *American Journal of Physics*, 75(5), 434–58.

Guillemin, V., and Sternberg, S. (1990). *Symplectic Techniques in Physics*. New York: Cambridge University Press.

Hansen, J.-P., and McDonald, I. R. (2013). *Theory of Simple Liquids* (4th edition). Oxford: Academic Press.

Hofer, H., and Zehnder, E. (1994). *Symplectic Invariants and Hamiltonian Dynamics.* Basel: Birkh auser.

Holland, P. R. (1995). *The Quantum Theory of Motion: An Account of the de Broglie-Bohm Causal Interpretation of Quantum Mechanics.* Cambridge: Cambridge University Press.

Horikoshi, A., and Kawamura, Y. (2013). Hidden Nambu mechanics: A variant formulation of Hamiltonian systems. *Progress of Theoretical and Experimental Physics,* 2013(7), 1.

Hydon, P. E., and Mansfield, E. L. (2011). Extensions of Noethers second theorem: From continuous to discrete systems. *Proceedings of the Royal Society of London A,* doi:10.1098/rspa.2011.0158. *II2007* Ivancevic, V. G., and Ivancevic, T. T. (2007). *Applied Differential Geometry: A Modern Introduction.* Hackensack: World Scientific.

Jackson, J. D. (1975). *Classical Electrodynamics* (2nd edition). New York: Wiley.

Jordan, A., and Libedinsky, M. (1997). Path integral invariance under point transformations. arXiv preprint hep-th/9703173.

Joyeux, M., and Sugny, D. (2000). On the application of canonical perturbation theory to floppy molecules. *The Journal of Chemical Physics,* 112(1), 31–9.

Jóźwikowski, M. (2013). Jacobi vector fields for Lagrangian systems on algebroids. International *Journal of Geometric Methods in Modern Physics,* 10(5), 1350011.

Kosmann-Schwarzbach, Y. (2011). *The Noether Theorems: Invariance and Conservation Laws in the Twentieth Century.* New York: Springer.

Lagrange, J. L. ([1811] 1997). *Analytical Mechanics* (translated and edited by A. Boissonnade and V. N Vagliente). Dordrecht, Springer.

Lanczos, C. (1970). *The Variational Principles of Mechanics* (4th edition). Toronto: University of Toronto Press.

Landau, L. D., and Lifschitz, E. M. (1976). *Mechanics* (3rd edition). New York: Pergammon Press.

Laplace, P. S. ([1814] 1951). A Philosophical Essay on Probabilities (translated by F. W. Truscott and F. L. Emory). New York: Dover Publications.

Lavin, R., and Pozo, J. (2011). The study and teaching of the vibrational modes of the triatomic molecule. *Revista Brasileira de Ensino de Física,* 33(1), 1–6.

Lovett, S. (2010). *Differential Geometry of Manifolds.* Natick: A. K. Peters.

Mandl, F. (1988). *Statistical Physics* (2nd edition). Chichester: Wiley.

Mangiarotti, L., and Sardanashvily, G. (1999). *Gauge Mechanics.* Hackensack: World Scientific.

Marcus, R. A. (1967). Analytical mechanics and almost vibrationally-adiabatic chemical reactions. *Discussions of the Faraday Society,* 44, 7–13.

Marcus, R. A. (1968). Analytical mechanics of chemical reactions. IV. Classical mechanics of reactions in two dimensions. *The Journal of Chemical Physics,* 49(6), 2617–31.

Marmo, G., Saletan, E. J., Simoni, A., and Vitale, B. (1985). *Dynamical System: A Dierential Geometric Approach to Symmetry and Reduction*. Chichester: Wiley.

Marsden, J. E., Patrick, G. W., Shkoller, S., and West, M. (2001). Variational methods, multisymplectic geometry and continuum mechanics. *Journal of Geometry and Physics*, 38(3–4), 253–84.

Marsden, J. E., and Ratiu, T. S. (1994). *Introduction to Mechanics and Symmetry*. New York: Springer.

Marsden, J. E., Ratiu, T. S., and Scheurle, J. (2000). Reduction theory and the Lagrange-Routh equations. *Journal of Mathematical Physics*, 41(6), 6.

Marsden, J. E., and Weinstein, A. (1974). Reduction of symplectic manifolds with symmetry. *Reports on Mathematical Physics*, 5(1), 121–30.

Martínez, E. (2005). Classical field theory on Lie algebroids: Variational aspects. *Journal of Physics. A, Mathematical and General*, 38(32), 7145.

Mauro, D. (2002). *Topics in Koopman-von Neumann Theory*. PhD thesis, University of Trieste.

McInerney, A. (2013). *First Steps in Differential Geometry: Riemannian, Contact, Symplectic*. New York: Springer.

McQuarrie, D. A. (2000). *Statistical Mechanics*. Sausalito: University Science Books.

Mimura, F., and T. Nono, T. (1986). Conservation laws derived from equivalent Lagrangian densities in continuum mechanics. *Bulletin of the Kyushu Institute of Technology. Mathematics, Natural Science*, 33, 21–35.

Misner, C. W., Thorne, K. S., and Wheeler, J. A. (1973). *Gravitation*. San Francisco: W. H. Freeman and Company.

Nambu, Y. (1973). Y, Nambu. Generalised Hamiltonian Dynamics. *Physical Review D: Particles and Fields*, 7(8), 2405–12.

Nayfeh, A. H. (1981). *Introduction to Perturbation Techniques*. New York: Wiley.

Nucci, M. C., and Leach, P. G. L. (2008). The Jacobi Last Multiplier and its applications in mechanics. *Physica Scripta*, 78(6), 065011.

Ortega, J. P., and Ratiu, T. S. (2004). *Momentum Maps and Hamiltonian Reduction*. Boston: Birkh auser.

Ott, E. (2002). *Chaos in Dynamical Systems* (2nd edition). Cambridge: Cambridge University Press.

Paufler, C., and Romer, H. (2002). De Donder-Weyl equations and multisymplectic geometry. *Reports on Mathematical Physics*, 49(2–3), 325–34.

Román-Roy, N. (2009). Multisymplectic Lagrangian and Hamiltonian formalisms of classical field theories. *Symmetry, Integrability and Geometry. Methods and Applications*, 5, 100.

Rothe, H. J., and Rothe, K. D. (2010). *Classical and Quantum Dynamics of Constrained Hamiltonian Systems*. Hackensack: World Scientific.

Rund, H. (1994). *The Hamilton-Jacobi Theory in the Calculus of Variations*. London: Van Nostrand.

Saunders, D. J. (1989). *The Geometry of Jet Bundles*. Cambridge: Cambridge University Press.

Schey, H. M. (1996). *Div, Grad, Curl and All That: An Informal Text on Vector Calculus*. New York: W.W. Norton & Company.

Siddiqi, K., Bouix, S., Tannenbaum, A. R., and Zucker, S. W. (2002). Hamilton-Jacobi skeletons. *International Journal of Computer Vision*, 48(3), 215–31.

Skinner, R., and Rusk, R. (1983). Generalized Hamiltonian dynamics. I. Formulation on $T^*Q \oplus TQ$. *Journal of Mathematical Physics*, 24(11), 2589–94.

Srikantha Phani, A., and Adhikari, S. (2008). Rayleigh quotient and dissipative systems. *Journal of Applied Mechanics*, 75(6), 061005.

Takhtajan, L. (1994). On foundations of the generalized Nambu mechanics. *Communications in Mathematical Physics*, 160(2), 295–316.

Talman, R. (2000). *Geometric Mechanics*. Nova Iorque: Wiley.

Tanner, D. J. (2007).*Introduction to Quantum Mechanics: A Time-Dependent Perspective*. Sausalito: University Science Books.

Thamwattana, N., Mccoy, J. A., and Hill, J. M. (2008). Energy density functions for protein structures. *The Quarterly Journal of Mechanics and Applied Mathematics*, 61(3), 431–51.

Thornton, S. T., and J. B. Marion, J. B. (2004). *Classical Dynamics of Particles and Systems* (5th edition). Belmont: Thomson-Brooks/Cole.

van der Waerden, B. L. (1935). Nachruf auf Emmy Noether [Obituary of Emmy Noether], *Mathematische Annalen*, 111, 469–74.

Walters, P. (2000). *An Introduction to Ergodic Theory*. Berlin: Springer-Verlag.

Weinberg, S. (1995). *The Quantum Theory of Fields*. Cambridge: Cambridge University Press.

Wells, C. G., and Siklos, S. T. (2006). The adiabatic invariance of the action variable in classical dynamics. *European Journal of Physics*, 28(1), 105–12.

Yoshimura, H., and Marsden, J. E. (2006). Dirac structures in Lagrangian mechanics Part I: Implicit Lagrangian systems. *Journal of Geometry and Physics*, 57(1), 133–56

Zangwill, A. (2013). *Modern Electrodynamics*. Cambridge: Cambridge University Press.

Index